T0135634

On Brine Entrapment in Sea Ice:
Morphological Stability, Microstructure and Convection

On Brine Entrapment in Sea Ice: Morphological Stability, Microstructure and Convection

Sönke Maus

Doktor Philosophiae Thesis

Geophysical Institute, University of Bergen, Norway

Bergen, Norway, March 2007

Tinfoil replica of a horizontal section of sea ice. Reproduced from E. v. Drygalski (1897), Grönlands Eis und sein Vorland. Vol. 1 d. Grönland-Expedition der Gesellschaft für Erdkunde zu Berlin 1891-1893, p. 424, orginal size.

Bibliographic information published by the Deutsche Nationalbibliothek

The Deutsche Nationalbibliothek lists this publication in the Deutsche Nationalbibliografie; detailed bibliographic data are available in the Internet at http://dnb.d-nb.de .

ISBN 978-3-8325-1523-2

Logos Verlag Berlin
Comeniushof, Gubener Str. 47,
10243 Berlin
Tel.: +49 030 42 85 10 90
Fax: +49 030 42 85 10 92
INTERNET: http://www.logos-verlag.de

Preface

The motivation for the present work on sea ice physics came from the depths of the polar ocean. During my work with the formation of cold, saline dense shelf waters, which form predominantly in areas of thin ice, I became aware of a lack in physical models to describe the salt release from thin ice properly. As I could neither find good empirical data sets, I was looking for an opportunity to take them on my own. When I spent a year in Longyearbyen, Svalbard, this possibility became realistic. The inner part of Adventbay, where Longyearbyen is situated, was an easily accessible ice laboratory. Due to the tides acting in waters of just a meter water depth, the ice cover was frequently changing. Every day i could observe new phenomena, like small ridges, nilas and frost flowers, opening and refreezing leads, the frazil-slush-pancake cycle and many ice forms growing and letting my interest grow. With the laboratory at hand in the University building (UNIS), and two weather stations in the vicinity, it was possible to obtain continuous data on newly growing ice, for which normally extensive field work must be performed.

During my second winter I had the opportunity to take almost daily observations and samples of sea ice growing in Adventbay, day by day on my 500-mters walk from my apartment to the university building. I got help from master student Cecilia Bennet and her friend Magnus Larsson to keep continuity in the sampling. By the mid of April we had obtained more than 200 thin ice cores and frazil samples, with well documented growth conditions, and cut most of them for appropriate subcore analysis.

When I returned from a conference by the end of April, I found the information that the door of the freezing room had been open over Eastern. The freezing aggregate had stopped working and most of our samples had been melted and refrozen. After having examined the cores we had to conclude that these samples were no longer appropriate for an analysis to link micro- and fine-structure to growth conditions. Neither was it possible to study vertical salinity distributions. Hence, with the exception of one data series, only bulk salinity determinations from whole cores packed in plastic-bags were finished. Knowing that this would have been my last winter on Svalbard, it was no luck to see the data were no longer of scientific value. However, when I left Svalbard some months later, I did it with the motivation to find a theoretical solution for the problem. The data were lost, but the days of observations and my interest were not.

For the past 6 years I have continued studying the microstructure of sea ice theoretically. Looking for data obtained by others was one thing, understanding the differences another one. Over the years my theoretical understanding of ice and sea ice and my empirical knowledge grew, but this went on slowly. Then I realised the need to study crystal growth and metallurgy and, last but not least, many molecular aspects of ice and water. Almost having given up the problem, I picked it up again and again until things became clearer. But even when they seemed clear, it was still one year to go. Perhaps I would not have come that far, if the polar night ice data on Svalbard had not been lost. Who knows?

This monograph has become quite long due to several reasons. First, the model approaches presented here are often new, and I felt it necessary to demonstrate, why earlier approaches should be replaced or were unsuccessful. Second, the problem to be solved here is rather complex and requires the coupling of a variety of physical models that all have to be critically discussed. Third, the model calcuations and predictions must reside on a solid fundament of thermodynamic properties of saline solutions, many of which were not available in the required form. Finally, I have to admit, that in the end the time was missing to put the present text into a shorter version. I will do so in the future.

Sönke Maus, Bergen, September 2006.

Ich bin die Nacht

Ich bin die Nacht. Meine Schleier sind
viel weicher als der weiße Tod.
Ich nehme jedes heiße Weh
mit in mein kühles, schwarzes Boot.

Mein Geliebter ist der lange Weg
Wir sind vermählt auf immerdar
Ich liebe ihn, und ihn bedeckt
mein seidenweiches, schwarzes Haar.

Mein Kuß ist süß wie Fliederduft -
der Wanderer weiß es genau ...
Wenn er in meine Arme sinkt,
vergisst er jede heiße Frau.

Meine Hände sind so schmal und weiß,
daß sie ein jedes Fieber kühlen,
und jede Stirn, die sie bedeckt,
muß leise lächeln, wider Willen.

Ich bin die Nacht. Meine Schleier sind
viel weicher als der weisse Tod.
Ich nehme jedes heiße Weh
mit in mein kühles, schwarzes Boot.

(Selma Meerbaum-Eisinger) [a]

[a] Most sea ice grows in high latitudes during the Polar Night, and also my motivation for the present work was born during this experience. Selma Meerbaum-Eisinger did not experience such a night. She died, in the age of 18 years, during december 1942 in the German concentration camp Michailowka.

Many of the observations, that served for me as a basis to validate the theories on sea ice presented here, have been conducted by researchers from the Cold Regions Research Engineering Laboratory (CRREL) of the US Army, within the scope of military programs. I aknowledge the importance of these observations, but I distance myself from the military actions being undertaken by the US Army all over the world. I hereby express my hope for a free world of education, where research is by no means driven by nationalism and fundamentalism or economics, but by the desire of human beings to understand nature and culture and to come and live together on planet earth.

With this ideal in mind, I would like to pass Selma's poem on patience, silence and peace to the reader. Sönke Maus, March 2007

Contents

1. Introduction

Ice that forms in the sea differs from ice that forms from freshwater in its microstructure. This difference implies, that sea ice contains considerable amounts of saline brine in pores and fluid inclusions, while lake ice does not. This microstructure and the resulting bulk salinity of sea ice are important variables in many geophysical systems and related fields of environmental sciences:

- *Engineering* applications, like traffic-ability of polar regions and risks to offshore structures, involve the mechanical properties of sea ice which essentially depend on its microstructure (Michel, 1978; Sanderson, 1988; Schulson, 2001; Cole, 2001).

- *The Ecosystem* of sea ice is essential to understand life cycles in polar regions. The microstructure near the ice bottom is particularly important for the biological activity and concentration of pollutants (Melnikov, 1997; Thomas and Dieckmann, 2003).

- *Remote Sensing* depends on the knowledge of dielectric and optical properties of sea ice which are intimately linked to salinity and microstructure (Carsey, 1992).

- *Geophysics of sea ice and Climate Research* involve a large number of mechanisms where the sea ice microstructure and salinity are relevant (Doronin and Kheisin, 1975; Untersteiner, 1986; Wadhams, 2001; Leppäranta, 2005). Some of them are (i) brine release from growing ice, driving rigorous convection in the ocean, (ii) propagation of sea ice melt signals in ocean gyres, (iii) large scale rheology of ice motion and deformation, linked to the sea ice's role as a thermally insulating lid between the sea and the atmosphere, (iv) the modification of CO_2 uptake by the ocean in ice-covered seas, (v) paleoclimatological aspects related to ice shelves. On average only 5 to 7 % of the world oceans are covered by sea ice, but its influence on polar climate is fundamental.

Most of these research topics and disciplines are young and the role of sea ice salinity and microstructure is, although evidently of high relevance, not well understood. The reason for this deficiency is that so far no validated theory for the formation of the microstructure exists. The present thesis attempts to close this gap. It deals with the formation of the microstructure of sea ice and the corresponding entrapment of saline

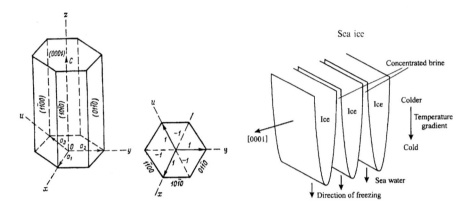

Fig. 1.1: Sketch of a hexagonal ice crystal indicating the c-axis and the three a-axis. From Doronin and Kheisin (1975).

Fig. 1.2: The preferred orientation of sea ice crystals with the c-axis or (0001)-direction being horizontal. From Petrenko and Whitworth (1999).

brine. This introduction shortly sketches the microstructural difference of sea and lake ice, gives a historical overview of research on sea ice salinity and microstructure, and formulates the main problems to solved. Then the present work is shortly introduced.

1.1 Pure ice versus sea ice

Crystal structure

Detailed descriptions of the molecular physics and crystal structure of pure single ice crystals are available in standard books and references therein (Shumskii, 1955; Fletcher, 1970; Hobbs, 1974; Petrenko and Whitworth, 1999). For the present study it is most important to recognize that ice has an approximate hexagonal geometry, with its main axis normal to the hexagonal planes [1]. The main axis is often termed c-axis or optical axis. The hexagonal planes are the basal planes, spanned by 3 equivalent hexagonal directions, the a-axes. The principal crystal structure is shown in figure 1.1.

The hexagonal ice lattice implies that the incorporation of atoms during freezing is different for the c-axis and a-axis directions. This in turn introduces an anisotropy in the growth behaviour in dependence on the supercooling below the equilibrium freezing point. During growth from the vapour the anisotropy results in a large variety of crystal forms and several shifts in the preferred growth directions (Shumskii, 1955; Fletcher, 1970;

---

[1] This is called ice Ih which forms under most natural conditions on earth.

Hobbs, 1974; Pruppacher and Klett, 1997). At supercoolings less than -4 to -5 K the growth is fastest in the direction of the basal planes. As the supercooling under natural conditions seldom exceeds a few degrees, the favoured growth direction in supercooled water is normal to the c-axis. In this regime the growth velocity anisotropy is very large, being one to two orders of magnitude (Hillig, 1958; Michaels et al., 1966; Ryan, 1969). The growth anisotropy is a molecular interfacial nucleation phenomenon and effective when the driving force for growth is the supercooling in the melt. It must not be confused with the thermal conductivity of ice. The latter is, if at all, very slightly larger in the c-axis direction (appendix A). However, when a certain amount of supercooling is present, then growth will be affected by the anisotropy and this is the basic cause for sea ice growing with horizontal c-axis orientation as shown in figure 1.2.

Formation of an ice cover

The formation of an ice cover on natural lakes starts by growth of needles and plates in a supercooled surface layer. It depends on the growth conditions like temperature, wind and precipitation, which principal crystal orientation develops to form the *primary ice skim* (Shumskii, 1955; Michel and Ramseier, 1971; Hobbs, 1974; Gow, 1986). Two limiting cases seem generally accepted: (i) during freezing of calm water plates will often grow in a thin surface film resulting in vertical c-axis-orientations, (ii) when the water is agitated by winds the orientation will be random (Shumskii, 1955; Michel and Ramseier, 1971). Shumskii (1955) was the first who pointed out that the primary ice formation is related to the crystal growth anisotropy, due to a competition of growth in the direction of maximum supercooling and the direction of maximum heat flow. Although crystal growth morphology in fresh water has been studied in some detail (Kumai and Itagaki, 1953; Arakawa, 1954; Williamson, 1967; Tirmizi and Gill, 1989; Shimada and Furukawa, 1997), an exact theoretical and quantitative test of these basic ideas is still outstanding. The further *secondary growth* below the primary skim most often retains the primary orientation, if it was *uniform*. The fact that a *random* primary orientation almost always evolves into a preferred horizontal orientation has been related to different processes. Being still a matter of debate, most scientists agree in the point that the principle process is some mode of geometric selection (Shumskii, 1955; Perey and Pounder, 1958; Knight, 1962b; Ketcham and Hobbs, 1967; Cherepanov and Kamyshnikova, 1971; Hobbs, 1974; Weeks and Wettlaufer, 1996).

In contrast to lake water, the primary freezing of seawater results always in a random c-axis orientation. This morphology of the top millimeters was already pointed out by Drygalski (1897). Later studies have shown, that even for calm freezing the initial crystal orientation is random in the upper centimeter (Weeks and Ackley, 1986). As stressed by Hamberg (1895), during secondary growth of sea ice the crystals with vertical c-axis are eliminated rapidly, and no exceptions to this rule were found since then (Weeks and Ackley, 1986). For primary sea ice, which contains considerable amounts of brine between the basal plane plates, an anisotropic effective conductivity is a relevant mechanism of geometric selection, because parallel heat conduction in a laminate of different conductivities exceeds serial heat conduction (Weeks, 1958; Harrison and Tiller, 1963). To explain, however, the plate structure of sea ice, the concept of supercooling must be viewed in combination with the latter.

Plate spacing of sea ice

Growth with horizontal c-axis is not sufficient to explain the salty brine that is included in sea ice. The latter relates to cellular growth of the ice and its lamellar microstructure, the main contrast to lake ice pointed out by scientists a long time ago (Parrot, 1818; Walker, 1859; Drygalski, 1897; Hamberg, 1895). Lake ice freezes with a flat, planar interface and retains little salt. Sea ice grows with a cellular interface and entraps considerable amounts of liquid brine in the cellular microstructure. This microstructure is lamellar, with plates being vertically oriented and parallel within each grain, see figure 1.4 below. It is the distance of these cells or *plate spacing*, which is the fundamental microstructural property of the sea ice medium.

Liquidus relationship

With the plate spacing as principal microstructure of sea ice, the second basic difference between sea ice and lake ice is at hand: the brine which is situated between the plates. While the ice cools brine has to adjust to its *liquidus* or equilibrium freezing temperature. The first observations of this kind were documented by Scoresby (1815), who analysed brine from pores of cold natural sea ice. He described that salty liquid brine of a density 1104.5 kgm^{-1} did not freeze above a temperature of -10 °C. This observation corresponds well to the present-day knowledge of seawater freezing point depression, which would yield a brine salinity of approximately 140 ‰ at -10 °C. Detailed observations of the liquidus relationship were first obtained for saline solutions. Blagden (1788) obtained the freezing point depression due to NaCl within 0.1-0.4 K of the present day standard, see appendix A.2.2. Using a simple elegant approach the liquidus relationship was determined for many salts by Ruedorff (1872): A stirred saline solution was first supercooled and then seeded with a snow crystal. Freezing then proceeded rapidly until the bath temperature stabilised at the freezing point of the actual solute concentration. Ruedorff (1861) obtained the liquidus of the NaCl-water system down to -13 °C. Considering his temperature accuracy limitation of 0.1 °C, the values obtained by him agree with the most accurate present day results.

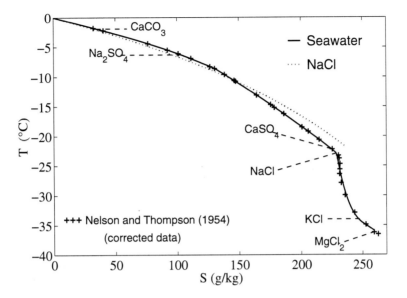

Fig. 1.3: Phase relationship of seawater. The ion data from Nelson and Thompson (1954) are shown as crosses and the eutectic temperatures of the precipitating salts are indicated (from appendix A.4). The dotted line shows the liquidus curve of aqueous NaCl from appendix A, based on the data from Zaytsev and Aseyev (1992).

Seawater is dominated by NaCl but contains several other salts, which implies a slightly different liquidus, with several eutectic points, where salts precipitate. First investigations by Buchanan (1887) were followed by a detailed analysis of the concentration of the major ions by Ringer (1906) for the temperature range -2 to -50 °C. On the basis of later studies with improved the accuracy and resolution (Gitterman, 1937; Nelson and Thompson, 1954) a number of authors have derived a seawater liquidus relationship (Assur, 1958; Anderson, 1960; Tsurikov and Tsurikova, 1972; Richardson, 1976). Figure 1.3 indicates the difference between seawater and NaCl.

From the liquidus relationship and the sea ice salinity the relative brine volume of sea ice can be obtained. It is critical in many applications and physical processes. It may be noted that the well known lever rule

$$v_b \approx \frac{S_i}{S_b},$$ (1.1)

where S_b is the liquidus salinity at a given temperature, and S_i the ice salinity, is often a reasonable approximation to the relative or fractional brine volume v_b.

1.2   Sea ice salinity and microstructure: historical overview

An overview of the history on sea ice research in general has been given by Weeks (1998b), with more detailed descriptions of the 19th and early 20th century available in the books from Zukriegel (1935) and Zubov (1943). A historical description of research on microstructure and salinity of sea ice has not been given so far and in the following an attempt is made to provide such an overview. With our present knowledge and growing understanding of the importance of sea ice salinity and microstructure in the environment, it is useful to trace the basic ideas back to the time where they entered the field of modern scientific analysis and physics. Any discussion of salt entrapment in sea ice must start from two basic physical questions: How is salt structurally incorporated and distributed in sea ice or saline ice? What are the physical processes influencing this distribution? The basic ideas evolved already 150 years ago during the first polar expeditions.

Ice purity and salt inclusions

It has long been known that the solid ice phase that forms during freezing of saline solutions is practically pure (Hobbs, 1974; Petrenko and Whitworth, 1999; Tyndall, 1858). Only a small fraction of the salt in a freezing solution is incorporated into the ice crystal, the majority being rejected and increasing the concentration of the solution. That fraction is often termed the *interfacial distribution coefficient* k. Its intrinsic value is difficult to determine from ice salinities alone, because the rejected solute always increases the interface concentration with respect to the bulk liquid. For planar freezing interfaces, modern studies indicate $k \approx 10^{-3}$ for aqueous NaCl and other alkali halides, with a slight dependence on concentration and growth velocity [2]. This k is two orders of magnitude lower than the fraction of seawater salts entrapped in natural sea ice. The discrepancy has been explained two centuries ago by Scoresby (1815). Scoresby pointed out that sea ice contains salt in form of liquid brine in the interstices between the ice-crystals, preventing it from becoming completely fresh, even in the presence of slow drainage processes of this brine.

   At the time of Scoresby's investigations there was no scientific agreement about the question, if ice crystals frozen from seawater contain salt. Nairne (1776) documented some early experiments. He first grew ice from seawater in a small vessel, and then performed a washing method during which the ice was warmed and partially thawed in a fresh water pail over several hours. From the shrunken sample he obtained melt water of similar density as measured for rain and river water. Other authors obtained apparently different results. Walker (1859) showed that, while naturally frozen seawater was not suitable as drinkable freshwater, the fractionation still allowed one to purify ice

---

[2] For a discussion see section (4.5)

by repeated freezing and melting. Parrot (1818) froze a NaCl solution of concentration comparable to seawater (3% by weight) partially in a vessel and washed the resulting ice several times in distilled water. Melting the samples and evaporating the water he obtained ice which contained one-fifth of the salts of the seawater from which it formed. Parrot (1818) also performed experiments of lateral freezing in vessels and noted that the salinity of the ice increased towards the center. His interpretation already contained the basic principles of growth of sea ice as an ice-brine mixture. He outlined that, due to the much faster diffusion of heat compared to salt, the solution would become more concentrated between growing crystals and that this concentrated solution would freeze at a lower temperature, leading to a subsequent freezing of brine between the propagating and cooling crystals. Parrot's error was to assume that at some fixed temperature also the salt would be incorporated into the crystals.

The general acceptance of the exclusion of impurities from ice crystals during freezing was established during the following years, when for example Faraday demonstrated it by freezing HCl solutions (Tyndall, 1858). The freezing point depression due to salt, indicated by the existence of presumably more saline liquid inclusions even in relatively fresh glacier ice, became a more obvious feature of the freezing process due to the detailed studies of, among others, Tyndall (1858). Tyndall described the internal melting of ice and subsequent expansion of such inclusions under exposure to the sun, termed today *Tyndall figures*. Faraday, in a response to Tyndall's work, suggested the entrapment process of salt in lake ice to be linked to liquid brine entrapment, proposing a complex interaction of crystal advance, salt rejection, diffusion and convective currents, driven by small perturbations (Tyndall, 1858). These intuitive ideas are still of interest, considering the incompleteness of theories of k and possible non-equilibrium convective effects during the freezing of relatively fresh water. In the present context it is interesting, that Faraday's sketch rather well represents the processes leading to salt entrapment in sea ice. Considering convection, internal melting and the description of sea ice by Scoresby (1815) it can be understood, why Nairne (1776) ended up with an almost fresh sea ice residuum while Parrot (1818) did not. The sample from Nairne floated for several hours in fresh water, probably warming it above the temperature at which it had originally frozen, thereby leading to internal melting and opening of the brine inclusions. The denser brine between the interstitials of solid crystals could then be exchanged by drainage against the fresh water in which the saline ice floated. In contrast, the bulk of the ice should not have changed its brine volume and fluid permeability during the short washing procedure applied by Parrot, keeping most brine inclusions disconnected.

In addition to the apparent contradiction between the works of Nairne (1776) and Parrot (1818), early in situ documentations of sea ice salt content, normally indirectly obtained via the melt water density, still must have appeared to be contradictory. Marcet (1819) summarised several floating ice observations from Polar expeditions, which implied lower ice salinities than normal tap water. With our present knowledge that Scoresby was right, it appears that the ice floe samples reported by Marcet (1819) must have originated from icebergs. A thin ice sample, upon which he had based his argumentation, likely formed from fresh melt water at the ocean surface, as it was described by Malmgren (1927) some century later, see figures (1.6) and (1.7) below.

Sea ice salinity: first quantitative observations

The question of the amount of salt remaining in natural sea ice was further addressed during the following years of polar expeditions. Melt water densities of naturally grown polar sea ice, sampled during an early scientific expedition, were in the range 1005-1007.8 kgm^{-3} (Walker, 1859), which typically corresponds to a salinity range 6 to 9 ‰. Walker discussed also the vertical distribution of salt. Based on subsections of ice cores Walker (1859) reported an increasing density (salinity) towards the surface. Moss (1878) provided one of the first reports of systematic chlorinity measurements obtained during the Arctic expedition with the 'Challenger' during 1875. The sampled first-year ice typically contained one-sixth of the chlorinity of seawater in which it floated, whereas in thicker multi-year ice the distribution ratio had dropped to one-fifteenth. With a representative water salinity of 32 ‰ based on the reported specific gravity measurements this corresponds to salinities of 5 and 2 ‰, respectively. Concerning the variability for annual ice, Moss (1878) found a decrease in bulk salinity with floe thickness. Moss also reported an observation of meltwater from rapidly formed thin ice: the latter roughly contained one-half of the salt of the seawater from which it had formed. A similar observation was documented by Weyprecht (1879) for 19 cm thick new ice which had an average salinity of 16 ‰. Weyprecht also gave the vertical distribution of salt in this thin ice, noting a salinity of 25 ‰ in the upper 5 centimeters. Also Rae (1874) described, more qualitatively, his observations during an Arctic expedition, noting that it was never possible to drink meltwater from sea ice formed in the same winter. He further pointed out, that ice which had formed in the previous winter and had been rafted above the water level, generally gave excellent drinkwater upon thawing. Rae suggested that such desalination should be related to the drainage of liquid brine.

The principle conclusions which could be drawn from these early observations were (i) ice formed from saline solutions is practically pure; (ii) the salinity of sea ice is governed by the fraction and salinity of structurally entrapped liquid brine; (iii) at high temperatures the liquid brine inclusions are interconnected and salt may be removed or exchanged by drainage or washing as in the experiments from Nairne (1776); inclusions become disconnected in cold ice, making it impossible to remove a certain brine fraction by washing, as in the experiments by Parrot (1818); (v) the salinity of floating ice in natural waters decreases strongly during summer, in particular in the free board; (vi)

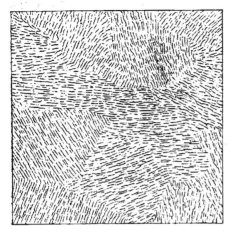

Fig. 1.4: Tinfoil replica of a horizontal section of sea ice from Drygalski (1897), showing crystals of 2 cm size with parallel plates within each crystal, spaced by 0.5 to 1 mm.

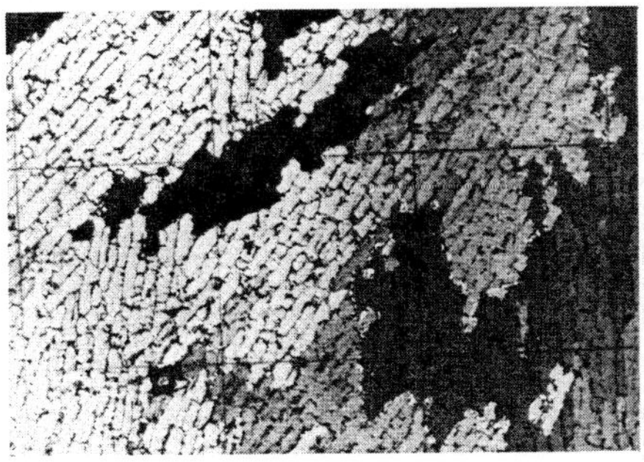

Fig. 1.5: Photomicrograph near the bottom of 30 cm thick young ice from Weeks and Hamilton (1962), showing individual crystals with parallel plates. The superimposed grid scale is 1 cm.

the salinity of first-year floating ice in natural waters decreases with thickness and age, indicating some slow desalination process.

Role of the microstructure for the ice salinity

The investigation of the microstructure of sea ice, concerning the distribution of the salt within the pores, also began at the time of the first scientific Arctic expeditions. The structure of ice forming from seawater and aqueous NaCl solutions had been described as lamellar with liquid brine being situated between the plates (Ruedorff, 1861). Walker (1859) characterized the sea ice as having a 'vertically striated structure'. The latter was most apparent near the ice-water interface. Later the cellular structure of the sea ice freezing interface was reported in studies by Drygalski (1897) and Hamberg (1895), pointing out the contrast to lake and glacier ice. In these pioneering studies the first graphical illustrations of vertical and horizontal sea ice sections were presented and may be compared to present day micrographs (figures 1.4 and 1.5) . Another detail pointed out was that for floating sea ice the orientation of the main crystal axis is horizontal (Hamberg, 1895).

Parrot (1818) and Ruedorff (1861) both found in experiments the salinity of ice grown from saline solutions to increase with growth velocity. Quincke (1905b,a) has given a detailed description of freezing experiments with solutions of salinities lower than 0.3‰ NaCl. He described the formation of lamellae and the milky ice with brine inclusions which he called *foam walls*. For such low salinities, the visibility was limited to raising the temperature of the ice close to the melting point, a fact related to the proportionality of melting temperature depression and brine salinity. The latter was already well established at that time due to the work of Blagden (1788) and Ruedorff (1861), see section 1.1 above. Quincke reported on a large number of experiments with different solutes and concentrations. Some of his numerous interesting findings were:

- The more rapidly ice freezes, the clearer its appearance and the harder the ice.

- The more rapidly ice freezes, the more lamellae or *foam walls* it contains.

- The more rapidly ice freezes, the higher its salinity.

Taken together, the finding from Ruedorff (1861) and Parrot (1818) that ice salinity increased with growth velocity was consistent with the field observations of thin ice by Weyprecht (1879). While Quincke experimented mostly with low salinity solutions, his analysis indicated that a relation between ice salinity and growth velocity was linked to the decreasing distance of brine inclusions for more rapidly growing ice.

First perennial salinity observations

Hence, already the observations at the turn of the 19th century could have allowed the suggestion of a basic principle of salinity entrapment in sea ice: its dependence on the number of solute inclusions within the microstructure. However, probably due to the lack of an ice research community joining the results from different fields and countries (Tyndall, 1872; Weyprecht, 1879; Hamberg, 1895; Drygalski, 1897; Quincke, 1905b), this connection was not realized. Its importance to predict sea ice salinities has still received little attention in present day discussions. At the beginning of the 19th century the salt entrapment in sea ice was no main principle question of Arctic Research and observations on sea ice salinity and growth conditions were only slowly increasing. The focus of Arctic expeditions was still mainly to understand the many other new aspects and describe sea ice and ocean on larger scales.

A more qualitative and detailed understanding of the salinity evolution of sea ice would also have made it necessary to perform time-dependent studies to remove an indeterminacy: On the one hand, as the growth velocity of ice normally decreases with its thickness, the observed decrease in ice salinity with bulk thickness relates to the lower average growth velocity. On the other hand, is the local ice salinity expected to decrease with time due to drainage, as evident from the comparison of salinities of first-year and multi-year floes (Rae, 1874; Moss, 1878; Weyprecht, 1879). To understand

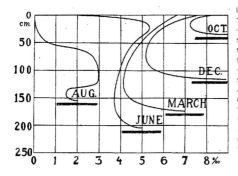 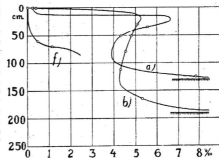

Fig. 1.6: Typical salinity profiles during seasonal growth and decay of arctic first-year ice reported by Malmgren (1927) from the Norwegian North Polar Expedition with the *Maud* 1918-1925.

Fig. 1.7: Typical salinity profiles of arctic first-year ice reported by Malmgren (1927). f) Almost molten first-year ice in July; a) and b) first-year ice during January and March which started growing from a fresh melt water layer.

and quantify both desalination mechanisms it was therefore in a first instance necessary to obtain continuous time series of vertical salinity distribution. A first step in this direction was made during the Norwegian North Polar Expedition with the *Maud* 1918-1925. The seasonal evolution of first-year salinity profiles obtained during the expedition was reported by Malmgren (1927)[3] and is shown in figures (1.6) and (1.7).

Although the data obtained by Malmgren (1927) were limited, the profiles gave a first quantitative indication of desalination changes of ice salinity at a fixed level, in comparison to the influence of the growth velocity on the initial salinity. The basic statements about the salinity of sea ice from earlier studies were thus confirmed in the first detailed analysis of Arctic sea ice samples by Malmgren (1927):

- *The entrapment of salt in sea ice increases with its freezing rate.*

- *The older the ice, the less salt it contains.*

Malmgren's figures also indicate the possibility of the formation of fresh surface ice in autumn from a surface melt water layer.

Malmgren's work remained unextended during the following decades. Research concentrated on the large-scale forms of sea ice (Zukriegel, 1935) and the influence of its

---

[3] In his dissertation Malmgren also provided a review of physical properties of sea ice. A particular and still actual result was a simple equation for the reduction in the effective latent heat of fusion due to the presence of brine

distribution on climate and ocean circulation (Pettersson, 1907; Helland-Hansen and Nansen, 1909; Koch, 1945). Hence, by the middle of the 20th century the information on sea ice salinity evolution was still very sparse (Zubov, 1943) and a theoretical description of the noted desalination mechanisms and the redistribution of brine within sea ice during drainage had not been attempted.

1.3   Sea ice salinity and microstructure: State-of-the-art

1.3.1   Salinity

In the second half of the 20th century observations of sea ice salinity increased. Beginning with multiyear-ice observations (Schwarzacher, 1959; Untersteiner, 1968; Cox and Weeks, 1974), now also the interest in thin ice was growing. First detailed observations of thin ice salinity profiles and their evolution in time were published by Weeks and Lee (1958), who also performed an analysis of the horizontal variability of salinity of thin samples (Weeks and Lee, 1962). Several authors addressed the horizontal variability of the salinity of first-year ice and discussed the role of brine inclusions on different scales (Bennington, 1967; Tucker et al., 1984; Cottier et al., 1999). Also classifications of salinity profiles in terms of their shape have been suggested (Eicken, 1992) for the Antarctic Ocean and bulk salinities were compared between Arctic and Antarctic (Kovacs, 1996).

Detailed field-work at a fixed location over the course of the winter has provided time-series of salinity profiles which are very important when physical modeling of ice growth is considered (Wakatsuchi, 1974b; Nakawo and Sinha, 1981). However, such time series of profiles are still sparse today. Nakawo and Sinha (1981) analysed their observations in connection with an ice growth model (figure 1.8). An important aspect of this study was, that it demonstrated a relation between the plate spacing and the salinity. This will be discussed next.

1.3.2   Plate spacing

Also the basic microstructural length scale of sea ice, the plate spacing, received increasing interest (Fukutomi et al., 1952; Anderson and Weeks, 1958; Schwarzacher, 1959; Tabata and Ono, 1962; Cherepanov, 1964). First models for the strength of sea ice treated the plate spacing as a constant (Anderson and Weeks, 1958; Assur, 1958). This assumption was in conflict with the large observed range of 0.2 to 0.5 mm for thin ice (Fukutomi et al., 1952; Anderson and Weeks, 1958) and 1 to 1.5 mm for thick ice (Assur, 1958; Schwarzacher, 1959; Cherepanov, 1964). It was rejected, when more observations became available and clearly showed an increasing plate spacing with decreasing growth velocity (Weeks and Hamilton, 1962; Assur and Weeks, 1963; Weeks and Assur, 1964). Lofgren and Weeks (1969b) aimed to establish this dependence quantitatively by freezing NaCl

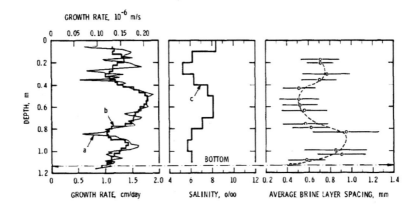

Fig. 1.8: Principal relation between freezing rate V, bulk ice salinity S_i and average brine layer or plate spacing. Adopted from Nakawo and Sinha (1984)

solutions in the laboratory, but the observations eventually differed by a factor of 2 to 3 from earlier field observations. This discrepancy remained when more systematic field studies of the plate spacing were performed (Nakawo and Sinha, 1984). The salinity dependence of the plate spacing was neither conclusive (Rohatgi and Adams, 1967b; Lofgren and Weeks, 1969b).

That plate spacing, growth velocity and ice salinity are related, was demonstrated by Nakawo and Sinha (1984) on the basis of the *a posteriori* analysis of a single ice core (figure 1.8). In a later study a similar relation between plate spacing and growth velocity was found (Sinha and Zhan, 1996). Observations obtained since then have confirmed the results for thicker ice (Lange, 1988; Cole et al., 1995). Due to variability and a lack in accurate growth rate measurements a quantitative evaluation has not been possible yet. The lack in systematic studies of plate spacing variability is even larger at higher natural growth velocities, where only a few field observations exist (Weeks and Assur, 1964; Nghiem et al., 1997). Some observations of microstructure and salinity have been obtained in artificial ice tank studies (Gow et al., 1987; Eicken et al., 1998, 2000; Haas, 1999), yet an analysis of the plate spacing has not been presented. Approaches to predict the plate spacing theoretically have been undertaken, but so far no solution has been found (Lofgren and Weeks, 1969b; Wettlaufer, 1992). To date, the natural variability of the basic microstructure of sea ice is still poorly understood.

1.3.3   Initial salt entrapment

In practical applications the effective distribution coefficient during freezing of seawater

$$k_{eff} = \frac{S_i}{S_\infty},$$
(1.2)

defined by the ratio of ice salinity S_i and seawater salinity S_∞, is important. The only theoretical attempt made so far to relate k_{eff} to the growth velocity of sea ice was once suggested by Weeks and Lofgren (1967). It is based on an analogy to solute entrapment at a planar interface in the presence of forced convection. As the approach could be reasonably correlated with laboratory observations (Cox and Weeks, 1975), it was extended to formulate a one-dimensional model that allows the prediction of salinity profiles during the growth phase (Cox and Weeks, 1988). Since then this concept of salt entrapment has been applied frequently (Nakawo and Sinha, 1981; Souchez et al., 1988; Eicken, 1998, 2003), although a clear validation by field data is still lacking. That some studies indicate a rather poor performance (Fertuck et al., 1972; Eicken, 2003), indicates that the analogy to planar freezing is not applicable in general. A critical discussion has not been given so far.

An alternative attempt has been made by Tsurikov (1965), who proposed a structural model for the entrapment of brine between the plates. He derived an empirical relation between the growth velocity and k_{eff} of very thin ice grown in a freezer. Shesteperov (1969) found agreement of the approach with field observations from Johnson (1943). However, the approximate agreement is restricted to thin bulk samples and tells little about the underlying physics: Tsurikov's formulation not even considered the increased *liquidus* salinities within the freezing ice-brine mixture. Thus, a physical justification is also lacking for this approach. Strict empirical correlations between bulk ice salinity and bulk growth rate have been suggested (Kovacs, 1996) and may be even more reasonable than the mentioned ambiguous parametric forms.

The main problem in the prediction of k_{eff} is its gradual transition. The bulk salinity of thin ice samples has been found to depend on thickness, growth velocity being the same (Wakatsuchi, 1974a). This implies a gradual change from the seawater salinity, near the cellular interface, towards much lower values, some centimeters within the ice. It is evident in nondestructive salinity profiles from Cox and Weeks (1975). Being aware of this problem, Cox and Weeks (1988) correlated k_{eff} at a fixed distance from the interface with growth velocity. In a proposed desalination model they still prescribed the further salt loss based on the growth conditions. Also this was done empirically. To date no physically consistent model to describe the entrapment of salt in sea ice exists.

1.3.4 Desalination and brine channels

The lack in a concise model for the salt entrapment is attributed to the difficulty to described the salt fluxes from sea ice theoretically. These take place by means of brine channels which are much wider than the plate spacings. In earlier work the concept of desalination of *warming* sea ice due to widening of brine channels had been outlined (Hamberg, 1895; Schwarzacher, 1959). The existence of these channels in *cooling* sea ice during the growth phase was first described by Bennington (1963, 1967). Lake and Lewis (1970) concluded from observations of brine channels and temperature fluctuations, that the basic mechanism of salt rejection at the ice-water interface implies downward convective motion of brine-rich streamers, balanced by upward moving waters around them. Far away from the ice the brine channels appear as in figure 1.9. Experiments in thin growth cells have confirmed this view (Eide and Martin, 1975; Niedrauer and Martin, 1979).

In the following years considerable effort was made by by Wakatsuchi and his colleagues to provide a quantitative description of brine channels. Observations included the falling velocity of streamers, their number density in the ice and water and their brine salinity (Wakatsuchi, 1977; Saito and Ono, 1977; Wakatsuchi, 1983; Wakatsuchi and Ono, 1983; Wakatsuchi and Saito, 1985). It became apparent that these parameters and the channel fluxes depended on growth conditions. Wakatsuchi and Kawamura (1987) further demonstrated experimentally that brine channels preferentially form at boundaries between crystals (figure 1.10). They suggested that the principle picture of desalination is a coupled mechanism. Expansion due to freezing in brine layers results in a flow towards the brine channels, where the resistance to flow is least and gravity drives the desalination.

Since Wakatsuchi's work little progress has been made with concern to the formation, distribution and convective activity of channels. Only a few additional observations of channel distribution were obtained (Tison and Verbeke, 2001). It is clear that brine channels are the key features in the primary desalination process. As locations of downward flow, they require a weak compensating upward flow. This mechanism exchanges persistently more saline brine against less saline seawater. Predictions of channel widths from convective stability scalings only yielded approximate agreement (Lake and Lewis, 1970; Eide and Martin, 1975; Niedrauer and Martin, 1979). A theory of their formation and a quantification of salt fluxes is not established yet. It is suggestive from figure 1.10 that the flow within brine channels will depend on their lateral feeding through the crystals, and thereby on the plate spacing of sea ice.

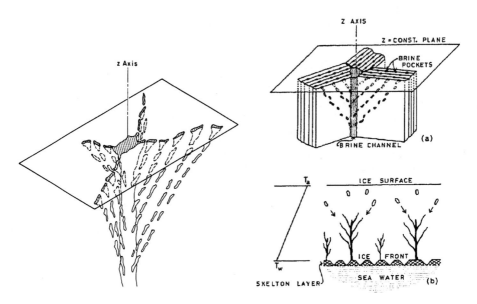

Fig. 1.9: Schematic drawing of a sea ice brine channel which is fed laterally. Adopted from Lake and Lewis (1970).

Fig. 1.10: Upper: drawing of a brine channel fed from the brine layers of surrounding crystals. Lower: Evolution of brine channel networks with distance from the ice-water interface. From Wakatsuchi (1983).

1.3.5 Convective instability

More recently, progress has been made with respect to the onset of convection during the growth of sea ice (Wettlaufer et al., 1997; Worster and Wettlaufer, 1997). The so called 'mushy layer' stability theory applied by the latter authors is a variant of convective stability in porous media (Nield and Bejan, 1999). It considers the convective stability in the porous medium sea ice, the 'mushy layer', and solutal convection in the liquid at the interface (Worster, 1992). The main salt fluxes are associated with the former internally driven convection (figure 1.11) which can be expected to set in when the Rayleigh number

$$Ra_p = \frac{\beta \Delta S_b g \Pi H_p}{\nu \kappa_b} \tag{1.3}$$

exceeds a critical value (Wettlaufer et al., 1997; Worster and Wettlaufer, 1997). Experimental results have been presented in a form shown in figure 1.12. It states that the onset of convection takes place when the thickness of the ice (H_p) exceeds a value that depends on its average solid fraction and the salinity difference ΔS_b between the seawater and the

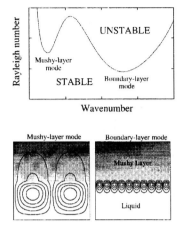

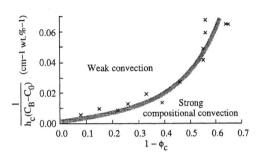

Fig. 1.11: Lower: convective mode ins sea ice (mushy layer) and a liquid boundary layer. Upper: Critical Rayleigh numbers for the onset of convection. From Worster and Wettlaufer (1997).

Fig. 1.12: Proposed marginal convective stability for internal convection from sea ice, in dependence on solid fraction Φ, ice thickness h_c and brine salinity difference between the cold boundary and the solution Worster and Wettlaufer (1997).

brine at the upper boundary. The curve may be interpreted as an empirical permeability law $\Pi(\Phi)$.

Wettlaufer et al. (2000) have compared the curve in figure 1.12 to the natural growth of lead ice and claimed an agreement with the onset of salt fluxes. No other confirmations of figure 1.12 have been published. Its generality may be questioned due to the many different salinities and temperatures on which the curve is based, while not considering the strong dependence of brine viscosity ν on temperature and concentration. Moreover does its derivation assume a constant solid fraction in the ice which is unrealistic.

Clearly, the application proposed by Wettlaufer and Worster to sea ice is a relevant step to improve the understanding of the desalination process, and the dynamics of the lower fraction of sea ice, its high porosity *skeletal layer*. However, in its to date applied form it is physically not fully consistent. Considerable progress could be made by predicting the permeability Π in dependence on growth conditions. The latter depends on the pore scale of a porous medium (Carman, 1956; Dullien, 1979; Nield and Bejan, 1999). This points also here to the plate spacing of sea ice as the key variable.

1.4   The present work

From the above summary it is clear that the plate spacing of sea ice is the fundamental microstructural variable in the process of early desalination and the establishment of a stable salt content. It is the formation of this fundamental microstructure that one has to understand in order to (i) model the salt entrapment during the early growth stages of sea ice properly, (ii) describe the microstructural evolution of sea ice of any age, (iii) treat sea ice physics properly in any application where microstructure and salinity play a key role. It is further expected to constrain the formation and activity of brine channels and thereby also should play a key role in understanding the permeability evolution of multiyear ice. It is the main goal of the present work to provide a theory which allows the prediction of the plate spacing of sea ice in dependence on growth conditions. To do so, this work is in principal structured into three parts, plus an appendix.

The chapters 2 through 6 review previously suggested theories and experiments on salt entrapment and plate spacing and attempt to establish a principal understanding of the desalination processes in terms of its vertical extent measured from the ice water interface. Chapter 2 compares sea ice bulk salinity observations available in the literature and proposes a conceptual view of salinity profiles, that differs from earlier classifications. The latter is further outlined in chapter 3, which considers the bottom regime of sea ice in terms of two types of boundary layers. The next chapter 4 makes a step back to a planar interface, to consider the possibility of modeling salt entrapment in sea ice as an analogy to the latter. An important aspect of this chapter is to provide an understanding of convective solute transport at a planar interface and critically discuss the difference between forced and free convection that so far has been largely ignored. Chapter 5 considers the initial salt entrapment for a cellular interface and discusses several flaws that must be reckoned with during laboratory experiments. This chapter provides a reference dataset of interfacial salt entrapment to be validated later. Observations of sea ice plate spacings for which the growth velocity is well known are rare, and chapter 6 compiles a reference dataset. Previous theories and approaches to predict the plate spacing are also critically reviewed.

The formulation, validation and application of a theory for the prediction of the plate spacing follows in chapters 7 through 12. The classical morphological stability theory for a planar interface is introduced and several applications to saline solutions are described to show its applicability, but also its limits, when a cellular interface is considered (chapter 7). The next chapter (chapter 8) introduces the theory of dendritic pattern formation as an important prerequisite to apply several scalings of dendrite spacings from the field of crystal growth and directional solidification to NaCl solutions. It is shown, that none of the theories predicts the plate spacings of sea ice properly. As a consequence a novel approach is formulated in chapter 9. It may be viewed as a macroscopic variant of morphological stability theory and is shown to give very promising predictions of plate spacings of NaCl solutions. The modification of the theory in the presence of natural convection, given in chapter 10 is fundamental to extend its applicability to sea ice. It is thereby shown that the theory gives proper predictions of the plate spacing for natural growth conditions.

The theory is readily applied to convective problems of the sea ice skeletal layer. This is done in chapter 11 and leads to a number of new relevant results. A heuristic approach to the convective stability of sea ice as an ensemble of packages of Hele-Shaw cells, provides a physically reasonable interpretation of experiments. The final chapter 12 provides three further applications of the combined theory of convection and morphological stability. (i) A solution to the problem of the planar-cellular transition during the freezing of brackish waters is presented. (ii) The constancy of the thickness of the skeletal layer of sea ice is explained. This further leads to an interesting new perspective on brine channel formation. (iii) Upper and lower bounds on the stable salinity of sea ice are proposed in dependence on its growth velocity.

The appendix provides additional material normally not available. Thermodynamic properties of ice and brine along the liquidus (appendix A), in particular the solute diffusivity and dynamic viscosity, are essential for a correct application. A discussion of the interfacial conditions during the growth of saline ice (appendix B) provides a simple view of nonlinear interfacial temperature gradients. Basic mechanisms of desalination of older sea ice are reviewed in (appendix C). Appendix (E) is a critical discussion of the reference sea ice salinity dataset from Cox and Weeks (1975), arguing why this dataset should be discarded or, at least, distinguished from natural sea ice observations. Appendix F is a detailed discussion of statistical datasets of microstructural observations in terms of the width of brine layers at the *bridging transition*, the moment where brine layers begin to transform to disconnected inclusions. Its basic result is an essential part of the prediction of the salt entrapment in sea ice, proposed in chapter 12. Finally, appendix D provides information on field observations, obtained by the present author and included in the reference data set on sea ice salt entrapment, discussed in chapters 5 and 12.

2. Sea ice salinity: observations and empirical relations

The present chapter gives an introduction of observations of sea ice salinity and their variation with vertical position in the ice. It begins with the relation between bulk salinity and ice thickness. In general, it is not unexpected, that variability in this relationship has been observed on the basis of growth conditions (Weeks and Ackley, 1986; Weeks, 1998a; Eicken, 2003). From a regional perspective, Kovacs (1996) has argued that the same empirical relations are applicable for Arctic and Antarctic first-year ice. This appears surprising, as there are considerable differences between the areas: (i) in the Antarctic Ocean the climate is not that cold as in the central Arctic; (ii) frazil ice growth is a more important growth mode in the Antarctic than in the Arctic; (iii) ice floes in the Antarctic consist frequently of rafted thin ice; (iv) water salinities are on average 5 to 10 % larger in the Antarctic. While (i) and (ii) should lead to larger Arctic ice salinities, (iii) and (iv) will have the opposite effect. The similar ice salinity thickness relationships thus may reflect a cancellation of several effects. Nevertheless, the ice may be microstructurally different.

2.1   The effective solute distribution coefficient

The following definitions of the redistribution of salt during the gradual transition from seawater to sea ice will be used in the present work:

- $\overline{k_{eff}}(t) = \overline{S_i}(t)/S_\infty$ is the bulk effective distribution coefficient, defined as the ratio of bulk sea ice salinity and bulk seawater salinity from which the ice forms. S_∞ is the seawater salinity far from the interface.

- $k_{eff}(t, z) = S_i(t, z)/S_\infty$ is the local effective distribution coefficient, normally a function of position and time in sea ice.

- $k(t, z = 0) = S_i(t, z = 0)/S_{int}$ is the interfacial distribution coefficient. This term is used for a planar interface during the freezing of fresh water *and* for a cellular macroscopic interface. It includes salt incorporated within and between grains and subgrains.

- $k_0(H_0) = S_{i0}(H_0)/S_\infty$ is the quasi-stable effective k_{eff} at some yet unspecified level H_0, where desalination almost drops. H_0 is measured upwards from the ice bottom.

The possible variation of S_∞ with time is neglected in this concept.

2.2 Empirical relations of bulk salinity and ice thickness

The rule of thumb that thick first-year ice contains one-sixth of the salinity of underlying water, $\overline{k_{eff}} \approx 1/6$, first suggested by Moss (1878) has been verified in many later studies (Zubov, 1943; Doronin and Kheisin, 1975; Weeks and Ackley, 1986). Another generally accepted aspect is that the colder the formation conditions or the faster the freezing rate of ice, the larger its salinity. Malmgren (1927) was the first who demonstrated, via time-series of salinity profiles, that the longterm salinity of thick ice is, besides the freezing rate, also related to its age: Thicker ice contains also less salt in the upper portions, due to the longer time available for desalination.

Both these processes are integrated in a relation between thickness and bulk salinity, which can be obtained in field studies without large effort. Among the empirical relations that have been suggested by different authors the most frequently used are reported as follows in a normalised form, the bulk distribution coefficient $\overline{k_{eff}}$ given as ice salinity divided by bulk salinity in the studies. Cox and Weeks (1973) suggested the following two relations, mainly based on the analysis of ice cores from the Arctic:

$$\overline{k_{eff}} = 0.445 - 6.06 \times 10^{-3} H_i \qquad (2.1)$$

for ice thinner than 40 cm and

$$\overline{k_{eff}} = 0.246 - 4.97 \times 10^{-4} H_i \qquad (2.2)$$

for thicknesses $H_i > 40$ cm. The original equations from Cox and Weeks (1973) have been divided by 32 ‰, which is most typical for the Arctic regions from which most of their correlated observations stem.

Ryvlin (1979) has proposed an equation which accounts for the growth conditions of the ice. In non-dimensional form it reads

$$\overline{k_{eff}} = \overline{k_\infty} + (1 - \overline{k_\infty})e^{(-bH_i^{1/2})}, \qquad (2.3)$$

where $\overline{k_\infty} = 0.13$ and the parameter b being slightly dependent on growth rate. According to Ryvlin, $b \approx 0.35$ cm$^{-1/2}$ for growth velocities above 4 cm d^{-1} and $b \approx 0.6$ cm$^{-1/2}$ for slow growth below 0.5 cm d^{-1}. For a further discussion it is interesting to approximate this equation by a form which only contains the thickness H_i. First, one may approximate the transitional behaviour of b between the bounds given by Ryvlin as $b \approx 0.25(1+V^{-1/2})$, V given in units of cm d^{-1}. Next, a typical ice surface temperature of -15 °C is used as a compromise between thin and thick ice. This leads, using appropriate thermal properties and assuming a linear temperature gradient in the ice, to the parametrization

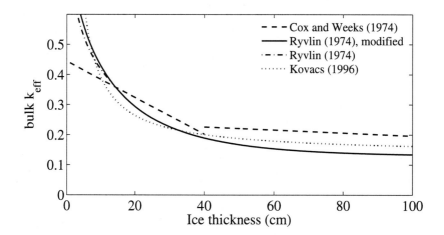

Fig. 2.1: Bulk relations between the average ice salinity $\overline{k_{eff}}$ and ice thickness from several authors. The relations from Cox and Weeks (1973) and Kovacs (1996) were normalized by 32 and 34 ‰, respectively.

$V^{1/2} \approx 9H_i^{-1/2}$, with H_i in centimeters and V in cm d^{-1}. Inserting these approximations into equation 2.3 gives

$$\overline{k_{eff}} = \overline{k_\infty} + (1 - \overline{k_\infty})e^{\left(-0.25\left(H_i^{1/2}+0.11H_i\right)\right)}. \tag{2.4}$$

It may be noted that this equation has the property that $\overline{k_\infty} = 1$ at very small thickness. At a thicknesses above 100 cm $\overline{k_{eff}}$ it approaches a constant fraction $\overline{k_\infty} = 0.13$ of the sea water salinity. Equation 2.4 formally gives a slightly larger $\overline{k_{eff}}$ than Ryvlins's formulations. However, as no physical model is underlying the approach from Ryvlin, equation 2.4 should rather be taken as an approach to approximately compare it to other empirical fits.

Kovacs (1996) has correlated a large amount of sea ice bulk salinity data from the Antarctic and suggested the form

$$\overline{k_{eff}} = 0.137 + 2.549H_i^{-1}, \tag{2.5}$$

with H_i in centimeters. Kovacs's original salinity equations have been normalised by 34 ‰, which is typical for samples from the Antarctic seas.

All three relations are compared in figure 2.1. It is seen that they deviate considerably from each other. Moreover, the data in each of the noted studies show a spread of ± 20 to 30 % around the respective relations, indicating the limitations of a simplified approach to estimate the sea ice salinity from its thickness. None of the above empirical relations is justified by a physical model. Their parametric form is arbitrary, and even if the scalings from Ryvlin (1979) and Kovacs (1996) appear to be more reasonable in being assymptotic to a constant salinity at large thickness, their skill may be inferior to the linear fits from Cox and Weeks (1973). The main aspects to be recognized in connection with figure 2.1 are

- At thicknesses of 100 to 200 centimeters the empirical relations all agree with the rule of thumb from earlier expeditions, that first-year ice typically contains one-sixth of the chlorinity of seawater on which it floats (Moss, 1878).

- None of the empirical relations is based on a physical model and the spread of ± 20 to 30% in the data on which the relations are based indicates the need for the latter.

- The uncertainties implied by a simplified parametric relation between thickness and bulk salinity are particularly large for thin ice less than 50 cm thick.

The above equations are in principal valid for first-year ice during its growth phase, or more exactly: during its cooling phase. When the ice is warming and melting, the salinity profiles become more complex, as seen in figure 1.7 adopted from Malmgren (1927). In first-year ice which has survived the first summer the salinity often increases towards the bottom and the salinity-thickness relationship becomes different. In this case one may find a slight increase of the bulk $\overline{k_{eff}}$ with thickness (Cox and Weeks, 1973; Weeks, 1998a). The latter is related to the combined effect of desalination, snow cover and melt, variable bottom accretion and surface and bottom ablation. These different mechanisms can produce much larger scatter in the bulk salinity and are no longer related to the entrapment process alone, the goal of the present study.

2.3   Thin ice growth sequences

The noted empirical thickness-salinity relationships show the largest differences for ice thinner than 40 to 50 centimeters. This is the regime where ice is subject to the largest variations in growth conditions. An analysis of the physical mechanism of salinity entrapment must be validated with respect to this variability. Of particular importance in this connection are time-series of salinity samples taken from ice growing at fixed locations. In figure 2.2 two such series of normalised bulk salinity versus ice thickness based on observations from Cox and Weeks (1973) and Melnikov (1995) are shown along with the above noted empirical relations.

Cox and Weeks (1973) reported, above an ice thickness of 10 cm, two salinity samples at each thickness increment. The range is indicated by bars in figure 2.2 and amounts approximately to 5 to 10 % of k_{eff}. A similar standard deviation has been found in other studies (appendices C.3 and D). It is apparent that variations from a smooth curve, connecting the data points from Cox and Weeks (1973) and Melnikov (1995), are larger than 5 to 10 %, also if some uncertainty in the thickness measurements is accounted for. Very basically, this scatter around a smooth curve may be related to several effects reported in the studies. First, it was shown by Melnikov (1995) that intrinsic fluctuations in the brine salinity take place on timescales of a few hours and are of the order of ±15 %. They may be interpreted in terms of in- and outflow of brine, being consistent with some limited laboratory studies (Eide and Martin, 1975; Niedrauer and Martin, 1979). Second, uncontroled loss of brine during sampling might have produced an artificial variability. The latter effect is expected to be most relevant for very thin ice. Finally, Melnikov (1995) also found a daily cycle, probably linked to radiative forcing variability during early growth, when sampling was performed on a four-hourly basis. Such variations appear also plausible for the dataset obtained by Cox and Weeks (1973) during Arctic spring or late winter. Considering these sources of intrinsic and artificial variability, one may deduce that the two time series might each be associated with a relatively smooth curve. It further can be concluded, that none of the empirical relations in figure 2.2 preferably captures the bulk ice salinity evolution during the two growth sequences.

Although the observed salinity evolutions from Cox and Weeks (1973) for the high Arctic and from Melnikov (1995) for the Antarctic are quite similar, a look on other published data indicates that this coincidence cannot be generalised. In figure 2.3 the two growth sequences are compared to other natural ice growth observations (Weeks and Lee, 1958; Wakatsuchi, 1983; Steffen, 1984; Gow et al., 1990) and a short time series obtained by the present author (appendix D). It is seen that bulk salinities at a certain thickness may differ by almost a factor of two. This result emerges also when comparing the individual growth sequences and thus is not related to sampling fluctuations. It must be related to the environmental conditions including air temperature, wind speed, snow cover and oceanic heat flux. Only for one of the noted series, taken from the study of Gow et al. (1990), these conditions have been properly documented. However, some of the studies provide information from which the growth velocity can be estimated (Weeks and Lee, 1958; Melnikov, 1995).

The only empirical relation which may be modified in terms of a dependence on growth velocity, see the previous paragraph, is equation 2.4 after Ryvlin (1979). For example, one may argue that the surface temperature of thin ice growing under typical high latitude conditions is normally higher than −15 °C, assumed to convert Ryvlin's empirical formula to equation 2.4. A more realistic thin ice surface temperature in the range −5 to −10 °C would change the relationship between V and H_i. Employing this modification leads to formula 2.4 with a slightly different exponent, but can only account for a small shift to a 5 to 10 % lower $\overline{k_{eff}}$ (not shown). It does neither improve the overall

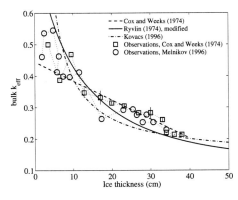

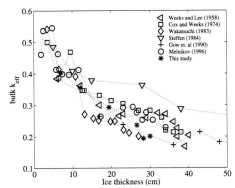

Fig. 2.2: Growth sequences of $\overline{k_{eff}}$ for thin ice from the Arctic (Cox and Weeks, 1973) and Antarctic (Melnikov, 1995), shown with suggested bulk relations.

Fig. 2.3: Growth sequences of $\overline{k_{eff}}$ for thin ice from several sources to be compared with figure (2.2).

agreement with the observations. This points again towards the arbitrary character of the empirical formula and the need for physical models to properly describe the salinity entrapment of young ice in dependence on growth conditions.

2.4 Salinity profiles

2.4.1 Profile shapes

Malmgrens discussion of vertical salinity profiles during the growth and aging of sea ice (figure 1.7) was already mentioned. He showed that the general C-shape of winter ice, with maximum salinities at the bottom and top, changes to a ?-shape in early summer, due to the warming and melting of ice from the surface. In mid-summer often warm ice tends to show an intermediate salinity maximum, as seen in figures 1.6 and 1.7 from the introduction. Eicken (1992) has proposed a classification of Antarctic sea ice salinity profiles, supplementing the C- and ?-shapes by classes of S- and I-shapes. He proposed polynomial expressions between salinity and normalized depth to characterise the profiles. Such a quantitative classification is problematic due to several reasons. Looking at figures 1.6 and 1.7 from Malmgren (1927), one may suggest that the I-shape, for example, probably does not represent a basically stable state, yet reflects the transition from the winter C-shape towards the summer S-shape. Summer warming and melting imply a permanent change of salinity profiles, due to internal redistribution driven by warming and melting from the top, but also due to bottom ablation in areas where oceanic heat fluxes are most relevant. In addition to this growth-cooling-warming-melt history, the rafting of thin winter ice may, as pointed out by Eicken (1992), produce rather irregular shapes. A polynomial classification of mentioned C-?-S-I-shapes makes, due to this transient character of salinity profiles, only sense, when it is related to the physical processes and thermal history. These redistribution processes have so far not been simulated successfully and their quantitative understanding must be rated as poor (Eicken, 2003).

A second critics relates to the use of *normalised* salinity profiles, as applied by Eicken (1992) to compute representative polynomials. This approach treats, for example, 20 and 100 cm thick ice as self-similar. As seen in Malmgren's profiles in figure 1.6 this is not the case for thicker ice. Detailed salinity profiles for even thinner ice obtained by Weeks and Lee (1958) show in more detail that the absolute values of k_{eff} and the asymmetry of C-shaped profiles change during the early growth. Here this is illustrated in figure 2.4 from a time-series of the present author and 2.5 for a laboratory study from Cox and Weeks (1975). These time series of growing thin ice demonstrate, that it is rather the physical extent of the bottom salinity layer and not the overall shape of the profile which is a stable feature in naturally growing sea ice.

Hence, if one considers typical sea ice during the growth season, it is more appropriate to characterise the latter by a salinity minimum between the bottom and the surface. The salinity increase towards the surface is principally explained by the decreasing growth velocity with depth. The increase towards the ice-water interface is related to the fact that the salt exclusion from the ice is gradual. The layers of increasing salinity in the upper and lower part of the ice are thus associated with different physical processes. Their extent is similar for thin ice, but not for thick ice. From these basic concepts, one may also conclude that a physical model for the bottom boundary of sea ice is essential: If, and only if the gradual salt entrapment in the bottom boundary layer is understood and predictable, then the upper part of the salinity profile may be computed as a consequence of the latter.

2.4.2   Bottom boundary layers

The informations that are generally accepted about the bottom boundary regime of sea ice are the following. The lamellar structure at the bottom of sea ice and saline ice has long been known (Walker, 1859; Hamberg, 1895; Drygalski, 1897). Its physical extent of 2 to 5 cm has been described in earlier studies to be relatively constant and independent of growth conditions (Assur, 1958; Weeks and Anderson, 1958). Assur introduced the term *skeletal layer* for this regime, where brine can circulate freely between vertical plates.

The vertical extent and microstructural changes above in this regime, necessarily linked to the entrapment process, have received little attention. From thin section analyses of the microstructure of first-year ice Stander (1985) reported several distinct regimes of microstructural development, measured from the bottom. Above a skeletal layer of 2 to 4 cm thickness he described a second regime of 6-8 cm vertical extension, reaching approximately 10 cm upwards from the ice-water interface. Here the brine is still situated in a sheetlike manner between the platelets, but the sheets are no longer perfectly connected. At a distance more than 10 cm from the interface, Stander noted the appearance of vertically elongated pockets. In addition, high lattice distortion features may indicate that the sheets of brine can no longer communicate. The transition layer between $\approx$ 3 and 10 cm from the ice-water interface is thus characterised by bridging between the plates. The expression *bridging layer* has been conceptually used in earlier work by Assur (1958) and Weeks and Anderson (1958), yet the latter authors did not analyse or discuss its vertical scale.

The vertical resolution of salinity cores during standard sample analysis is normally from 2 to 10 cm, and the salinity of some high resolution studies may be compared to the scales indicated by the microstructure analysis from Stander (1985). A closer look at the thin ice salinity profiles in figures 2.4 and 2.5 indicates that the 10 cm lengthscale of the bridging layer in general corresponds to a local minimum in the salinity. In thin ice the bottom boundary layer has, of course, still to evolve. However, all profiles indicate a salinity minimum 6 to 10 cm from the ice-water interface, once the ice has reached a thickness of 15 cm. To be discussed below, the study by Cox and Weeks (1975), from which the profiles in figure 2.5 are taken, has provided observations with a non-destructive method. In the latter figure the distinguishment of the 3 cm skeletal layer from the bridging layer reaching to 10 cm from the interface, is clearest, as sampling is not affected by brine drainage from the open skeletal layer. The 3 cm level, indicated by a dot in all profiles, separates the skeletal layer with a much steeper salinity gradient from the upper levels of the ice.

Some more observations of the noted regimes will be discussed in the following chapter 3 for both the *skeletal layer* and the *bridging layer*. For the moment it suffices to say that these lengthscales are (i) quite distinct from each other, (i) that their typical values for growing sea ice are 3 and 10 cm, respectively, and (iii) that the vertical extent of the bridging layer is similar for thin and thick ice.

2.4.3 Time series of salinity profiles

Similar to the general understanding of the seasonal evolution of sea ice from seasonal profile shapes elucidated by Malmgren (1927), one must obtain profiles on a shorter timescale to understand the desalination process. For example, both figures 2.4 and 2.5 indicate, that the salinity above the minimum changes indeed very little in subsequent profile. An exception is the change between March and April in figure 2.4. As it is described in appendix D, this re-initiation of salt loss is associated with a severe warming

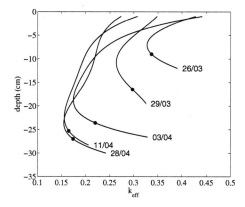

Fig. 2.4: Normalized salinity profiles during the growth of thin ice in the field described in the appendix (D). Sampling dates for the profiles are denoted. The dots indicate the position of the 3 cm distance from the macroscopic ice-water interface.

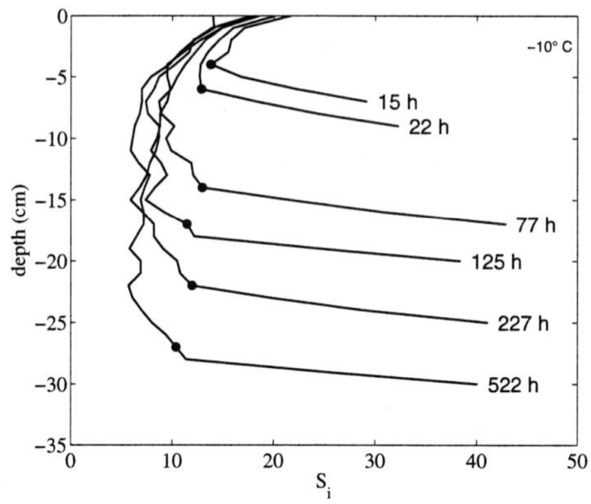

Fig. 2.5: Salinity profiles during the growth of thin ice in a laboratory experiment with a constant upper boundary temperature of -10 °C Cox and Weeks (1975). The hours since the onset of growth are denoted for each profile. Dots: 3 cm distance from the ice-water interface. Salinities have not been normalised, as the water salinity S_∞ increased during the experiment.

event.

The desalination due to a warming event demonstrates, that simultaneous time series of salinity and temperature profiles in the ice are of particular importance. These give the possibility to evaluate the desalination of the ice in terms of the brine volume. Such field observations, which also allow a reasonable prediction of the growth velocity of the ice, are still rare. An exemplary study is the work from Wakatsuchi (1974b), for which ice salinity and temperature are shown as contourplots in figures 2.6 and 2.7. Although the contour resolution is limited, it may be deduced that the ice, during a prolonged phase of constant growth rate, attained an almost constant salinity near the interface, changing very little at a distance of more than ≈ 10 cm. In practice, due to the linear temperature gradient, it may often be sufficient to obtain ice surface temperature measurements. When ice is very thin, or temperature fluctuations are large, this is no longer reasonable.

Figure 2.6 also clearly demonstrates the small extent of the bottom regime that is so important for the understanding of sea ice salinity. It is noted, that the observations from Wakatsuchi (1974b) refer to the re-initiation of growth in second-year ice. A similar study, yet presented in a somewhat different form, has been performed by Nakawo and Sinha (1981) for the entire growth season of first-year ice. Cox and Weeks (1975) observed, for

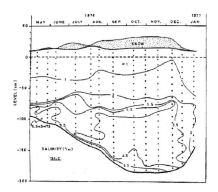
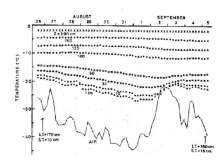
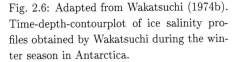

Fig. 2.6: Adapted from Wakatsuchi (1974b). Time-depth-contourplot of ice salinity profiles obtained by Wakatsuchi during the winter season in Antarctica.

Fig. 2.7: Adapted from Wakatsuchi (1974b). Time-depth-contourplot of ice temperatures corresponding to the salinity profiles in figure 2.6.

the profiles in figure 2.5, also the ice temperatures with reasonable resolution. In a later chapter 5 it is illustrated, how this dataset thereby can be explored to obtain a basic understanding of aspects of desalination in the bottom regime of sea ice.

Modeling

Salinity profiles have in principal been simulated (Cox and Weeks, 1988; Eicken, 1992) on the basis of the following model: (i) prescribe the growth velocity dependence of k_{eff} 3 cm from the interface, (ii) define a bottom desalination regime by a threshold brine volume v_{b0}, (iii) prescribe a local desalination rate dS_i/dt in the desalination regime in dependence on the local brine volume and temperature gradient. However, as pointed out in the appendix C.4.2, the critical desalination parameters have not been properly validated. As may be deduced from the graphs shown by Cox and Weeks (1988), their proposed model algorithm produces a bottom boundary layer of 10 to 20 centimeters under typical rapid growth conditions, because this is the length scale to reach v_{b0}. However, the predictability of the bottom regime is not warranted by this algorithm and it failed in more more recent simulations by Eicken (2003).

The present day (empirical) model approaches are thus incapable to represent the bottom desalination in a reasonable manner. Their prediction of the upper part of the quasi-C-shape is no proof for their consistency. Approximate agreement can also be acquired by replacing k_{eff} at 3 cm by a somewhat smaller stable k_0, relating to positions farther away from the interface (Fertuck et al., 1972; Petrich et al., 2006). A recent refinement of the empirical relations from Cox and Weeks (1988) has been suggested by

Petrich et al. (2006), on the basis of a more sophisticated numerical model of fluid and solute transport through the pore space. However, this model involves new adjustable parameters and its permeability function is questionable, see appendix C.4.2. It is also based on the rather scattered desalination observations from Cox and Weeks (1975). The latter study will, in chapter 5, be shown to be only representative in a limited way for natural sea ice growth.

2.5   Bulk salinity versus growth velocity

As mentioned above, among the discussed bulk salinity-thickness relations, only Ryvlins's approach 2.3 formally allows to account for a growth velocity dependence. It turned out that the parametrisation is not capable to produce the larger natural variability in $\overline{k_{eff}}$ at some given thickness. Other authors have proposed relations between average growth velocity $\overline{V}$ and bulk salinity $\overline{k_{eff}}$ for thin ice (Wakatsuchi, 1983; Kovacs, 1996). These approaches are inasmuch problematic, as they imply integrated values of growth velocity and ice salinity in a regime where the entrapment may be nonlinear and the thickness fraction of the bottom boundary layer decreases. Such relations, as they are based on rather different thermal histories, and thus combinations of growth velocity and ice thickness, cannot elucidate the physics of the entrapment problem. To understand the latter, one must compare the salinities in the ice locally to the local growth velocity. However, as will be seen in chapter 4, a similar problem applies to the evaluation of ice salinities at different distances from the interface, as performed in an earlier study by Weeks and Lofgren (1967) for 10 to 20 cm thin ice. The reason is that, according to the different position in the bottom boundary layer, the samples represent different desalinations stages. This has been realised by the latter authors, who later evaluated k_{eff} at a fixed distance from the interface (Cox and Weeks, 1975). The latter dataset has become a frequently cited reference in sea ice studies (Nakawo and Sinha, 1981; Weeks and Ackley, 1986; Cox and Weeks, 1988; Weeks, 1998a; Eicken, 2003). It will be discussed in detail in chapter 5.

The analysis of bulk salinity-growth velocity observations may be useful when comparing the early growth stage to a small and similar thickness of a few centimeters. Some experiments of this kind (Johnson, 1943; Tsurikov, 1965; Wakatsuchi, 1974a) are also discussed in a later section 4 to test simple models of convective salt transport. In the latter case, however, another problem is faced. Due to the high mobility of brine in the bottom fraction, conventional ice core sampling almost certainly implies the loss of brine. Kusunoki (1957) has documented some quantitative comparisons of this effect, frequently reported as a regular problem (Cox and Weeks, 1975; Nakawo and Sinha, 1981; Williams et al., 1992). An approximate estimate of the magnitude may be obtained from sea ice density observations: If ρ_* and ρ_{si} are the expected and observed densities of sea ice, and H is the sample thickness, then approximately $H \times (\rho_*/\rho_{si} - 1)$ water equivalents should have been lost. The thin ice sample analysis by Kusunoki (1957) then indicates a maximum loss of 5 to 8 mm water equivalent, presumably associated with the very bottom fractions. The drainage might be less pronounced, when the ice is very thin and crystals are still inclined, but it can hardly be controlled. A recent nondestructive experiment by Notz et al. (2005) indicates that brine loss can be larger when the ice is warming and therefore is inherently unstable. However, the method from the latter authors is based on freezing an instrument, which consists of wires that have a diameter comparable to the sea ice plate spacing, into the ice. It would be surprising, if this does not affect the microstructure and thus the desalination of the ice. Observations must clarify this question.

2.6 Summary

Empirical relations of the effective distribution $\overline{k_{eff}}$ of salt in bulk sea ice in dependence on its thickness have been compared. For thick ice all these relations agree in that the salinity of first-year ice approaches $\approx 1/6$ of the salinity of the seawater on which it floats. This rule-of-thumb has been pointed out much earlier by Moss (1878) and the relations do not give additional information. In the thinner ice regime of less than 40 cm all algorithms differ considerably, indicating the much more variable range in growth conditions. Here their prediction of $\overline{k_{eff}}$ may be in error by more than 50 %. None of the algorithms is based on a physical model or scaling. Therefore, in addition to absolute uncertainties, also the predicted dependencies of $\overline{k_{eff}}$ on the thickness are arbitrary. A validation of a reasonable parametric form can only be provided by a physical model.

It is clear that a physical model of salt entrapment in sea ice must solve two problems: The initial entrapment at the interface and the delayed desalination. Principal ideas can be obtained by identifying distinct regimes in vertical salinity profiles of growing sea ice. These profiles provide a reasonable estimate of the bottom regime, where the main desalination takes place. Its extent appears to be ≈ 10 cm upwards from the ice water interface and it is revealed by a salinity minimum ($dS_i/dz = 0$). This is a general result for ice of any thickness during the growth phase under not too large fluctuating cooling conditions. It implies that a similarity classification of ice of different thickness is not appropriate. It is rather the bottom boundary layer than the shape of the profile which is similar for ice of different thickness.

Considering the bulk ice above the salinity minimum, desalination is still expected under warming events. Otherwise only a slight salt flux due to expulsion may be conjectured. The increase of the ice salinity towards the surface is natural due to the inverse relation between growth velocity and thickness, if one accepts in principal an increase in k_{eff} with growth velocity. It should be mentioned that the interpretation in these simple terms is no longer valid close to the surface, where other factors increase the complexity: (i) during early stages of growth the ice is still permeable at the top and expansion flow of brine may be expelled upwards; (ii) temperature fluctuations are largest and brine may be redistributed by intermediate opening of pores; (iii) In the freeboard of the order of $(\rho_w/\rho_{si} - 1) \approx 1/8$ times the ice thickness disequilibrium pressure effects may occur; (iv) interaction with the snow cover is likely. However, the present 'single-minimum' classification, which is literally preferable over the term 'C-shape', makes sense for the freezing season and when the ice does not experience considerable periodic warming cycles.

Before the entrapment problem is addressed, the bottom regimes of the skeletal and bridging layers will be discussed in some more detail in the following chapter 3.

3. Skeletal layer, bridging and percolation transition

Having discussed observations of sea ice bulk salinity and profiles in some detail, the present chapter is meant to given an overview of observations in the bottom regimes of sea ice. To do so it is useful to start with a principal classification that separates the bottom region from the overlying bulk sea ice. The principle regimes, already pointed out in section 2.4, are sketched in figure 3.1. To illustrate this classification, observed non-destructive salinity profiles from a study by Cox and Weeks (1975), which will be discussed in detail in the following chapter, are shown. When moving vertically upwards from the ice-seawater interface these are

- The *convective layer* of height H_{pc}, wherein rigorous convective motion is limited by high mobility of the ice matrix. Below this regime, on the liquid side of the interface, convection may also occur. This layer is termed H_c but belongs by definition not to the ice.

- The *skeletal layer*, extending to a distance of $H_{sk} \approx 3$ cm from the freezing interface. In this regime brine may circulate freely in vertical planes parallel to the plates and a direct vertical connection to the seawater below exists. The bulk ice salinity decrease in this layer is still dominated by convective exchange between the interstitial brine and the seawater.

- The *bridging layer* or *percolation transition*. Above the skeletal layer, brine is situated in horizontally and vertically elongated sheetlike inclusions. These are supposed to be still interconnected, presumably in a mode of finger-like pattern. Permeability and free convection are strongly reduced, as the convective exchange with the seawater below is no longer possible along direct paths. It occurs via the Felike brine-sheet network. This regime reaches up to a distance of $H_0 \approx 10$ cm from the interface. Concerned with sea ice strength models, Assur (1958) has called the latter layer the *bridging layer*, within which the sea ice strength increases strongly from its low skeletal value.

- The *bulk sea ice*, where the salinity attains a *quasi-stable* value. Above the bridging layer the sheetlike inclusions have lost their connectivity. Desalination is limited by the small effect of downward expulsion of brine due to expansion. The freez-

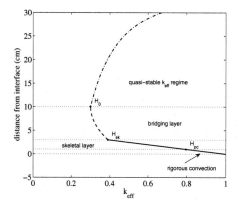

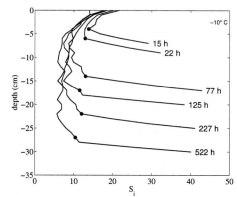

Fig. 3.1: Normalized salinity (k_{eff}) profile typical for rapidly growing thin ice. The four distinct regimes are described in the text. The distances H_{pc}, H_{sk} and H_0 are similar for any growing ice sheet.

Fig. 3.2: Salinity profiles during the growth of thin ice, based on data from Cox and Weeks (1975), see figure 2.5. The vertical resolution was 1 cm and the 3 cm distance from the ice-water interface is indicated. The hours refer to the time since the onset of growth.

ing of disconnected inclusions may now also lead to lattice distortions noticed by Bennington (1963) and Stander (1985).

This classification has not been proposed in earlier studies. The convection layers are very difficult to observe. Temperature fluctuations of thermistors during the passage of the freezing front may be considered to indicate a H_{pc} regime of only a few millimeters (Niedrauer and Martin, 1979; Haas, 1999). The boundary layer in the liquid appears to be even thinner. Due to a few observations of perturbation wavelengths, a thickness of less than a millimetre can be expected (Foster, 1972). The detailed nature of these convection processes will be discussed in later chapters. The liquid boundary layer H_c is considered by Worster and Wettlaufer (1997), yet their 'mushy layer' thickness is much thicker than the thin internal convection layer H_{pc}. Conceptually, the 'mushy layer' may eventually be identified wit the skeletal layer, but it must be warned, that the theory from Worster and Wettlaufer (1997) only refers to the onset of convection.

As already mentioned, the definition of distinct length scales of bottom boundary layers also differs from the classification of salinity profiles via self-similar C-shapes, independent of thickness. The latter was proposed by Eicken (1992). To date the distinguishment between a *skeletal layer* and a *bridging layer* has been conceptually made only by Assur (1958) but neither been analysed quantitatively, nor received much attention since then. The structural difference between the skeletal and bridging regimes has been pointed out by Stander (1985) from an analysis of first-year ice cores. The following critical discussion of these two regimes in terms of a variety of observations illustrates the importance of this distinguishment in quantitative modeling.

3.1 Skeletal layer thickness H_{sk}

3.1.1 Optical and mechanical observations

A high salinity bottom layer is documented in historical vertical profiles of ice salinity, f.e., Malmgren (1927). Its principle identification as a *skeletal layer*, above which the transition from brine layers to small inclusions takes place, is a consistent idea first proposed by Assur, Anderson and Weeks in connection with its effect on sea ice strength (Anderson and Weeks, 1958; Weeks and Anderson, 1958; Assur, 1958). Observations by Weeks and Anderson (1958) for ice of 6 to 40 cm thickness gave H_{sk} in the range 2 to 5 cm, determined by scraping off the underside of ice of low strength. Weeks and Anderson (1958) also reported on Assur's unpublished measurements of a fairly constant skeletal layer of 2.4 to 2.8 cm thickness at the bottom of thick arctic sea ice, while Assur himself has suggested a characteristic value of 2.5 cm for thick ice []Assur (1958). Stander (1985) has given a structural classification based on vertical and horizontal thin section analysis. He noted the entrapment of brine in well-defined sheet structures, beginning 2 to 4 cm from the interface of 50 to 100 cm thick first-year ice. Bennington (1963) interpreted optical observations near the ice-water interface in terms of stress concentrations that must occur during expansion of freezing brine that is no longer interconnected or connected with the seawater below. He noted that the transitions, which he called 'corrosion bands', were found always at a distance of 3 to 6 cm from the ice-water interface, for 160 cm thick first-year ice. It is not clear, if these transitions correspond to the top of the skeletal layer H_{sk} or, as discussed below, if they represent the first stages of the percolation transition at distance H_0. Langhorne (1979) described that the skeletal layer increased from 0.5 cm for 3 cm thick ice to 2 cm when the ice thickness reached 10 cm. The scraping measurements obtained by Weeks and Anderson (1958) for ice of 6 to 40 cm thickness did not appear to show a systematic dependence when plotted against the temperature gradient. However, a plot of these skeletal layer measurements against thickness (figure 3.3) indicates that the skeletal layer reaches some limiting value of 3 to 4 cm at 20 cm ice thickness.

3.1.2 Indirect estimates of H_{sk}

The above reported values of H_{sk} are mainly based on optical and mechanical observations. Anderson and Weeks (1958) noted that for thin ice it is not always easy to define the skeletal layer. In the present work it is suggested to define the *skeletal* layer H_{sk} by the transition from laminar brine layer state at H_{sk}, identifying the *skeleton* of sea

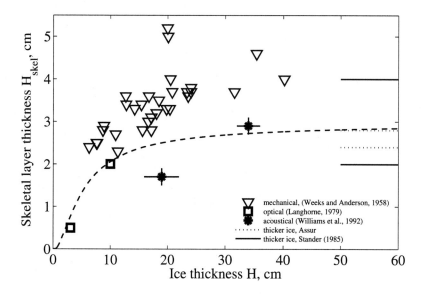

Fig. 3.3: Observations of the skeletal layer of young ice obtained by (i) mechanical scraping (Weeks and Anderson, 1958), (ii) visual inspection (Langhorne, 1979) and (iii) sound velocity measurements (Williams et al., 1992). The dotted lines span the range reported for thicker by Assur (Weeks and Anderson, 1958) and Stander (1985).

ice with these brine layers. As the bridging is a smooth process, one may expect some scatter problems to define the value from mechanical scraping and, in principal, interpret the observations from Weeks and Anderson (1958) as an overestimate of H_{sk} defined here. Also, its optical definition is no clear objective procedure, and Langhorne (1979) noted difficulties to distinguish it for thicknesses above 10 cm. An objective definition of the transition requires a microstructure analysis as described by Stander (1985), to make progress in evaluating the thickness and growth dependence. However, some indirect observations based on the ice permeability and the skeletal layer's different physical properties are of interest in this context. Valuable alternative observations have been obtained by Kawamura (1988) from X-ray computed tomography measurements. He reported a change in the signal distribution at a distance of 1.5 cm from the bottom layer of laboratory ice grown to a thickness of 9 cm, also consistent with the above reported values. An optical cross section for this ice indicates a similar length scale as the CT image. Another useful approach are acoustic (sound velocity) measurements performed by Garrison and Francois (1991) and Williams et al. (1992), which invoke the porosity dependence of the elastic modulus to profile the ice bottom layers. The former authors obtained a value of $H_{sk} \approx 3$ cm for thick Arctic sea ice, above which a sharp transition was observed. The latter suggested lower bounds of the skeletal layer thickness, changing from 1.7 ± 0.2 cm to 2.9 ± 0.2 cm, when thin lead ice grew from 16 to 22 and 32 to 36 cm thickness. These estimates appear consistently somewhat below other observations in figure (3.3). In their further analysis Williams et al. (1992) suggest an average value of 3 cm, obtained when all observations of ice between ≈ 10 and 40 cm were analyzed together. Another indirect source are dye and oil release experiments (Wolfe and Hoult, 1974; Eide and Martin, 1975; Niedrauer and Martin, 1979; Eicken et al., 1998), but as the ice also stays permeable above the skeletal transition they are discussed in the next section on the percolation transition level H_0 .

Finally, Cole et al. (2002) extended the qualitative description of banding features from Bennington (1963). They performed a quantitative analysis of a larger number of first-year ice cores from two fast ice sites near Barrow, Alaska, and found that the first-year ice consisted of alternating bright and less bright horizontal layers. A microscopic comparison showed that high porosity bands were associated with an increased fraction of larger brine tubes, reflecting a larger permeability. Most cores indicated a median between 2 to 3 cm in the spacing of these layers. It is tempting to interpret the spacing in terms of intermittent buildup and collapse of a skeletal layer of similar height. On the other hand, Cole et al. (2002) also reported timescales of 3 to 5 days associated with the distance of the bands and suggested a possible connection to local sea level variations and under-ice pressures. Due to the general difficulties to reconstruct an accurate growth history, these ideas await further studies.

3.1.3   Variation of H_{sk} with thickness

Most observations are consistent with a sharp initial increase in the skeletal layer with ice growth, reaching a limiting value at 20 to 30 cm ice thickness. To describe this asymptotic behaviour the equation

$$H_{sk} = H_{sk\infty} e^{\left(-\frac{H_{sk\infty}}{H - H_{sk\infty}}\right)} \tag{3.1}$$

has been tentatively chosen and is shown as the dashed curve in figure 3.3. The form depends on one parameter, a saturation skeletal layer thickness $H_{sk\infty}$. If the latter is taken as 3 cm, it reflects a compromise between most observations, considering the upper bound character of mechanical scraping. However, equation 3.1 is not based on any physics, and is just an ad-hoc scaling to describe the observations. While the noted results from different methods are consistent with each other, the scatter clearly points to the necessity of a systematic study of the skeletal layer's variation with growth conditions. As all observations have been reported for normal seawater salinities of 30 to

35 ‰, no information about the salinity dependence of H_{sk} is available at present. Some observations indicate the influence of under-ice flow. Eicken et al. (1998) reported that the skeletal layer of sea ice grown in a tank to similar thickness was strongly reduced from several centimeters to only 0.5 cm under the presence of a strong under-ice current.

3.2   Bridging layer and percolation transition H_0

Stander (1985) described that the transition from brine planes to sheets by bridging began at a distance of 2 to 4 cm from the interface. Here the bridging layer begins. Stander further pointed out that the sheetlike brine layers in the bridging layer start to split into elongated pores at a distance 10 cm from the interface. He reported no systematic difference in this length scale for 50 and 100 cm thick first-year ice. Anderson and Weeks (1958) also noted such a behaviour but did not distinguish the skeletal and bridging regimes. They neither reported length scales or ice thickness, but associated the behaviour in principle with an increase in sea ice strength. The splitting into pores may also be associated with a large change in the horizontal permeability. Therefore the level H_0, above which brine tubes start to dominate, may be called the *percolation transition*.

3.2.1 H_0 and the salinity minimum

The permeability drop at the level H_0 can be expected to limit the further salinity loss, in such a way that, at the corresponding distance from the freezing interface, k_{eff} should attain a quasi-stable value. As k_{eff} increases with growth velocity, and the latter normally decreases with thickness, the percolation transition should show up as a minimum in ice salinity profiles. Indeed, as already mentioned, such a minimum appears to be a general feature of first-year ice during the growth period. Profile data presented in section 2.4 showed the transition in thin ice at a distance of 6 to 10 cm from the interface. A similar distance is evident in profiles of laboratory grown NaCl ice from Cox and Weeks (1975), see figure 3.2. Other published data for thin ice confirm the critical distance of approximately 10 cm from the ice-water interface, once the ice has exceeded 20 cm in thickness (Weeks and Lee, 1958; Cottier et al., 1999). The length scale does not appear to depend strongly on ice thickness. Most salinity profiles of up to 150 cm thick first-year sea ice published by Nakawo and Sinha (1981) show the minimum at a distance 9 to 11 cm from the interface, while in thinner ice samples the minimum was situated between 5 and 11 cm. The contour plot from Wakatsuchi (1974b) shown as figure 2.6 above, is also noted. It indicates, in the limits of its resolution, a minimum near 10 cm from the interface, as long the growth rate is stable. Salinity profiles of 60 to 120 cm thick first-year ice shown by Bennington (1967), based on a somewhat coarser 5 cm resolution, also support a transition 10 cm from the interface. Profiles of thick first-year ice from Meese

(1989) are at least, although not giving further proof due to their coarse 10 cm resolution, not inconsistent with the scale. Interestingly, the minimum also appears during columnar ice growth in low-salinity brackish water (see figure 5c from Granskog et al. (2004a)).

In very thin ice the transition level H_0 appears to be smaller, if identified with the salinity minimum. In the profile of 12 cm thick sea ice, shown in figure 2.4, and comparable early growth profiles in figures 2.4 and 2.5 from the laboratory runs of Cox and Weeks (1975), the minimum appears at about 5 to 6 cm from the interface. Further evidence is found in several other studies, where salinity profiles of ice thinner than 15 cm have been sampled (Weeks and Lee, 1958; Bennington, 1963; Nakawo and Sinha, 1981). Notably, for thin ice also the skeletal layer scale H_{sk} has been found to be smaller. As will be discussed next there are several other indications of the percolation transition H_0, which are consistent with the above scales.

3.2.2 Corrosion features

Stander (1985) reported another interesting observation for the level H_0. He noted that the shift from brine sheets to brine layers was accompanied by lattice distortions observable at tips of brine inclusions. This may also indicate that brine finger pattern are no longer interconnected, implying that pressure build-up due to expansion on freezing can no longer be released by fluid movement through the brine network. Bennington (1963) found another, eventually related feature. He described pressure figures between 3 and 6 cm distance from the ice-water interface, which appeared as ensembles of minute brine inclusions. Bennington interpreted them in terms of expansion and termed them 'corrosion bands'. One may then ask, if the transition level could be determined from the evaluation of those lattice distortions.

Cole and Shapiro (1998) described similar features at positions of 2 to 8 cm and 20 cm from the interface of different first-year ice cores. They questioned a fracture mechanism as the origin of corrosion bands, arguing that close to the ice-water interface the temperature decrease is too small. This argument appears not convincing, because the change in the brine volume due to expansion is actually largest at high temperatures. It may be important, under which conditions the micrographs of cross sections were obtained. Stander (pers. comm.) performed the preparation of sections mostly within several minutes and at temperature near -10 °C. Bennington (1963) applied a different procedure. First, dye and brine were used to seal a thin section, after which the ice was drained for some hours at -25 °C, to finally elucidate the microstructure. The latter procedure probably has a larger potential to change the in-situ microstructure. Also must the lack in generality of corrosion features in some ice cores, reported by Cole and Shapiro (1998), not necessarily exclude an expansion fracture mechanism: The oscillatory nature of localised brine channel convection and air temperatures, eventually also the sampling procedure, may create large variability in forcing conditions. Bennington never observed the 'corrosion bands' farther than 6 cm from the interface, giving rise to the idea that these bands may disappear again (or fundamentally change their appearance) by slow migration or thermal cycling processes during aging. Cole et al. (2002) recently described corrosion features in the upper portion of a first-year ice core, corresponding to a position in the freeboard, eventually involving different physics. In their opinion, the 'corrosion' bands reflect local areas of recrystallization and microscopic slush formation. If this is correct, the question remains how such processes can be triggered. As the formation of small crystals requires large supercoolings, fast convective transport due to expansion and fracturing would appear as a more plausible candidate than slow changes in hydrodynamic conditions in the seawater as suggested by Cole et al. (2002).

The reports by Stander (1985), Bennington (1963) and Cole et al. (2002), pointing to the potential of stress indicators to reveal permeability transitions, involved different observational methods. The influence of the method of sample preparation on the observable microstructure was pointed out by Sinha (1977). A valuable discussion of aspects of sea ice microstructure evolution due to internal stresses has also been given by Knight (1962a). Hence, an analysis of different sampling conditions appears necessary to decide, if cooling and disequilibrium during sampling may affect the microstructure, and for which temperature change or cooling rate this might happen. At present an objective description of expansion and lattice distortion features is lacking.

3.2.3 Dye experiments and percolation

From dye release experiments onto the cold sea ice surface one may learn that the main body of young ice is permeable, if liquid is added at its surface, yet that the penetration is limited to the freeboard (Bennington, 1967). To evaluate the possibility of internal brine drainage, the flow through the bottom of ice, driven by the actual pressure gradients due to brine salinity gradients, is more critical. Eide and Martin (1975) released dye at the bottom of growing laboratory ice and observed a rapid penetration to a depth of ≈ 1 cm (5 cm initial ice thickness, yet still growing). For thicker ice of 20 cm (almost stagnant) thickness penetration was deeper, 3 to 4 cm. In the latter case, the rapid penetration was followed by a slower mode, during which the dye reached a level 5.5 cm from the interface after 24 hours, after which the signal propagation almost ceased. Niedrauer and Martin (1979) found a similar dye penetration depth of 1.5 to 2 cm for ice thinner than 10 cm. Wolfe and Hoult (1974) reported on the upward penetration of oil inserted at the bottom of 12 to 16 cm thick ice. During 12 to 24 hours the oil penetrated to 2.5 cm depth. Similar oil release experiments under 13 cm thick sea ice grown in the INTERICE tank study showed an upward penetration of 3.5 to 4 cm during a period of 4 days (Eicken et al., 1998). All these penetration distances are less than the percolation transition level suggested from salinity profiles. However, it is uncertain if the dye experiments performed by Eide and Martin (1975) and Niedrauer and Martin (1979) represent conditions of natural sea ice, because the ice had been grown in very thin (1.6-3 mm) growth cells. Considering the low average brine volume of 0.17 reported by Niedrauer and Martin (1979) for the lower 2 cm, it appears probable that the skeletal layer H_{sk} in these experiments was less than normal, probably more like 1 cm. This would likely also imply a smaller H_0. With respect to the oil entrapment results from Wolfe and Hoult (1974) and Eicken et al. (1998), it seems also plausible that the more viscous oil would intrude to a lesser degree than seawater into the ice. It is therefore suggested, that the noted dye and oil intrusion experiments, for different reasons, underestimate the level to which seawater percolates under natural growth conditions of thin ice.

Williams et al. (1992) reported on results from acoustic methods interpretable in this context. They presented some results of sound velocity measurements before and after complete submersion of a complete 42 cm sea ice core in seawater. After submersion, the porosity and salinity attained a minimum at a distance of 10 cm from the bottom, while the sound velocity reached a maximum. This signal distribution has been explained by the authors due to exchange of saline brine against less saline seawater in which the ice was immersed, followed by additional internal freezing. The minimum at the 10 cm level may indicate that the circulation of brine was restricted to this regime and supports it as being the level of the percolation transition. Another indicator for H_0 are the X-ray computed tomography measurements performed by Kawamura (1988) on 9 cm thick ice, which had been interpreted in terms of a skeletal layer of thickness $H_{sk} \approx 1.5$ cm. The vertical thin section images shown by Kawamura also indicate a larger transition layer, extending upwards 3 to 4 cm from the interface. The smaller scales for both H_{sk} and H_0 appear consistent with the ice being very thin.

3.2.4 Brine salinity sampling

Melnikov (1995) has obtained brine samples from rapidly grown 28 cm thick lead ice. Samples were taken via bore holes approximately 6 to 8 cm from the ice-water interface. The layer below this level was described as 'wet' by Melnikov (1995), while the upper layer was described as 'almost dry'. From half-hourly measurements of brine salinities Melnikov suggested a 1.5 to 2 hourly cycle during which the brine salinity fluctuated between 60 and 90 ‰. This 'wet' ice observation agrees well with H_0 related to the salinity maximum discussed above.

An observation of similar brine salinity had been already reported by Malmgren (1927). Also Eicken (1998) collected brine systematically from bore-holes drilled into first-year ice, interpreting sampled brine salinities in terms of a 'close-off' brine volume. It would be tempting to estimate the 'close-off' level from simultaneous ice temperature profiles. However, there are many problems with such an approach. First, the brine sampled from boreholes will often be a mixture of brine from different levels, the relative contribution being unknown (Eicken, 1998). While this may be avoided by taking samples at fixed levels with plastic needles, as it was done by Melnikov (1995), such an approach is not nondestructive. The collected brine certainly derives from larger brine channel features, which makes an allocation to a certain level uncertain. Also, the draining brine will reflect the 'close-off' of the channels, which may be quiet different from the effective horizontal percolation limit of brine layers. The most serious objection in this context is probably, that the lateral drainage of brine into a borehole does not represent the convective limit or 'close-off', because the lateral pressure gradient created by the open hole is much larger than the gradient related to internally driven convection in undisturbed ice.

If careful and well-suited sampling techniques could be devised, to avoid some of the mentioned problems, one would still be left with the noted fluctuations and the non-equilibrium character of the brine concentration in channels, as it was was further corroborated by Melnikov (1995) by a brine salinity profile from 80 cm thick ice. The latter indicated a 30% undersaturation of brine salinities for the whole ice sheet. Similar non-equilibrium conditions have already been pointed out by Ono (1967) based on an laboratory experiment. The noted complications indicate that one needs additional informationh than brine temperature and salinity to infer the permeability behaviour of sea ice. An ad-hoc model of oxygen isotope fractionation has been applied by Eicken (1998) in this connection, yet the signifance of the results suffers from the above problem of borehole destruction. The potential of oxygene isotope or other chemical fractionation processes between brine and ice, linked to growth rates, diffusion and brine mobility, as highlighted in the discussion by Eicken (1998), needs to be investigated further, but probably requires sophisticated sampling, theoretical model approaches which have not been realized so far.

3.3 Microstructural transitions of brine layers

Two very simple geometrical models have been suggested for the microstructural transition in the bottom regimes of sea ice:

- Anderson and Weeks (1958) proposed the splitting of brine layers into vertical cylinders above the skeletal layer.

- Assur (1958) suggested an alternative model, where first sheet-like channels inclusions form by disintegration of brine layers, keeping their width at the original brine layer width. These sheets are assumed to shrink in the basal planes until they become circular cylinders.

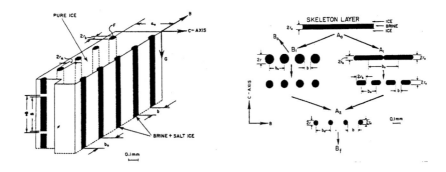

Fig. 3.4: Adapted from Assur (1958). Two modes of brine layer transition near the bottom of sea ice. The right figure is a view from the bottom: Skeletal brine layers transform first to circular inclusions (B_1) or quasi-rectangular channels (A_1). Left: Side view of the ice matrix, eventually accounting for vertical disconnections.

The geometrical concept is shown in figure 3.4 adapted from Assur (1958). Both Assur (1958) and Anderson and Weeks (1958) were mainly concerned with the predictability of sea ice strength, and its increase by the bridging mechanism during cooling. The basic approaches neglected vertical disconnections, but the latter could be included by some simple geometrical factors.

Observations obtained since these early discussions support the view of sheetlike inclusions which are slightly elongated in the vertical (Stander, 1985; Eicken et al., 2000). A detailed review and discussion is given in the appendix F. The following note is concerned with the mechanism that leads to the transition.

3.3.1 Misconception of surface energy minimisation

Both Assur (1958) and Anderson and Weeks (1958) associated the splitting of brine layers with the principle of minimisation of solid-liquid surface energy. Since then this principle has been included by many authors in different model parameterisations (Weeks and Assur, 1963; Tsurikov, 1965; Weeks and Ackley, 1986; Petrich et al., 2006). The concept was adopted by Anderson and Weeks from Nakaya (1956) and may be sketched as follows. One first considers the specific surfaces s_{d0} and s_{cy} for a thin long brine layer and an array of cylinders as

$$s_{d0} = \frac{2(L_{gra} + d_0)}{d_0 L_{gra}} = \frac{2}{d_0} + \frac{2}{L_{gra}} \qquad and \qquad s_{cy} = \frac{\pi d_{cy}}{\frac{\pi}{4} d_{cy}^2} = \frac{4}{d_{cy}}, \tag{3.2}$$

where L_{gra} is the horizontal extension of the brine layers of width d_0, and d_{cy} the cylinder diameter. Further, if l_{cy} is the spacing between the splitted cylinders, conservation of volume requires $d_0 l_{cy}/d_{cy}^2 = \pi/4$. If one now introduces the surface energy minimisation condition that the specific surface area must decrease or stay constant, $s_{d0}/s_{cy} \geq 1$, and neglects the contribution $2/L_{gra}$ to s_{d0}, it follows

$$l_{cy} \geq \pi d_0 \qquad and \qquad d_{cy} \geq 2d_0. \tag{3.3}$$

This requires the least spacing of cylinders to be π times the brine layer width. Similar considerations can be made by assuming that the brine cylinders split further into equally distributed spheres. The diameter of these spheres will then be at least $d_{sp} \geq 3/2d_{cy} \geq 3d_0$ and their spacing l_{sp} will be larger than $9/2d_0$. The simplified concept implies that a film of thickness d_0 first transform into cylinders spaced by πd_0 and finally splits into spheres spaced by $9/2d_0$, the ratio of these length scales being $9/(2\pi) \approx 1.43$.

Although Anderson and Weeks (1958) reported observed scales $d_0' = 0.07$ mm, $l_{cy} = 0.23$ mm and $d_{cy} = 0.136$ mm, in apparent agreement with the prediction, the concept and numbers must be questioned due to the following reasons:

- Anderson and Weeks (1958) did not observe the splitting *in situ*, yet the scales were derived from horizontal sections at different vertical positions in the ice. No statistics of the analyzed micrographs and levels was presented, and the temperatures during processing were not reported. The scale $d_0' = 0.07$ mm may be viewed as the smallest brine layer width d observed in samples before the appearance of predominantly circular vertical tubes. It is no objective estimate. In view of more detailed microstructural analysis, performed since then and showing variable distributions of inclusion length scales (Perovich and Gow, 1996; Cole and Shapiro, 1998; Light et al., 2003), the values for d_0', l_{cy} and d_{cy} suggested by Anderson and Weeks (1958) can only be regarded as approximate estimates.

- The assumed physical mechanism of surface energy minimisation is a misconception. The latter should lead to a coarsening of a structure via *merging* of inclusions rather than *splitting*, e.g., Voorhees (1992). Moreover, are the basal planes the low energy surfaces of ice (see appendix A.2.10), and there is no driving force from surface energy principles to deform this geometry.

- The concept was adopted from Nakaya (1956), who reported reasonable agreement with the above scalings during the migration of *vapour figures* in glacier ice. These had separated from liquid *Tyndall figures* (Tyndall, 1858). The vapour figures were highly volatile films that changed in response to thermal history. Their dynamics cannot be expected to be comparable to brine liquid inclusions at the equilibrium freezing point. Even for vapour figures, Maeno (1967) has observed very different scales and ratios than Nakaya (1956).

- If at all, then brine inclusions should be compared to Tyndall figures. The latter, however, are also highly dynamical disequilibrium phenomena and not fully understood (Nakaya, 1956; Kaess and Magun, 1961; Knight and Knight, 1972; Mae, 1976). What can be stated on the basis of all observations is that Tyndall figures do not split *during cooling*.

- Nakaya (1956) observed that, *during warming*, Tyndall figures mostly increase in area, keeping an almost constant thickness of the order of 0.1 mm. A similar observation has been made by Anderson and Weeks (1958). It indicates the problem, that cooling, storage and warming of samples changes the microstructure considerably (appendix F).

The noted aspects indicate that a theory and reliable quantitative observations of microstructural transitions of brine layers are lacking. One may, however, consider some basic ideas obtainable from first principles. A more plausible explanation for the bridging of brine layers is the interaction of lateral freezing with hydrodynamic instabilities due to flow through these layers. A classical problem is the stability of a liquid jet due to capillarity (Rayleigh, 1879). Lord Rayleigh showed that instability requires an axisymmetric perturbation wave length π times the radius of the cylinder, while the wave length of maximum perturbation growth rate, which is of interest in many applications, is $\pi/0.697 = 4.508$ times the cylinder diameter d_{cy}. Chandrasekhar (1961) has discussed many similar problems of liquid jets. The gravitational instability of a cylinder, disregarding the effect of capillarity, has its maximum growth rate at a slightly larger wavelength of $\pi/0.0.580 = 5.417$ times the diameter, whereas for the capillary instability of a hollow jet the nondimensional wavelength is $\pi/0.0.484 = 6.491$. As pointed out by Levich (1962), the problem of the capillary instability of a cylindrical liquid jet is very general and can be extended to arbitrary shapes. It has been shown that instabilities are expected for wave lengths larger than 2π times the layer thickness d, the wave length of maximum growth rate being $2^{3/2}\pi d = 8.976d$ (Ruckenstein and Jain, 1974; Prevost and Gallez, 1986). The latter value is, for example, in agreement with volatile vapour film splitting observed by Maeno (1967). An alternative classical problem is the thermocapillary instability of a horizontal liquid layer with a free surface (Pearson, 1958). Nield (1964) obtained coupled solutions of surface tension and gravitational instabilities and showed that, for different boundary conditions and combinations of the effects, the cell sizes are typically 2 to 3 times the layer width.

While none of the noted problem is directly applicable to the problems of brine layer freezing with throughflow, they point out that most instability processes produce length scales proportional to the layer or channel width. When freezing sea ice brine layers are considered, one must not even have a hydrodynamic problem to obtain similar scales. In directional solidification one sometimes observes a solute droplet pinch off similar to what one might term a brine layer transition. For this problem Brattkus (1989) found a characteristic equation very similar to Rayleigh's problem for inviscid liquid jets, with essentially the same wavelength of maximum growth rate of $\pi/0.697 = 4.508$ times the diameter. The formal difference in the problems was that the transport in the channel was not dynamical yet diffusive. The prediction from Brattkus are in agreement with experiments (Kurowski et al., 1989) and numerical simulations (Conti and Marconi, 2000). The pinch off depends, in essence, on the creation of disequilibrium by fluctuations in the growth velocity. It is interesting that the behaviour is qualitatively similar to what Harrison (1965b) has described as 'solute transpiration pores' in ice. These cylindrical pores were often stable, but started to break up into droplets when the motion of the temperature gradient was halted.

For saline solutions, a few studies allow an estimate of the relevant scales for pinch-off of inclusions. Light et al. (2003) have analysed the statistics of brine inclusions from vertical sea-ice thin sections. They found that below a length scale of 0.03 to 0.04 mm the inclusion aspect ratio approaches unity. Micrographs taken by Light et al. (2003) during a cooling sequence (their figure 13) do not manifest a splitting behaviour of the thinnest ($\approx$ 0.03 to 0.04 mm) elongated inclusions. In some of their figures strings of very small inclusions have sizes $\approx$ 0.02 to 0.03 mm, which is similar to the break-up size observed by Harrison (1965b) for very thin pores.

The noted scales refer to transitions of closed pores and the brine layer transition problem is certainly a different one that must involve the fluid flow through the brine layers. An exact theoretical solution of the problem of a laterally freezing brine layer may moreover involve aspects of viscous fingering and thin boundary layers and can be expected to be very difficult. A heuristic model approach and its further discussion is delayed to the final chapter of this work (chapter 12), when a more clear picture on freezing, convection and length scales has been formulated. It should be clear that the observation $l_{cy}/d_0' \approx 3$ reported by Anderson and Weeks (1958) cannot be associated with the minimisation of surface tension, but with instabilities introduced by fluctuating heat and fluid transport as well as salt rejection.

3.3.2   Permeability changes

It is useful to consider the transition into cylinders as a bound of the permeability change upon a geometrical transition. It is straightforward to compute the ratio of vertical permeability of an array of cylinders of, $\Pi_{v,cy}$, to the brine layer permeability $\Pi_{v,d0}$. The former is related to Poiseuille flow through cylinders, the latter is given by the Hele-Shaw permeability. From classical hydrodynamic lubrication theory (Lamb, 1932) their ratio is

$$\frac{\Pi_{v,cy}}{\Pi_{v,d0}} = \frac{v_b d_{cy}^2/32}{v_b d_0^2/12} = \frac{3}{8}\left(\frac{d_{cy}}{d_0}\right)^2, \tag{3.4}$$

where d_0 is the width of the brine layers, d_{cy} the diameter of the cylinders and v_b their volumetric fraction. For a splitting into an array of touching cylinders of smallest possible diameter $d_{cy}/d_0 = 4/\pi$ one finds a ratio of $6/\pi^2$ and thus a *decrease* in the vertical permeability. For $d_{cy}/d_0 = (8/3)^{1/2} \approx 1.63$ the vertical permeability would be conserved, while for wider cylinders the effective permeability is increased. As mentioned, such a jump from brine layers to considerably wider channels has been proposed by Anderson and Weeks (1958) on the basis of a few observations, but is not consistent with more recent work on microstructures (appendix F).

One may therefore conclude, that realistic changes in the brine layers above the skeletal layer can be expected to *decrease* the vertical permeability, if not very wide channels form. Such wide brine channels are known to be present also in young sea ice at spacings of the order of a few centimeters (Wakatsuchi, 1977; Niedrauer and Martin, 1979; Wakatsuchi and Saito, 1985). While they imply a much larger local permeability, the overall transition effect of brine layers, referring to bulk ice away from these brine channels, is, in contrast, a larger resistance to *vertical* throughflow. The primary effect of the horizontal separation of brine layers into cylinders is, of course, a radical drop of the *horizontal* permeability. The real transition is, as mentioned, sheet-like and still allows horizontal percolation at its beginning. Such a transition allows a formulation of the horizontal permeability in terms of twodimensional percolation theory and a simple solution will be discussed at the end of this work in appendix F. At the moment it suffices to say, that the conceptual understanding of the percolation transition is the transition of brine layers by a bridging mechanism of ice plates, rather than a splitting mechanism into cylinders.

3.4 Summary

A classification of sea ice into different regimes measured from the ice-water interface has been proposed. Available observations on the bottom regimes of the *skeletal layer* and *bridging layer* have been described to provide a basic understanding of these regimes in terms of their salinity, modes of desalination and percolation. The observations support the classification in figure 3.1 and give rise to a simple preliminary interpretation. In the skeletal layer of $H_{sk} \approx 3$ cm, measured from the ice-water interface, the main desalination takes place. Desalination is reduced within the bridging layer. This is not simply a consequence of the decreasing brine volume, but follows from the microstructural transition. The top of the bridging layer, at a distance $H_0 \approx 10$ cm from the ice-water interface, may be thought to represent the level were bridging is almost complete and the original brine layer planes can no longer be percolated.

The fact that the location of the salinity minimum, at $H_0 \approx 10$ cm, shows little variability with growth conditions and age of the ice is interesting. Only for thin ice, H_0 appears to be smaller, which parallels the saturating behaviour of the skeletal layer thickness. Both scales appear, at any ice thickness, to scale as a ratio of approximately $2 < H_0/H_{sk} < 4$. The constancy of the bottom boundary layers H_{sk} and H_0 is an important aspect of growing sea ice that to date has received almost no attention. Apparently it is this regime which must be understood, to describe the desalination of sea ice towards a quasi-stable k_{eff}. Consequently, the present work will almost solely deal with the description of the bottom boundary regimes and focus on the understanding of the noted transitions.

Two aspects have not been considered so far and shall only be partially addressed in the present work. These are the slow desalination processes of bulk sea ice, presumably pocket migration and redistribution on expansion, and the formation of brine channels and their networks. For an overview of these processes and features the reader is referred to appendix C. Brine channels extend farther into the ice than the skeletal layer or bridging transition (Bennington, 1963, 1967; Niedrauer and Martin, 1979; Wakatsuchi, 1983; Wakatsuchi and Saito, 1985). This does not mean that they are not coupled to the physics in these layers. However, while the skeletal layer reaches a quasi-stable height, the brine channels evolve into a more mature network system, apparently being fed by brine from the surrounding crystals (Wakatsuchi, 1983; Wakatsuchi and Kawamura, 1987). The first step to their understanding is the principal plate microstructure. From observations, it is known that brine channels transport brine downwards (Niedrauer and Martin, 1979; Wakatsuchi, 1977, 1983; Wakatsuchi and Saito, 1985). This must be compensated by upward flow through the brine layers in the rest of the ice. When bridging takes place, it leads to a state in which suction from the larger channels can no longer drive a circulation between ice and seawater. The *percolation transition* level H_0 may then be viewed as the limit for gravity-driven salt-fluxes from growing sea ice. These fluxes always have their origin in downward flow through the channels and they cease, when the brine layers, through which they are fed, offer a two large flow resistance. The brine channels are background features that always can become reactivated. Considering only the cooling and growth phase of ice, this activation process is also excluded from the present work.

The transition of the microstructure at the bridging condition will be discussed at the end of this work. Before this can be done in a meaningful manner, one must understand the formation of the microstructure and the exclusion of salt in the skeletal layer by convection. The following chapters establish a theoretical framework of crystal growth in seawater, convective processes at the freezing interface and their interaction.

4. Solute redistribution at the ice-solution interface

The present chapter considers physical processes during solute redistribution at a freezing ice-solution interface. As the solid ice phase is almost pure, solute is rejected at the interface and will be redistributed by diffusion and convection. Before this problem can be formulated for a cellular sea ice interface it is useful to review the scalings to date suggested for planar freezing. This *planar* freezing solute distribution is discussed in detail for two reasons. First it is critical to understand the discussion of the planar interface breakdown in later sections of this work. Second, most applications published to date on the problem of solute entrapment in *cellular* sea ice and saline ice have been based on an analogy of *planar* solute distribution models (Weeks and Lofgren, 1967; Weeks and Ackley, 1986; Cox and Weeks, 1988; Eicken, 2003).

The planar freezing solute distribution coefficient k of NaCl and most other salts is of the order of 10^{-3}. While its direct influence on the solute budget of sea ice is negligible, it is essential to understand basic elements of solute redistribution during planar freezing. Two aspects of the planar k may be noted at this point. First, due to the smallness of k, the solute flux at a planar interface is simply related to the growth velocity. With such a simple boundary condition, the solutal convection problems become tractable and of a different nature as if k would depend on V. Second, it is essential to know the true value of k in order to interpret vertical solute distribution in the ice in terms of diffusive and convective solute fluxes on the liquid side of the interface. The first part of the present section reviews earlier work on k and solute redistribution during planar freezing. In particular, a relatively simple framework of free solutal convection and solute redistribution is formulated and validated with observations. Finally, some published observations of freezing salt solutions are reviewed to outline some fundamental differences between a planar and a cellular interface.

4.1 Equilibrium diffusion

A freezing interface where the fraction $(1 - k)$ of the solute in the solution is rejected is considered. In aqueous solutions $k < 1$ and solute enriches at the interface. The effective solute entrapment coefficient $k_{eff} = C_s/C_\infty$ is normally defined as the ratio of concentration in the solid to the bulk concentration in the liquid far from the interface. It depends on the interfacial solute distribution coefficient k and the concentration at the solidification interface via

$$k_{eff} = \frac{C_s}{C_\infty} = \frac{C_{int}}{C_\infty} k. \tag{4.1}$$

One may, for the sake of simplicity, assume that k is a material constant. Then C_{int} at the interface must be known to predict k_{eff}. In the absence of fluid motion solute is transported away from the interface by diffusion only, until an equilibrium between diffusion and rejection is reached. The problem has been first discussed in some detail by Tiller et al. (1953). Conservation of solute in the liquid is expressed by

$$\frac{\partial C}{\partial t} = D_s \frac{\partial^2 C}{\partial z^2}, \tag{4.2}$$

where D_s is the solute diffusivity. Assuming unidirectional growth of a planar interface at velocity V the solute balance may be written for a coordinate system moving at velocity V. In the steady state it becomes

$$D_s \frac{\partial^2 C}{\partial z^2} + V \frac{\partial C}{\partial z} = 0 \tag{4.3}$$

Let the freezing interface be fixed at $z = 0$ and the positive z-direction extend into the liquid. Using the boundary conditions for the solute flux at the interface and constant solute concentration at infinity

$$
\begin{aligned}
C &= C_\infty & at \quad z &= \infty \\
D_s \frac{\partial C}{\partial z} &= V(C_{int} - C_s) & at \quad z &= 0
\end{aligned}
\tag{4.4}
$$

one obtains the steady-state solute distribution in the liquid

$$C(z) = C_\infty + (C_{int} - C_s) \, exp\left(-\frac{V}{D_s} z\right). \tag{4.5}$$

The steady state requires the concentration C_s in the solid and C_∞ in the liquid to be equal and hence an interfacial distribution coefficient k which obeys

$$C_\infty = C_s = C_{int} k \tag{4.6}$$

Equation (4.5) may also be written as

$$C(z) = C_\infty \left(1 + \frac{1-k}{k} exp\left(-\frac{V}{D_s} z\right)\right). \tag{4.7}$$

The solute concentration profile in a purely diffusive solutal boundary layer is described by an exponential distribution with 'decay constant' V/D_s. The steady state implies $k_{eff} = 1$. The transient to approach this steady-state has been discussed approximately by Tiller et al. (1953) and more exactly by Smith et al. (1955).

4.1.1 Practical restrictions

Validation attempts of the diffusive equilibrium and/or the transient modifications have faced some practical problems in the case of aqueous saline solutions. For dilute solutions the solute fluxes are small and thermal convection effects may be relevant during *upward growth* (Jaccard and Levi, 1961; Gross, 1967; Gross et al., 1977). Gross (1967) performed freezing experiments with convection 'suppressed as far as possible', but the estimates of k were rather scattered and not conclusive. To suppress the effect of thermal convection during upward freezing the salinity must be increased above 23 ‰ for NaCl (appendix A.2.4). However, as shown theoretically in the course of this monograph and observed by many previous investigators, under such conditions the interface is no longer planar. It becomes cellular, with a fundamental change in the mode of solute entrapment. Due to these aspects the use of the diffusion equation (4.7) as a standing-alone analysis of solute redistribution during ice growth is limited.

Under natural conditions of *downward* ice growth molecular solute transport is enhanced by fluid flow. The most important driving processes of such flow are (i) *external stirring* or *forced convection*, (ii) laminar or turbulent fluid flow in a boundary layer and (iii) *natural free thermal or solutal convection*. Most previous work to account for fluid flow has been based on a concept of a stagnant diffusive boundary layer. This concept will be reviewed next.

4.2 Stagnant boundary layer approach

The most widely used approach to correlate ice salinities with a theoretical model is based on the work of Burton et al. (1953a) in the field of metallurgy. It was first applied by Jaccard and Levi (1961) to ice, evaluating the planar freezing process of aqueous hydrofluoric solutions. Weeks and Lofgren (1967) introduced it into the analysis of sea ice salinity data. Since then it has been adopted by many authors (Weeks and Ackley, 1986; Cox and Weeks, 1988; Souchez et al., 1988; Eicken, 1998; Killawee et al., 1998; Ferrick et al., 2002), yet has seldom been discussed critically. For a discussion of its applicability to several situations of ice growth the theory is summarised as follows.

4.2.1 Growth of a rotating crystal

Burton et al. (1953a) were interested in the particular situation of a rotating crystal of infinite radius solidifying directionally into a liquid volume of infinite height. For this case the classical relation between the fluid vertical velocity W and stirring rate has been given by Cochran (Levich, 1962; Schlichting, 1965; Batchelor, 1967). Burton et al. (1953a) then investigated a modified diffusion equation (4.3), growth velocity V being replaced by $V + W$, with the same boundary conditions (4.4) as given above. The result was an equation which only could be solved numerically. Next, for practical reasons, a simplified approach to the problem was made, solving the diffusion equation (4.3) under the assumptions that (i) in a stagnant layer of thickness δ_{bps}, the solute transport is by molecular diffusion only, (ii) at the distance $z = \delta_{bps}$, strong mixing in the liquid keeps the concentration at C_∞. Hence, the boundary conditions to be applied are

$$C = C_\infty \qquad at \quad z \geq \delta_{bps} \qquad (4.8)$$

$$D_s \frac{\partial C}{\partial z} = V(C_{int} - C_s) \qquad at \quad z = 0 \qquad (4.9)$$

and lead to the solution

$$C(z) = C_s + (C_\infty - C_s) \, exp\left(-\frac{V}{D_s}(z - \delta_{bps})\right) \qquad (4.10)$$

within the stagnant film $z < \delta_{bps}$. This implies a concentration at the interface

$$C_{int} = C_s + (C_\infty - C_s) \, exp\left(\frac{V}{D_s}\delta_{bps}\right). \qquad (4.11)$$

The next point made by Burton et al. (1953a) was that, using $k_{eff} = C_s/C_\infty$ and $k = C_s/C_{int}$, equation (4.11) may be written as

$$k_{eff} = \frac{k}{k + (1 - k)exp\left(-\frac{V}{D_s}\delta_{bps}\right)} \qquad (4.12)$$

The usefulness of equation 4.12 depends on the question, if δ_{bps} can be found by experiment or theory and if it is constant. By comparing the approximation to the numerical computations, Burton et al. (1953a) showed that δ_{bps}, required to match the solute fluxes, depends on the growth rate, but that this dependence becomes weak for

$$Pe_\delta = \frac{V\delta_{bps}}{D_s} < 1 \qquad (4.13)$$

which defines the applicability of the δ_{bps}-approach[1]. The interpretation of the length scale δ_{bps} is less clear. For a *rotating disc*, Levich (1962) has shown that at the distance

$$\delta_{rot} = 1.61 Sc^{1/6} \left(\frac{D_s}{\omega}\right)^{1/2} \qquad (4.14)$$

the diffusive and convective fluxes are the same. The length scale δ_{rot} depends solely on the Schmidt number $Sc = \nu/D_s$ and rotation rate ω. At low growth velocities the solution for a growing *rotating crystal* approaches that of a *rotating disc* given by Levich (1962) and $\delta_{bps} \approx \delta_{rot}$.

---

[1] In an exact expression the density change upon solidification must be included, which can be done by rescaling the velocity to $V' = V\rho_i/\rho_b$ in equation (4.12), where ρ_i and ρ_b are the solid and liquid densities. In natural water the effect translates into the lifting of the freeboard of ice and may be ignored

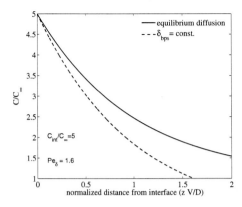

Fig. 4.1: Normalized concentration profile (4.10) from the δ_{bps}-approach (dashed line) and the diffusive equilibrium (4.5) (full line) for an interface concentration $C_{int} = 5C_\infty$.

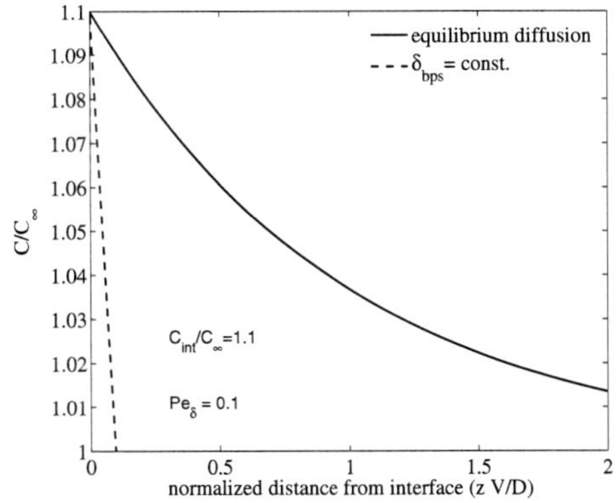

Fig. 4.2: Contrasting figure (4.1), for low $C_{int} = 1.1C_\infty$ the concentration profile from the δ_{bps}-approach is almost linear.

Burton et al. (1953b) found agreement of equation 4.12 with a number of observations. Its attractivity is due to its potential to use measurable quantities k_{eff} and V in order to estimate the interfacial k. Written in the form

$$ln\left(\frac{1}{k_{eff}} - 1\right) = ln\left(\frac{1}{k} - 1\right) - \frac{V\delta_{bps}}{D_s}, \tag{4.15}$$

one finds that, in a graphical presentation of $ln(1/k_{eff} - 1)$ versus V, one may eventually identify a constant slope $-\delta_{bps}/D_s$ and a zero-intercept $ln(1/k - 1)$. In case that D_s is known, it is also tempting to interpret δ_{bps} as the width of the forced diffusive boundary layer and compare it to the diffusive case. In figures 4.1 and 4.2 the concentration profile (4.10) from the δ_{bps}-approach is compared to the diffusive equilibrium (4.5) for two different ratios C_{int}/C_∞. For $C_{int}/C_\infty \approx 1$, corresponding to $k_{eff} \approx k$, the gradient in the hypothetical boundary layer is close to linear. With increasing C_{int}/C_∞ the concentration distribution approaches the exponential form of the diffusive equilibrium. The profile shape depends on C_{int}/C_∞ and, whereas the equilibrium diffusion concentration profiles are similar for any interface salinity ratio, the profiles from the δ_{bps}-approach are not. This shape-change is critical when interpreting δ_{bps} as a boundary layer scale, to be discussed next.

4.2.2   Misconception of δ_{bps} as a boundary layer scale

When suggesting the approach of constant δ_{bps}, Burton et al. (1953a) pointed out that δ_{bps} is a 'somewhat arbitrary quantity', as it just gives a similar dependence of the interface composition on growth parameters as a physical model, and warned to interpret it in general as a physical length scale. They showed that δ_{bps} corresponds closely to δ_{rot} computed from Levich's boundary layer model, provided that the crystal growth rate is small and $Pe_\delta < 1$. However, the evaluation of δ_{bps} as a boundary layer scale has to consider the model of interfacial fluxes. Levich (1962) termed the boundary layer scale δ_{rot} as the distance where convective and diffusive flux contributions are comparable. He defined it via $\delta_{rot} = D_s(C_{int} - C_\infty)/F_s$, where F_s is the total solute flux towards the interface of the crystal. With pure diffusion at the interface one has $F_s = D_s(\partial C/\partial z)_{int}$ and δ_{rot} is related to the interfacial solute gradient and the overall concentration difference across the boundary layer. Levich's work is valid for δ_{rot} of a fixed *rotating disk*. Wilson (1978) has outlined that applying this definition to the *rotating growing crystal*, the flux boundary layer scale δ_{rot*} is modified as

$$Pe_{rot*} = 1 - exp(-Pe_{rot}), \tag{4.16}$$

where the Peclet number $Pe_{rot*} = V\delta_{rot*}/D_s$ represents the flux boundary layer of a growing rotating crystal. Notably, this interfacial flux definition implies that at $Pe_{rot} = 1$ the growing crystal boundary layer thickness δ_{rot*} is just 63 % of δ_{rot} for a rotating disc obtained with Levich's boundary layer model. At $Pe_{rot} = 0.2$ the difference is still 10 %. Hence, while for $Pe_{rot} < 1$ δ_{bps} closely agrees with δ_{rot}, it misinterprets the interfacial flux scale δ_{rot*} by the above quoted numbers at larger Pe_{rot}. Most important is, that it becomes completely invalid and misleading as a length scale, because the growth compresses the boundary layer.

Wilson (1978) noted that the δ_{bps}-approach appears to give 'the right answer for the wrong reason'. It is constructed by forcing some arbitrary mathematical equation to match the interfacial solute gradient, disregarding the physical mechanisms of the solute transport. This may be illustrated in a different way. Formally, the Peclet number is fixed in the δ_{bps}-model via

$$Pe_\delta = ln\left(\frac{C_{int}}{C_\infty}\left(\frac{1-k}{1-\frac{C_{int}}{C_\infty}k}\right)\right), \tag{4.17}$$

which is equation (4.15) written in terms of $C_{int}/C_\infty = k_{eff}/k$. It is seen that $Pe_\delta \approx ln(C_{int}/C_\infty)$ for small k and moderate C_{int}/C_∞. As seen in figures 4.1 and 4.2, the increase in Pe_δ is reflected by the shape change in the concentration boundary layer profile. This shape change is not based on a realistic physical model. When Pe_δ becomes large, the formalism δ_{bps}=const. is no longer valid and leads to unrealistic scalings. It does not reflect the reality, where the equilibrium diffusion solution is approached and the scale based on the interfacial solute flux becomes equal to D_s/V.

Due to its simplicity, the δ_{bps}-approach from Burton et al. (1953a) has for a while been used as a basic boundary layer concept in metallurgical applications, where the method of rotating crystal growth from Czochralski has been popular (Tiller, 1962; Chalmers, 1964; Flemings, 1974). Today it is known that it is not only the concept δ_{bps}=const. at larger growth velocities which is questionable. Results are modified considerably by unsteady oscillating growth (Wilson, 1980), convection and finite-size effects (Tiller, 1991a; Vartak et al., 2005). In conclusion, the applicability of the δ_{bps}-approach is limited. One must at least distinguish the following regimes

- $Pe_{\delta} < 0.1$. Both the calculation of k from k_{eff} via the approach δ_{bps}=const. and the implied boundary layer δ_{bps} are within 5 % of the exact solutions.

- $0.1 < Pe_{\delta} < 1$. The relation between k and k_{eff} is properly resolved by the simplified δ_{bps}-approach. The constant boundary layer scale δ_{bps} already disagrees with a consistent interfacial gradient scaling by a factor 0.63 at $Pe_{\delta}= 1$.

- $Pe_{\delta} > 1$. The correlation of k_{eff} with V leads to an underestimate of k, if the simplified equation 4.15 is applied. This underestimate increases with Pe_{δ}. The scale δ_{bps} is no appropriate boundary layer scale at all.

4.2.3   The δ_{bps}-approach for other geometries

The δ_{bps}-approach resembles a stagnant film model to describe interfacial fluxes. Such models have been suggested already by both Nernst and Langmuir at the beginning of the 20th century (Burton et al., 1953a; Levich, 1962). Being eventually applicable under rotating, slow crystal growth conditions, the δ_{bps}-approach was later applied to crystal growth configurations in general (Tiller, 1962; Chalmers, 1964; Flemings, 1974). However, already Levich (1962) pointed out many aspects, due to which the extension of the method to other geometries than the rotating disc is incorrect. The main points of critics (Levich, 1962, 1967; Wilcox, 1993) are:

- The assumption of no fluid velocity in the boundary layer is wrong. While, for the rotating disc, the fluid velocity can be computed from molecular properties, the treatment of other 'nonuniformly accessible surfaces' is more difficult. A correct procedure would require a model for δ_*, which for nonrotating geometries will also depend on position x and velocity u due to $\delta_* \sim (\nu x/u)^{1/2}$.

- As shown for the rotating crystal, the interpretation of δ_{bps} as a length scale is also here problematic.

- The picture of transport processes implied by the $\delta_{bps} = const.$ approach is misleading. The solution of equation (4.10) is not valid for $z > \delta_{bps}$, because it implies a concentration $C < C_{\infty}$. In this context the boundary condition $C = C_{\infty}$ at $z = \delta_{bps}$

is problematic. The assumption of molecular diffusion for $x < \delta_{bps}$ and infinite diffusion at $x > \delta_{bps}$ implies a concentration gradient discontinuity at $x = \delta_{bps}$ (figures (4.10) and (4.5)).

- The solute transport away from the phase boundary is often enhanced by free convection which may depend on the growth velocity.

The general approach $\delta_{bps} = const.$ lacks a model of a *boundary layer* but just forces δ_{bps} to fulfill the interfacial solute balance, if there was no flow in this layer. Physically the approach corresponds to a sharp interface at the distance δ_{bps} and reflects an insulating, infinite flux boundary condition. It can, at most, be imagined to parametrise the solute transfer - see discussion above - but not the physical scale of the boundary layer.

4.3   Convective solute transfer

Convecto-diffusive solute fluxes depend in general on the systems' geometry as well as laminar or turbulent flow conditions. They can eventually be approximated by simplified hydrodynamic boundary layer models (Levich, 1962; Grigull et al., 1963; Schlichting, 1965; Batchelor, 1967). Some success in predicting k_{eff} from equation 4.12 has been obtained in the field of metallurgy, both for rotating and nonrotating crystal growth (Tiller, 1962; Chalmers, 1964; Flemings, 1974; Wilcox, 1993). This is likely related to the fact that a study at a specific geometry will often result in specific and constant flux conditions, at least over limited parameter ranges. It does not imply that the δ_{bps}-approach is a consistent method.

   To better understand the predictability of k_{eff} it is important to compare the solute transfer for different experimental geometries, stirring or flow conditions. In its most general the problem can be discussed by defining an effective solute transfer rate V_* and a corresponding Peclet Number $Pe_* = V_*/V$, rather than considering the detailed transfer process. Instead of the diffusive condition (4.9) the interfacial solute flux is parametrised via

$$V_*(C_{int} - C_\infty) = V(C_{int} - C_s) \qquad at \quad z = 0 \tag{4.18}$$

It is recalled that, expressing V_* as D_s/δ_*, δ_* only eventually represents a boundary layer scale. Rather one should think of $V_* = Nu_* D_s/L_*$ being given by the Nusselt Number Nu_* and characteristic length scale L_* of the system. Examples of expected scalings (Levich, 1962; Grigull et al., 1963; Schlichting, 1965) are

$$V_* \sim \omega^{1/2} \qquad \text{rotating crystal}$$
$$V_* \sim \omega^{2/3} \qquad \text{stirred solution}$$
$$V_* \sim U^{2/5} \qquad \text{rectangular channel,}$$

where rotation rate ω or fluid velocity U reflect the flow conditions. V_* normally further depends on the molecular transport properties, their combinations like the Schmidt number $Sc = \nu/D_s$ and geometric factors. The solute transfer is related to forced flow with a source far from the interface and is normally termed *forced convection*.

A second class of solute transport configurations is related to the Rayleigh number Ra and V_* typically scales as

$$V_* \sim Ra_c^{1/4} \qquad \text{laminar free convection}$$
$$V_* \sim Ra_c^{1/3} \qquad \text{turbulent free convection.}$$

Under these *free convective* conditions the fluid flow is driven by solute release at the interface.

It is desirable to predict V_* from semi-empirical relations that parametrise the flow in free or forced convective boundary layers. The principal difference compared to the δ_{bps}-approach is that diffusion is no longer assumed to control the solute distribution in a boundary layer. The assumption is rather that the solute flux due to equation 4.18 is linearly related to the difference between interface and bulk concentration. Equation (4.18) may then be written in the form

$$k_{eff} = \frac{k}{1 - \frac{V}{V_*}(1 - k)} \approx \frac{k}{1 - \frac{V}{V_*}} \tag{4.19}$$

, where the right hand side holds for $k \ll 1$ and the case of planar freezing. This approximation implies an interfacial concentration

$$C_{int} = C_\infty \frac{k_{eff}}{k} \approx C_\infty \frac{1}{1 - \frac{V}{V_*}} \tag{4.20}$$

which for $V/V_* \ll 1$ may be further approximated as

$$C_{int} \approx C_\infty (1 + \frac{V}{V_*}). \tag{4.21}$$

4.3.1 Planar freezing: V_* versus δ_{bps}

The difference between k_{eff} from equation (4.19) and the δ_{bps}-approach is demonstrated in figure 4.3 for $k = k_{eq} = 0.002$. Free convection model computations, also shown in the figure, will be discussed later. The curve for the bulk solute transfer approach (dashed) appears no longer as a solid line in $ln(1/k_{eff} - 1)$ versus V space with slope V_*. The latter is only asymptotically reached at low growth velocities. Towards $V = V_*$ the curves diverge. At low growth velocities, approximately $Pe_* = V/V_* < 0.1$, the approaches almost agree, simply because k_{eff} is close to the equilibrium k. In the solute transfer model the value $Pe_* = 1$ represents the diffusive limit and implies $k_{eff} = 1$ at

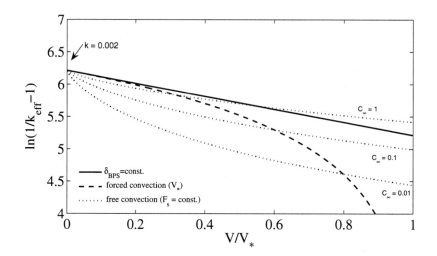

Fig. 4.3: Effective distribution coefficient k_{eff} in dependence on the Peclet number V/V_*. The solid line is the δ_{bps}-approach using D_s/δ_{bps} instead of V_*, as adopted by Weeks and Lofgren (1967) from Burton et al. (1953a). The dashed line the simple mass flux from equation (4.19). Dotted lines show free convection V_{f*} based on equations (4.24) and (4.30) for $C_\infty = 1$ (upper), $C_\infty = 0.1$ (upper) and $C_\infty = 0.01‰$ (lower). $V_* = 1.5 \times 10^{-4}$ cm s^{-1} was chosen somewhat arbitrarily to indicate the difference between free and forced convection. An interfacial $k = 0.002$ is typical for aqueous NaCl at $C_\infty = 0.1‰$, see below.

equilibrium. The solute transfer is bounded by $k_{eff} = 1$ at $Pe_* = 1$, and $k_{eff} = k$ at $Pe_* = 0$. In the δ_{bps}-approach the diffusive equilibrium $k_{eff} = k$ is not reached below $Pe_\delta \approx 5$. This is, as mentioned, the consequence of forcing the diffusive boundary layer to a constant width δ_{bps}.

The comparison demonstrates, that a convective solute transfer model leads to a different relation between k_{eff} and V than implied by the δ_{bps}-approach. If the solute transfer depends linearly on the concentration difference across the boundary layer, then the V_*-approach is correct. This needs not to be the case. Laminar or turbulent solute transport parameterisations are mostly valid if the convective transport dominates (Grigull et al., 1963; Schlichting, 1965; Batchelor, 1967). However, near the transition from diffusive to convective mass transfer, here at $Pe_* = 1$, the Nusselt number changes often in a nonlinear manner. This may imply V_* that is not independent of V. One may argue that the solute transfer approach is, in the present simplified form, most useful and trustworthy in the regime where $k_{eff} \approx k$, where it almost coincides with the δ_{bps}-approach. However, while V_* can be estimated *a priori* by empirically incorporating a boundary layer model, the δ_{bps}-approach has no clear physical basis. In general, it is essential to first obtain an estimate of V_* for the experimental setup in question, before a refined analysis of the solute transfer can be attempted. The δ_{bps}-approach applied in most analyses to date (Jaccard and Levi, 1961; Weeks and Lofgren, 1967; Weeks and Ackley, 1986; Cox and Weeks, 1988; Souchez et al., 1988; Eicken, 1998; Killawee et al., 1998; Ferrick et al., 2002) does, considering figure 4.3, not provide a correct description of the transport processes.

If the solute transfer V_* is by free convection similar arguments apply. In this case the mass flux velocity V_* is related to the salinity flux at the interface and can be expected to depend on the growth velocity. As mentioned above, often the Rayleigh number is the appropriate parameter in a scaling law for V_*. In the following, some empirical relations for the solute transfer during forced and free convection for frequently used geometries are summarised, before the results of the studies are analysed within this framework.

4.4 Practical scaling laws of solute transfer

4.4.1 Rotating crystal

The rotating crystal has been discussed in section 4.2.1 above. It is recalled that for sufficiently small $Pe_* = V/V_* < 0.2$ the solute transfer velocity V_* is found by dividing D_s by δ_{rot} given by equation (4.14). However, ice is seldom grown under such idealised conditions. The scalings which may evolve at larger Pe_* and nonideal boundary conditions are more complex (Tiller, 1991a).

4.4.2 Vertical cylinder with stirring

A frequently used setup in ice growth experiments is unidirectional growth in a cylinder under stirring (Weeks and Lofgren, 1967; Wintermantel, 1972; Gross et al., 1975, 1977; Ozum and Kirwan, 1976; Gross et al., 1987; Cragin, 1995). Among these authors only Wintermantel (1972) determined experimentally the solute transfer velocity V_* of the system. Some aspects of his procedure are outlined here.

Wintermantel (1972) first derived an empirical heat transfer relation for his geometry and the Prandtl number regime $7 < Pr < 230$, which then was applied to the solute transfer problem by replacing the Prandtl number Pr by the Schmidt Number Sc. According to his observations V_* at the bottom of the cylinder may be presented in the form

$$V_* = c1 \frac{D_s}{L_*} \left(\frac{\omega L_{st}^2}{\nu} \right)^{c2} Sc^{c3}, \tag{4.22}$$

where ω and L_{st} denote rotation rate and diameter of the stirrer. The term in brackets may be thought as a Reynolds number. The constants $c1$ through $c3$ were obtained for two different stirring devices, being ($c1 = 0.29$, $c2 = 0.67$, $c3 = 0.43$) for a propeller and ($c1 = 0.37$, $c2 = 0.65$, $c3 = 0.43$) for a blade stirrer. Wintermantel used the diameter L_* of the cylinder as length scale, which was almost the same as the distance H_* between the stirring devices and the bottom. Increasing H_* by 40 % left the heat transfer almost unchanged for the blade stirrer, but led to an increase of V_* by 25 % for the propeller device. Hence, for the propeller also H_* should be included in the scaling law. A tentative modification in agreement with his limited experimental data is to introduce the ratio of cylinder diameter to stirrer distance from bottom, L_*/H_*, into the Reynolds Number. The modification of equation 4.22 by the factor $(L_{st}/H_*)^{2/3}$ is consistent with scaling laws in general (Grigull et al., 1963; Schlichting, 1965), and with turbulent exchange in the related thermosyphon problem (Japikse et al., 1971). Unfortunately, in most of the noted studies on ice growth under stirring (Weeks and Lofgren, 1967; Wintermantel, 1972; Gross et al., 1975, 1977; Ozum and Kirwan, 1976; Gross et al., 1987; Cragin, 1995) the provided information is not adequate to compute V_*.

It should be kept in mind that the scaling 4.22 has been determined in a particular laboratory configuration. Several factors may affect the constants of this equation. For example, also in Wintermantel's study deviations from a radially uniform heat transfer at the bottom of the container were found upon increasing the stirring rate.

4.4.3   Forced flow through a channel

Freezing experiments of NaCl solutions with forced flow through a vertical channel, to be discussed below, have been performed and analysed by Dschu (1968). The scaling law of the Nusselt number for turbulent flow through a pipe (see for example Grigull et al., 1963, p.232 ff.) may be expressed in the form

$$V_* = c1 \frac{D_s}{L_{hyd}} \left(\frac{U L_{hyd}}{\nu} \right)^{c2} Sc^{c3} \left(\frac{L_{hyd}}{L_{pip}} \right)^{c4}. \tag{4.23}$$

Appropriate values for the constants have also been summarised by Grigull et al. (1963). These are $c_1 \approx 0.032$ and $c_2 \approx 4/5$ and $c_3 \approx 1/3$ for the exponents of the Reynolds and Schmidt numbers. The last factor on the right of equation 4.23 accounts for a weak dependence on the ratio of hydraulic radius L_{hyd} and pipe length L_{pip}. The value $c_4 \approx 0.054$ is typically valid for the range $10 < L_{pip}/L_{hyd} < 400$.

4.4.4   Free solutal convection

For free turbulent convection the generally accepted scaling for the mass transfer V_{f*} is

$$V_{f*} = c_{Nu} \frac{D_s}{L_*} Ra^{1/3}, \tag{4.24}$$

where the solutal Rayleigh number is

$$Ra = \frac{\beta \Delta C_{int} g L_*^3}{\mu D_s}. \tag{4.25}$$

In these equations β is the haline density contraction coefficient, μ the dynamic viscosity and g gravity acceleration. A detailed discussion of solutal free convection follows in section 10. Here it suffices to say that a considerable body of theoretical and experimental work justifies the use of a constant c_{Nu}. It relates to solute flux in a turbulent boundary layer driven by free convection for large enough $Ra \approx 10^5$. Due to the thin solutal boundary layers the latter condition is fulfilled in most laboratory configurations as during natural ice growth. c_{Nu} is normally in the range 0.15 to 0.19 for free convection at high Schmidt or Prandtl numbers. The scalings imply that V_{f*} is independent of sample dimensions L_*, but depends on the salinity difference ΔC_{int} between the bulk and the interface due to

$$V_{f*} \sim \Delta C_{int}^{1/3}. \tag{4.26}$$

Hence, under free solutal convection the form 4.19 cannot be used with a constant V_{f*} to validate observations of k_{eff}.

The change in the dependence between k_{eff} and V may be illustrated for $V/V_{f*} \ll 1$. With a reference ΔC_{int0}, at which $V_{f*} = V_{f*0}$, equation 4.21 may be written as

$$\frac{k_{eff}}{k} \approx (1 + \frac{V}{V_{f*}}) \approx \left(1 + \frac{V}{V_{f*0}} \left(\frac{\Delta C_{int0}}{\Delta C_{int}}\right)^{1/3}\right). \tag{4.27}$$

Now, from equation 4.21, one has $\Delta C_{int} \approx C_\infty V/V_{f*}$. If ΔC_{int} changes solely due to variation in C_∞ and the higher order terms in V_{f*} are neglected, one gets

$$\frac{k_{eff}}{k} \approx \left(1 + \frac{V}{V_{f*0}} \left(\frac{C_{\infty 0}}{C_\infty}\right)^{1/3}\right) \tag{4.28}$$

and k_{eff} is found to decrease slightly with C_∞. On the other hand, for constant C_∞ and the growth velocity V varying with respect to a reference value V_0, one has

$$\frac{k_{eff}}{k} \approx \left(1 + \frac{V^{2/3} V_0^{1/3}}{V_{f*0}}\right). \tag{4.29}$$

These examples illustrate the difference between free convective scalings of V_{f*} and forced convection V_*, the former depending on C_∞ or yielding a weaker V-dependence. When V approaches V_* one has to use 4.24 with 4.25 instead of the approximation 4.27. For the case of $k \ll 1$ equations 4.19, 4.24 and 4.25 yield

$$\frac{\Delta C_{int}^{4/3}}{C_\infty} - c_{1/3} \frac{\Delta C_{int}}{C_\infty} - c_{1/3} = 0, \tag{4.30}$$

where

$$c_{1/3} = \frac{V}{c_{Nu}D_s} \left(\frac{\mu D_s}{\beta g} \right)^{1/3} \tag{4.31}$$

is a function of growth velocity.

Although equation (4.30) looks promising due to its solvability, it must be viewed with caution. It assumes that the critical ΔC_{int} is instantaneously given. This implies a kind of insulating (constant salinity step) boundary condition. However, if the solute-buildup is intermittent this condition is not correct. One might argue that a constant salt flux ($F_s = const.$) boundary condition is a better choice. The difference has been discussed by Foster (1968b) and is further outlined in chapter 10.2.1 on solutal convection. A comparison of k_{eff} for the two boundary conditions is illustrated for $k = 0.0015$ and $C_\infty \approx 0.3$ ‰ in figure 4.4 at a later stage of the present chapter. It is seen that, for $\Delta C_{int} = const.$, k_{eff} increases sharply at low growth velocities and reaches a similar slope as the $F_s = const.$ model with further increasing V. It is noted that, at low concentrations and low growth velocities, k_{eff} for a $F_s = const.$ model is less than k_{eff} for ΔC_{int}=const. This behaviour changes at large C_∞ and large V (not shown). As long as there is no mechanism which controls the salinity step, the $F_s = const.$ condition is the physically more reliable approach[2]. It will be denoted as the *free convection model* in the rest of this chapter.

The solution of the *free convection model* is shown in figure (4.3) for three concentrations C_∞ and V normalised by (an arbitrary) $V_* = 1.5 \times 10^{-4}$ cm/s. The dependence of k_{eff} on growth velocity depends strongly on C_∞ and is very different from forced convection scalings. This is also seen in figure 4.4, where k_{eff} decreases with increasing bulk concentration and the slope of k_{eff} versus V differs from the δ_{bps}-approach. One must conclude that convection produces a relation between k_{eff} and V which only coincidentally may coincide with forced convection scalings.

4.4.5 Free thermal convection

For water and dilute salt solutions the density has a maximum at a temperature above the melting point (see the appendix A.2.4 for NaCl). Therefore free thermal convection may influence the solute transport during the upward freezing of unstirred dilute solutions. Considerable effects of thermal convection have been reported in upward freezing experiments (Kvajic and Brajovic, 1970; Gross et al., 1977). The resulting double-diffusive solute transport is complicated and depends on experimental details and molecular properties (Turner, 1973). For the rotating crystal, some aspects of the complexity of thermosolutal convection in confined geometries have been discussed by Tiller (1991a). An

---

[2] Once the salinity increases considerably at the interface, also $F_s = const.$ is no longer fulfilled. The equations that must be used in this case are discussed in the final chapter, section 12

ad-hoc approach applied by Gross et al. (1977) to account for thermal convection therefore gave only the correct order of magnitude for k, derived from k_{eff}.

During downward freezing at salinities higher then 23 ‰ NaCl (25 ‰ for seawater) both the thermal and solutal stratifications are vertically unstable and two convective modes will occur. However, while for solutal convection often a parametrisation based on infinite Rayleigh number may be used, the thermal convection pattern for typical freezing situations may interact with the laboratory geometry, as the thermal boundary layer is much larger. The formulation of scaling laws is more complex (Siggia, 1994; Grossmann and Lohse, 2000; Ahlers et al., 2002). When unknown heat fluxes through the side walls come in addition, one often has to find empirical scalings.

4.5 Applications of solute transfer models to planar freezing

It is now possible to review several studies on the freezing of aqueous NaCl solutions with respect to the above described solute transport concepts. First of all, k derived in several studies from observations of k_{eff} is discussed critically. The question, if the interfacial k resembles the true equilibrium value, has been discussed by several authors on the basis of different theoretical approaches of equilibrium solute incorporation (Drost-Hansen, 1967; Gross, 1968; Seidensticker, 1972; Hobbs, 1974; Tiller, 1991b; Thibert and Domine, 1997). However, a fundamental and generally validated theory is still lacking. Many processes may influence k, for example (i) electric interfacial processes (ii) hydrodynamic and mechanical influences on interface structure and kinetics, (iii) interstitial entrapment at grain and subgrain boundaries, (iv) interface stability and, (v) growth rate fluctuations, (vi) differential diffusion of anions and cations, (vii) crystallographic orientation. In the following, k is denoted as an effective interfacial solute distribution at a planar freezing interface, distinguishing it from the true, unknown equilibrium value k_{eq}.

Most studies cited below are based on Cl$^-$ measurements. The concentrations of NaCl by weight discussed in the present chapter are obtained by assuming that the natrium ion is incorporated in ice at the same fraction, although there are indications that the bulk distribution of chloride ions in ice is slightly larger (Gross, 1968; Killawee et al., 1998). Assuming an equivalent behaviour of anions and cations is necessary here for a tractable computation of diffusive and convective fluxes. For small k in saline solutions this approach is justified. Of course, differential distribution needs to be considered in connection with a more sophisticated model of the incorporation process itself.

4.5.1 Forced convection due to stirring

The distribution of the chloride ion in ice has received attention in several recent studies. Seidensticker (1972) reported a concentration dependence of $k \sim C_{int}^{-1/3}$ for freezing of aqueous HCl, which he explained in terms of a plausible simple model of interface potential. Gross and his colleagues performed experiments for many halides for stirred and unstirred conditions (Gross et al., 1975, 1977) and suggested that the concentration dependence reported by Seidensticker might be related to a detection limit, or to inappropriate estimates of C_{int}. Although Seidensticker (1972) noted that thermal convection was not apparent in dye experiments, his results scatter over one order of magnitude, casting some doubt on the applicability of his diffusion-only approach. Gross et al. (1975) discussed the influence of thermal convection on C_{int} and, on the basis of their own data, suggested a simple bulk model to account for the latter. The later experiments by Gross et al. (1977) are now frequently used as a reference for the chloride ion's k. These observations are discussed below in some detail.

Discussion of experiments by Gross et al. (1977)

The conclusions drawn by Gross et al. (1975) and Gross et al. (1977) from considering the upward freezing of stirred solutions of hydrogen and alkali halides including NaCl were:

- k_{eff} was, for the range of growth velocities, almost independent of the stirring rate once the latter exceeded a typical limiting value. Hence, Gross et al. (1977) suggested $k \approx k_{eff}$ for the experiments.

- k_{eff} for different halides is almost constant and its average is $k_{eff} \approx 0.0027$.

- k_{eff} is almost independent of concentration

- k_{eff} is not affected by electrical interface potential and the ph of the solution

With emphasis on the freezing of aqueous NaCl solutions, several aspects of the studies from Gross and coworkers have been reexamined critically. The following points, to be discussed below, are most important:

1. In figure 3 from Gross et al. (1975) for the freezing of HCl there appears to be an increase of k_{eff} by approximately 20 to 50 % when the growth velocity is doubled, at least for concentrations above 10^{-5} M.

2. Figure 10 from Gross et al. (1977), reflecting the stirred growth of NaCl solutions, shows a significant decrease from $k_{eff} \approx 0.0034$ at 0.0006 ‰ NaCl to $k_{eff} \approx 0.0012$ at 6 ‰.

3. This concentration dependence may be approximated to within 10 % by $k_{eff} \approx 0.0016C_\infty^{-0.1}$, with C_∞ given in ‰. That this average dependence of k_{eff} on C_∞ is not constant but increases with concentration, is also apparent for other halides.

4. At high NaCl concentrations the data shown by Gross et al. (1977) give a less clear picture. In one run shown in their figure 10 a sudden decrease in k_{eff} is indicated near 1 ‰, followed by a minimum in $k_{eff} \approx 0.0007$ at 2 to 3 ‰.

5. The initial values of k obtained in unstirred experiments in the same figure do not reveal a concentration dependence, yet indicate a constant $0.002 < k < 0.003$. These estimates of k, reduced from the observed k_{eff} by a convection model, are more scattered and restricted to concentrations below 0.06 ‰.

6. A reduction from k_{eff} to k, due to the finite Peclet number of the system, was not performed.

Concentration dependence of k_{eff}.
Gross et al. (1977) attributed the slight decrease of k_{eff} with C_∞ to the fact that the distance between crystal and stirrer decreased during growth, being associated with an increasing solute concentration in all the runs. According to the above considerations this would in fact be a plausible explanation when considering V_* given by equation (4.22) for a propeller device. However, it cannot explain that also the initial values of k_{eff} showed the observed relationship $k_{eff} \approx 0.0016 C_\infty^{-0.1}$. Also in a later study by Gross et al. (1987) with relatively low C_∞ a larger $k_{eff} \approx 0.0032$ was reported, giving support to the concentration dependence. The decrease of k with C_∞ may even have been masked by the following hydrodynamic effect: Increasing C_∞ at constant growth rate likely implies an increasing interfacial salinity difference ΔC_{int} which in turn should retard the stirring strength V_* and imply an increase in k_{eff}. A crude estimate of this effect can be made by assuming a similar form as the unstable free convection scaling. It would lead to a reduction of V_* by the order of $Pe_* V_{*f}$ with V_{*f} being calculated from equation 4.24. One may, between $C_\infty = 0.001$ and $C_\infty = 1‰$, and for $Pe_* \approx 0.2$ in the experiments (see next paragraph), expect an increase in k_{eff} by ≈ 10 %. It is finally noted, that the data shown by Seidensticker (1972) for HCl indicate a lower bound on the concentration dependence similar to $k_{eff} \sim C_\infty^{-0.1}$. It can be concluded that a concentration dependence of k is very likely.

Peclet number and correction of k_{eff}.
In earlier studies Gross (1968) has noted some dependence of k_{eff} on stirrer shape and distance. Implications were not discussed in later studies and the stirrer dimensions not reported (Gross et al., 1977). Here the Peclet number $Pe_* = V/V_*$ in the experiments from Gross et al. (1977) has been roughly estimated from the variation in k_{eff} with stirring rate shown by Gross (1968) for aqueous HF (his figure 17). If V_* in the experimental setup is approximately given by equation (4.22) with $L_{st} = 4$ cm (cylinder diameter) and $L_* = 3$ cm (stirring device diameter) one may estimate $V_* \approx 4.6 \times 10^{-4}$ cm s^{-1} and $Pe_* \approx 0.21$ for the stirred freezing runs reported by Gross et al. (1977) with aqueous NaCl

[3]. This value is also consistent with the slight velocity dependence of k_{eff} for HCL - see note 1. above. Using equation 4.19 for the solute transfer and $V \approx 9.6 \times 10^{-4}$ cm s^{-1} in the experiments, it seems appropriate to reduce k_{eff} given by Gross et al. (1977) by a factor ≈ 1.26 to obtain the interfacial k. In conclusion, a revised dependence of the intrinsic k on NaCl concentration

$$k \approx 0.0013 C_\infty^{-0.1}$$
(4.32)

may be suggested according to the present reanalysis of the data from Gross et al. (1977).

Proposed k at lake water salinities

Based on the likely concentration dependence one may define some reference value of k at concentrations typical for natural lake water. For $0.1 < C_\infty < 0.3$ NaCl equation 4.32 based on the data from Gross et al. (1977) gives $0.0016 < k < 0.0015$. In a recent laboratory study, Ferrick et al. (2002) reported on freezing experiments of 0.16 to 0.2 ‰ chlorine aqueous solutions (0.26 to 0.32 ‰ in an NaCl system) under 'well-stirred' conditions, where both free convection and an under-ice flow existed. The observed k for the three lowest freezing rates were within 10 % of 0.0015. Only for rapid growth a slightly larger $k \approx 0.0025$ was found. These results are thus in good agreement with the data from Gross et al. (1977). Below it will be seen that unstirred experiments from Ferrick et al. (2002), when solute transport is due to free convection only, also validate this estimate.

Comparison with other studies

For a typical brackish water salinity of 1.65 ‰ NaCl a value $k_{eff} \approx 0.00105$ has been reported by Cragin (1995) using a similar growth apparatus as Gross et al. (1977). Growth velocities were similar as in the experiments by Gross yet stirring conditions have not been documents by Cragin (1995) who refers to Gross et al. (1977). The uncorrected salinity dependence from Gross et al. (1977), note 3. above, gives almost the same $k_{eff} \approx 0.0011$, whereas the actual experimental data from Gross, note 4. above, indicate a lower value $k_{eff} \approx 0.0007$ at this concentration. The differences are indicative of an intergranular and an intragranular solute distribution mode. This was also suggested by Cragin (1995) from a qualitative comparison of k_{eff} with grain size variations.

After correction due to V_* the smallest k observed by Gross et al. (1977) appears to be $k \approx 0.0005$ at a salinity of ≈ 2 ‰ NaCl. Also Tiller (1991b) has, for KCl, given a value of $k = 0.0005$ at a comparable concentration, while Kvajic and Brajovic (1971)

---

[3] In section 7.2.2 below it will be seen that this estimate of Pe_* is also consistent with interface instability observed in these experiments

observed minimum distribution coefficients in the range $0.0005 < k < 0.001$ during the freezing of KOH, yet based on K^+ measurements. The experiments from Thibert and Domine (1997), who doped single ice crystals with gaseous HCl for several weeks, may be most relevant in terms of an intragranular k_g, or even an equilibrium k_{eq} value. These authors report $0.00012 < k < 0.00025$ decreasing with mole fraction for a temperature range which correspond to liquidus mole fractions of approximately 0.002 to 0.02 M HCl. For higher solute concentrations k increased again to ≈ 0.0006 at a mole fraction of 0.1 M HCl, the upper limit of observations.

Weeks and Lofgren (1967) reported a much larger $0.005 < k_{eff} < 0.009$ at even smaller growth velocities for two stirred experiments at 3 ‰ NaCl in the bulk solution. Th stirring level might have been lower in the latter experiments, implying a larger correction factor from k_{eff} to k. However, due to downward freezing here also free convection was present. The investigation of the non-stirred free convection runs from the same authors in section 4.5.2 below indicates, that free convection at low growth velocities is sufficient to limit k in a similar way. The most plausible explanation of the larger k seems to be a history-dependent effective solute incorporation. In the experiments from Weeks and Lofgren the cellular-planar transition took place under decreasing growth velocities. One may speculate that the preceding cellular growth created smaller grain sizes which then survived during only a few centimeters of planar growth.

Grain size and crystal orientation

Hence, all mentioned studies are consistent with a grain size dependence of the bulk k in unconfined growth. It is unlikely that the intrinsic intragranular k has been observed by Gross et al. (1977) and most other authors.

The values of k obtained by Gross et al. (1977) probably relate to crystals of a maximum grain size of 2 cm, embedded in an ice sample of 4 cm diameter as described in earlier studies (Gross, 1968). Similar grain sizes are normally observed in the upper centimeters of lake ice (Ragle, 1963) and may also be expected for the freezing runs by Ferrick et al. (2002). The noted agreement between k from Gross et al. (1977) and Ferrick et al. (2002) should be viewed in this context. It was already mentioned that k_{eff} in the experiments of Weeks and Lofgren (1967) was much larger. The scatter in the latter observations, which was also present in other experiments (Kvajic and Brajovic, 1971; Seidensticker, 1972), supports the conjecture of a large contribution of intergranular solute entrapment to k_{eff}. Cragin (1995) pointed out that, during initial crystal growth, k_{eff} appeared to decrease strongly with increasing grain size. It is worth a note, that such distribution of solute within grains and grain boundaries has been reported for glacier ice samples by Harrison and Raymond (1976), although growth, aging and coarsening of the latter are certainly different from unidirectional growth on lakes. The importance of grain boundaries may be traced to earlier descriptions of solute in veins and liquid figures revealed by internal melting (Quincke, 1905b; Nakaya, 1956). Quincke (1905b) noted that the solute content of samples grown from dilute salt solutions was the higher the more internal melt figures were found in a sample. Recent microscopic studies clearly show the preferred location of impurities in different ice types at grain-, but also at subgrain boundaries (Cullen and Baker, 2002; Illiescu et al., 2002; Baker et al., 2003).

Concepts of intermittent solute incorporation had already been proposed by Tyndall and Faraday (Tyndall, 1858). Later also Tiller (1962) pointed out the role of grain boundaries for solute incorporation and proposed some simple ideas which have received little attention. In principal one can expect that the value of k will decrease with increasing grain or subgrain size (sample sizes being the same).

The crystallographic orientation with respect to the growth direction is likely another important factor, concerning the incorporation process into single crystals and the selection of grain sizes. Some observations of k_{eff} for different crystal orientations have been obtained by Kvajic and Brajovic (1971) during freezing of aqueous NaOH solutions. However, the presented results appear too scattered in the planar growth regime to identify a difference. Proper measurements of k for single grains and subgrains, different grain sizes and orientations are an important future task and still outstanding. The studies noted above do principally indicate an intrinsic intragranular $0.0001 < k0.0005$ for the chloride ion.

4.5.2   Free convection

The values of k estimated from the study by Gross et al. (1977) should be understood as being valid for growth velocities and grain sizes typical for natural lakes. The value $k \approx 0.0015$ at $C_\infty = 0.3$ ‰ (equation 4.32) is in agreement with stirred experiments from Ferrick et al. (2002). The latter authors also performed unstirred downward freezing runs, when haline free convection can be assumed to be the main agent of solute transport. In figure (4.4) these observations of k_{eff} are compared to different solute transport relations. Note that for the δ_{bps}-approach and the mass transport an arbitrary $V_* = 1.5 \times 10^{-4}$ cm s^{-1} has been used in connection with $k = 0.0015$. It is recalled that such an approach is physically unreasonable, while the free convection solutions, shown as dotted and dash-dotted lines, reflect self-sustaining turbulent boundary layer models. It is illustrated that equation (4.30) for a constant salinity step leads k_{eff} which slightly larger than observed. Using here in advance the constant heat flux boundary condition in a convection scaling given in section 10.2.1 a reasonable agreement with the observations is obtained, being shown as the dash-dotted curve.

The constant salt flux convection model underestimates k_{eff} by ≈ 30 % at higher growth velocities ($V/V_* \approx 0.35$), which might be related to the following factors. First,

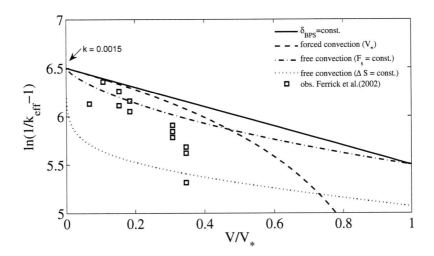

Fig. 4.4: Effective distribution coefficient k_{eff} observed by Ferrick et al. (2002). For comparison, the solid line is the $V_* = D_s/\delta_{bps}$-approach, the dashed line the simple mass flux from equation (4.19), for which $V_* = 1.5 \times 10^{-4}$ cm s^{-1} was chosen somewhat arbitrarily. Self-standing free convection models for constant salinity step (dotted) and constant salt flux (dash-dotted) boundary conditions. An interfacial $k = 0.0015$ is assumed for $C_\infty \approx 0.3\%_0$NaCl, see below.

the initial ice skim may entrap more salt between the smaller grains, taking some time until quasi-equilibrium is attained. Indeed, the fractionation measurements of ^{18}O shown by Ferrick et al. (2002) for this growth sequence indicate a somewhat different entrapment mode in the upper 5 cm, corresponding to the four points of largest deviation in figure (4.4). This is further corrobated by the fact that the ice salinities from a melt-regrowth cycle shown by the authors also resulted in smaller k_{eff}. Second, Ferrick et al. (1998) doped the mother solution with additional Cl$^-$. This change in the ph value value might have influenced the nonequilibrium evolution of k (Malo and Baker, 1968). Third, k is known to decrease slightly with concentration. As the free convection model is physically more reasonable, these influences on k need to be investigated.

The agreement of the free convection model at lower velocities not only lends further credence to $k \approx 0.0015$ at $C_\infty \approx 0.3 \%_0$ NaCl. It also indicates that the self-standing free convection model correctly describes the variation in k_{eff}. It makes future analysis of non-equilibrium processes more reliable. Finally, figure 4.4 demonstrates, that a linear relationship, as given by the δ_{bps}-approach, may eventually be correlated to data of k_{eff}

over a limited velocity range. However, slope estimates and interpretations in terms of a boundary layer thickness (Ferrick et al., 1998, 2002) are likely to be misleading.

The mentioned two times larger $0.007 < k_{eff} < 0.009$, reported by Weeks and Lofgren (1967) for two unstirred freezing runs at similar growth velocities, is slightly out of the range covered by figure 4.4. A different solute incorporation, due to the cellular-planar transition that took place earlier in the experiments from Weeks and Lofgren (1967), has already been suggested as a possible explanation.

4.5.3   Forced convection due to shear flow

Dschu (1968) performed freezing experiments of NaCl solutions flowing through a rectangular channel. For such a geometry the turbulent heat or solute transport V_* has been obtained empirically in many studies (see for example Grigull et al., 1963, p. 232 ff.). Dschu (1968) pointed out that in a few experiments the ice initially was very clear if the growth velocity exceeded a certain value depending on V_* and C_∞.

Figure 4.5 shows the k_{eff} for $k = 10^{-3}$, typical for high salinities, computed with the forced convection model and the δ_{bps}-approach. Dschu's observations are shown in figure 4.6 on a reduced scale. They are from freezing runs with solute concentrations between 20 and 60 ‰ where $k \approx 10^{-3}$ is expected. The values of V_* were also given by Dschu. The initial k_{eff} observations denoted by Dschu as 'clear' ice are plotted as squares, the ice with 'filamental structures' as triangles. All observations were apparently obtained in a Peclet number regime, where k_{eff} should be very close to the interfacial k. The initial planar value of k_{eff} was between 3.4 and 4.4×10^{-3} which corresponds to $3.0 < k < 4.0 \times 10^{-3}$. Upon a slight decrease in the growth velocity the observed k_{eff} rose rather abruptly.

The range $3.0 < k < 4.0 \times 10^{-3}$ for 'clear' ice is considerably larger than $k \approx 10^{-3}$ observed at approximately 6 ‰ by Gross et al. (1977). As already mentioned, this might be explained by a dependence of k on crystal size. For example, Matsuoka et al. (2001) grew ice from saline NaCl solutions under strong fluid flow in a rotating chamber and presented photographs of the microstructure with crystal sizes of several μm. As in the experiments from Matsuoka et al. (2001) the smallest k observed was 0.01, the ice growth in the latter experiments might always have been unstable. However, it is also plausible that crystals growing in a channel under strong fluid flow are smaller in general. Although there is a lack in theories of grain size control, one can expect a connection to thermal, solutal and viscous boundary layers. For rapid flow the latter may become relevant and create smaller grain sizes. The range in observed k points to the importance to formulate solute incorporation models in terms of the microstructure, even for planar freezing.

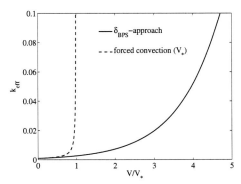

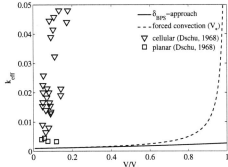

Fig. 4.5: Effective distribution coefficient k_{eff} for forced convection V/V_*, here shown as a linear plot (compare to figure 4.3). Curves are drawn for a realistic $k = 0.001$ for high salinities. Again $V_* = 1.5 \times 10^{-4}$ cm s^{-1} was chosen somewhat arbitrary

Fig. 4.6: Figure (4.5) on a smaller scale. The squares and triangles show observations from Dschu (1968) for planar and milky ice appearance, respectively. Solution salinity range $20 < C_\infty < 60$.

4.6 Solute transport during cellular freezing

The solute redistribution k at a planar interface is apparently influenced by several mechanisms, which are not well understood and have the potential to change k by a factor of two. Grain size and growth history seem to play an important role. With this uncertainty in mind one may now extend the above discussion to solute incorporation during cellular growth of ice.

Planar-cellular transition

The easiest measurable indicator of a change in interface morphology from planar to cellular is, as mentioned above, an abrupt change in k_{eff}. If this is observed at a V/V_* value far below one, as seen in figure 4.6 for the experiments by Dschu (1968), this can no longer be explained by convective solute transport alone. It indicates a change in the mode of solute entrapment due to a transition from planar to cellular growth. Dschu (1968) and Wintermantel (1972) reported the dependence of this transition on stirring and growth conditions. Kvajic and Brajovic (1971) documented k_{eff} during freezing of KOH solutions in the case of free convection. Their observations show the noted jump in k_{eff} in dependence of the solute concentration, for constant growth velocity. Also Weeks and Lofgren (1967) presented transition observations from freezing of aqueous NaCl solutions. Their stirred experiments cannot be compared to other studies because (i) stirring rates and device properties were not documented, (ii) very probably both forced and free convection were present. All mentioned authors described that ice grown with a planar interface had a 'clear' appearance, cellular ice grown with much larger k_{eff} being 'milky'. In the following observations of such 'milky' ice are compared.

Proposed analogy between planar and cellular growth

Weeks and Lofgren (1967) applied the δ_{bps}-approach to a number of unstirred free convection experiments with downward freezing of solutions at different NaCl concentrations. A uniquely estimated D_s/δ_{bps} was suggested to be valid for cellular growth under natural conditions. Weeks and Lofgren (1967) proposed that also the entrapment in cellular ice may be described by the δ_{bps}-approach. k was now simply treated as an effective interfacial value for cellular growth. The observations from Weeks and Lofgren (1967) showed rather large scatter, which was reduced in a later study at a smaller solution salinity range (Cox and Weeks, 1975). However, the observations have never been critically discussed in terms of the question, if such an analogy is reasonable.

Much later work on the predictability of sea ice salinities has been based on accepting the analogy hypothesis of Weeks and Lofgren (1967), and applying the δ_{bps}-approach (Cox and Weeks, 1975; Weeks and Ackley, 1986; Cox and Weeks, 1988; Weeks, 1998a; Eicken, 1998, 2003). From the discussion in the previous sections it is clear that the δ_{bps}-approach does not capture the physics of the problem for a planar interface. One may therefore, at most, expect agreement with an empirical δ_{bps}-fit for a limited growth regime. The least condition for such an agreement is that, analogous to the planar interface, a quasi-constant interfacial k exists during cellular freezing. In the following this condition will be examined by focusing on the the relation between k_{eff} and V/V_* for freezing runs where V_* is approximately known due to forced convection scalings. This avoids the more complex changes during free convection indicated in figure (4.4).

Cellular k and structural entrapment

It is useful to point out some aspects of the interfacial k during cellular freezing. The planar interfacial value of k consists, as discussed so far, very probably of an *intergranular* and an *intragranular* component. Although detailed observations are lacking, the intragranular component of k is, for salinities of 1 ‰ and higher, probably less than 0.5×10^{-3}, see section 4.5.1 above. It follows that during cellular growth, when the cellular effective k is generally of order 0.1 or larger, the contribution of the solid ice phase to the bulk salinity content may be neglected. Disregarding a small correction factor due to the difference between ice and brine densities, the interfacial k during cellular freezing may then essentially be understood as the volumetric fraction of brine. Hence, the *cellular k* at the macroscopic freezing interface is a structural property closely reflecting the *relative brine volume*.

4.6.1 Forced convection

To test the analogy hypothesis from Weeks and Lofgren (1967) the following forced convection results from Wintermantel (1972) and Dschu (1968) in the cellular growth regime were analysed:

- In the experiments from Wintermantel (1972) freezing was upwards and forced convection created by the stirring device described above in 4.4.1. Wintermantel reported cellular freezing runs for different values of V and V_* at a fixed concentration of 79.5 ‰NaCl. The observations taken from his figures (5.40 and 5.42) have been interpolated to three growth velocities in order to illustrate the variation of k_{eff} with varying V_* in figure 4.8. Due to the slightly different thermal properties derived in the present work the values for V_* are approximately 25 % smaller than originally given by Wintermantel (1972) for the nominal concentration 79.5‰.

- The observations by Dschu (1968) for lateral freezing in a vertical rectangular channel with throughflow are shown in figure (4.7). Throughflow velocities were of the order of 100 cm s^{-1}. Only values for the largest V/V_* are plotted to exclude effects near the cellular-planar transition, which has been shown figure (4.6) for these experiments. Data points are shown for three salinities. For each salinity the dependence on V_* is emphasized by plotting only a narrow growth velocity regime. However, the V_*-range covered in each run is limited to 2 and 3 different values differing by a factor of 2.

Cellular k. Dependence on V and V_*

Both datasets of cellular ice growth have been plotted along with the δ_{bps}-approach and the V_* relation in figure 4.7 and 4.8, $k = 0.1$ chosen arbitrarily. The first aspect to note is that, although both datasets cover approximately the same V/V_* range, the k_{eff}-values from Wintermantel (1972) are a factor of 5 to 10 larger than those from Dschu (1968). The solute incorporation is apparently rather different. In the runs from Dschu, figure 4.7, the change of k_{eff} with V exceeds the slope of the V_*-relation. In figure (4.8) the differences between the data from Winzermantel and the V_*-relation are even larger. It is evident that the variation of k_{eff} with V and V_* is very different from a theoretical prediction based on any constant k, which is shown for $k = 0.1$ and $k = 0.7$. Eventually at $V/V_* \approx 0.5$ the theoretical slope for a constant $k \approx 0.7$ is reached, yet the coincidence may be accidental.

Comparing k_{eff} at approximately constant growth velocities, the symbol groups in figure 4.8 also indicate that k is not constant yet increases with growth velocity. The effect of the stirring V_* is apparently much stronger than the weak slope suggested by the scaling (4.19) or the δ_{bps}-model. Too mechanisms may be considered as explanations.

First, the stirring may influence the structural k at the interface by changing the distance of cells. Second, stirring may enhance the circulation in the ice matrix. As the average brine salinity within the dendritic ice layer is set by the average liquidus temperature, any delayed exchange of the bulk liquid with C_∞ against a fraction of interdendritic brine with $C > C_\infty$ effectively decreases the solute content of the dendritic ice. The process involves intermittent freezing and exchange of saline interstitial brine and k_{eff} is thus changed by an *internal* desalination process driven by *external* forced flow.

In the studies from Dschu (1968) and Wintermantel (1972) a limited range of V and C_∞ was covered and V_* only varied by a factor of three. A larger range of forced convection conditions was tested by Chen et al. (1998) for sucrose and NaCl solutions. These authors did not compute mass transfer velocities V_* for their systems yet correlated k_{eff} with fluid velocities. The relations between k_{eff} and V or fluid velocity U were principally similar to those obtained by Dschu and Wintermantel and also inconsistent with a constant k assumption.

Cellular k. Dependence on C_∞

An increase of k_{eff} with C_∞ is indicated in the observations from Dschu (1968) shown in figure 4.7. It has also been found in other forced convection studies of freezing NaCl and KCl solutions (Kirgintsev and Shavinskii, 1969b; Wintermantel, 1972; Chen et al., 1998; Matsuoka et al., 2001). Conceptually it may be related to the following two processes:

- The ratio of interdendritic and bulk concentration decreases with salinity. This leads to a less pronounced decrease in k_{eff} due to interdendritic exchange induced by forced fluid flow for high salinities.

- The velocity of the planar-cellular transition increases with salinity (Dschu, 1968; Kirgintsev and Shavinskii, 1969b; Wintermantel, 1972). The transition generally leads to larger k_{eff}.

- The grain size may depend on concentration

It has to be taken into account that during the planar-cellular transition the mechanisms occur simultaneously. Free convection produces, as mentioned above, an opposite concentration dependence, while the V_*-approach alone is concentration invariant.

Conclusion: Analogy of planar and cellular k with forced convection

It is concluded that for a cellular freezing interface (i) no values of k and V_* can be derived that do not depend strongly on growth velocity or concentration and therefore (ii) an analogy to planar freezing is not valid. Several mechanisms at a cellular interface

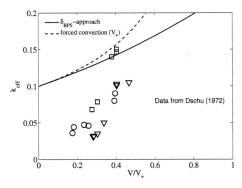

Fig. 4.7: k_{eff} in the cellular growth regime observed by Dschu (1968) for three experimental conditions, varying mainly V_*. As circles: $C_\infty = 20$ ‰ and $7.7 < V < 8.5 \times 10^{-4}$ cm s^{-1}; squares: $C_\infty = 15$ ‰ and $12.5 < V < 13.3 \times 10^{-4}$ cm s^{-1}; triangles: $C_\infty = 10$ ‰ and $12.3 < V < 15.4 \times 10^{-4}$ cm s^{-1}. The narrow velocity regimes appear as groups. The δ_{bps}-approach and mass transfer solutions are shown for $k = 0.1$ and $k = 0.7$.

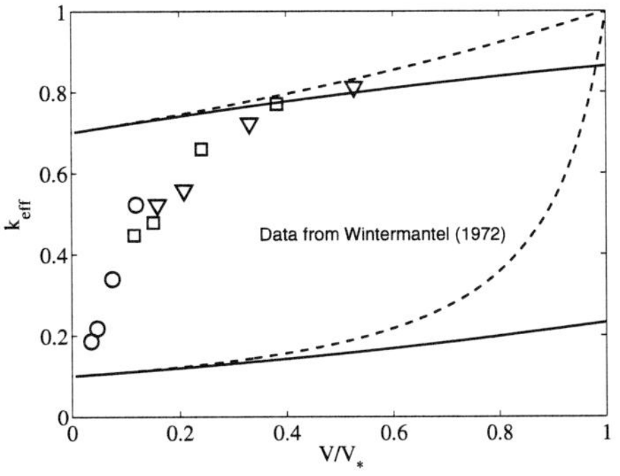

Fig. 4.8: k_{eff} in the cellular growth regime observed by Wintermantel (1972) for $C_\infty = 79.5$ ‰ and three growth velocities: $V \approx 2.5 \times 10^{-4}$ cm s^{-1} (circles), $V \approx 8 \times 10^{-4}$ cm s^{-1} (squares) and $V \approx 11 \times 10^{-4}$ cm s^{-1} (triangles). The δ_{bps}-approach and mass transfer solutions are shown for $k = 0.1$ and 0.7.

appear to cause a much more complex variation of k_{eff} with V and V_*. Furhermore, k depends strongly on the mode of freezing and turbulent solute transport. Also for a fixed experimental cellular freezing configuration k_{eff} does not vary in accordance with the simple V_*-relation for solute transport. The variation of k may either be explained by a substantial change of the interfacial k with V/V_*, or by a delayed change of k_{eff} due to forced interdendritic fluid flow. The least condition $k = const.$, to justify an analogy of planar and cellular solute redistribution, is not fulfilled.

4.6.2 Free convection

Laboratory experiments by Weeks and Lofgren (1967)

Weeks and Lofgren (1967) froze solutions with initial salinities $1 < C_\infty < 100$ ‰ NaCl, which increased typically by $10 - 20$ % during the downward growth of 10 to 20 cm ice in a 60 cm long vertical cylinder. k_{eff} was related to growth velocities estimated for segments of final ice cores. Results shown by Weeks and Lofgren (1967) have not been published for particular salinity runs. They therefore have been inverted from the solution salinity below the growing ice by conservation of mass and solute. The inversion was done from tables of water salinity C_∞ versus ice thickness and growth velocity given by Lofgren and Weeks (1969a). As these values were originally calculated from measured ice salinity data (Weeks and Lofgren, 1967), the latter should be reproducable in this way. However, less data points then based on the orginal 1 cm resolution are reconstructable, and some scatter is introduced due to the limited precision of tabulated data. A freezing run at 1 ‰ and some values from a 3 ‰ run were omitted here for these reasons. The inversion precision implies that k_{eff} in the low salinity runs at 3 and 5 ‰ and above $V/V_* \approx 1$ can only be reconstructed within 10 to 20 %, as indicated by the scatter in figure 4.9 below. Three other runs performed at cooling temperatures of -70 °C have also been omitted. The reconstructed k_{eff} is plotted in figure 4.9. Each group of symbols is connected by a dotted line and resembles, as growth velocity was generally decreasing monotonically with thickness, a vertical salinity profile. The vertical spacing of the values in a profile increases tyically from 1 cm at the top to 3 cm at the bottom.

Figure 4.9 also shows the δ_{bps}-relation as fitted by Weeks and Lofgren (1967) to the complete data set with $k = 0.26$ and $D_s/\delta_{bps} = 1.96 \times 10^{-4}$ cm s^{-1}. Due to their inclusion of the rapid growth runs at -70 °C (not shown here), the relation does not represent the data well. Notably, Lofgren and Weeks (1969a) only used data above $V \approx 0.5 \times 10^{-4}$ cm s^{-1} in the fitting procedure. The rapid growth runs all resulted in k_{eff}-values above the shown δ_{bps}-relation. Extending the growth velocity range to $10V/V_*$, these runs shifted the centre of the data in the fit by Weeks and Lofgren (1967) to a regime $V/V_* > 1$. However, in this regime the δ_{bps}-approach is no longer valid at all (see section 4.2.1 above). Therefore it must be concluded that the δ_{bps}-relation given by Weeks and Lofgren (1967) and shown in figure (4.9) is by no means preferable over an arbitrary empirical fit. It rather conveys an incorrect view of transport processes, and an empirical relation which is likely to perform worst at low growth velocities, the case of interest in nature.

Expected concentration dependence of k_{eff}

In the next figure 4.10 the free convection model with constant solute flux, equations 4.24 and 10.9, has been applied in connection with the same $k = 0.26$ as in the δ_{bps}-approach. It is recalled that k is the only free parameter in the free convection model. For the sake of clarity, observations from only two salinity runs at $C_\infty \approx 5.5$ and ≈ 105 ‰ are shown with the predictions. Comparing the runs with $C_\infty \approx 5.5$ and ≈ 105‰ the data appear to show a slight decrease of k_{eff} with concentration, as the free convection model predicts. However, when all runs in figure (4.9) are considered, the data are too scattered to clearly demonstrate such a dependence. The chosen $k = 0.26$ is arbitrary, yet it appears to represent the low salinity run quite reasonably.

Although the free convection model appears promising, the data are insufficient for

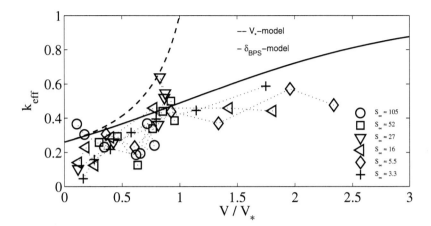

Fig. 4.9: Effective distribution coefficient k_{eff} based on the experiments from Weeks and Lofgren (1967). For each run the average C_∞, which typically increased by 10%, is given and the observed k_{eff} values are connected by a dotted line. The solid line is the δ_{bps}-approach with $k = 0.26$ and $V_* = D_s/\delta_{bps} \approx 1.96 \times 10^{-4}$ cm s^{-1} as proposed by Weeks and Lofgren (1967) for the same but more detailed dataset. The dashed line is the more correct solute transfer relation, expected if V_* were constant.

its validation. The expected concentration dependence of k_{eff} during free convection is opposing the effects of forced convection. Quantitatively any dependence may result from coupling between convection in the liquid and interdendritic flow and an eventual concentration dependence of the interfacial k. Experiments with KCl solutions, frozen laterally in horizontal bottles, indicate a complex relation between k_{eff}, V and C_∞ (Kirgintsev and Shavinskii, 1969a). Furthermore do all runs indicate the classical C-shape salinity profile of thin ice. The subsamples from the cores thus reflect different stages of desalination to be discussed in later chapters. Hence, rather than representing a normalised dependence between k_{eff} and V, the observations from Weeks and Lofgren (1967) resemble the vertical salinity profiles of thin ice in a stage of convective evolution.

Dependence of k on growth velocity

Due to the above arguments one should disregard the bottom observations in figures 4.9 and 4.10 (lowest V). If this is done, the observations of k_{eff} eventually indicate an increase of k with growth velocity, when compared to the free convection model. Any effects of delayed drainage are expected to be least for the runs at lowest C_∞,

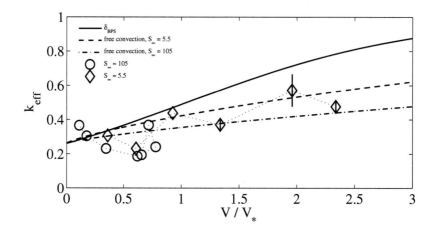

Fig. 4.10: Effective distribution coefficient k_{eff} based on two selected freezing
runs from Lofgren and Weeks (1969a), a low $C_\infty \approx 5.5$ contrasting the highest
$C_\infty \approx 105$. The corresponding free convection model solutions are shown as
dashed and dash-dotted lines, the δ_{bps}-approach as a solid line, with $k = 0.26$
and $V_* = D_s/\delta_{bps} \approx 1.96 \times 10^{-4}$ cm s^{-1}as derived by Weeks and Lofgren
(1967). For $C_\infty \approx 5.5$ error bars due to the inversion procedure are indicated
at high growth velocities.

because brine volumes and mobilities are lowest. In fact, the runs with $C_\infty \approx 3.1$ and
≈ 5.5 ‰ indicate most clearly an increase in the interfacial k with V (figure 4.10). Hence,
also for the case of free convection Weeks and Lofgren's hypothesis of a constant k needs
to be be rejected on the basis of their own data. Two main questions have emerged:

- What determines the dependence between the interfacial k and the growth velocity?

- The instantaneous effective k_{eff} at the macroscopic interface is the consequence of
  some structural k and free convection. How is this k_{eff} reduced due to the effect
  of delayed drainage processes within the ice?

So far it is probably safe to say, that the free convection solution will, provided that k
is known, give an upper bound for k_{eff}, before the onset of any drainage processes from
within the ice. This section will close with two supplementing datasets to discuss the
concentration dependence of k_{eff} and the noted drainage processes.

4.7   Thin bulk ice samples

Another category of observations that are useful to obtain a better understanding of solute redistribution during cellular freezing are thin bulk ice samples, grown at different initial velocities. Studies of this kind have been obtained by several authors (Johnson, 1943; Tsurikov, 1965; Wakatsuchi, 1977).

4.7.1   Field data from Johnson (1943)

It appears that the first quantitative observations of the relationship between growth velocity and ice salinity were obtained by Johnson (1943). Weeks and Lofgren (1967) discussed Johnson's work to introduce into the problem, and as a motivation for their experiments to clarify the growth velocity dependence. They noted Johnson's observations to be inconsistent with their own study, where k_{eff} increased with growth velocity. The systemntic problems to interpret the study from Lofgren and Weeks have been noted above. Before the data from Johnson are discussed, it is important to point out the difference between the sampling methods: Johnson reported bulk salinities of the initial ice skim after 14 hours of growth. The bulk samples were only a few centimeters thick and thus only correspond to one or two samples of the sectioned cores from Weeks and Lofgren (1967). The sampling procedure by Johnson has advantages and disadvantages to be discussed below. The values of k_{eff} from the studies are, however, not comparable without a critical discussion.

Johnson (1943) obtained ice observations during the early freezing of seawater from holes within fjord ice and in a water tank. In the following, to compute the free convection scalings, any differences between NaCl and seawater are neglected. The observations from Johnson are shown in figure 4.11 for three salinity regimes. For the free convection model $k = 0.26$ is assumed to facilitate the comparison with previous figures.

It is first noted that the field observations of k_{eff} are significantly higher than the observations from Weeks and Lofgren (1967). It appears that the use of a higher $k \approx 0.4$ to 0.45 in the free convection model could reasonably predict most observations. The free convection model implies only a slight difference in k_{eff} for high and low salinity regimes, near 5.5 and 30 ‰. The observations indicate the opposite dependence. They do not indicate a difference between the 20 and 30 ‰ regimes. It should be mentioned that the low salinity ice observations were obtained from freezing of artificial leads, while for the high salinity regimes a tank with a saline solution was placed on the fjord ice.

The contradiction between the free convection model and observations suggests that delayed interdendritic exchange is also important during free convection conditions. A second explanation might be that under-ice currents were present in the field and not in the tank experiments from Johnson. However, Johnson noted no difference between

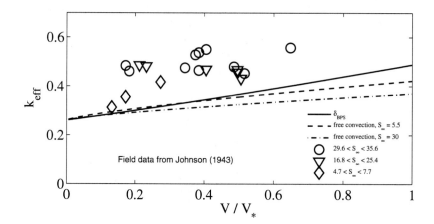

Fig. 4.11: Effective distribution coefficient k_{eff} based on the field observations from Johnson (1943), with three salinity regimes distinguished. The corresponding free convection model solutions with $k = 0.26$ are shown as dashed and dash-dotted lines for $C_\infty \approx 5.5$ and 30‰ . The growth velocity is again normalised by $V_* = D_s/\delta_{bps} \approx 1.96 \times 10^{-4}$ cm s^{-1}, to facilitate the comparison to the δ_{bps}-approach from Weeks and Lofgren (1967).

growth in artificial leads and tanks. The question of the salinity dependence can apparently not be answered from the limited number of observations.

Delayed interdendritic exchange appears to be the most likely cause for the differences between the data from Johnson (1943) and Weeks and Lofgren (1967). As already mentioned, the ice samples grown by Johnson were only between 1.7 and 6.5 cm thick. They likely and contained a considerable fraction of brine within a skeletal layer, discussed earlier. Also Fukutomi et al. (1952) determined the average brine volume fraction of similar thin ice samples grown in the laboratory and estimated high brine volumes of approximately 0.35 to 0.45. As already noted in connection with the data from Weeks and Lofgren (1967), thin ice profiles generally show a decreasing salinity within a decay distance of 5 to 10 cm from the bottom of sea ice - see section 2.4. Ice from from normal seawater therefore does not appear to reach a stable k_{eff} as long as it is thinner than 5 to 10 cm. In fact, also Johnson (1943) noted an increase in salinity from surface to bottom of a thin sample. The thin bulk sample salinity will generally exceed the practical values of k_{eff} for mature saline ice grown at the same velocity, if sampling is careful and the skeletal layer brine is included. This was likely the case in the samples from that Johnson.

Growth velocity dependence of k_{eff}

Ice grown from the lowest solution salinity range of 4.7 to 7.7‰ is expected to be influenced least by the noted skeletal layer sampling bias and other uncertain brine drainage or interdendritic flow processes, simply because the brine volume and permeability are lowest. The physical explanation, to be further discussed in later chapters, is that the decay scale of bottom desalination depends on the permeability and the brine volume. This decay distance is for salinity of 5.5 ‰ expected to be about one fifth of the 5 to 10 cm distance estimated so far for normal seawater of $\approx$ 30 to 35 ‰. Therefore Johnson's observations for the lowest 5 to 8 ‰ salinity regime should be closer to a longterm stable k_{eff}. Accepting this suggestion the three datapoints at this salinity can be interpreted as an increase in interfacial k with growth velocity. This result had already been proposed in the reanalysis of the data from Weeks and Lofgren (1967), but is most significant in Johnson's low salinity observations.

4.7.2 Laboratory experiments by Tsurikov (1965)

Tsurikov (1965) performed freezing experiments of seawater and NaCl solutions in small vessels in the laboratory, approximately for the same salinity regimes as Johnson (1943). These data are shown figure 4.12, confirming most of the conclusions drawn from Johnson's data in figure 4.11. In particular one finds that (i) there is no detectable difference between k_{eff} when comparing freezing runs with salinities of $\approx$ 35 and 25 ‰, (ii) k_{eff} during low salinity runs is, with out exception, significantly lower, (iii) the high growth velocity data are more consistent with $k \approx 0.4$ to 0.5.

Also the dataset from Tsurikov indicates an increase in k with growth velocity. The samples from Tsurikov (1965) had thicknesses of 1 to 1.5 cm. The above considerations of sampling of skeletal layer brine apply also here to explain the larger k_{eff}, which should not be interpreted as longterm values of mature ice sheets.

4.7.3 Concluding remarks

Essentially the same arguments as for sampling methods from Johnson (1943) and Tsurikov (1965) apply to the laboratory observations from Wakatsuchi (1974a). His data may be further analysed in this context, as he obtained ice salinities after 1.5, 3, and 5 centimeters of growth. The conclusion that can be drawn from the bulk thin ice sampling studies is thus: These observations cannot be associated with the stable salinity in the ice and represent some unknown stage of delayed interdendritic desalination. Due to these reasons the bulk salinities are not comparable to the ice core study by Lofgren and Weeks (1969a). The latter, however, also represents different stages of desalination, in dependence on the vertical position in the thin ice core.

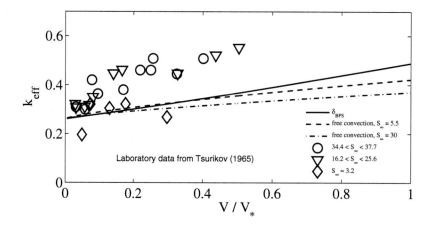

Fig. 4.12: Effective distribution coefficient k_{eff} based on the field observations from Tsurikov (1965), with three salinity regimes distinguished. Notations correspond to figure (4.11).

4.8 Summary

4.8.1 Planar ice growth

The motivation of the present chapter was to formulate a framework of physics of solute distribution at a planar freezing ice interface. Such a framework appears necessary before the more complex problem of a cellular interface can be analysed. Most work done so far had been based on very crude semiempirical approaches to describe solute transport near a planar ice interface. The present work has summarised and compared physically reasonable parameterisations of solute transport forced and free convection solute transport. Several aspects of planar freezing of aqueous solutions have been reexamined by considering the predictability of solute transfer in specific geometries. The overall goal was to obtain proper scalings of solute fluxes under natural growth condition, where haline free convection is the dominant transport process.

Rotating crystal

First, the general limitations of the stagnant boundary layer δ_{bps}-approach from Burton et al. (1953a) for a *rotating crystal* have been summarised: (i) The approach describes solute fluxes only in a limited parameter regime of $Pe_\delta = V\delta_{rot}/D_s < 1$. (ii) The interpretation of δ_{bps}-as a boundary layer scale is already a misleading physical concept for the case of the *rotating growing crystal*.

Forced convective solute transfer

Furthermore, the δ_{bps}-approach is literally not applicable to other geometries than the rotating crystal. Instead, a general approach of solute transfer velocity V_*, equation (4.19), was suggested as a more reliable basic concept of simple solute transfer scalings. Empirical relations for V_* have been summarised for several classical stirring and fluid flow conditions. Only for very strong stirring the δ_{bps}-approach becomes similar to V_*-scalings and may be applicable (Jaccard and Levi, 1961; Gross et al., 1977).

Free convection

Based on simple model concept it has been shown that free convection leads to a fundamentally different relation between the growth velocity V and the effective solute entrapment k_{eff}. In this sense a parametric correlation based on the δ_{bps}-approach becomes arbitrary and can not be recommended as a model for solute incorporation. The application of the δ_{bps}-approach to free convective solute redistribution introduced by Weeks and Lofgren (1967) and since then adopted by many authors (Weeks and Ackley, 1986; Cox and Weeks, 1988; Souchez et al., 1988; Eicken, 1998; Killawee et al., 1998; Ferrick et al., 2002) is physically questionable.

Planar interfacial distribution coefficient k

The formulation of simplified solute transfer approaches allowed for an improved estimate of k given by Gross et al. (1977) for a limited range of NaCl concentrations and growth velocities. At typical lake water concentrations of $C_\infty \approx 0.1$ ‰ the value $k \approx 0.0015$ is more consistent with observations than $0.0027 < k < 0.0030$ from Gross et al. (1977, 1987). This derived k is, however, only valid for a limited growth velocity regime and needs to be reexamined in terms of its *intergranular* and *intragranular* contributions. A proper treatment of k during freezing of dilute solutions (freshwater) likely requires a model for subgrain and grain boundary solute distribution. Solute inclusions within fresh ice have been described a long time ago by Tyndall (1859) and Quincke (1905b,a, 1906). Also concepts of the principle mechanisms of entrapment by intermittent convective small-scale motions were already suggested by Tyndall (1858), yet have received little attention since then. The closure of this principal gap in the understanding of freshwater and brackish water ice growth and impurity incorporation is outstanding. Recent progress in microstructural observation methods (Illiescu et al., 2002; Baker et al., 2003) is promising in terms of validation.

Application to field data

The free turbulent convection model of solute transfer at a planar interface, to be discussed in a later section, has been compared to a properly determined ice growth series from Ferrick et al. (2002). It gave reasonable agreement with observations, also supporting the revised value of $k \approx 0.0015$ for typical freezing conditions of natural lake water. The comparison of the predictions with actual measurement of k further supported a grain size dependence of k.

4.8.2 Cellular ice growth

Cellular ice growth experiments have been discussed primarily with respect the question: Is there a simple analogy between planar and cellular interface solute distribution ? The analogy hypothesis was once put forward by Weeks and Lofgren (1967) and since then used as a basic model of salt entrapment in sea ice (Cox and Weeks, 1975; Nakawo and Sinha, 1981; Cox and Weeks, 1988; Weeks, 1998a; Eicken, 2003). In the present work the question was addressed critically by analysing the appearance of (i) the V_*-approach for forced convection, (ii) the free convection model, and (iii) the δ_{bps}-approach in connection with observations. Albeit rejected for the planar interface, the analysis of the δ_{bps}-approach was included to eventually emphasis a reinterpretation of earlier studies based on the latter.

Forced convection

It was found that the simplified V_*-model (equation 4.19) *cannot* fit observations of cellular ice growth over larger ranges of V/V_* in case of forced convection. Several factors like interdendritic mixing and a cell size dependence of k, analogous to the grain size dependence for planar growth, appear responsible. A number of experimental studies suggest that the interfacial k is not constant but depends on growth velocity. The free convection model predictions indicate that this dependence is masked by the unphysical δ_{bps}-approach. The comparison demonstrates that the latter creates a misleading physical picture of processes determining both k and k_{eff}.

Free convection

The free convection model implies a slight decrease in k_{eff} with concentration. Observations do not agree with such a prediction (Johnson, 1943; Tsurikov, 1965). Instead k_{eff} in these datasets was found to increase slightly with C_∞. This result is qualitatively but not quantitatively similar to observations during forced convection (Matsuoka et al., 2001). It has been outlined in principle that solutal convection in the melt leads to an increase of k_{eff} with C_∞, while interdendritic mixing has the opposite effect. Some limited freezing experiments indicate that the competition of the effects leads to a complex relation between k_{eff}, V and C_∞ (Kirgintsev and Shavinskii, 1969a).

Delayed desalination

Some interpretation of salinity profiles of ice grown at normal seawater salinities was suggested in section 2.4. It implies a distance of the order of 5 to 10 cm from the ice-water interface, within which delayed desalination is active. This distance is expected to depend on the fractional brine volume and thus to approximately scale with the salinity C_∞ of the melt. A consequence is, for example, that the k_{eff} data obtained by Weeks and Lofgren (1967) at several levels of 15 to 20 cm thick ice cores represent mature ice when grown from low C_∞, yet ice in a transitional stage of desalination when grown at large C_∞. The question, how the coupling of this scale to forced and free convective fluid flow may affect k_{eff}, is important but could not be addressed so far.

Thin bulk samples

The large difference between k_{eff} based on very thin ice samples from Johnson (1943) and Tsurikov (1965) and thicker ice from Weeks and Lofgren (1967) can be related to the same process of interdendritic exchange which creates the dependence of k_{eff} and C_∞. Thin samples include liquid brine and are in a state of transition with drainage still taking place. The considerable change of k_{eff} by almost a factor of two between thin ice (Johnson, Tsurikov and Wakatsuchi) and the thicker, older ice (Weeks and Lofgren) implies that k_{eff} is controlled by interdendritic drainage processes. It was concluded, that the free convection model alone, being reasonable for planar interfaces, does not capture the main principles of desalination during cellular ice growth. The δ_{bps}-approach does, of course, neither do so. It's prolonged use rather has suppressed a better understanding of the entrapment process, let it be the change in k with growth velocity evident from freezing at low C_∞, or the general picture of delayed interdendritic desalination.

The relevance of interdendritic exchange during cellular freezing highlights the role of two factors that control the solute redistribution:

- The fractional brine volume near the interface

- The microstructure or spacing of the cells

In fact, it will be discussed in later sections, that these are the key parameters in the permeability relationship of porous media. The following chapter introduces and evaluates a microstructural model of solute entrapment that considers the basic principles of delayed desalination on the basis of solute redistribution near a cellular freezing interface.

5. Solute redistribution during growth of saline ice

The discussion in the previous chapter has demonstrated that one serious problem in unambiguously interpreting ice salinity observations is the delayed desalination due to interdendritic flow. Also salinity profiles (section 2.4) show that observations of k_{eff} depend on the distance from the ice-water interface. To proceed in understanding the desalination process it is desirable to obtain timeseries of ice salinity profiles or perform at least observations at a fixed level. The destructive method of sectioning an ice core implies, due to the mobility of liquid brine in sea ice, almost unavoidable the loss and redistribution of salt.

These problems led Cox and Weeks (1975) to devise an experiment of nondestructive in-situ observations of ice salinities, a dataset which will be denoted as $CW75$ in this chapter. The results of this study were less scattered then the prior observations by Weeks and Lofgren (1967) and have become a frequently used standard reference in the sea ice community, see chapter 4. The outcome of the study by Cox and Weeks (1975) also serves as an empirical boundary condition for sea ice growth models (Weeks and Ackley, 1986; Cox and Weeks, 1988; Eicken, 1998; Weeks, 1998a; Wadhams, 2001; Eicken, 2003). However, as Cox and Weeks (1975) used the δ_{bps} approach to correlate their data, and as experiments were performed at non-constant solution salinity, the proposed empirical relations need to be critically reexamined. This will be done in the present chapter. Another purpose is, of course, the formulation of a more consistent brine entrapment approach which accounts for interdendritic and convective processes. Special attention will also be paid to the observations near the ice interface and the experimental protocol from Cox and Weeks (1975).

5.1   Nondestructive experiments by Cox and Weeks

The experiments from Cox and Weeks (1975) were performed with a similar laboratory configuration as in earlier studies from Weeks and Lofgren (1967). However, using the radioactive isotope ^{22}Na as a tracer, the local salinity entrapment could be derived during the growth process, and at any level within the ice. Results were documented as discrete profiles of ice bulk salinity and temperature at 1 cm resolution. Some limitations of the method at growth velocities higher than 1×10^{-4} cm s^{-1} were pointed out by Cox and Weeks. However, the geophysically most relevant natural growth velocity regime

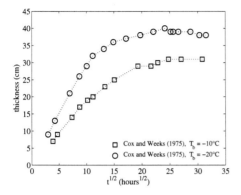
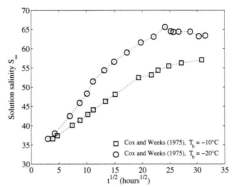

Fig. 5.1: Thickness time series of the laboratory freezing runs from Cox and Weeks (1975) experiment with constant upper boundary temperatures of −20 and −10 °C. Each point corresponds to the snapshot of a nondestructive salinity profile.

Fig. 5.2: Solution salinity S_{infty} increase during the laboratory freezing runs from Cox and Weeks (1975) experiment with constant upper boundary temperatures of −20 and −10 °C. Each point corresponds to the snapshot of a nondestructive salinity profile.

was reasonably covered. The ice thickness reached 30 to 40 cm in these experiments. In two performed freezing runs the solution salinity C_∞ increased from initially 35 to finally 57 and 65 ‰ and thus exceeded natural conditions by a factor two at the end of the experiments. Time series of ice thickness and solution salinity to be discussed in the present chapter are shown for the freezing runs with fixed boundary (upper ice) temperatures of −20 and −10 °C in figures 5.1 and 5.2.

Evaluation of k_{eff} above the bottom skeletal layer

Selected profiles from the freezing runs are shown in figures 5.3 and 5.4, the number of hours since the beginning of freezing indicated. k_{eff} was, for each profile, evaluated by Cox and Weeks (1975) at a fixed distance of 3 cm from the interface. Although somewhat arbitrary, this distance was chosen by Cox and Weeks as it corresponds to the typical scale H_{sk} of the high mobility *skeletal layer* measured from the bottom of sea ice (chapter 3). In figures 5.3 and 5.4 this 3 cm distance from the freezing interface has been marked. Cox and Weeks (1975) obtained H_{sk} graphically and reported a rather stable value of 3 cm throughout their experiments. This was the empirical motivation for Cox and Weeks to choose 3 cm as the determination level of k_{eff}. Questions about the assumption $H_{sk} = const.$ will emerge and be discussed in the coarse of this chapter. The corresponding k_{eff} will be henceforth termed k_{sk}.

Correlating k_{sk} at 3 cm with the δ_{bps}-approach, Cox and Weeks (1975) found slightly

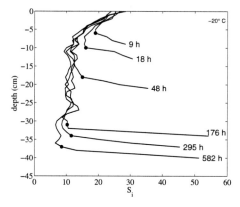

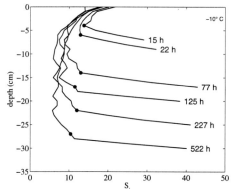

Fig. 5.3: Salinity profiles during the growth of thin ice in a laboratory experiment with a constant upper boundary temperature of -20 °C Cox and Weeks (1975). The hours since the onset of growth are denoted for each profile. Dots denote the 3 cm distance from the ice-water interface. Salinities have not been normalized, as the water salinity S_∞ increased during the experiment.

Fig. 5.4: Salinity profiles during the growth of thin ice in a laboratory experiment with a constant upper boundary temperature of -10 °C Cox and Weeks (1975). The hours since the onset of growth are denoted for each profile. The dots denote the 3 cm distance from the ice-water interface. Salinities have not been normalized, as the water salinity S_∞ increased during the experiment.

larger values for k_{eff} than obtained in the earlier study from Weeks and Lofgren (1967), see figure (5.5). This was noted to be consistent with the expectation, that the initial k_{sk} should decrease further due to slow drainage processes, a change which can also be seen in the series of salinity profiles at any fixed level. A further comparison of these curves is not done here, as the fit from Weeks and Lofgren (1967) was based on both high and low solution salinities (section 4.6.2). Also, after the critical discussion in the preceding chapter, it appears not useful to further evaluate the δ_{bps}-approach as proposed by Weeks and Lofgren (1967). However, due to the noted popularity of the approach, some of the results are documented in the following. It is recalled that any relation between k_{eff} and V based on the δ_{bps}-approach has no physical justification, but must be considered as an inferior empirical fit.

Comparison to other observations

The data points from $CW75$ are plotted in figures 5.5 and 5.6 with their proposed δ_{bps}-fit ($D_s/\delta \approx 1.38 \times 10^{-4}$ cm s^{-1}, $k = 0.26$) as a solid line. This is slightly above the earlier relation obtained from the discussed more scattered data (figure 4.9). Wilcox (1967) reported $\delta_{bps} \approx 2$ mm at 10 to 16 ‰for free convection, which corresponds to

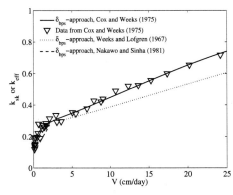

Fig. 5.5: Experimental results presented by obtained by Cox and Weeks (1975) for k_{sk} (as discussed mostly, but not always 3 cm from the ice-water interface) during freezing of NaCl solutions. The solid line is the δ_{bps}-fits to these data, the dotted line the fit to earlier experiments from Weeks and Lofgren (1967).

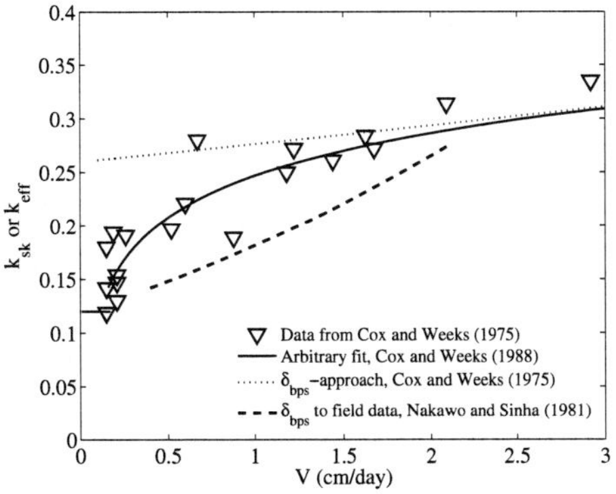

Fig. 5.6: Data from Cox and Weeks (1975) as in figure (5.5) depicting the regime of low growth velocities. The dotted line is the δ_{bps}-fits to the data, the solid line an arbitrary relation later suggested by Cox and Weeks (1988). The dashed line is a δ_{bps}-fit to field data of freezing sea ice from Nakawo and Sinha (1981).

$D_s/\delta \approx 3.3 \times 10^{-4}$ cm s^{-1}. The large difference compared to other studies further supports the lack in generality in the δ_{bps}-fit. The low growth velocity regime is emphasised in figure 5.6, showing the curve $D_s/\delta \approx 2.38 \times 10^{-5}$ cm s^{-1}, $k = 0.12$ proposed by Nakawo and Sinha (1981) from a large number of sea ice salinity profiles in the field. While the latter relation was favoured by Weeks and Ackley (1986) in a later review, Cox and Weeks (1988) suggested a set of equations to model the salinity evolution in first-year ice. They now distinguished three different growth velocity regimes fitted to the data from Cox and Weeks (1975) by the relations

$$k_{eff} = \frac{k}{k + (1-k)exp\left(-\frac{V}{D_s}\delta_{bps}\right)}, \qquad V > 3.6 \times 10^{-5} cms^{-1} \qquad (5.1)$$

$$k_{eff} = 0.8925 + 0.0568 ln V, \qquad 3.6 \times 10^{-5} > V > 2.0 \times 10^{-6} cms^{-1} \qquad (5.2)$$

$$k_{eff} = 0.12, \qquad V < 2 \times 10^{-6} cms^{-1} \qquad (5.3)$$

Equations 5.1 to 5.3 are at present viewed as the state-of-the-art by many sea ice researchers (Weeks, 1998a; Eicken, 1998, 2003).

In figure 5.6 a considerable difference between k_{sk} from Cox and Weeks (1975), equation 5.2 and the fit to the field data from Nakawo and Sinha (1981) is evident. This may be partially related to the fact that the values from Nakawo and Sinha (1981) refer to a winter-averaged k_{eff} at a given level, whereas the k_{sk} from Cox and Weeks (1975) gives the initial salinity 3 cm from the interface. The exponential dependence (equation 5.2) assumed by Cox and Weeks (1988) is hardly evident, when considering the scatter in the experimental values at low growth velocities. The proposed zero-growth constant $k = 0.12$ is arguable from δ_{bps} -theory, but if the latter is rejected, there is neither a theoretical nor a clear observational basis for such an assumption. Below it will be seen, that also the data at high velocities need to be reevaluated.

5.2   Macroscopic interfacial brine volume

Most of the variability in k_{eff} observations for sea and saline ice emerging in the previous section 4 has been related to two unknowns, the intrinsic k at the cellular interface, and the delayed desalination due to intercellular fluid flow. The nondestructive measurements from Cox and Weeks (1975) resulted in temperature and salinity profiles close to the interface and thus give the opportunity to calculate the interfacial brine volume brine v_{int}. As, neglecting the density difference between ice and solution, $v_{int} \approx k$ (recall section 4.6), a first step to a better understanding of the interfacial process is to discuss the variation of v_{int} with growth conditions.

The definition of v_{int} at a macroscopic freezing interface implies already some assumption about the modes of solute transport one aims to describe. At first glance a natural definition would be to define the interface by the level $v_{int} = 1$ beyond which no ice dendrite or spongy excursion reaches and the medium is strictly liquid. For a planar interface one will have a sharp front across which v_{int} must jump to a very small value, eventually resembling the contribution of grain boundaries to the solute distribution k. The interface is then almost sharp but v_{int} depends on the question if one chooses it on the solid ($v_{int+} \approx k$) or on the liquid ($v_{int-} = 1$) side. Notably, if one describes the incorporation k_g into single crystals, v_{int} jumps from 1 to 0 at the phase transition, but also in this case the transition process layer has a finite width on the molecular scale (Hobbs, 1974; Petrenko and Whitworth, 1999). For a cellular interface, as will be seen in later sections, it can be useful to distinguish v_{int} in a similar manner and use the radius of curvature at the tip of cells as a vertical criterion for the transition between ($v_{int-} = 1$) and $v_{int+} = k$. Assuming cells with a circular cap as a simple model, this radius may not be larger than the plate spacing a_0, also to be discussed in later sections, and the transition scale typically less 0.5 mm. Hence, the following values of v_{int} reconstructed from the phase diagram may either represent $v_{int+} \approx k$ or the liquid limit $v_{int-} = 1$, which is impossible to distinguish with a crude optical estimate.

The interfacial brine volume v_{int} based on the profiles from Cox and Weeks are shown in figures (5.8) and (5.7) for the two freezing runs. The variation in v_{int}, and some occasional unphysical values of $v_{int} > 1$, indicate the difficulties in determining the exact position of the freezing interface noted by Cox and Weeks (1975). Another restriction

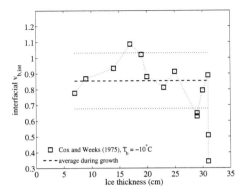

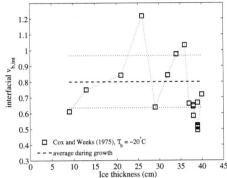

Fig. 5.7: Brine volume v_{int} at the cellular freezing interface versus ice thickness. Freezing run with $T_b = -10$ °C. Approximate changes due to a 0.5 cm uncertainty in the freezing interface are indicated.

Fig. 5.8: Brine volume v_{int} at the cellular freezing interface versus ice thickness. Freezing run with $T_b = -20$ °C. Approximate changes due to a 0.5 cm uncertainty in the freezing interface are indicated.

is that the temperature had to be interpolated between sensors spaced by 5 to 10 cm. The eventually systematic fluctuations in v_{int} might be related to the passage of these temperature sensors and the implied uncertainties in the calculation of v_{int}. In both runs v_{int} dropped at the end of the experiment, when the growth ceased. If only the growth period is considered, one may obtain average interfacial brine volumes $v_{int} = 0.80 \pm 0.19$ for $T_b = -20$ °C and $v_{int} = 0.85 \pm 0.13$ for $T_b = -10$ °C, the respective median values being 0.73 and 0.88. The approximate uncertainty in v_{int} corresponding to an error of ± 0.5 cm in the freezing interface has been obtained by extrapolation and is shown as dotted lines in the figures. With respect to the question of interface definition, it is of interest, that an a priori assumption of an intrinsic interfacial brine volume of $v_{int-} = 1$ would, by linear extrapolation from the two bottommost values, require that the interface positions for the runs at $T_b = -20$ and $T_b = -10$ °C to be moved by 0.60 and 0.62 cm into the liquid. The variability in v_{int} could thus be related to observational uncertainties in the interface position, but the latter could also have been systematically underestimated. Its objective determination by optical methods can be expected to be difficult, if the ice interface is very spongy.

Despite the noted uncertainties, the nondestructive measurements from Cox and Weeks (1975) can be used to highlight that the interfacial brine volume during growth is likely $\approx 0.8 < v_b < 1$. One may suspect that there is a tendency of v_{int} to decrease with growth velocity, a question which will not be addressed here. The main point is that the data show v_{int}, and thus an interfacial distribution coefficient k, close to unity. Some other investigations may be noted as a support.

First, the results from a recent study based on nondestructive measurements of ice salinities by Notz et al. (2005) show a similar interfacial brine volume close to unity. Even if their instrument, as pointed out in chapter 2, can be expected to influence the sea ice microstructure, it seems unlikely that it fundamentally changes the interfacial solid fraction. Second, Wettlaufer et al. (1997) measured the onset of salt fluxes from growing thin ice indirectly by monitoring the solution salinity in the laboratory. In their experiments with similar boundary temperatures initially most salt was retained in the cellular ice layer, until a critical thickness, depending on temperature and concentration and being in the range 4 to 6 cm for similar growth conditions as in the experiments from Cox and Weeks (1975), was reached.[1]. The initial retainment of most salt implies necessarily that the cellular interface corresponds to a brine volume fraction, and thus k, near unity. In other words: The effective interfacial solute rejection, and thus the solid fraction in the cellular interface regime, is small. For upward freezing, in the absence of solutal convection in the liquid, this has been demonstrated for dilute (0.29 to 5.8 ‰ NaCl) solutions of NaCl (Terwilliger and Dizio, 1970) and higher concentrations of up to 100 ‰ KCl by (Kirgintsev and Shavinskii, 1969a). For the downward freezing of high salinity solutions and the presence of solutal convection, the case investigated here, the situation is apparently similar. This implies, that a free solutal convection model alone, as applied to the planar interface in the previous chapter, cannot be used to predict an effective solute distribution coefficient for sea ice.

The comparison does *not* imply, that solutal convection in a thin interfacial layer does not take place, but that it is not strong enough to transport salt faster than the interface advances. Furthermore should the above numbers for the interfacial brine volume not be interpreted in terms of the true interface solid fraction. The resolution in the interface position, temperature and salinity, are insufficient to resolve any detail.

5.3   A simple scaling concept for k_{eff}

A simple concept to describe the evolution of interdendritic k_{sk} is now formulated. It replaces the rejected δ_{bps}- and free convection approaches and is meant to serve as a basis to analyse observations of k_{eff}. Let the value of k_{eff} at a distance H_{sk} from the ice-water interface be $k_{sk} = S_{i,sk}/S_\infty$, where $S_{i,sk}$ is the bulk ice salinity at the sk-level and S_∞ the bulk salinity of the solution at infinity. Denoting the brine salinity and brine volume at this level with S_{sk} and v_{sk}, respectively, one may compute k_{sk} from the following equations:

$$k_{sk} = r_\rho v_{sk} \frac{S_{sk}}{S_\infty} \tag{5.4}$$

---

[1] These experiments will be addressed in more detail in a later section 11

$$r_\rho = \left((1 - v_{sk}) \frac{\rho_i}{\rho_{sk}} + v_{sk} \right)^{-1} \tag{5.5}$$

where r_ρ is slightly larger than unity and related to the ratio of brine and fresh ice densities, and the brine volume v_{sk}. On the basis of the large thermal to solutal diffusivity ratio $\kappa_b/D_s \approx 300 \gg 1$, it is reasonable to assume that the local brine salinity S_b within the ice corresponds to the liquidus S_{liq} and that at the sk-level

$$S_{sk} = S_{liq}(T_{sk}). \tag{5.6}$$

The temperature T_{sk} at any sk-level may be written as the sum of three terms

$$T_{sk} = T_\infty - \Delta T_{int} + \left(\frac{\Delta T_{sk}}{H_{sk}} \right) H_{sk} \tag{5.7}$$

$$\left(\frac{\Delta T_{sk}}{H_{sk}} \right) = \left(\frac{dT}{dz} \right)_{sk} = r_{sk} \frac{V L_f \rho_i}{K_i}, \tag{5.8}$$

with the introduced parameters explained as follows:

- r_ρ depends only on the density ratio of pure ice and brine and is slightly larger than unity.

- ΔT_{int} is the interfacial supercooling within the solutal diffusive boundary layer in the liquid. For the planar interface it was given by the free convection model which has been rejected above. Taking in advance results from later chapters, it may be said that the cellular plates grow in a way, that ΔT_{int} is not important for k_{eff} at large salinities[2]. It corresponds typically to $\Delta S_{int} \approx 1.3$ ‰, which is used here for simplicity.

- r_{sk} relates the growth velocity V to the skeletal layer bulk temperature gradient $(dT/dz)_{sk} = \Delta T_{sk}/H_{sk}$. It may be viewed as a *reduction* factor of the temperature gradient of an ice-brine mixture with respect to pure ice growing at the same velocity. The mechanism behind the reduction is to first order the reduced effective latent heat of fusion in the skeletal ice-brine layer, in connection with a slightly lower reduction in the effective conductivity. The temperature profiles from Cox and Weeks (1975) were coarsely resolved but at growth velocities above 3 cm d^{-1} they yield r_{sk} in the range 0.6-0.8, shown in figure (5.9). A similar reduction has been observed in other studies at comparable solution salinities (Niedrauer and Martin, 1979; Wettlaufer et al., 2000). Therefore $r_{sk} = 0.7$ is a reliable value on an empirical basis. A physical justification is given in appendix B.6. It is further seen in figure 5.9, that the temperature gradient near the interface deviates increasingly from 0.7 at growth velocities below 3 cm d^{-1}. This is discussed in appendix E.

---

[2] It will be seen in chapter 7 that $\Delta T_{int} \approx 0.08 K$ is reasonable when assuming a cellular interface stability limit

- $H_{sk} = 3$ cm will be used also here to be consistent with Cox and Weeks' evaluation of k_{sk} at this level. The particular choice of the distance $H_{sk} = 3$ cm would be rather arbitrary, if it were not associated with a transition of physical regimes: the cutoff of convectively driven desalination. However, as described in chapter 3, this is not quite correct. Instead the *onset of bridging* and a drop in the rate of desalination is more reliable. The problems that arise from the assumption $H_{sk} = const$ will become clear below, which makes it useful to use the term H_{sk}^*, where the *-sign denotes the intrinsic skeletal layer thickness.

- The brine volume v_{sk} at the H_{sk}-level represents the main degree of freedom in the formulation. Figure 5.9 shows v_{sk} at the 3 cm level from the profiles from Cox and Weeks. Its average is $v_{sk} \approx 0.17$, if only the profiles during considerable growth are included ($V > 0.3$ cm d^{-1}). This value is apparently uncertain. It eventually differs for the two freezing runs, and is not constant. It is noted that in the observations from Cox and Weeks an interface position error of 0.1 cm is typically related to a change in v_{sk} by 0.01, while the precision reported by the authors was $\approx \pm 0.5$ cm .

Equations (5.4) and (5.8) are the exact formulation for k_{sk} at any level H_{sk} from the freezing interface and its dependence on growth velocity. They may be written in the following simple illustrative form:

$$k_{sk} \approx r_\rho v_{sk} \left(1 + \frac{\Delta S_{int}}{S_\infty} + \left(\frac{\Delta S_{sk}}{H_{sk}}\right)\frac{H_{sk}}{S_\infty}\right). \tag{5.9}$$

The average brine salinity gradient in the skeletal layer, linked to the average temperature gradient via the liquidus slope, wherein now $(\Delta S_{sk}/H_{sk}) = (dS/dz)_{sk}$. Because at water salinities larger than 35 ‰ the contribution of $\Delta S_{int}/S_\infty$ is less then 4 %, see above, equation 5.9 is still almost exact when the latter term is omitted. This relation for solute redistribution in the vicinity of the cellular ice-water interface will be refered to as the *skeletal model*.

To what degree k_{sk} may be predicted by the skeletal model, depends on two conditions. First, one needs to know, how v_{sk} depends on growth velocity, solution salinity S_∞ and other growth conditions that might influence the relation between $(\Delta S_{sk}/H_{sk})$ and V. The second relevant question is, of course, the level H_{sk}. According to the review in chapter 3 it is unlikely that the intrinsic skeletal layer thickness H_{sk}^* is a constant of 3 cm. A change with thickness, in particular for thin ice, is suggested by figure 3.3. However, as a first order estimate a $H_{sk}^* = const.$ model is reasonable and will therefore be discussed in the reanalysis of the data from Cox and Weeks, appendix E.

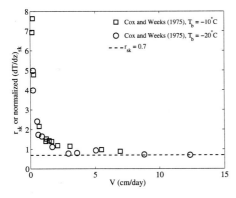
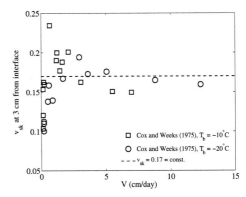

Fig. 5.9: Observed normalized temperature gradient r_{sk} in the bottom 3 cm layer, from equations (5.8). The dashed line $r_{sk} = 0.7$ is the proposed value for the skeletal layer. Note that the observed r_{sk} becomes more uncertain at low velocities due to a coarse spacing of the temperature sensors from Cox and Weeks (1975).

Fig. 5.10: Brine volume v_{sk} at 3 cm distance from the ice-water interface based on the salinity profiles from Cox and Weeks (1975). The dashed line for $v_{sk} \approx 0.17$ is the average value considering velocities above 0.3 cm d^{-1}. The growth-velocity dependence is similar to the normalized k_{sk} shown in figure (E.2).

5.4 Caveats of the laboratory protocol from Cox and Weeks (1975)

The observations from Cox and Weeks (1975), and also the earlier study from Weeks and Lofgren (1967) obtained with a very similar apparatus, are laboratory results. Possible differences with respect to natural growth conditions have not been discussed by the authors. The following particular aspects of the laboratory protocol from Cox and Weeks (1975) are noteworthy in this context and will be discussed below:

- The temperature resolution was not very accurate. Sensors were placed at distance of $0, 5.3, 11.2, 20.1, 33.8, 46.6$ and 58.5 cm distance from the top. In addition, the ice-water interface provides a 'virtual' temperature sensor in connection with the freezing temperature of the solution. Towards the end of the experiment the coarser resolution may have led to considerable uncertainties of brine volumes and temperature gradients near the interface.

- The observations of k_{eff} were standardized by Cox and Weeks (1975) to a level 3 cm from the ice-water interface. However, the timescales needed for the radioactive tracer measurements restricted the observations to thicknesses larger than 7 to 9 cm. Most data points at growth velocities above 8 cm d^{-1} in figure 5.5 therefore refer to k_{eff} in the upper portion of the first profile of each run (emphasised as triangles in figure E.1 to be discussed below). They are not strictly comparable to the data at an otherwise fixed 3 cm distance. Because desalination by gravity drainage and expulsion is clearly evident at distances farther than 3 cm from the interface, easily visible by comparing later subsequent profiles in figures 5.3 and 5.4, a correction towards higher initial k should have been applied to these data.

- In both experimental runs the growth stopped, when the ice reached a thickness of 30 to 40 cm (figure 5.1). This is presumably related to thermal convection and sidewall effects in the laboratory container. The corresponding thermal interface conditions, distinct from natural sea ice growth, may influence k_{eff} and k.

- The confined laboratory setup may have other consequences for the desalination behaviour. Even if, as claimed by Weeks and Cox (1974), the lateral insulation does not affect the macroscopic planarity of the ice-water interface if the room temperature is close to the freezing point of the solution, this does not exclude an influence of the lateral boundaries on the threedimensional temperature and solute fields in the ice. From the insulator thickness of ≈ 2 cm shown by Weeks and Cox (1974), assuming an insulator thermal conductivity of the order of 2 % with respect to ice, one may estimate that in the colder upper half of the ice the linear lateral heat flow may already reach 10 to 20 % of the vertical heat flow, once the ice becomes 20 cm thick. This heat flow will not only effect the overall growth of the ice, yet it may also trigger internal melting and brine channel flow pattern which are not created under natural conditions.

- The salinity profiles at 1 cm resolution were obtained by Cox and Weeks (1975) in a procedure which involved deconvolution, filtering and convolution of salinity observations obtained with a coarser probing method, taking advantage of the known endpoints. Near the ice-water interface and in the skeletal layer this method implies largest uncertainties of 5 to 10 % in k_{eff}.

- During the experiment the liquid salinities increased from 35 to almost 70 ‰ (figure 5.1). This may have influenced the relation between k_{eff} and growth velocity V. It may also have created differences in the brine circulation and the length scales of convective regimes like the intrinsic skeletal layer thickness H_{sk*}.

- Weeks and Cox (1974) report that the insulation and room temperature were sometimes adjusted to produce slightly thicker ice at the edge of the container, while in the analysis only the centered ice was used. Also if the differences in growth velocity are properly corrected, it is likely that such a state changes the internal brine circulation in the ice, in connection with horizontal temperature gradients and convection pattern in the liquid.

- Some sequences of more detailed growth observations (figure 22. in Cox and Weeks (1975)), indicate growth velocity fluctuations in response to the observational procedure, involving the removal of the insulation for several to up to 16 hours, depending on thickness. For the intermediate growth velocity regime the disturbance can be estimated to have lasted for 10 to 30 % of the time span between the sampling dates, indicating a considerable potential for velocity perturbations.

5.4.1 Summary: Reanalysis of the dataset

The nondestructive experiments by Cox and Weeks (1975) have yielded a presently unique dataset which allows a discussion of interfacial properties. As mentioned above, this dataset is frequently used as a reference for salt entrapment in natural sea ice. Therefore a detailed analysis of this dataset has been performed in the appendix E. The results are summarised in the following. It will be seen that the joint effect of the mentioned caveats, especially the first four aspects, is substantial and requires a reinterpretation.

The predictive equations 5.1 to 5.3 for the salt entrapment in sea ice, derived in the original analysis (Cox and Weeks, 1975, 1988), should no longer be used due to the following arguments:

- From a theoretical point of view the δ_{bps}-approach, suggested more than 30 years ago (Weeks and Lofgren, 1967; Cox and Weeks, 1975) and still in use (Weeks, 1998a; Eicken, 2003), is physically inconsistent and creates a misleading picture of solute transport. This inconsistency has been discussed for a planar interface in chapter 4.

- k_{sk} documented by Cox and Weeks (1975) at a distance 3 cm from the interface is affected by several previously unrecognised physical processes and systematic errors, not expected under natural growth conditions.

- Considering the observations two aspects of the original representation of the data by Cox and Weeks (1975) are noteworthy: (i) almost all observations above a growth velocity of 8 cm/day in figure 5.5 are not from a 3 cm evaluations level and brine drainage corrections to adjust them were omitted; (ii) reported growth velocities are 20 to 30 % lower than the actual growth velocities of the 3 cm distance when it passed the interface.

The mentioned observational inaccuracies can be corrected in a reanalysis, but several caveats of the laboratory protocol cannot. Among them thermal convection is best quantifiable. It considerably changed the relation between growth velocity V and k_{sk}. As this is a laboratory effect, it makes the obtained relations not representative for most natural conditions. Natural oceanic heat fluxes, if present, are normally less than in the laboratory where growth ceased at 30 to 40 cm thickness, while the cold interface was still at -10 and -20 °C. This problem is likely to be found in most other laboratory studies and suggests, as pointed out already by Fertuck et al. (1972), that the interfacial or skeletal layer temperature gradient $(dT/dz)_{sk}$ is a better parameter to describe k_{sk} during laboratory growth conditions. Observations of $(dT/dz)_{sk}$ in the laboratory are more elaborate but essential.

In the $CW75$ study the temperature resolution was limited, yet sufficient to demonstrate that the interfacial gradient in the skeletal layer during saline ice growth under the absence of thermal convection is typically a factor $r_{sk0} \approx 0.6$ to 0.7 lower than the gradient of pure ice growing at the same velocity. This result is important, as it allows a conversion between $(dT/dz)_{sk}$ and V when comparing laboratory with natural growth data. Theoretical arguments for the reduction are given in the appendix B.6.

Some empirical observations of the skeletal thickness originally motivated Cox and Weeks to evaluate the ice salinity at a distance 3 cm from the bottom, where it should take a quasi-stable value. Cox and Weeks reported an almost constant $H_{sk}^* \approx 3$ cm while the ice was growing. Investigating the hypothesis of a skeletal layer of constant thickness $H_{sk}^* = 3$ cm, it was found that such a level is, during rapid growth, consistent with a critical brine volume of v_{sk} in the range 0.15 to 0.20. However, the compilation of limited field and laboratory observations of the skeletal thickness H_{sk}^* suggests that the latter is not a constant, yet generally depends on ice thickness and growth conditions. This is supported by the analysis of several near-interface properties and parameters derived from the $CW75$ dataset. Due to the limited resolution and mentioned effects like thermal convection and the changing solution salinity, this interpretation must remain qualitative.

In the discussion of the relation between k_{eff} and $(dT/dz)_{sk}$ other discrepancies in the $CW75$ data have been detected, for example an interfacial temperature above the freezing point during the -10 °C-run, indicating problems in the determination of the ice-water interface position. These in turn translate to uncertainties of the actual vertical position with respect to H_{sk}^* which adds to the unresolved intrinsic variability of the latter. Therefore the concept to determine k_{sk} at a constant 3 cm level from the interface gives uncertain results. Several aspects of delayed desalination above the skeletal layer have confirmed that the data for the two freezing runs at different boundary temperatures -10 and -20 °C do no longer collapse when plotting k_{sk} against $(dT/dz)_{sk}$.

The quantitative validation of theoretical desalination concepts requires reference data of k_{eff} at reference levels. From a phenomenological standpoint the most important level is H_{sk}^*, as here the abrupt change in salinity gradients (figures 5.4 and 5.3) indicates a physical transition. Due to the practical determination problems of a comparable k_{eff} in the vicinity of $H_{sk}^* \approx 3$ cm, the reference k_0' at $H_0' = 6$ cm has been determined and was analysed in dependence of $(dT/dz)_{sk}$. This 6 cm value, however, no longer represents the sharp skeletal transition and it involves a difference in the desalination above the skeletal layer. The lower k_0' at -10 compared to -20 °C may be explained by the fact that, the initial $(dT/dz)_{sk}$ being equal, the temperature at a fixed position decreases less rapidly in the warmer ice. The desalination in the warmer ice is enhanced and the difference already becomes significant at 6 cm from the interface.

5.5 Field observations of k_{eff}

Having discussed the dataset from Cox and Weeks (1975) in detail, some other field and laboratory observations of ice salinities will be compared to these results. In spite of the problems in the $CW75$ data, it is now compared to field observations to emphasise the difference. Two aspects emerging from its detailed reanalysis are recalled as a basis for the following comparison:

- The principle conjecture of an intrinsic H_{sk}^* and corresponding k_{sk}^* as a unique function of the temperature gradient, or growth velocity in the field, seems consistent with the observations. This level is approximately 2 to 3 cm from the interface. In the present view it corresponds to the onset of bridging between plates.

- Above the skeletal layer the further desalination depends on the local cooling rate.

For a level of 6 cm a difference of only 5 to 10 % in k_{eff} was found between the warm and the cold freezing runs from Cox and Weeks (1975). This level appears to be a compromise to compare observations. It will be noted with H_0' and k_0' henceworth.

There are, of course, also some caveats during field sampling of sea ice. An analysis of sea ice field data is normally performed on the whole vertical core obtained from an ice sheet, implying that the measured k_{eff} is always influenced by unknown differential desalination. In addition, the reconstruction of the growth velocity history is often subject to larger uncertainties from unmodeled natural variability. This contrasts laboratory experiments, where the growth velocity may be traced more accurately, but instead the effects of thermal convection can hardly be avoided. A combination of careful modeling with a high frequency of thickness and snow cover observations may reduce the uncertainties (Nakawo and Sinha, 1981). Another drawback during field measurements is that, due to natural variability of brine channels, one must reckon with a related variability of 10 to 20 % of k_{eff}, and a few centimeters in the ice thickness (Weeks and Lee, 1962; Tucker et al., 1984; Cottier et al., 1999). This can only be avoided by sampling of a number of cores which is seldom performed (appendix D).

In a few field and laboratory studies, to be compared with the dataset from Cox and
Weeks (1975), the mentioned problems were not too severe or handled to some degree.
These observations are summarized in figure 5.14 below. To facilitate a comparison of
the datasets a distance $H'_0 = 6$ cm from the interface is used as a reference level. So far,
no other authors than Cox and Weeks have presented their observations of k_{eff} in terms
of a fixed level from the interface and the growth velocity. Observations of k_{eff} far from
the interface may also be compared, if alternative cores indicate, how k_{eff} has decreased
due to desalination with respect to the 6 cm level. Figure 5.11 illustrates this delayed
desalination in the runs from Cox and Weeks (1975). It shows, how the salinity k_{min} of
the final two profiles is reduced with respect to the 3 cm (upper) and 6 cm (lower symbols
of each bar) levels. The bars themselves thus reflect the desalination between 3 and 6
cm. The final salinities are typically reduced to 70 to 95 % of the values at the reference
level $H'_0 = 6$ cm,. The reduction is, as pointed out above, larger for the warm run at
-10 °C. Figure 5.12 indicates a similar measure for the field observations from Nakawo
and Sinha (1981), yet plotted against the ice thickness, to be discussed next.

In most following studies only the growth velocity has been documented and $(dT/dz)_{sk}$
has been computed from V via equation 5.8, assuming $r_{sk} = 0.66$. This value is theoretically plausible (appendix B.6) and ensures comparability with the $CW75$ observations
rather well.[3]

5.5.1   Field data from Nakawo and Sinha (1981)

A sea ice salinity field study frequently cited in connection with the $CW75$ laboratory
observations (Weeks and Ackley, 1986; Cox and Weeks, 1988; Eicken, 1998, 2003) is the
work by Nakawo and Sinha (1981). These authors obtained a whole winter series of
vertical salinity profiles at 2.5 cm vertical resolution, ice thickness increasing from 17 to
160 cm. Nakawo and Sinha (1981) did not present their observations as initial values
near the interface, but as $\overline{k_{eff}}$, averaging all profiles at fixed levels. They published the
details of half of their 26 cores. Due to the large number of cores this $\overline{k_{eff}}$ is closer to
the longterm desalination limit than to k'_0 at 6 cm distance. The timeseries presented
by Nakawo and Sinha (their figures 7 and 8) indicate that k_{eff} in general stabilised
one month after the passage of the freezing front. The reported $\overline{k_{eff}}$ reflects this stable
stage within a few percent, because towards the end of the winter, when the ice started
warming, a slight decrease in k_{eff} is seen and in the computations compensated largely
for the initially larger values k'_0 and k_{sk}. Therefore the reported $\overline{k_{eff}}$ from Nakawo and

---

[3] For normal seawater, assuming a $r_{sk} = 0.66$, the conversion $V = 9.43(dT/dz)_{sk}$, with units
cm d^{-1} and K cm^{-1} is valid at -3 °C, a typical skeletal layer bulk temperature; a temperature change of
± 1 K shifts the proportionality by approximately $\approx \mp 1$ %, which is of little relevance for natural growth
conditions.

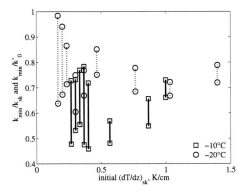

Fig. 5.11: Longterm decrease in k_{eff} in the $CW75$ data. For each level the ratio of the average $\overline{k_{eff}}$ from the last two profiles to k_0' at 6 cm (upper symbols) and to k_{sk} at 3 cm are shown. The height of the bars thus represents the desalination between these levels.

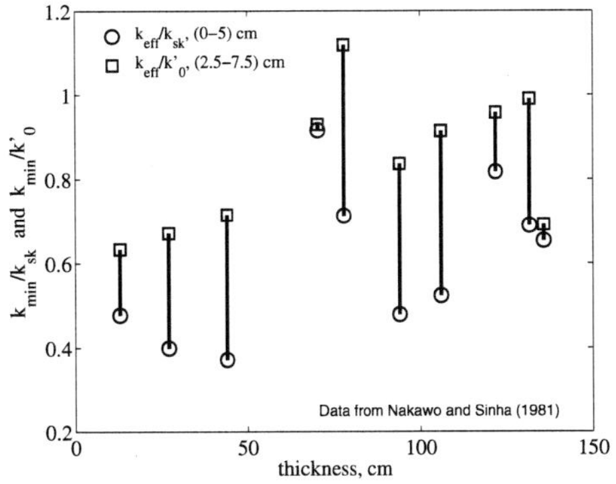

Fig. 5.12: Longterm decrease in k_{eff} based on the profiles published by Nakawo and Sinha (1981). Shown is the ratio of the average $\overline{k_{eff}}$ from the three final profiles in relation to k_0' (here: initially $2.5 - 7.5$ cm from the interface, squares), and k_{sk} (initially $0-5$ cm, circles.

Sinha (1981) should be comparable to k_{min} from Cox and Weeks (1975) shown in figure E.24. Nakawo and Sinha, however, only compared their data with k_{sk} documented by by Cox and Weeks 3 cm from the interface (figure 5.13) and found, as one expects, significantly lower $\overline{k_{eff}}$.

To facilitate a comparison of the data from Nakawo and Sinha with levels comparable to H_0' and H_{sk} from Cox and Weeks the following procedure was applied. Nakawo and Sinha's k_0' at the 6 cm level was estimated from their published salinity profiles by averaging the two bottommost samples (0 to 5 cm), assuming it to represent roughly the H_{sk} level, while the H_0' level was associated with an average of the two samples between 2.5 and 7.5 cm. Next, the longterm k_{min} at each level was computed from the three profiles at the end of the winter, disregarding the very last one, obtained after the onset of considerable warming. Figure 5.12 shows, that the ratios of the final k_{min} to k_0' obtained in this way, are similar to those for the cold run from Cox and Weeks (1975) in figure 5.11, being mostly between 70 and 90 %. Also the 3 cm ratios k_{min}/k_{sk} are similar, between 50 and 70 %. The scatter in the data from Nakawo and Sinha (1981) is, according to the discussion in appendix D, not unexpected: 10 to 20 % variability in k_0' may be related to horizontal variability of larger brine features, not accounted for by sampling of single cores. Due to the similarity of the plots, one may conclude that the crosses from Cox and Weeks (1975) shown in figure 5.13 should be reduced by $k_{min}/k_{sk} \approx 0.5$ to 0.7 to justify a comparison with $\overline{k_{eff}}$ or k_{min}. This in turn lets one deduce a smaller k_{min} in the $CW75$ experiments, compared to the field data from Nakawo and Sinha, if looking at the growth velocity plot in figure 5.13. Hence, in contrast to previous conclusions, the ice corres from Nakawo and Sinha reflect the higher salt entrapment. If the data from Cox and Weeks are corrected for thermal convection, or presented in terms of $(dT/dz)_{sk}$, as discussed in appendix E, this difference is enhanced. The approximate ranges are shown in figure 5.14, yet for k_0' at 6 cm distance, estimated as follows.

Due to the scatter in k_{min}, an approximate conversion of $\overline{k_{eff}}$ from Nakawo and Sinha to k_0' was deemed preferable and two aspects have been considered to do so. First, based on figures 5.11 a ratio $k_0'/\overline{k_{eff}} \approx 1.2$ to 1.3 seems to be most appropriate in the $CW75$ data. On the other hand, the 2.5 to 7.5 cm average from the profiles from Nakawo and Sinha is likely an overestimate of k_0', both due to a smaller (5 cm) average distance and steeper salinity gradients towards the interface. The profiles from Cox and Weeks (1975) and Nakawo and Sinha (1981) have been used to estimate that this, for moderate growth velocities, yields on average a difference of ≈ 10 to 15 %, when comparing k_{eff} at 5 and 6 cm from the interface. Therefore a factor of 1.1 was used to convert Nakawo and Sinha's $\overline{k_{eff}}$ to k_0'. In figure 5.14 the data range from the original $\overline{k_{eff}}$ in figure 5.13 is increased by 1.1 and shown as parallel solid lines. The comparison indicates, that k_0' from Nakawo and Sinha (1981) is significantly different from the reanalysed Cox and Weeks results even if, due to the limited growth velocity regime, the datasets no longer overlap.

Due to this large difference, the procedure from Nakawo and Sinha, reconstructing the growth velocity from meteorological forcing data, has been examined critically. It was found that it is likely, that the modeled growth velocities of much of the upper 70 cm of ice were underestimated by 20 to 30 %. The underestimate may be explained by the fact that Nakawo and Sinha (1981) used an average snow thickness in a thermodynamic ice growth model. When the ice was thin and the true snow thickness was much less, this considerably underpredicts growth rates, an effect showing up mainly in the regime of larger k_{eff} and temperature gradients. From the open circles, 'using growth rates measured' in figure 5.13, the relation

$$\overline{k_{eff}} \approx 0.14 + 0.4V \approx \frac{k_0'}{1.1} \qquad (5.10)$$

was estimated, with V in cm d^{-1}. This equation, converted to $(dT/dz)_{sk}$ by $r_{sk} = 0.66$, is shown in figure 5.14, denoted as 'Nakawo and Sinha, most plausible', the dash-dotted line. It is still above the observations from Cox and Weeks, yet in closer agreement with a second dataset to be discussed next.

5.5.2 Field data of k_0' from Adventfjord, Svalbard

Results from a field study by the author, obtained in Adventfjord(Svalbard) during the winter season 2000 and described in appendix D, are also presented in figure 5.14 along

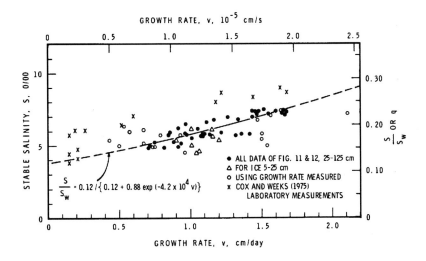

Fig. 5.13: Adopted from Nakawo and Sinha (1981). Field determinations of an average $\overline{k_{eff}}$ at fixed levels in the ice against modeled (dots) and measured growth velocities (open circles). The data from Cox and Weeks (1975) refer to k_{sk} at a distance of 3 cm from the interface.

with the studies noted so far. As described in the appendix, these values have been obtained by averaging over several cores, thereby reducing the uncertainties to less than 10 % for k_{sk} and 20 % for the growth velocity, while for single ice cores the uncertainties would have been a factor 2 to 3 larger. As already noted this may explain much of the scatter in the observations discussed so far. The observations, covering the growth velocity range 1 to 6 cm d^{-1}, have been converted to the skeletal temperature gradient by assuming $r_{sk0} = 0.66$, and are shown as crosses connected with a stippled line.

As described in appendix D, the ice growth slowed down during the field work, presumably due to increasing solar radiation and/or oceanic heat flux. This effect, equivalent to a temperature gradient reduction $(dT/dz)_{sk}$ in the range ≈ 0.04 to 0.05 K cm^{-1}, has been roughly estimated during a period of stable meteorological conditions, and is indicated by lateral bars emerging from the crosses. Due to increasing solar radiation during the growth phase, it is possible, but not certain, that these bars relate primarily to the lowest temperature gradients. These observations, due to averaging of 5 to 7 profiles least biased by natural variability, also give k'_0 clearly above the reanalysis of the $CW75$ data.

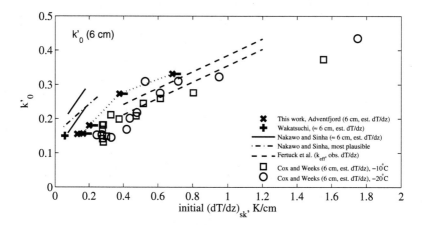

Fig. 5.14: Comparison of observations of k'_0. For all observations $r_{sk0} = 0.66$ has been assumed to convert V to dT/dz. The $CW75$ data, corrected for thermal convection, are thus presented as in figure E.5. Crosses: present work Adventfjord; lateral bars indicate a possible oceanic heat flux effect. Solid lines: field data range from Nakawo and Sinha (1981) increased by 10 % to convert $\overline{k_{eff}}$ to k'_0. Dash-dotted line: 'most plausible' evaluation of observations from Nakawo and Sinha (1981) discussed in the text. Dashed line: laboratory study with observed temperature gradient from Fertuck et al. (1972) 1 to 2 cm from interface in a with variable $10 < S_\infty < 30$ and k_{eff} from different levels. Plus sign: Field data from Antarctic (Wakatsuchi, 1974b).

5.5.3   Field data of k'_0 from Wakatsuchi (1974b)

Wakatsuchi (1974b) obtained time series of salinity profiles at Syowa Station, Antarctica (see figure 2.6 in an earlier section). The sample resolution was between 5 and 3 cm, and the ice was second-year ice which started regrowing at a thickness of ≈ 100 cm. At two stations the environmental conditions implied rather constant growth velocities of 0.5 to 0.6 cm d^{-1} over the course of three months. The contour plots shown by Wakatsuchi (1974b) appear sufficient to estimate $k'_{sk} \approx 0.15 \pm 0.01$ at about 6 cm from the interface. This point is shown as a plus in figure 5.14. It may be viewed as a reasonably validated reference at low growth velocities and compares to the lower bound and lowest temperature gradients from Nakawo and Sinha (1981).

5.5.4 Laboratory data from Fertuck et al. (1972)

Another comparable study has been documented by Fertuck et al. (1972). While these authors froze ice in the laboratory, they installed a high-resolution chain of thermocouples and reported a linear relationship of the form $k_{eff} = 0.1 + 0.293 (dT/dz)_{int}$, where $(dT/dz)_{int}$ was referred as the measured temperature gradient at the interface. The regime to which the temperature gradient relates is variable, yet due to the reported spatial resolution it likely corresponds to an interfacial distance of 1 to 2 cm during most of the growth phase and should be comparable to the skeletal gradient $(dT/dz)_{sk}$. Observations based on core sampling after the experiment were obtained for initially lower solution salinities of $2.5, 5$ and 10 ‰. The solution salinity was not constant but increased by approximately a factor of three during each experiment. The results from Fertuck et al. (1972) indicate an initially 20 % larger k_{eff} for the 2.5 compared to the 10 ‰ runs. Therefore only the limits for the run at 10 ‰ have been depicted from their figure 2 and are shown in figure 5.14.

The k_{eff} values reported by Fertuck et al. (1972) relate to a variable distance from the interface, based on 1.3 cm slices obtained from final cores of ≈ 20 cm thickness. One would therefore expect that the values are most representative for the 6 cm k'_0 at intermediate temperature gradients (corresponding to this distance in the final core). Upper levels (and observations at higher temperature gradients) should be affected by desalination, lower levels by the increasing k_{eff} in the high-salinity skeletal layer. Due to the lower solution salinity the skeletal layer is expected to have been less than for seawater, yet also this condition likely changed due to the threefold increase in S_∞. As runs were obtained at fixed boundary temperatures of -18 and -7 °C, the upper half of the covered temperature gradient regime represents the colder runs. Taking these factors into account the detailed data appearance or linear slope from Fertuck et al. (1972) is not meaningful. However, one may suggest that the high temperature gradient data are likely least problematic. These also fall above the reanalysed points from Cox and Weeks.

5.5.5 Rejected data sources

It is recalled here that several data sources of initial sea ice growth are less meaningful in connection with a proper analysis of the salt entrapment process. Bulk samples of thin ice may include either considerable parts of the bottom skeletal layer, be biased by uncontrolled loss of brine, and suffer from a lack in vertical resolution. The studies from Tsurikov (1965) and Johnson (1943), affected by these problems and thus giving considerably larger k_{eff} when related to their average growth rate, have already been noted and discussed in the previous chapter. Another study of this kind, performed by Wakatsuchi (1974a) in the laboratory, most clearly illustrates the caveats: Growth rates being the same, $\overline{k_{eff}}$ for 1.5 cm thick ice exceeded the value for 5 cm thick ice by up to 50 %. Relations between growth rate and bulk $\overline{k_{eff}}$, suggested on the basis of these data sources (Tsurikov, 1965; Wakatsuchi, 1983; Kovacs, 1996) are of some practical importance when the rough evolution of bulk salinities of an ice sheet is considered. Bulk salinities, however, do not provide direct information about the entrapment and desalination process near the interface considered here on the basis of k'_0 and k_{sk}. Such bulk relations are, as already discussed in chapter 2, also much more uncertain for thin ice. Observations of bulk salinity $\overline{k_{eff}}$ may have the capability to indirectly validate a theory, and the systematic growth data from Wakatsuchi (1974a) and Johnson (1943) are of particular value, because thickness and growth velocity were documented. Such a theory, a framework of interfacial brine entrapment and delayed desalination, apparently still needs to be formulated.

Any evaluation of field ice core samples must take into account that the stability of the ice salinity and its delayed change from k'_0 during a winter season depends on the thermal evolution. In a warming ice sheet stable k_{eff} values can no longer be expected. This has been most clearly shown by Sinha and Zhan (1996), comparing the salinity of a first-year ice core with earlier determinations by Nakawo and Sinha (1981) from the same area. While still being done under cold winter conditions, sampling by Sinha and Zhan (1996) was performed two months after the onset of warming, and the observed salinities were almost a factor of two smaller than those from Nakawo and Sinha (1981). Also the significantly smaller k_{eff} reported by Souchez et al. (1988) should be interpreted in this manner, their derived values of k_{eff} being no longer related to the actual growth conditions.

5.5.6   Discussion

The field study from Adventfjorden and the work by Wakatsuchi (1974b) are likely the least uncertain field observations of k_{eff} at a fixed level $H'_0 \approx 6$ cm from the interface, because growth velocities and k'_0 are based on a large number of cores or stable growth regimes. At low temperature gradients and growth velocities a range of $0.15 < k'_0 < 0.16$ is approached in these studies. The estimate of k'_0 from the observations by Nakawo and Sinha (1981) is more uncertain both in terms of k'_0 and V. It is important to recall that the steeper slope, in the data proposed by the latter authors, is likely a systematic bias due to the underestimate of the modeled growth velocities during early growth.

For the study from Nakawo and Sinha a most plausible relation 5.10 for k'_0 has been estimated. This dashed-dotted line in figure 5.14 indicates a larger k'_0, with respect to the noted two studies, the lowest k'_0 from Nakawo and Sinha being ≈ 0.18. The difference could have several causes. One may consider that also the field ice sampled by Nakawo and Sinha (1981) was subjected to an oceanic heat flux, as indicated for the Adventfjord data. Two aspects shown in a later study (Nakawo and Sinha, 1984) indicate this possibility: (i) growth was faster at a station further offshore and (ii) the crystal alignment at the coastal station indicates considerable alongshore currents. A comparison of offshore and coastal growth rates by the end of the season may be used for a rough estimate of an oceanic heat flux corresponding to $V \approx 0.3$ cm d^{-1}or $(dT/dz)_{sk} \approx 0.03$ K cm^{-1}. A correction would be slightly less in magnitude than indicated by the lateral bars for the Adventfjorden data in in figure 5.14. It thus seems incapable to explain the difference compared to other observations. On the other hand, it is also plausible that not only the growth, but also k'_0 from Adventfjorden was decreased by the significant tidal current activity and its induced enhancement of mixing in the skeletal layer. The data from Nakawo and Sinha would then eventually resemble k'_0 under less intensive flow conditions. Another aspect of interest in this context could be the different growth history of the ice types. The ice from Nakawo and Sinha grew under persistent cooling, and temperature conditions were rather stable during the whole winter. The ice from Adventfjorden was warmer as a whole and temperature-related brine drainage is already expected to show up in k'_0 (as evident from the comparison of the warm and cold freezing runs from Cox and Weeks (1975) in figure 5.14). Moreover, the Adventfjorden ice was rather early subjected to thermal cycling which might be resembled by the three observations near 0.2 K cm^{-1}, obtained after a natural temperature rise. Enhanced desalination may be related to direct brine loss due to freezing-expansion hysteresis or to an indirect mechanism: a different population of larger brine channels due to thermal cycling may in turn influence the interfacial transport processes. A final aspect to be considered is, that a slow downward transport of salt in the thick ice from Nakawo and Sinha, might have increased k_{eff} at lower levels, and thus at lowest temperature gradients. This redistribution source is neither available in thin ice from Adventfjorden, nor in the regrowing ice from Wakatsuchi (1974b), because the upper levels were much less saline.

In spite of these differences a larger k'_0 than derived from the laboratory data of Cox and Weeks emerges clearly from all field studies. This holds also for the ice from Adventfjorden at high growth velocities. Next, some quantitative estimates related to differences between seawater and NaCl solutions are made.

Expected difference between seawater and NaCl solutions

Due to the considerable difference between seawater observations of k'_0 and the NaCl ice from Cox and Weeks, one must ask for the principle difference between these aqueous solutions. The molecular properties of NaCl and seawater are very similar and a crude estimate may be obtained as follows. As shown in figure 5.16 below, the slight difference in the liquidus relationships alone may, at most, account for a 5 % smaller k_{sk} for an NaCl solution (dashed curve) compared to seawater (solid curve) at the same concentration. On the other hand, the haline contraction coefficient is typically 5 % lower for

NaCl solutions than for seawater (appendix A.2.5), and one might expect an opposite effect due to enhanced gravity drainage, and only a small difference is expected. This conclusion, of course, requires the same skeletal distance H_{sk} from the interface and the same microstructure or plate spacings. As will be discussed in the following chapters the plate spacing mainly depends on the diffusivity D_s of the main ions or salts in solution. Due to the dominance of Na^+ and Cl^- ions in seawater, it is likely that plate spacings are very similar, in particular because NaCl is the most rapidly diffusing major salt (Caldwell and Eide, 1981; Zaytsev and Aseyev, 1992). From these considerations alone it seems likely that k'_0 or k_{sk} for NaCl solutions and seawater will not differ by more than 5 %. For colder ice, where brine channels dominate the fluid flow, the precipitation of Na_2SO_4 ($\approx$ 11 % of all salts by weight) can change these figures slightly, as NaCl brine stays more mobile. How this in turn will influence the brine channel distribution, and eventually trigger differences near the interface, requires detailed studies of microstructure and chemical constitutents and cannot be adressed here. Thinking in simple terms, desalination of NaCl may eventually be 10 % more intense, if a considerable fraction of the ice reaches temperatures below the onset of precipitation of Na_2SO_4 near ≈ -7 °C (see appendix A.4). This is, however, only half the truth, as larger haline contraction of seawater may compensate this effect partially, and as diffusion coefficients may be involved in a complex manner.

Dependence on C_∞

A prominent difference, compared to the field observations, is the increasing solution salinity in the experiments from Cox and Weeks, reaching almost two times the seawater values at the end of the runs. Two effects have been noted in the previous chapter: (i) free solutal convection in the liquid leads to a slight increase in k_{eff} with S_∞, but is likely of small relevance when the interfacial brine volume is close to one; (ii) forced flow through the skeletal layer leads to an increase in k_{eff} with S_∞, yet this effect is hard to quantify without a hydrodynamic model. In the present case the main mechanism is free convection from within the ice and the dependence in a simple constant skeletal layer model (resembling this convection) would be an increase in k_{eff} with S_∞, as indicated in figure E.1. A constant skeletal layer, however, seems questionable. The following aspects can be summarised for the studies considered so far.

In the skeletal approach applied in the present chapter k_{eff} decreases with S_∞ at a fixed level (e. g., figure E.8). From equation 5.9 one would expected a rather strong dependence of k_{sk} on S_∞. This is evidently not the case, as the observations from Fertuck et al. (1972) for an initial $S_\infty = 10$ ‰ indicate, if at all, only a slight difference in k_{eff}. Notably, Fertuck et al. (1972) also performed experiments at $S_\infty = 2.5$ ‰, which indicate an $\approx$ 10 % larger k_{eff} than in their runs at initially 10 ‰. Detailed experiments

at low salinities are rare, yet recently Granskog et al. (2004a) have reported continuous
salinity profile series at $\approx$ 5 cm resolution from three winters in the Gulf of Finland,
Baltic Sea. Their figure 5 indicates a rather stable growth velocity of 0.6 cm d^{-1}over
the course of three weeks during 2001, and the drawn contours may be used to estimate
$1.0 < S_i < 1.2$ ‰ at $\approx$ 5 cm from the interface. Divided by the water salinity of
$\approx$ 5.8 ‰ one may obtain a typical bound $0.17 < k_0' < 0.21$ on the salt entrapment.
This is 5 to 20 % larger than the values for seawater at comparable growth velocities.
Although the data basis is sparse, the available observations indicate a slight decrease
in k_0' with salinity under natural growth conditions, when comparing brackish water of
$S_\infty \approx$ 5 ‰ with seawater. Some experiments reported by Ryvlin (1979) for seawater
with salinities between 6 and 34 ‰ support these numbers.

Considering other work, the freezing runs from Lofgren and Weeks (1969a) have been
discussed in the previous chapter, being too uncertain for an evaluation of the problem
(figure 4.9). Interesting information was provided by (Kirgintsev and Shavinskii, 1969a)
on the basis of freezing experiments of aqueous KCl solutions normal to the direction of
gravity. These authors found a complex dependence of k_{eff} on growth velocity. Their
figures show $dk_{eff}/dS_\infty < 0$ at high growth velocities and moderate concentrations, yet
the opposite behaviour at low growth velocities or large solute concentrations. Although
their comparably smaller k_{eff} indicates that freezing normal to gravity in small ampoules
is accompanied by stronger convective salt fluxes, one may eventually compare their
results to the observations from Fertuck et al. (1972) at similar k_{eff}, where both indicate
a 10 to 20 % decrease in k_{eff} between $S_\infty \approx$ 10 and 30 ‰. While it seems that most
observations are consistent for the salinity range 5 to 35 ‰, for a further doubling in S_∞
the results from Kirgintsev and Shavinskii (1969a) are not clear. The change can be in
any direction but is less than 20 %.

It may be noted that, as the possible variation of k_0' with S_∞ is not more then 20 to
30 %, this restriction allows a different conclusion. Considering equation 5.9, it can be
concluded that k_{sk} is largely independent of S_∞, and one hence must have $H_{sk} \sim S_\infty$. It
can be further argued that the empirically supported skeletal thickness value of $H_{sk} \approx 3$
cm, and the variation shown in figure E.16, is only valid for normal seawater ice or similar
solution salinities of 30 to 35 ‰. In principle the results indicate that the skeletal layer
likely scales with the salinity of the solution, provided the growth velocity is similar.

Keeping the simple standpoint one may then conclude that two opposing effects are
in effect. The constant skeletal layer model results in a decrease of k_{sk} with S_∞, but the
convective exchange of liquid works against this dependence. Due to the lack in empirical
data for vertical growth, and recalling the discussion from section 4.6, the change in k_{eff}
or k_{sk} for an intermediate salinity increase from 35 to 60 ‰, as in the $CW75$ experiments,
remains an open question. While figure 5.14 supports a significant difference between k_0'
in the $CW75$ experiments and field observations, it cannot be decided by simple means

to what degree this is created by the laboratory setup, creating hydrodynamical changes
due to thermal convection or lateral temperature gradients, or to the increasing salinities.

On the reinterpretation of studies based on Cox and Weeks (1975)

Another aspect that needs to be considered is the question, how the data appearance
changes when changing variables from $(dT/dz)_{sk}$ to V. In the analysis of the $CW75$ data
$(dT/dz)_{sk}$ has been suggested as the relevant variable, because thermal convection was
severe.

In figure 5.15 the reviewed observations of k_0' are alternatively shown in dependence
on the growth velocity the ice had when this level was the freezing front. Also in this
presentation the field observations of k_0' are considerably above the data from Cox and
Weeks. It is important to note that Cox and Weeks had presented their data by correlating k_{sk} at 3 cm distance with the momentary interface velocity, arguing that the skeletal
layer is constant. This introduces a considerable inaccuracy in the case of decelerating
growth. An evaluation indicates a relatively constant ≈ 20 and 30 % underestimate of
V for the runs at -20 and -30 °C, respectively, with increasing deviations towards the
end of the runs. It is interesting, that this inaccuracy also changed the appearance of
the runs at -10 and -20 °C relative to each other. As the inaccuracy actually moves the
points in figure 5.15 by ≈ 20 and 30 % to the left, it creates the impression of reasonable
agreement with natural growth conditions, as for example reported by Nakawo and Sinha
(1981) in a comparison study. The predictive equations 5.1 through 5.3, suggested by
Cox and Weeks (1988), are therefore relatively close to natural sea ice salt entrapment.
However, this is due to a cancellation of errors, and in a correct analysis their dataset
reveals considerable disagreement with natural ice observations.

INTERICE tank observations and fluid flow

Figure 5.16 also evaluates some additional observations from the INTERICE ice tank
studies (Eicken et al., 1998), indicated by an ellipse. The following estimates enable a
distinction between ice growth with and without a forced under-ice current. To compute
k_0' from published profiles an increasing S_∞ from the initial tank salinities, based on brine
release into a 1 m water column, was accounted for. This correction was only applied in
the runs with under-ice flow, assuming that in the quiet runs the brine fluxes will mainly
stratify the water column. The change in the estimates of k_0' is small, only 4 to 8 %, yet
likely relevant for the following discussion.

Several results from the INTERICE II experiment have been presented by Tison et al.
(2002) and k_0' at a distance of 6 cm from the interface was determined from the salinity
profiles in figure 5 from the latter authors. The freezing run without currents (upward
triangle) can be compared with two runs under fluid flow of ≈ 6.5 to 8.5 cm s^{-1} (lateral

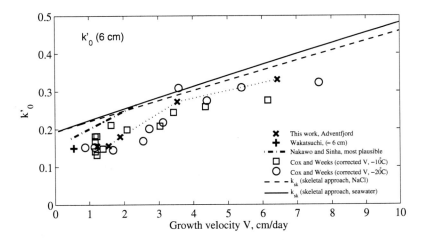

Fig. 5.15: Comparison of observations of k_0'. Observations correspond to 5.14, yet are plotted against the observed growth velocity. Note that k_0' at 6 cm from the interface from Cox and Weeks (1975) is related to its earlier interfacial freezing velocity, which implies 20 to 30 % larger V values than associated by Cox and Weeks (1975) with these levels in their original work. The simplistic skeletal approach (for k_{sk} with $H_{sk} = 3$ cm and $v_{sk} = 0.17$) is shown for seawater (solid curve) and NaCl (dashed curve) assuming the same concentration and skeletal layer microstructure.

triangles). A second observation was taken from Tison et al. (2002) for the run with under-ice current, yet after regrowth (lateral triangle). Observations and growth conditions from the earlier tank study INTERICE I had to be picked from several publications. A single observation of k_0' has been taken from Eicken (2003) and another one from Cottier et al. (1999). Respective growth velocities are from Eicken (2003) and Eicken et al. (1998) and k_0' is shown as a diamond and a pentagram, respectively.

The estimates of k_0' for 5 profiles agree reasonably with the observations from Adventfjorden. Most important is that, both for INTERICE I and II, k_0' is larger under absence of fluid flow. The results from from Nakawo and Sinha (1981) remain above the quiet runs, yet an above mentioned heat flux correction of $(dT/dz)_{sk} \approx 0.03$ K cm^{-1} would decrease the difference to less than 10 %. It is worth a note that during a three day period of warming the salinity in the tank experiments decreased by typically 20 % (Haas, 1999; Tison et al., 2002). As the ice from Adventfjorden was subject to a similar thermal cycling, this may serve as an explanation for the remaining difference between the latter and the INTERICE II observations.

It is further seen that k'_0 from INTERICE I is systematically below the values from IN-TERICE II, although grown without a weaker under-ice current. A possible explanation is, that during INTERICE I the growth was much more decelerating near the evalua-tion levels than during INTERICE II. In the particular case denoted with 'regrowth' for INTERICE II it was even accelerating after a short melting period. For this sample the related increase in the temperature gradient during further growth is indicated by a lateral bar. This may indicate why its k'_0 was larger than for all other comparable obser-vations. This aspect is also a plausible explanation for the larger k'_0 in the field work by Nakawo and Sinha (1981), during which the growth rate in the early winter was almost constant. This indicates, that under a constant growth velocity a 10 to 20 % larger k'_0 may be realised. The possibility, that accelerated growth or a steepening of temperature gradients can change k_{eff} by downward redistribution of salt was also pointed out by Tison et al. (2002). It indicates the difficulty to interpret profiles *a posteriori* in terms of interfacial processes alone. The consideration also applies to the data from Nakawo and Sinha. As the desalination in upper levels will always create a slow downward redistri-bution of salt, this may explain that k'_0 estimated from $\overline{k_{eff}}$ only appears larger in the lower levels. The changing $k'_0/\overline{k_{eff}}$ in figure 5.12 supports this mode of redistribution.

After all it seems that, while a number of growth conditions can affect k'_0, all consid-ered datasets collapse quite reasonably. Fluid motion, oceanic heat flux and decelerating growth may create variability in k'_0. Downward redistribution of salt may bias the latter, if obtained *a posteriori* from thicker samples. The variation due to these processes seems to create not more than ±15 % in k'_0. To understand the noted mechanisms of vari-ability there is still a need for systematic high resolution studies with properly controled boundary conditions.

Proposed reference data

Although the empirical basis is somewhat sparse, with respect to reference data for k'_0 it is concluded:

- The k_{eff} laboratory observations performed by Cox and Weeks (1975) differ con-siderably from field data for natural sea ice growth.

- The exact dependence of k'_0 on S_∞ needs to be known to interpret the $CW75$ observations. It cannot be decided if the different k'_0 due to much larger solution salinities, thermal convection, or lateral boundary effects on the ice growth and desalination.

- k'_0 is reasonably bounded by the observations from Nakawo and Sinha (1981) and observations from Adventfjorden (present work). If these studies are corrected by an oceanic heat flux corresponding $0.03 < (dT/dz)_{sk} < 0.05$ K cm^{-1} they also match the range in INTERICE results.

- Differences in a relation of k_0' versus $(dT/dz)_{sk}$ appear to be related to warming periods and under-ice flow, to the question if growth is accelerating or decelerating. Redistribution of brine has likely introduced another bias in the thick ice from Nakawo and Sinha.

Field data are lacking at high temperature gradients and freezing rates. Only those $CW75$ observations with moderate solution salinity ($C_\infty < 45$ ‰ and small thermal convection effects, limited by $r_{sk} < 1$, were retained in the reference data plot figure 5.16. Under these constraints the observations from Cox and Weeks (1975) are more reliable, as the solution salinities were not far from normal seawater values and the relative influence of thermal convection was small. However, also the few retained data points from Cox and Weeks, of which the two lowest values are seen in figure 5.16, must be viewed with caution, as the growth rates from which $(dT/dz)_{sk}$ was computed are rather uncertain during rapid initial growth. Alternative estimates of the interfacial $(dT/dz)_{sk}$ were obtained by considering simply the cold boundary temperatures in the study and reducing the linear gradient by 0.66. Doing so, up to 20 % different and mostly smaller values of $(dT/dz)_{sk}$ are obtained and indicated in figure 5.16 as lateral bars. In the overlapping regime these values are 5 to 15 % below the Adventfjorden field data.

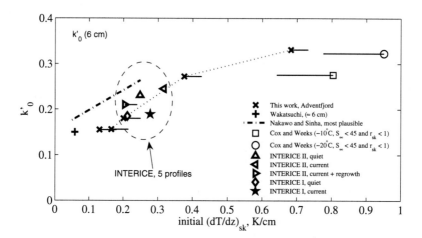

Fig. 5.16: Proposed reference data of k_0' for growing sea ice, selected from figure 5.14. For the study from Nakawo and Sinha (1981) only the 'most plausible' evaluation is shown. INTERICE data are added here. They were taken from figures by Eicken (2003) (pentagram) and Cottier et al. (1999) (diamonds) and Tison et al. (2002) (triangles), all corresponding to a distance of 5 to 6 cm from the interface. The NaCl data from Cox and Weeks (1975) have been restricted to $r_{sk} > 1$ and $S_\infty < 45$ ‰ to exclude unrealistic growth conditions in the laboratory. The lateral bars indicate the uncertainty in the temperature gradients.

5.6   Summary and outlook

The nondestructive experiments by Cox and Weeks (1975) have yielded a presently unique dataset which allows a discussion of interfacial properties. Moreover, this dataset is frequently used as a reference for salt entrapment in natural sea ice.

Taking up the questions which emerged from chapter 4 the interfacial brine volume in the experiments was considered and found to be close to one. This implies, that desalination at salinities comparable to seawater, is primarily controlled by salt fluxes from a *skeletal layer*, within which brine freely circulates between the plates. It further implies, that the δ_{bps}-, but also the free convection approaches discussed in section chapter 4, must be rejected. A simple concept for the calculation of the interdendritic k_{sk} has been formulated.

The analysis of the dataset from Cox and Weeks (1975), given in appendix E, has led to the conclusion that the empirical relations 5.1 through 5.3, deduced from these data,

represent natural sea ice growth only to a limited degree. This conjecture is based on
(i) the lack in a reasonable physical model or parametric form, and (ii) the discussion of
many inconsistencies and caveats during the experiments. It was further confirmed by
a comparative analysis with several studies of natural sea ice growth. The dataset from
Cox and Weeks (1975) has to date been a standard reference for sea ice salt entrapment
(Cox and Weeks, 1975, 1988; Weeks, 1998a; Eicken, 2003). Although it contains, due its
nondestructive character, very important information, one must be aware of the particular
laboratory conditions when interpreting it.

The analysis of the data from Cox and Weeks has motivated the choice of $H_0' = 6$
cm as the evaluation level of k_{eff}. The choice of a fixed level from the bottom ensures
the comparability of observations. The level of 6 cm is normally far enough from the
interface to exclude brine loss during sampling. It is still close enough to the interface to
be situated mid in the bridging layer. It thereby resembles a position near the salinity
minimum in a profile, being interpretable in terms of initial entrapment and interfacial
process.

Only a few field data series were found to be appropriate to validate a relationship between k_0', at a fixed interfacial distance, and growth conditions (figure 5.16). Considering
previously published work these are the time series of salinity profiles from Wakatsuchi
(1977) with rather stable growth velocities, and the dataset from Nakawo and Sinha
(1981). An analysis of field work performed by the present author, documented in appendix D, has provided another set of reference values of k_0' for growth velocities up to
6 cm d^{-1} $((dT/dz)_{sk} \approx 0.7$ K cm$^{-1})$. Ice tank data from INTERICE I and II Eicken
et al. (1998) were also included in the intercomparison of relevant field studies. The
analysis resulted in the following conclusions:

- k_0' in the laboratory observations from Cox and Weeks (1975) is typically 20 to 40
  % smaller than in field observations of natural sea ice growth. Possible causes are
  the much larger solution salinities, thermal convection, or warm lateral boundary
  effects on the ice growth and its desalination.

- From simple considerations it is likely that the maximum expectable difference
  between k_0' for seawater and aqueous $NaCl$ solutions at the same concentration is
  less than 5 to 10 %. Aspects related to differential diffusion, Na_2SO_4 precipitation,
  and brine networking deserve further studies.

- k_0' is reasonably bounded by the observations from Nakawo and Sinha (1981) and
  observations from Adventfjorden (present work). If these studies are corrected by an
  oceanic heat flux corresponding $0.03 < (dT/dz)_{sk} < 0.05$ K cm^{-1} they also match
  the range of critically discussed results for several INTERICE ice tank experiments.

- Differences in a relation of k_0' versus $(dT/dz)_{sk}$ can be qualitatively related to thermal cycling, under-ice flow and changes in the growth velocity.

- The critical review of field observations and laboratory data was concluded by suggesting reference values of k_0', shown in figure 5.16. In spite of very different conditions the spanned range to be matched by an entrapment model seems not to be larger than ± 15 % in k_0'.

- Low growth velocity $\overline{k_{eff}}$ from Nakawo and Sinha (1981) most clearly indicates a redistribution bias due to downward migration of brine, making *a posteriori* interpretation of thick winter ice samples, even for rapid steady cooling conditions from Nakawo and Sinha (1981), more problematic than assumed.

5.6.1 Outlook: A structural model for k_{sk}

While desalination also takes place above the skeletal layer, it appears from figures 5.11 and 5.12, that under normal conditions k_{eff} drops to typically 50 to 70 % of its 3 cm skeletal value. To predict the longterm salinity of ice one needs to predict k_{sk}. However, from figure 5.16 it is now clear, that a simplified skeletal approach using $H_{sk} = const.$ and $v_{sk} = const.$ does not account properly for the dependence of the initial salt entrapment on growth conditions.

With the outcome that the skeletal layer thickness H_{sk} can hardly be assumed to be constant one may modify the earlier equation 5.9, and suggest the following conceptual form:

$$k_{sk} = r_\rho r_{gra} \frac{d_{sk}}{a_0} \left(1 + \frac{\Delta S_{int}}{S_\infty} + \frac{\Delta S_{sk}}{S_\infty} \right) \tag{5.11}$$

In this equation

- $$r_\rho = \left((1 - v_{sk}) \frac{\rho_i}{\rho_{sk}} + v_{sk} \right)^{-1} \tag{5.12}$$

  is an expansion parameter slightly larger than unity and related to the ratio of brine and fresh ice densities, and the brine volume v_{sk}. It is relatively straightforward to compute and ≈ 1.10 for brine of normal seawater salinity and $v_{sk} \approx 0.2$, increasing to ≈ 1.16 at a temperature of -8 °C. A lower temperature is unrealistic for the skeletal layer even for very rapid growth.

- $$r_{gra} = 1 + \frac{a_0}{L_{gra}} \left(1 - \frac{d_{sk}}{a_0} \right) \tag{5.13}$$

  r_{agr} is a structural correction factor which accounts for the fact that the brine layers have no infinite length but are limited by the grain length L_{gra}, assuming rectangular grains.

- ΔS_{int} is the salinity increase across the liquid solutal boundary layer, which so far has been assumed negligible for high salinities, because of the high liquid fraction at the interface.

- $\Delta S_{sk}/S_\infty$ is the salinity increase across the skeletal layer. It depends on the internal desalination due to fluid flow and indirectly contains the so far discussed skeletal layer thickness temperature gradient.

- a_0 is the plate spacing at the top of the skeletal layer or skeletal transition.

- d_{sk} is the width of brine layers at the top of the skeletal layer or skeletal transition, at the onset of bridging between the plates.

Grain size parameter r_{gra}

The parameter r_{gra} depends mainly on the ratio of grain length to plate spacing. Typical columnar sea ice grain sizes have been reported to increase with depth from ≈ 3 mm at the surface to ≈ 20 mm at 60 cm depth (Tabata and Ono, 1962; Weeks and Hamilton, 1962; Weeks and Ackley, 1986; Wakatsuchi and Saito, 1985). The increase is strongest in the upper layer and then becomes approximately linear. Wakatsuchi (1983) has, for thin ice of 2 cm thickness, reported a strong growth rate dependence. The average grain size ranged between ≈ 3.5 mm at 12 cm d^{-1} and ≈ 13 mm at 4 cm d^{-1}. Most studies show, that the crystal dimension in the direction of the plates is typically two times larger than normal to it. therefore the grain sizes should be multiplied by $2^{1/2}$ to obtain the relevant grain length L_{gra}. As will be seen later in more detail, the natural range in a_0 is typically 0.3 to 1.5 mm. The ratio L_{gra}/a_{sk} can be expected to range between ≈ 15 for very thin ice (Wakatsuchi, 1983) to 30 for thick ice (Cole et al., 1995). Neglecting d_{sk} this results in $1.03 < r_{gra} < 1.1$. Some parallel observations of a_0 and L_{gr} from Tabata and Ono (1962) indicate, assuming a grain aspect of 2, a ratio between 20 and 30 and thus a rather constant $r_{gra} \approx 1.04$.

In addition to the additional brine at grain boundaries one may suspect a brine volume increase due to the presence of brine channels which, considering the statistics from Wakatsuchi and Saito (1985), is not expected to exceed 1 %. Larger second-generation channel systems, evolving by brine redistribution, may be relevant for thicker ice (Bennington, 1967) and may ocassionally reach area fractions of 5 % (Lake and Lewis, 1970; Cole et al., 2004). Their contribution can be quite substantial at low brine volumes, but this is far away from the freezing interface, when they are to be associated with secondary features. These are not considered in the present approach of the primary structure a_0.

Main dependence: ΔS_{sk}, a_0 **and** d_{sk}

Under natural growth conditions the joint effect of r_ρ and r_{gra} likely can account for an increase in k_{sk} by ≈ 15 %. As mentioned above, the salinity increase ΔS_{int} in the convective boundary layer below the ice, results in a further increase of less than 4 % for seawater. From the detailed discussion of the data from Cox and Weeks (1975) it is clear, that the most important properties to predict the variation of salt entrapment in sea ice are ΔS_{sk}, the brine salinity increase across the skeletal layer, and a_0 and d_{sk}, which give the brine volume v_{sk} at its top. Finally, a few earlier conceptional approaches in the lines of equation 5.11 may be mentioned here.

The brine layer width d_{sk} is to a certain degree comparable to d_0 suggested in in connection with strength studies (Anderson and Weeks, 1958; Assur, 1958; Weeks and Ackley, 1986). However, the definition of d_0 by the latter authors was the 'thinnest observable brine layer width before splitting' and not the onset of separation into sheets. d_0 has always been assumed constant, yet a convincing statistical analysis of microstructural data or a physical explanation have not been published yet. As noted earlier, Anderson and Weeks (1958) discussed the role of the ratio d_0/a_0 in sea ice strength models, yet did not make an approach to compute the salinity entrapment based on this parameter. This was attempted by Tsurikov (1965) who once suggested the form

$$\overline{k_{eff}} = r'_\rho \frac{d_0}{a'_0 + d_0} \tag{5.14}$$

to describe the salt content of sea ice. However, Tsurikov did not determine the microstructure, yet only made an indirect approach to determine a_0, correlating the salinity of very thin, porous ice with the growth velocity. He adopted $d_0 = 0.07$ mm from Anderson and Weeks (1958) and determined the constant γ in the assumed equation $a'_0 = \gamma V^{-1/2}$, to be discussed in the next chapter. Notably, the scaling 5.14 does not account for the skeletal layer salinity increase due to the temperature gradient. In addition Tsurikov's formulation is inaccurate, as he uses a'_0, which is not the plate spacing a_0 as he claimed, yet the difference $(a_0 - d_0)$. The formulation suggested by Tsurikov was therefore incomplete and his equations, if at all, are only valid for the bulk $\overline{k_{eff}}$ of 1 to 2 cm to thick ice, as already pointed out above. However, Tsurikov's *concept of structural entrapment* is physically more reasonable than the δ_{bps}-approach from Weeks and Lofgren (1967), which was formulated at the same time and paradoxically has survived many years as a parametric form to describe salt entrapment in sea ice (Weeks and Ackley, 1986; Cox and Weeks, 1988; Weeks, 1998a). In spite of emerging evidence of a correlation between sea ice salinity and a_0 in a study by Nakawo and Sinha (1984), to be discussed in the next section, Tsurikov's structural approach is still not considered at all in present state-of-the-art reviews of the salt entrapment problem (Weeks, 1998a; Eicken, 2003).

Having reanalysed the available data and basic formulations against which a structural approach needs to be validated, the present work reopens Tsurikov's ideas in a more valid form, on the basis of equation 5.11. The sections which follow demonstrate that the parameters ΔS_{sk}, a_0 and d_{sk} represent key mechanisms of the sea ice growth and salt entrapment problem, and propose a theoretical framework of their prediction.

6. The plate spacing of sea ice

The first descriptions of the phenomenon, that ice grown from saline solutions comparable to seawater entraps cold brine between very fine plates, date almost 2 centuries back (Parrot, 1818; Walker, 1859; Ruedorff, 1861). The first micro-graphical documentations of the microstructure of sea ice in the field were performed during expeditions by the end of the 19th century (Drygalski, 1897; Hamberg, 1895). Drygalski (1897) described the microstructure of sea ice, as consisting of crystals of the order of 20 mm diameter, each of them being separated by parallel vertical plates spaced by 0.5 to 1 mm. Another peculiarity in contrast to glacier ice was pointed out by Hamberg (1895), who noted that for floating sea ice the orientation of the main crystal axis is always horizontal. This basic description of the skeleton of unidirectionally growing sea ice has since then been confirmed without exception. Figures 1.4, 1.5 and 1.2 from the introduction are recalled to illustrate the structure.

6.1 Observations

Quantitative descriptions of the microstructure of sea ice, in particular the plate spacing a_0 began half a century later. Fukutomi et al. (1952) determined an average $a_0 \approx 0.39$ mm by analysing the structure of 1 to 2 cm thick laboratory grown sea ice. The range during several experiments with slightly different seawater salinities was between $0.23 < a_0 < 0.50$ mm. Anderson and Weeks (1958) determined an average $a_0 \approx 0.46$ mm, with an even larger range $0.2 < a_0 < 0.8$ mm for ice of different thickness. They suggested, in an average sense, a_0 to be constant. Investigations during the following years have shown that this is not the case. Tabata and Ono (1962) reported an increase in a_0 from 0.65 to 0.87 mm with depth in 18 cm thick sea ice sheet. Weeks and Hamilton (1962) observed an increase in a_0 with decreasing growth velocity, during field work and in a laboratory study. While all these observations were obtained for ice thinner than ≈ 30 cm, the plate spacing of thick Arctic sea ice was found to be larger and close to 1 mm (Assur, 1958; Schwarzacher, 1959; Muguruma and Higuchi, 1963), while observations of exceptionally thick sea ice indicate plate spacings of up to ≈ 1.5 mm (Muguruma and Higuchi, 1963; Cherepanov, 1964).

Lofgren and Weeks' laboratory study

The noted observations strongly indicated that a_0 increases with decreasing growth velocity. Motivated by the role of a_0 in sea ice strength models its dependence on the growth velocity was sought by Weeks and Assur (1964, 1967). To establish this dependence Lofgren and Weeks (1969a) performed a detailed laboratory study of aqueous NaCl solutions, extending the earlier limited measurements from Weeks and Hamilton (1962). The main results of this work are shown in figures 6.1 and 6.2 and may be summarized as follows:

- A presumed growth velocity dependence of the form

$$a_0 \propto V^{-b} \tag{6.1}$$

  with a constant power law exponent b could not be established. For most runs b appeared to decrease with growth velocity (figure 6.1).

- A rather complex dependence of a_0 on the concentration C_∞ of the solution was found (figure 6.2).

Lofgren and Weeks (1969a) admitted that no explanation for the complex behaviour was available, in particular for the apparent minimum in a_0 at NaCl concentrations of 10 to 25 ‰, depending on the growth velocity. They claimed this minimum to be a fundamental difference compared to an earlier study by Rohatgi and Adams (1967b). This is not correct, because the lowest solute concentration from Rohatgi and Adams (1967b) was $0.2N \approx 12$ ‰, and thus would not have resolved the minimum in a_0, if it were a general result.

An increase in a_0 with C_∞ at higher concentrations, as indicated by most of the freezing runs from Lofgren and Weeks, was also observed by Rohatgi and Adams (1967b). However, in their experiments a_0 was ≈ 60 % larger for $C_\infty \approx 100$ ‰, compared to the values obtained at ≈ 20 ‰, while Weeks and Lofgren only found an increase half of this value. The behaviour in the study from Lofgren and Weeks appears to depend on growth velocity. Concerning the study by Rohatgi and Adams (1967b), it needs to be pointed out that these authors observed the substructure of NaCl droplets, frozen while falling through a cold organic liquid. Thus, the mode of freezing was very different from directional soldification. Due to the much more rapid growth in the latter study there is also little overlap in the experimental regimes.

Sea ice field work by Nakawo and Sinha

A systematic study of plate spacing observations in natural sea ice was published by Nakawo and Sinha (1984). These authors determined the plate spacing at different

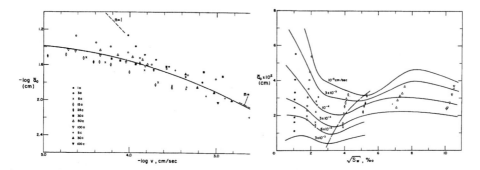

Fig. 6.1: Adopted from (Lofgren and Weeks, 1969a). Plate spacing of laboratory ice grown from aqueous solutions with different NaCl concentrations, denoted in ‰ in the figure.

Fig. 6.2: Adopted from (Lofgren and Weeks, 1969a). Plate spacing of laboratory ice grown from aqueous solutions with different NaCl concentrations, plotted against the square root of the solute concentration ‰.

levels in a core obtained in mid winter (figure 6.3) and related these to growth velocities estimated in a parallel study (Nakawo and Sinha, 1981). Although their observations, shown in figure 6.4, only covered a growth velocity regime from 1 to 2 cm d^{-1}, the main results were very different from the study by Lofgren and Weeks (1969a) in that

- a_0 was a factor of 2 larger than at comparable growth velocities and NaCl concentrations from Lofgren and Weeks.

- No constant a_0 was approached at low growth velocities.

In a later study Sinha and Zhan (1996) have repeated the procedure and confirmed the earlier results. The second study yielded somewhat less scattered observations (figure 6.5). It is likely that the larger scatter in figure 6.4 is related to the fact that the authors related a_0 to growth velocities from an ice growth model, which agreed less well with actual thickness observations than in the later study by Sinha and Zhan (1996). The likely error in the growth velocities has already been discussed in connection with the ice salinity observations from the same field study (5.5.1).

To evaluate the effect of this modeling uncertainty in figure 6.6 the observations from Nakawo and Sinha (1984) have been plotted against the observed growth velocity, which was simply derived from the thickness observations (Nakawo and Sinha, 1981). a_0 between the observed thickness increments was averaged, which reduces the number of observations. For comparison, from the work by Sinha and Zhan (1996) only observations above 170 cm are shown, because at later stages the modeled growth velocities start diverging from the observations, while they fit the thickness observations rather well

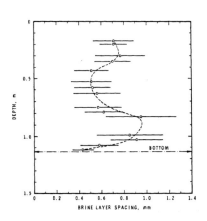

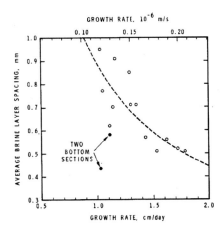

Fig. 6.3: Adapted from Nakawo and Sinha (1984). Variation of the plate spacing with depth in a first year ice sheet.

Fig. 6.4: Adapted from Nakawo and Sinha (1984). Dependence of the plate spacing on growth velocity calculated from a calibrated model.

before. It is seen that the scatter is reduced and the datasets collapse better. While of course also some information is lost two aspects are noteworthy. Whereas Nakawo and Sinha (1984) and Sinha and Zhan (1996) have pointed pout that the bottom samples in both studies gave a significantly smaller a_0 and emphasised this in their figures, a critical check shows that, accidentally, at these levels the modeled growth velocities show the largest deviation from the observed growth velocities. The bottom values from Sinha and Zhan (1996), omitted in figure figure 6.6 , might easily have grown at two-fold velocities. The bottom regime from Nakawo and Sinha (1984), depicted with a star in figure figure 6.6 appears still at the lower end of the observations, yet much less pronounced than indicated by figure 6.4. Two other points have been depicted with a plus sign, resembling events of short drops in the growth velocity over just 3 to 5 centimeters. Their outlying behaviour indicates that adjustment of a_0 to a new growth regime takes some time.

Although the work by (Nakawo and Sinha, 1981) indicated that the observed growth velocities were also limited in their accuracy, the collapse of the datasets in figure 6.6 is promising. The comparison indicates that a more detailed discussion of these, and of future observations requires more accurate growth velocity data and modeling. However, due to the large number of observations the studies from Nakawo and Sinha (1981) and Sinha and Zhan (1996) can be expected to bound the relation between a_0 and V correctly.

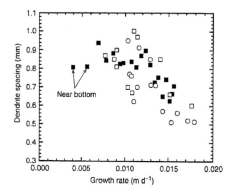

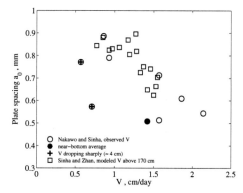

Fig. 6.5: Adapted from Sinha and Zhan (1996). Comparison of plate spacing versus growth velocity from two more recent studies with the observations from Nakawo and Sinha (1984).

Fig. 6.6: Observations from 6.5 with modifications. Instead with modeled growth velocities values from Nakawo and Sinha (1984) are shown against observed growth velocities taken from Nakawo and Sinha (1981).

6.1.1 Compilation of plate spacing studies

No other systematic studies on the dependence of a_0 on the growth velocity are known to the author, yet some work is worth noting. Due to the limited data also single observations are important and the following sources are summarised in figure 6.7, if growth velocities and a_0 are available. Lange (1988) obtained a profile of plate spacing a_0 for Antarctic sea ice, shown a similar depth dependence as found by Nakawo and Sinha (1984), with $a_0 \approx 0.5$ mm in the upper layers and ≈ 1 mm near the bottom. Growth velocities were not reported, yet the inverse correlation of a_0 with ice salinity was indirect evidence of a strong linkage between these properties.Cole et al. (1995) obtained a profile of the plate spacing at 10 cm vertical resolution for a site near Barrow, Alaska. Although in the latter study the growth velocity was only very coarsely resolved, growth was rather stable near 1 cm d^{-1} during a period of two months, with correspondingly stable plate spacings between 0.6 and 0.7 mm.

Nghiem et al. (1997) have reported 3 observations of V and a_0 for thin lead ice, extending the natural growth observations to the growth velocity regime up to 16 cm d^{-1}, even if their largest V appears somewhat high compared to other studies of, for example, (Wettlaufer et al., 2000). Some limited observations were reported for ice tank experiments (Arcone et al., 1986; Gow et al., 1987). In the runs from 1984, Arcone et al. (1986) give $a_0 \approx 0.4$ mm at 4 cm from the interface, to which ice grew at ≈ 3.3 cm d^{-1}. The report by Gow et al. (1987), while discussing some more observations of a_0, contains insufficient information on the growth rate, but at least supports this latter value indirectly due to the reproducibility of the microstructure in the following year.

In addition to the noted single observations the plate spacing reported by Weeks and Hamilton (1962) are shown in in figure 6.7. These include both a field study of sea ice and observations from an NaCl freezing run discussed by Weeks and Assur (1963). Considering the two observations at lowest growth velocities, it is worth a note that they also stem from levels 1 to 2 cm close to the interface of 30 cm thick young ice which had recently retarded growing. It would be plausible, that also here the plate spacings did not have enough time to adjust to the slow growth, as discussed for the observations from Nakawo and Sinha above.

The observations from the very different studies are seen to differ by $\approx$ 30%, yet provide a reasonable overview of the overall increase in the plate spacing. Considering the figures discussed above, there is certainly room for a $\pm$15 % systematic undertainties considering joint errors in a_0 and V. However, figure 6.6 based on the work from Nakawo and Sinha also indicates the adjustment of a_0 to new growth conditions as an intrinsic source of variability. The only dataset which significantly deviates from all other observations is the laboratory experiment from Lofgren and Weeks (1969a). Recalling the experiment by Cox and Weeks (1975) discussed in the previous chapter, where a similar freezing apparatus was used and growth at low velocities ceased due to thermal convection freezing, the growth velocities from Lofgren and Weeks (1969a) have been tested in a similar manner. Observed growth velocties were simply compared to expected growth due to the known bulk temperature gradient. The results was similar to the finding based on figure 5.9: below a growth velocity of $\approx 10^{-4}$ cm/s the growth velocities from Lofgren and Weeks (1969a) reveal a fundamental increasing effect of thermal convection (not shown here). The dataset therefore is not representative of most growth conditions in the field and will not be included in the present compilation, to be discussed later.

An interesting aspect of the field data compilation is a relatively constant a_0 over a wide growth velocity regime above $\approx$ 5 cm d^{-1}. This is corroborated by a study by Dykins (1967), who obtained a limited range $0.30 < a_0 < 0.40$ mm for laboratory grown NaCl ice. Although growth velocities were not reported the growth conditions and salinity profiles make an initially rapid growth and decay to values near 5 cm d^{-1} or less likely. Of particular interest in the problem to determine the relation between a_0 and V is the adjustment rate of a_0 upon a change in the growth velocity. If adjustment takes at least several centimeters, as it does for alignment of crystals under fluid flow (Langhorne, 1986; Stander and Michel, 1989), then this might at least partly explain the large difference in the observations from Lofgren and Weeks (1969a) at lowest growth velocities, because the evaluation levels in this study were just separated by 1.5 to 2.5 cm. To distinguish the possible systematic errors from intrinsic differences in growth conditions and adjustment, more field observations and laboratory studies are essentially needed.

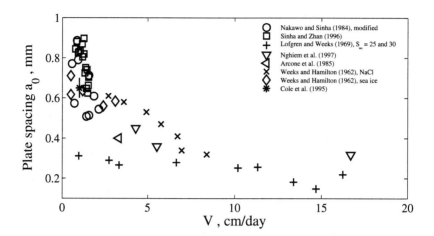

Fig. 6.7: Observations of plate spacings from different field and laboratory studies of seawater and NaCl solutions at similar salinity. The data from Lofgren and Weeks (1969a), shown as crosses, fall considerably below all other measurements.

6.2   Early explanations and models of plate spacings

At the same time when the first quantitative observations of sea ice plate spacings were made (Fukutomi et al., 1952; Anderson and Weeks, 1958; Tabata and Ono, 1962), the problem of crystal substructure received increasing interest in the field of metallurgy (Rutter and Chalmers, 1953; Tiller et al., 1953), resulting in the formulation of basic theories of the cellular solidification process (Chalmers, 1964). Rutter and Chalmers (1953) formulated qualitatively the theory of *Constitutional Supercooling* to explain (i) why cellular growth occurs at all and (i) why the plate or cell spacing in solidification processes normally decreases with increasing growth velocity. Qualitatively their arguments are as follows:

- (i) Solute, rejected at the freezing interface, lowers the freezing point and a solutal boundary layer forms by diffusion. (ii) Due to much faster thermal compared to solute diffusion, the thermal boundary layer of width $\propto \kappa_b$, will have a much larger extension than the solute boundary layer, $\propto D_s$. (iii) Therefore the liquid will be *constitutionally supercooled* near the freezing interface and (iv) a perturbation moving into this supercooled liquid will have the tendency to grow spontaneously. This is the basic mechanism of cellular instability.

- The wave length of the most rapidly growing perturbations is expected to increase with the solutal boundary layer width and is thus $\propto D_s$.

- The more rapid the growth, the less time is allowed for lateral diffusion normal to the solidification direction. This smaller *circle of influence* leads to smaller cell sizes.

- For larger solute concentrations more solute is rejected and the boundary layer will normally be larger. Therefore an increase in the cell spacing with solute concentrations is expected.

Rutter and Chalmers (1953) supported all their qualitative arguments with observations.

As noted above, observations of cellular substructures during the freezing of saltwater and seawater are much older than for metals (Parrot, 1818; Scoresby, 1815; Walker, 1859; Ruedorff, 1861). Interestingly, it had been suggested more then a century prior to Rutter and Chalmers (1953), in connection with freezing salt water solutions, that the cellular solidification is the necessary consequence of differential diffusion of heat and salt (Parrot, 1818)[1] Also the first documentations of an increase in the number of lamellae with growth velocity date back to observations of the freezing of dilute salt solutions by Quincke (1905b), who then extended his descriptions to metals (Quincke, 1906). Quincke (1905b) described a decrease in the distance of lamellae with increasing solute content, in contrast to the later proposal by Rutter and Chalmers (1953). Respectively, the observations by Lofgren and Weeks (1969a) in figure 6.2 indicate that both relations between a_0 and C_∞ may occur . As Quincke only performed experiments with dilute solutions of initial $C_\infty <$ 0.1‰, his results need to be may be compared, and are consistent with the results from Weeks and Lofgren (1967) at low concentrations. At higher concentrations a_0 increases with C_∞ for aqueous NaCl solutions as proposed and observed by Rutter and Chalmers (1953) for some metal alloys. Interestingly the increase in a_0 with C_∞ has since then been confirmed in many studies of metal alloys, yet the decreasing a_0 at low concentrations has only been documented for saline solutions. This shows that the concept of constitutional supercooling, as qualitatively outlined by Rutter and Chalmers (1953), is modified in a complex manner by mechanisms which are not intuitively accessible.

6.2.1 Constitutional supercooling quantified

After Rutter and Chalmers (1953) had formulated *constitutional supercooling* as a qualitative consistent concept, Tiller et al. (1953) postulated an appropriate quantitative

---

[1] In the same paper Parrot suggested a theory of glaciation, concluded that the Arctic sea ice due to its low reported salt content cannot have formed from seawater. While overlooking gravity drainage in this argumentation, his double-diffusive theory of cellular freezing was remarkable.

solution of the problem. They proposed that the necessary condition for cellular growth is

$$mG_c = -mC_\infty \frac{1-k}{k} \frac{V}{D_s} \geq \frac{\partial T_b}{\partial z} = G_b \qquad (6.2)$$

The left hand side gives the interfacial gradient mG_c in the liquidus temperature, based on the steady state salinity gradient at the interface [2]. If mG_c exceeds the temperature gradient G_b in the liquid, one has constitutionally supercooling near the interface and cellular perturbations can grow.

In industrial applications condition 6.2 offers the possibility to influence the onset of cell formation by superimposing a large temperature gradient G_b, and the following years brought, in principle, many experimental validations (Chalmers, 1964). Considering the freezing of natural waters, the applicability of condition 6.2 is limited, because temperatures are often close to the freezing point and G_b is either small or unknown. Under such conditions constitutional supercooling would always be realised, if not other processes are operating. Tiller (1962) therefore proposed a second condition relevant in this context. From considerations of solute and heat conservation he proposed that for cellular growth to occur it is also necessary that steep grain boundary grooves are stable. According to Tiller the condition for their stability is obtained by replacing G_b in equation 6.2 by the interfacial temperature gradient G_i in the solid ice, while the interfacial k must be replaced by the true value k_0 for single crystals. G_i may, see equation 5.8, be written as $G_i = r_{tz} \frac{V L_f \rho_i}{K_i}$. Assuming no temperature gradient in the liquid and a negligible brine volume in the solid one has $r_{tz} \approx 1$ and obtains

$$C_\infty \frac{1-k_0}{k_0} \geq -\frac{L_f \rho_i D_s}{mK_i}. \qquad (6.3)$$

According to the discussion in chapter 4 one must distinguish k_0 for single crystals from k for an interface including grain boundary segregation of solute. Although there seems a lack in measurements of k_0, a value 5×10^{-4} seems reliable for the freezing of halides. Applying equation 6.2 to such dilute NaCl solutions one finds that steep grain boundary cells can be stable at C_∞ above ≈ 0.001 ‰, which is smaller than the typical concentration $C_\infty \approx 0.1$ ‰ of natural lake water (Millero and Sohn, 1992). Molecular properties of other solutes than NaCl are not likely to change this result by more than a factor of 2 or 3. The reason that lakes freeze in general with a non-cellular interface must therefore, as pointed out by Weeks and Ackley (1986), be related to the absence of a diffusive equilibrium. Under normal growth conditions the left hand side in equation 6.3 will approach a much lower value $\approx C_\infty$, – for quantitative examples see section 4.5.2, such that for aqueous NaCl the critical C_∞ becomes ≈ 1.6 ‰. .

---

[2] The solutal gradient G_c simply follows from the state of equilibrium diffusion (section 4.1) by differentiation of equation 4.5

6.2.2 Basic scalings for a_0

The concept of constitutional supercooling qualitatively illustrates the expected dependence of a_0 to be a decrease with growth velocity, yet leaves the nature of this dependence unspecified. Also equations 6.2 and 6.3, being least criteria for the formation of unstable cellular growth, do not provide information about the sizes of perturbations that will develop under different freezing conditions. Early attempts were therefore based on principal arguments. For example, considering the wave lengths of instabilities in other perturbed system problems being proportional to the layer width (Rayleigh, 1916, 1879), one might expect that the perturbation is limited by the width D_s/V of the solutal boundary layer and an inverse dependence $a_0 \sim V^{-1}$ would emerge. Notably, such a result would be the consequence of a *frozen time* system moving at constant velocity. On the other hand it is expected that the formation of perturbations is a time-dependent process where lengthscales evolve proportional to the diffusion length $(Dt_D)^{1/2}$. If the time t_D allowed for diffusion is limited by the growth velocity and a different spatial scale z_D via $t_D \sim z_D/V$ one might expect a dependence $a_0 \sim V^{-1/2}$.

Bolling and Tiller's approach

Bolling and Tiller (1960) were the first who considered the above ideas more quantitatively by a tentative equation of the form

$$a_0 \approx \frac{4D_s}{V}\left(\frac{1}{4\pi} + \frac{z_D}{a_0}\right),\tag{6.4}$$

where z_D/a_0 may be imagined to represent the time-dependent perturbation growth in terms of t_D. The formulation reflects that that the latter may be dominant, if z_D, the distance by which a perturbation can propagate during the critical time t_D of its formation, is comparable to the plate spacing. Bolling and Tiller (1960) proposed further the idea that z_D should be related to the curvature-related freezing point depression at the cell tips. They suggested $z_D \approx (0.6\Gamma/G_{eff})^{1/2}$, where G_{eff} is an effective temperature gradient at the cell tips, given by either G_s or G_l or both. Γ is the Gibbs-Thompson parameter, which is defined as the ratio of solid liquid surface energy and the entropy of fusion and has the unit length times temperature (appendix A.2.10). Although Bolling and Tiller (1960) found agreement with some crystal growth experiments, their approach contained several arbitrary approximations and has not been substantiated by further studies. However, the principle ideas and relevant parameters in the problem, in particular the key role of $(\Gamma/G_{eff})^{1/2}$, have later emerged to be important in the understanding of cell formation, to be discussed in the following chapters (Mullins and Sekerka, 1964; Langer et al., 1978).

Lateral freezing of cells

Another simplified approach, quoted in connection with plate spacings of sea ice (Lofgren and Weeks, 1969a; Weeks, 1998a) and metal alloys (Flemings, 1974), has been suggested by Rohatgi and Adams (Rohatgi and Adams, 1967b,a), who considered the situation of *existing plates* of spacing a_0 freezing *laterally*. They assumed that the temporal increase in brine concentration between the plates is independent of position or lateral distance, which implies two symmetric boundary layers of constant brine salinity gradient dC_b/dt. By solution of Fick's law their equation, in the notations from the present work, is

$$a_0^2 \frac{dv_b}{dt} = -8D_s \frac{\Delta C_{lat}}{C_\infty} \tag{6.5}$$

where v_b is the fractional macroscopic brine volume and ΔC_{lat} is the difference in brine concentration in each lateral boundary layer. The fractional freezing rate dv_b/dt was determined by an empirically validated simplified heat conduction approach (Rohatgi and Adams, 1967a). The authors found that the maximum $(dv_b/dt)_{max}$ was proportional to a_0^2 in a large number of experiments with different saline solutions.

As equation 6.5 implies the fractional freezing rate, its comparability to natural growth conditions is better evaluated from a different form. Using $dv_b/dt = dv_b/dT_b dT_b/dt$ and $dv_b/dT_b \approx 1/T_b$ with T_b the brine freezing temperature in °C, and a linear liquidus slope $dT_b/dC_b = m$, Rohatgi and Adams (1967a) derived

$$a_0^2 \frac{dT_b}{dt} = 8D_s m \Delta C_{lat} \frac{T_b}{T_\infty}. \tag{6.6}$$

Now $\Delta T_{lat} = m\Delta C_{lat}$ may be viewed as the constitutional supercooling in the intercellular liquid boundary layers. While equation 6.6 relates a_0 to the temperature change dT_b/dt it provides no clear information about the dependence of a_0 and the growth velocity V. From, for example, figure 7 in the article from Rohatgi and Adams (1967b) it can be further deduced, that the experimental results do not validate equation 6.5, because ΔC_b is not found to be constant, but increases with dv_b/dt. Bower et al. (1966) have used a similar criterion to ensure that the deviation from the equilibrium temperature between the plates is small, which allows macroscopic simplifications of the system. It did, however, not lead to a predictive equation for spacings, as also in the latter study a non-constant ΔC_b was found. Consequently, Rohatgi and Adams (1967b) were not able to relate their observations to quantitative, generally applicable predictions of a_0.

Even in the case of constant ΔC_b the approximations by Rohatgi and Adams (1967b) would, at most, give the possibility to *deduce* a_0 indirectly from the changing local temperature field or, as used by the latter authors, the 'local solidification time' it takes for the temperature to drop to its eutectic value. Such a relation would not have a general predictive skill for a_0, because it considers existing plates beyond the interface and operating level of plates or dendrites. The formation mechanism of a_0, and the question how

ΔC_{lat}, dT_b/dt and T_b evolve at their tips to influence the cell spacing, is not taken into account by the lateral solute budget.

6.2.3   Correlation attempts of a_0 for seawater

A dependence $a_0 \sim V^{-1/2}$ was suggested by Assur and Weeks (1963), mainly based on the above noted qualitative argument of a diffusional distance scaling $a_0 \sim (D_s t_D)^{1/2}$. At that time the simultaneous observations of a_0 and the growth velocity, summarised by Weeks and Hamilton (1962), were limited and insufficient to validate such an exponent in a power law. Weeks and Assur (1963) correlated the same experimental results in a different form $a_0 \sim z^{1/6}$, z being the depth in the ice sheet, in some correspondence to the observations by Tabata and Ono (1962).

The systematic laboratory experiments by Lofgren and Weeks (1969a) were meant to elucidate these questions, but yielded, as shown above in figure 6.1, no single exponent. The theoretical approaches discussed by Lofgren and Weeks in connection with the results were limited to the noted conceptual studies by Bolling and Tiller (1960) and Rohatgi and Adams (1967b). However, both studies were to a certain degree misinterpreted by Lofgren and Weeks. First, the experimental and theoretical work by Rohatgi and Adams (1967b) *does not*, as interpreted by Lofgren and Weeks, indicate a dependence $a_0 \sim V^{-1/2}$, as these authors did not consider directional but fractional freezing rates. Second, Lofgren and Weeks suggested that equation 6.3 from Bolling and Tiller (1960) implies $a_0 \sim V^{-1}$ at large growth velocities, which is also incorrect. At large growth velocities one will find z_D in equation 6.4 to be determined by the temperature gradient in the solid, via $z_D \sim G_s^{-1/2}$. For directional sea ice growth this implies normally $z_D \sim V^{-1/2}$, and therefore the weakest dependence possible on the basis of equation 6.4 is $a_0 \sim V^{-3/4}$. The noted misinterpretations have not been corrected since then and appear in later reviews of the subject (Weeks and Ackley, 1986; Weeks, 1998a).

It may be noted that, inserting observed values of a_0 and $z_D \approx (0.6\Gamma/G_s)^{1/2}$ from the study by Lofgren and Weeks into equation 6.4 from Bolling and Tiller (1960), then the second term in the bracket on the right hand side is not more than 20 to 30 % of the first term $1/4\pi$. Hence, according to the approach from Bolling and Tiller, a_0 should be close to $D_s/\pi V$. While this gives the correct order of magnitude for natural growth rates, the implied proportionality $a_0 \sim V^{-1}$ is not was is observed. To illustrate this the compiled data in figure 6.7 are plotted on a log-log scale in figure 6.8. Disregarding the very different laboratory results from Lofgren and Weeks it is seen that the average exponent in a dependence of the form $a_0 \sim V^e$ corresponds typically to $-0.4 < e < -0.3$. However, as discussed above, also some of the revised observations are uncertain in terms of their growth velocity and the adjustment to its changes. Therefore figure 6.8 does not provide a clear picture of the dependence of a_0 on V.

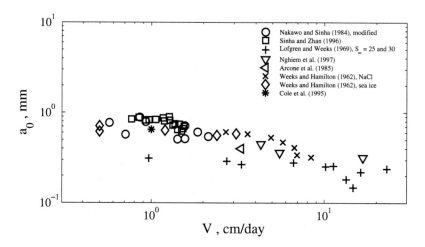

Fig. 6.8: Compiled plate spacing observations versus the ice growth velocity. The crosses are the laboratory observations from Lofgren and Weeks (1969a). All other points, with an exception from Weeks and Hamilton (1962) refer to field observations.

6.3   Stability theory by Mullins and Sekerka

The early cell spacing prediction approach suggested by Bolling and Tiller (1960) was an approximate analysis, but indicated, how the principle ideas of diffusive wave pattern at an interface, and the role of the solid-liquid interfacial energy, may interact to determine the cell spacing during solidification under constitutional supercooling. Since then a lot of observations and theoretical progress have been made in the field of crystal growth and cellular solidification being summarized in subsequent reviews (Langer, 1980b; Kurz and Fisher, 1984; Coriell and McFadden, 1993; Billia and Trivedi, 1993). The classical theory on which many of the theoretical and semi-empirical approaches are still based, has been formulated by Mullins and Sekerka (1964). Mullins and Sekerka presented a rigorous theory of the formation of cellular pattern during directional solidification of dilute solutions. They determined analytically, how in the case of constitutional supercooling an infinitesimal perturbation of arbitrary shape will grow with time. The formulation also implies a prediction of the wave-length which grows fastest, leading to $\lambda_{max} \sim V^{-1/2}$.

Morphological stability theory implies a modification of the constitutional supercooling criterion, but for the ice-water system the difference is small, even for one or two order of magnitudes larger growth velocities than found under most natural conditions (Sekerka, 1968; O'Callaghan et al., 1980). Therefore the estimates noted above apply and, under the absence of convection, also natural lake water should freeze with a cellular interface. First confirmations for the ice-water system were presented for ice cylinders in supercooled liquid (Hardy and Coriell, 1969). Later the onset of instabilities during directional solidification during quasi-twodimensional growth has also been found in reasonable agreement with predictions (Körber and Scheiwe, 1983). Several related studies of freezing of aqueous solutions were summarised by Koerber (1988).

Koerber (1988) also reported, that the transitional wavelength was in reasonable agreement with predictions. Mullins and Sekerka (1964) had pointed out, that the theory is, in a strict sense, only valid for small disturbances during the transition from a planar to a cellular interface, and might not give correct values for the spacings of deep cells. Indeed, the situation becomes less clear when sea ice with deep cell grooves is considered. Weeks and Ackley (1986) noted the theory from Mullins and Sekerka (1964), but did not report quantitative estimates. Such an application was attempted by Wettlaufer (1992, 1998), who compared the above noted datasets from Lofgren and Weeks (1969a) and Nakawo and Sinha (1984) to predicted wave lengths from morphological stability theory. He found that the observations were typically one order of magnitude larger than the predicted spacings. Wettlaufer's conclusion was, that a prediction of plate spacings from the *linear* theory is not possible because the problem requires a *nonlinear* treatment. The principal value $k_{eff} = 0.3$ for the distribution coefficient of sea ice, used by Wettlaufer (1992) in his computations, is, however, questionable. As it was seen in Chapter 6, $v_b \approx k_{eff}$ is of order unity near the interface, and it is this level, and not the delayed entrapment at the top of the skeletal layer, which determines the spacing. This aspect will become clear in the following chapters.

To consider the potential of morphological stability theory, with respect to the prediction of plate spacings in natural sea ice, first the principles of the classical solution derived by Mullins and Sekerka (1964) are outlined in the following chapter, and applications to saline solutions are summarised. In chapter 9 it will then be shown, how morphological stability theory may be modified in a simple consistent manner, to be applicable to the solidification of sea ice.

7. Morphological stability

The original morphological stability analysis from Mullins and Sekerka (1964) has later later been presented with slight modifications (Sekerka, 1968; Langer, 1980b; Coriell and McFadden, 1993). The following short description refers to the derivation by Coriell et al. (1985) and Coriell and McFadden (1993).

7.1 Linear stability theory for a planar interface

7.1.1 Basic theory

The directional solidification of a saline solution at constant velocity V in the z-direction is considered in a (x-y-z) coordinate system. In the following the limit of large ratios of thermal diffusivity to solutal diffusivity (for ice and brine: $\kappa_i/D_s \gg 1$ and $\kappa_b/D_s \gg 1$) is considered. In the steady state the diffusion equations to be fulfilled for the temperature T_i and T_b in the (liquid) brine and (solid) ice are

$$\kappa_b \nabla^2 T_b + V \frac{\partial T_b}{\partial z} = 0, \tag{7.1}$$

and

$$\kappa_i \nabla^2 T_i + V \frac{\partial T_i}{\partial z} = 0. \tag{7.2}$$

The solute concentration C in the liquid obeys

$$D_s \nabla^2 C_b + V \frac{\partial C_b}{\partial z} = 0. \tag{7.3}$$

In the solid solute diffusion is neglected and C_i

$$C_i = k C_{int} \tag{7.4}$$

is given by C_{int} at the interface in the liquid and the interfacial solute distribution coefficient k. The other boundary conditions at the interface are

$$(\mathbf{v} \cdot \mathbf{n}) L_v = (K_i \nabla T_i - K_b \nabla T_b)) \cdot \mathbf{n} \tag{7.5}$$

for heat and

$$(\mathbf{v} \cdot \mathbf{n}) C_{int} = D_s \nabla C_b \cdot \mathbf{n} \tag{7.6}$$

for solute, where $\mathbf{v}$ is the local solidification velocity and $\mathbf{n}$ the unit vector normal to the interface. L_v is the volumetric latent heat of fusion, K_b and K_i are the thermal conductivities of the brine and ice and m is a linearised local liquidus slope. Equilibrium at the interface implies

$$T_b = T_i = T_f + mC_{int} - \Gamma\mathcal{K}, \tag{7.7}$$

with melting temperature T_f of the pure liquid (water). Γ is the Gibbs-Thomson parameter

$$\Gamma = \frac{T_f\gamma_{sl}}{L_v}, \tag{7.8}$$

which together with the local mean curvature $\mathcal{K}$ of the interface gives the freezing point depression due to surface tension. For an interface given by $z = h(x,y)$ the linearized mean curvature has the form

$$\mathcal{K} = \left(\frac{\partial^2 h}{\partial x^2} + \frac{\partial^2 h}{\partial x^2}\right). \tag{7.9}$$

All transport coefficients are assumed constant for a certain concentration range C_{int} to C_∞. As they, as well as the local liquidus slope m, depend on concentration, they are computed at C_∞ throughout the present work. Fluid flow due to the change in density upon solidification is neglected. In practice its effect in freezing experiments depends on the capability of the ice to raise its freeboard in compensation to expansion. During freezing from a solid cooling interface in the laboratory it will be present, during natural freezing likely not. In principle it can be accounted for by rescaling the growth velocity V to $\rho_i/\rho_b V$. As the implied change is not very large for ice-brine mixtures it is neglected and eventually reintroduced later in comparison to observations.

The assumption of a thermal steady state is justified in the limit of large ratios of thermal to solutal diffusivities and is satisfied for the ice-brine system ($\kappa_i/D_s \approx 1.6 \times 10^3$ and $\kappa_b/D_s \approx 2 \times 10^2$ near 0 °C). In a linear stability analysis approach the temperature and concentration fields are written as the sum of an unperturbed part only depending on z and a perturbed part $exp(\Sigma t + i(\omega_x x + \omega_y y))$. For example, the freezing interface, defined as $z = 0$ in the steady state, takes the form

$$z = h(x,y,t) = \delta exp\left(\Sigma t + i\left(\omega_x x + \omega_y y\right)\right), \tag{7.10}$$

in the perturbed state, where Σ describes the time-dependent behaviour of an infinitesimal perturbation and ω_x and ω_y its horizontal wave number vector. As shown in several similar treatments (Mullins and Sekerka, 1964; Coriell et al., 1985; Coriell and McFadden, 1993) this analysis leads to a the following dispersion relation

$$\Sigma = \left(-G_{eff} - \Gamma\omega^2 + mG_c\frac{\omega_* - \frac{V}{D_s}}{\omega_* - (1-k)\frac{V}{D_s}}\right)\left(\frac{L_v}{(K_b + K_i)\omega} + \frac{mG_c}{V(\omega_* - (1-k)\frac{V}{D_s})}\right)^{-1} \tag{7.11}$$

, which characterises the onset of instability from the thermal steady state. The growth rate Σ of an instability has the unit time^{-1}. The parameters therein are

$$\omega_* = \frac{V}{D_s} + \left(\left(\frac{V}{D_s}\right)^2 + \omega^2\right)^{1/2}, \tag{7.12}$$

a wave number

$$\omega = (\omega_x^2 + \omega_y^2)^{1/2}, \tag{7.13}$$

while

$$G_{eff} = \frac{G_i K_i + G_b K_b}{K_i + K_b} \tag{7.14}$$

is an effective temperature gradient at the interface, based on the temperature gradients G_i and G_b in the liquid and ice, respectively. Under equilibrium freezing the solute gradient G_c at the interface is given by

$$mG_c = C_\infty \frac{1-k}{k} \frac{V}{D_s}. \tag{7.15}$$

Equation 7.11 is valid in the thermal steady state, while a more complete general analysis for finite diffusivity ratios is given by, for example, Coriell and McFadden (1993). The general meaning of the dispersion relation 7.11 is that for wave lengths $\lambda = 2\pi/\omega$ with $\Sigma > 0$ the interface is unstable to infinitesimal perturbations, while it is stable for wave lengths with $\Sigma < 0$. For a better understanding and discussion of the influence of different terms in equation 7.11 it is useful, with help of equation 7.5, to express the effective temperature gradient G_{eff} in the form

$$G_{eff} = \frac{L_v V + 2G_b K_b}{K_i + K_b} = \frac{L_v V}{K_i} \left(\frac{1 + \frac{2G_b K_b}{L_v V}}{1 + \frac{1}{n_k}}\right), \tag{7.16}$$

with $n_k = K_i/K_b$, emphasising the role of the thermal conductivity ratio of liquid and solid and of the relative dominance of latent heat L_v in the problem. One may further define two other nondimensional parameters which are, as already pointed out by Mullins and Sekerka, of relevance in the theory. These are the absolute stability parameter

$$\mathcal{A} = \frac{k\Gamma V^2}{mG_c D_s^2} = \frac{k^2 \Gamma V}{m(1-k)C_\infty D_s}, \tag{7.17}$$

and the ratio

$$\mathcal{G} = \frac{G_{eff}}{mG_c} = \frac{L_v D_s}{K_i} \frac{k}{mC_\infty(k-1)} \left(\frac{1 + \frac{2G_b K_b}{L_v V}}{1 + \frac{1}{n_k}}\right), \tag{7.18}$$

which characterises the degree in constitutional supercooling.

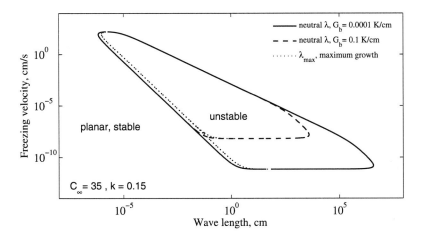

Fig. 7.1: Morphological neutral stability balloon for an NaCl solution with $k = 0.15$ and $C_\infty = 35$ ‰, freezing at two different temperature gradients G_b the liquid brine.

7.1.2   Principle results of morphological stability theory

In figure 7.1 the neutral stability curve of the dispersion relation has been plotted for a two sets of parameters. It may be noted that the use of $C_\infty = 35$ ‰ restricts the possible k to a value of 0.132, two orders of magnitude larger than typical NaCl values for a planar interface, because the interface concentration is limited by the eutectic composition $C_\infty \approx 230$ ‰. The choice of C_∞ thereby becomes somewhat arbitrary, and the dispersion curve for $C_\infty = 0.5$ ‰ and a more realistic planar $k = 0.0021$ would look essentially the same if all thermal properties are evaluated at the interface. Such large interface concentrations are seldom reached during natural freezing. Therefore figure 7.1 serves in the first sense as an illustration of a theory which still needs to be modified in account of natural growth conditions. The basic qualitative interpretation of the graph is: (i) Inside the closed polygon in figure 7.1 a cellular interface is unstable, outside it is stable; (ii) an upper and lower limit of the freezing velocity V exists, beyond which the interface is always stable; (iii) within the velocity regime interfacial instability is limited to a band $\lambda_\Gamma < \lambda < \lambda_D$; (iv) the wave length λ_{max} of most rapidly growing perturbations is slightly larger than the lower limit of neutral stability and shown as a dotted line within the stability balloon.

Constitutional supercooling modified

Another implication of the stability analysis is that the constitutional supercooling condition (6.2) discussed earlier and recalled here,

$$\frac{G_b}{mG_c} = \frac{G_b D_s}{V} \frac{k}{mC_\infty(k-1)} \leq 1 \tag{7.19}$$

takes a sightly different form when considering the linear stability analysis. Now the condition for instability may be written as

$$\frac{G_{eff}}{mG_c} = \frac{L_v D_s}{K_i + K_b} \frac{k}{mC_\infty(k-1)} \left(1 + \frac{2G_b K_b}{L_v V}\right) < \mathcal{S}(\mathcal{A}, k), \tag{7.20}$$

which was pointed out already by Mullins and Sekerka (1964). The function $\mathcal{S}(\mathcal{A}, k)$ was introduced by Sekerka (1965), being very useful to describe the effect of the surface tension on interface stability. At the moment it suffices to note that $\mathcal{S}$ is mostly in the range 0.9 to 1 under natural ice growth conditions, delaying its discussion to paragraph 7.1.4 below. If the liquid temperature gradient G_b is dominant in equation 7.20, the latter stability condition becomes

$$\frac{G_{eff}}{mG_c} = \frac{G_b D_s}{V} \frac{k}{mC_\infty(k-1)} \leq \frac{K_i + K_b}{2K_b} \mathcal{S}(\mathcal{A}, k). \tag{7.21}$$

Neglecting a slight deviation of $\mathcal{S}$ from unity, one finds that, with $K_i/K_b > 1$ for ice and brine, the interface stability condition 7.21 is less stringent than equation 7.19. In the case of ice and dilute brine or water with $K_i \approx 4K_b$ the condition for the onset of cellular freezing is shifted considerably, by a factor $\approx 5/2$.

If $2G_b K_b$ is small compared to $L_v V$, then equation 7.20 simplifies to

$$\frac{G_{eff}}{mG_c} = \frac{L_v D_s}{K_i + K_b} \frac{k}{mC_\infty(k-1)} < \mathcal{S}(\mathcal{A}, k), \tag{7.22}$$

, where the growth velocity is no longer directly present, yet only due to its influence on the surface tension stability function $\mathcal{S}$. One realises that interface stability is expected to be independent of V, if latent heat dominates the phase transition. Equation 7.22 is most important when considering the onset of cellular ice growth from saline solutions under natural growth conditions. Neglecting still the influence of $\mathcal{S}$ and using a bulk interface value $k \approx 10^{-3}$ for dilute NaCl solutions (chapter 4) one finds that cellular instabilities should occur at C_∞ above ≈ 0.0013 ‰[1]

---

[1] The question if one should, instead of the bulk k for a planar interface, better use the single crystal value of eventually $k_0 \approx 5 \times 10^{-4}$, is essential and will be considered below. A related problem, the stability of *existing* grooves, was addressed by Tiller (1962) and already mentioned in connection with equation 6.3.

7.1.3 Wave lengths and neutral stability

The instability band of wave lengths in figure 7.1 is illustrated by a quantitative example of the growth rate Σ of perturbations in figures 7.2 and 7.3. These figures resemble a different situation than in figure 7.1 above, where a maximum interfacial concentration close to the eutectic value was assumed. Now C_∞ has been chosen very close to the value ≈ 0.0013 ‰, below which a planar interface will be stable for directional freezing of aqueous NaCl solutions[2]. To compute Σ in both figures G_b has been set to zero.

In figure 7.2 the dependence of Σ on the perturbation wave length is shown for $C_\infty = 0.0015$ ‰, slightly above the critical concentration for the onset of cellular perturbations. Moving from the left to larger wave lengths, the perturbation growth rate becomes positive above λ_Γ, reaches a maximum at $\lambda_{max} = 3^{1/2}\lambda_\Gamma$, and becomes negative again above λ_D. This is the behaviour of Σ which leads to the neutral stability regime of possible wave lengths in figure 7.1.

In figure 7.3 the behaviour of Σ is illustrated very close to the critical concentration for the onset of instability. Under this condition the wave lengths λ_Γ, λ_{max} and λ_D are still very close to each other. In figure 7.1 this *bifurcation* of the wavelengths appears as the wide basis of the stability balloon. At the critical concentration close to the onset of instabilities the $\Sigma = 0$ line is touched at a single wave length, which is terned λ_{mi}.

Based on the summary given so far a change of the neutral stability curve in figure 7.1 is qualitatively related to the following boundary conditions and properties in equation 7.11:

- A positive temperature gradient (temperature increasing in the growth direction) in the liquid increases G_{eff} and always stabilises the interface. This is the classical compensation of constitutional supercooling and raises the bottom of the stability balloon to higher growth velocities.

- Latent heat stabilises the interface as it has to be removed before freezing can proceed. The larger the solid and liquid conductivities K_i and K_b, the more rapid the removal of the latent heat, and the less stable the interface.

- A faster removal of solute due to a larger solute diffusivity D_s stabilizes the interface.

- Constitutional supercooling increases with concentration C_∞ and the local liquidus slope m, which both are destabilising.

- Surface tension always stabilises the planar interface. This effect is only relevant at small wave lengths (high growth velocities) and implied in the decrease of $S < 1$ with growth velocity.

---

[2] With the properties in the present paper a value ≈ 0.001325097 ‰ gives λ_Γ and λ_D differing by only 2 %

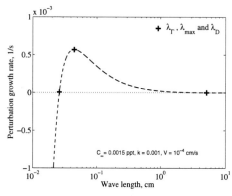

Fig. 7.2: Perturbation growth rate Σ at the planar interface of an unidirectionally freezing aqueous NaCl solution under diffusive equilibrium and $G_b = 0$ in the liquid brine brine. Positive growth rates (above the dotted line) imply an unstable interface, for which λ_Γ and $lambda_D$ span the possible wave lengths. Instability growth is most rapid at λ_{max}.

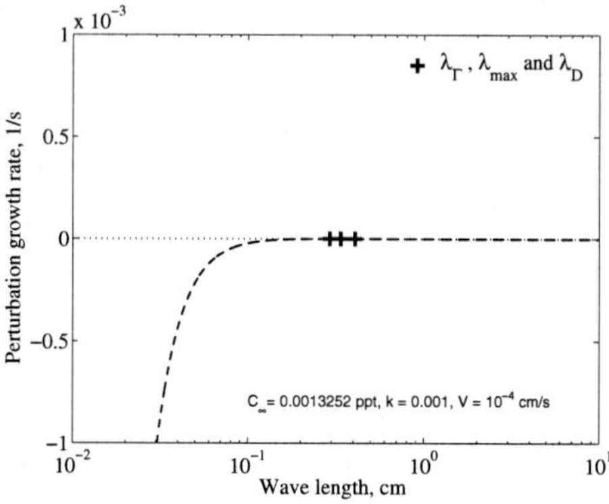

Fig. 7.3: Perturbation growth rate Σ at the planar interface as in figure 7.2, marginally above the critical concentration for the onset of cellular growth, where the neutral stability curve bifurcates. λ_Γ and λ_D are still very close to each other.

7.1.4 Margins of stability

From the preceding discussion it is clear that for each set of k and the temperature gradient G_b in the liquid, there is a minimum concentration at which the *onset of instability* is expected, corresponding to the lowest possible freezing rate V for cellular growth. As shown by Sekerka (1965) it is possible to obtain a solution of the dispersion relation 7.11 for the onset of stability in the form

$$\mathcal{S}(\mathcal{A}, k) = \mathcal{G} = 1 + \frac{\mathcal{A}}{4k} - \frac{3\mathcal{A}^{1/2}r}{2} - \frac{\mathcal{A}(1 - 2k)r^2}{4k}, \tag{7.23}$$

when $\mathcal{S}(\mathcal{A}, k) = \mathcal{G}$ also defines the noted stability function. The variable r is defined as

$$r^4 = 1 + \left(\frac{2D\omega}{V}\right)^2, \tag{7.24}$$

and in turn is determined by the one real root of

$$r^3 + (2k - 1)r - \frac{2k}{\mathcal{A}^{1/2}} = 0 \tag{7.25}$$

greater then zero.

Negligible temperature gradient in the liquid

For the case $G_b = 0$, when latent heat controls the interface stability, the situation is further simplified. Now, for each k there is one and only one concentration $C_{mi}(k)$, at which the neutral stability curve in figure 7.1 is a closed polygon. This value is defined here has

$$C_{mi} = C_{mi}(k) = C(C_\Gamma = C_{max} = C_D, G_b = 0). \qquad (7.26)$$

As noted above and indicated in figure 7.3, for the dilute NaCl-water system one has $C_{mi}(k = 0.001) \approx 0.0013$ ‰, when using the thermodynamic properties from appendix A. As only a weak growth velocity dependence enters equation 7.22 via the stability function $\mathcal{S}(\mathcal{A}, k)$, which is very close to unity at low growth velocities, it does not seem useful to speak of a lower growth velocity limit. Most important is that, for C_∞ larger than C_{mi}, the neutral stability curve does not close at low velocities, but the instability wave length can in principle grow to infinity. This is also the case, if some small G_b exists but is simply proportional to G_i in the ice, as one will usually find if freezing starts with some amount of superheat in the water. As an example one may consider superheat during the slow freezing of lake or brackish water that has not overturned and mixed after reaching its density maximum. Assuming idealized conditions of maximum natural superheat of 4 K in the water and a similar temperature difference through the ice, one may derive from the classical Stefan solution for diffusion (B.1) that the liquid temperature gradient contributes ≈ 25 % of the latent heat term in equation 7.20 and would increase the critical concentration by this amount. This calculation is, however, of little practical importance, because under real conditions thermal and/or haline convection will alter the picture more severely, to be discussed later.

Absolute stability

The other margin of cellular instabilities apparent in figure 7.1 is the limit at large growth velocities. Here the stability balloon always closes and the condition sufficient for stability, once given by Mullins and Sekerka (1964) as $\mathcal{A} > 1$, has later been refined to $\mathcal{A} + \mathcal{G} > 1$ (Coriell et al., 1985), with a more general discussion given by Coriell and McFadden (1993). From the definition of $\mathcal{A}$, equation 7.17, one sees that absolute stability strongly depends on k.

Absolute stability arises when the length scales in the problem become so small that capillarity and the Gibbs-Thomson effect completely control the freezing point depression. The transition to this stage, which occurs at high velocities, is characterised to a certain degree by the stability function $\mathcal{S}(\mathcal{A}, k)$. It is obtained by solving equations 7.23 and 7.24 and shown in figure 7.4. For the specific case of aqueous NaCl solutions, $\mathcal{A}$ has been computed in dependence on V for several sets of k and C_∞, to obtain $\mathcal{S}$ as a function of the

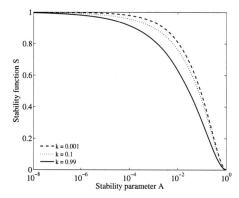

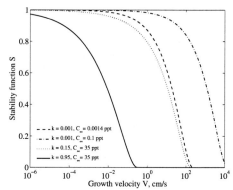

Fig. 7.4: Stability function $\mathcal{S}$ defined by Sekerka (1965) in dependence on $\mathcal{A}$ for three values of k. $\mathcal{S}$ approaches zero when $\mathcal{A}$ approaches unity, which is the point of absolute stability.

Fig. 7.5: Stability function $\mathcal{S}$ in dependence on the freezing velocity V for several sets of freezing parameters of aqueous NaCl solutions.

freezing velocity in figure 7.5. Assuming, for example, a NaCl-water system resembling lake and river compositions with $C_\infty = 0.1$ ‰ and $k = 0.001$, absolute stability is obtained near $V = 10^4$ cm s^{-1}, while for the eutectic interface concentration system in figure 7.1 it takes place at somewhat lower growth velocities. For natural growth conditions, in particular the freezing of lake and sea water, absolute stability is of little importance, when k is small.

7.1.5   Transient versus equilibrium

The linear stability analysis by Mullins and Sekerka is based on a time-independent basic state. As for the onset of cellular growth constitutional supercooling normally is established within a finite time after the onset of freezing, the problem in reality can be expected to be time-dependent. The limitations of the theory must therefore be discussed, when the condition of a steady state is not given. The aspects which need to be discussed are:

1. The time independence of G_c enables the derivation of the dispersion relation 7.11. What is the effect of G_c changing with time?

2. Equation 7.15 gives the steady-state equilibrium concentration gradient at the interface under absence of convection. Its use leads, for example, to the simplified constitutional supercooling equation 7.22. How is this limit affected by a transient G_c ?

The short discussion which follows again only applies to the onset of cellular growth and infinitesimal perturbations.

Sekerka (1967) has performed an analysis of the time-dependent problem and found that under the condition of thermal equilibrium, only the solute field being time-dependent, equation (7.20) can still be applied. He proposed that it is likely that the classical analysis will also hold in general. Numerical simulations of the initial transient by Coriell et al. (1994) for a tin-bismuth alloy have confirmed this hypothesis to some degree. These authors tested a scenario of V and C_∞ increasing with time from a basic state of zero velocity. It was found that, as long the critical interface concentration for instability was close to the equilibrium value, the planar interface breakdown from the transient took place at an interface concentration within a few percent of the steady state prediction, provided that the instantaneous interfacial values were used in the stability analysis. However, if instability was attained already at 50 % of the equilibrium value $C_\infty \frac{1-k}{k}$, the Mullins-Sekerka analysis underpredicted the critical solute concentration by 10 %. In other words, the numerically predicted interfacial solute gradient at the time of instability was 10 % larger than predicted by steady-state theory. Coriell et al. (1994) suggested that this effect may be related to the increasing velocity and solute gradient at the interface, the instantaneous values not accurately representing the situation over some infinite previous time period, when the system is more stable. Interestingly, as discussed below, a comparable overshoot of the critical interface concentration has been observed during the freezing of NaCl solutions with constant interface velocity but a solute gradient increasing with time (Körber and Scheiwe, 1983). This delay appears plausible. When a perturbation advances into a regime towards which, due to the increasing solute gradient, also the solute flux increases, then the local constitutional supercooling to which the perturbation travels, is reduced, with respect to a steady state.

Critical interface concentration

One may conclude that, in the transient problem, the equilibrium value C_∞/k in the general equation 7.20 has to be replaced by the time-dependent interface salinity $C_{int}(t)$ to get the condition.

$$mC_{int}(t)(k-1) > \frac{L_v D_s}{K_i + K_b}\left(1 + \frac{2G_b K_b}{L_v V}\right)\frac{1}{S(\mathcal{A}, k)} \qquad (7.27)$$

for the interface concentration. Considering the case of small relative temperature gradient in the liquid, $2G_b K_b/(L_v V) \ll 1$ and moderate growth velocities for which the effect of surface tension can be neglected ($S \approx 1$) one obtains a critical interface concentration that must exceed

$$C_{int}(t) > \frac{L_v D_s}{K_i + K_b}\frac{1}{m(k-1)}, \qquad (7.28)$$

For dilute NaCl solutions with a small k it takes the value $C_{int} \approx 1.33$ ‰. It must be noted that equation 7.27 has been validated numerically for the case when the critical interface concentration is approached from below. The wave length at instability is then found by inserting

$$G_c = C_{int}(t)(1-k)\frac{V}{D_s} \tag{7.29}$$

in the dispersion relation 7.11. Below it will be seen, that equation 7.27 is only sufficient for interface instability, if the interface concentration is reached from below. Under conditions where C_∞ already exceeds the critical concentration, the time-dependent onset must be described in a slightly different form, accounting for the establishment of not only the interfacial gradient, but also a certain degree in constitutional supercooling.

7.2 Confirmations of morphological stability theory for water and aqueous solutions

The two basic limits emerging from the preceding discussion of freezing of aqueous NaCl solutions in the case of a relatively small temperature gradient $G_b \simeq 0$ are

- Under freezing conditions where a *diffusive equilibrium* is reached cellular instabilities may be expected at $C_\infty \approx 0.0013$ ‰, if $k = 0.001$ is assumed.

- Under absence of diffusive equilibrium instability occurs when $C_{int} > 1.33$ ‰.

Although a derivation of these critical concentrations is straightforward, they have not been presented or discussed in earlier work. Experimental verification is complicated because (i) also during upward freezing, when the diffusive limit may be realised, thermal convection due to the density maximum above the freezing point disturb the strict diffusive solute transport and (ii) interface concentrations are not easily measurable. The following summary of experimental studies which contain information about the concentration-dependent interface breakdown is chronological.

7.2.1 Early observations by Quincke

An interesting early confirmation of morphological stability can be traced back one century to the observations by Quincke (1905b,a), who performed a large number of experiments with saline solutions freezing under different conditions. Quincke's observations are the first documents of cellular freezing of rather dilute saline solutions. He called the cellular ice *foam cells* surrounded by *foam walls* of *oily liquid* of concentrated saline solutions. He found these foam walls to be very thin and often invisible, but made careful descriptions of their evolution during the warming and remelting of ice.

Quincke described the macroscopic appearance of saline ice as either lamellar or milky. Freezing forced by coolants from the bottom upwards always created milky ice. The number of lamellae increased with freezing velocity and solute concentration, which is the expected qualitative dependence in morphological stability theory. Quincke reported some quantitative tests. During upward freezing of NaCl solutions with initial salinities 0.01 and 0.001 ‰ he found, that the final ice block was more milky in the upper part, while for a larger concentration of 0.1 ‰, corresponding to typical lake water, the lower part was more milky. This indicates either, that during upward growth of very dilute solutions the onset of stability was apparently delayed, until the solute in the container was sufficiently concentrated, or that milkiness increases with solute concentration, which is also plausible. Quincke noted that a remarkable milkiness was already produced by a concentration of 0.001 ‰ NaCl. This value is close to the diffusive transition of 0.0013 ‰. The results reported by Quincke are not detailed enough to associate them as a quantitative confirmation, yet his observations reflect the transition to cellular growth at reasonable values of C_∞. A quantitative comparison will also have to consider that the interfacial distribution coefficient for single crystals is likely only half of the effective value $k \approx 10^{-3}$ assumed here. The question, how the crystal and grain boundary modes of solute entrapment interact, will then also be important for an exact prediction.

Quincke also performed experiments where small containers filled with dilute saline solutions (0.001 to 0.1 ‰ NaCl) were frozen in still air, leading to mixed lateral and downward solidification. In these experiments he generally observed an outer shell of very clear ice. These findings are consistent with the mechanism shown quantitatively in the earlier section 4.5.2, solutal convection effectively prohibiting the buildup of the critical interface concentration $C_{int} > 1.33$ ‰. Dilute solutions of natural tap water with $C_\infty \approx 0.1$ ‰ therefore likely need to be concentrated by a factor of approximately 10 to produce ice with a milky appearance during downward freezing, and simple experiments to illustrate this may be performed with tap water and a conventional refrigerator. It may be finally noted that Quincke already pointed to the importance of surface tension in the problem and described cellular substructures not only for ice but also for metals and alloys (Quincke, 1906).

7.2.2 Planar-cellular transition: quantitative confirmations

Perturbed ice cylinder

Mullins and Sekerka's theory of interface instability for binary alloys has also been formulated for interfaces or particles growing into a pure supercooled melt (Mullins and Sekerka, 1963; Langer, 1980b). The dynamical aspects of the theory were first quantitatively confirmed in experiments of ice cylinders growing into pure supercooled water (Hardy and Coriell, 1969; Coriell and Hardy, 1969). This arrangement has the advantage that free growth velocity of the cylinder, its radius and the supercooling are related to each other in a unique manner. In their experiments, Hardy and Coriell found good agreement between predictions and observations, considering both the growth rates of perturbations and the wave lengths. The measured wave lengths corresponded closely to λ_{max} for the most rapidly growing perturbations. The theory has been refined by a nonlinear analysis, necessary when the perturbation growth rate was comparable to the radial cylinder growth rate, accounting for the coupling of perturbations along and around the axis of the cylinder (Coriell et al., 1971). In later experiments with dilute HCl and NaCl solutions a similar confirmation was obtained (Hardy and Coriell, 1973).

Unidirectional freezing

Kvajic et al. (1971) reported morphologically unstable growth during upward freezing of NaCl solutions at salinities as low as 0.1 ‰. Similar to Quincke (1905b), they described the appearance of the ice as milky, with lines and steps. Kvajic et al. (1971) compared their observations of the first appearance of lines and steps to equation 7.27 from morphological stability theory. They found that the transition to a cellular interface took generally place at concentrations gradients 20 to 60 % of the predicted critical G_c. The disagreement can partly be explained by their use of a 50 % too large solute diffusivity D_s. Another cause for the deficiency seems to be the uncertainty in the interface concentration. As only the temperature gradient was measured, the solute gradient had to be predicted from a diffusion model, for which a steady state approximation was used. Such a model is, during the initial transient, often inappropriate, and may underestimate the interface concentration by a factor of 2, as shown quantitatively in figure 12.1 of chapter 12. Considering these inaccuracies the work from Kvajic et al. (1971) is consistent with morphological stability theory.

Freezing under the cryomicroscope

The first detailed experimental confirmation of morphological stability theory during directional solidification was carried out by Körber and his colleagues (Körber and Scheiwe, 1983; Körber et al., 1983). These authors obtained accurate observations of the planar-cellular transition during freezing of aqueous 21.6 ‰ $NaMnO_4$ solutions under a cryomicroscope[3]. The interfacial temperature gradient in these studies was calculated from a parabolic diffusion law, while the solute concentration profile was measured with a spectrophotometer by means of transmitted light. Photo-micrograph series of the interface breakdown were also provided.

---

[3] It is noted here that the diffusion coefficient of $NaMnO_4$ in aqueous solutions is similar to NaCl (Körber et al., 1983), while the liquidus slope m of $NaMnO_4$ is 50 % smaller. Therefore morphological stability theory simulations with 10 ‰ NaCl resemble the system of 21.6 ‰ $NaMnO_4$

The onset of morphological instability was in most cases observed when the liquidus temperature gradient (based on the solute gradient) exceeded the externally imposed temperature gradient by 10 to 50 %, with largest differences at lowest growth velocities. Capillarity was neglected, which is reasonable due to soldification velocities less than 10^{-3} cm s^{-1}, as one recognises from $S \approx 1$ in figure 7.5. Also the difficulties to obtain concentration gradients directly at the interface, discussed by Körber et al. (1983) in connection with a transient solute diffusion model, seem insufficient to account for an uncertainty of up to 50 %. Here it is suggested that the approach from Körber and Scheiwe (1983), to replace G_{eff} from stability theory by an effective interfacial thermal gradient, is more problematic. The latter was estimated from the interface velocity and assuming a parabolic temperature distribution through both phases assuming $G_i = G_b$ at the interface (Körber et al., 1983). Such a temperature distribution has not been verified experimentally and is not expected during unidirectional freezing of water with a high Stefan Number (appendix B). Assuming instead a linear temperature distribution in the ice would increase G_{eff} in general. This correction increases with the distance between interface and cold boundary, and would be largest at lowest freezing rates in the experiments from Körber and Scheiwe (1983). An approximate evaluation of the simulated temperature profiles shown by (Körber et al., 1983) indicates that the maximum underestimnate of G_{eff} may well have been 30 %.

It seems therefore that the results on the onset of instabilities presented by Körber et al. (1983) require a slight correction. If at all, they may be interpreted as an overshoot of the Mullins-Sekerka stability criterion by 10 to 20 %. Later results obtained with the same method (Koerber, 1988) are likely to be interpreted in a similar way. A slight overshoot would, as discussed above in paragraph 7.1.5, be plausible in connection with the transient evolution of the system, and has been simulated by Coriell et al. (1994). In view of these refinements it can be concluded that the experiments from Körber and Scheiwe (1983) are a reasonable confirmation of the Mullins-Sekerka stability criterion for the planar-cellular transition.

Thin growth cell interferometry

Recently several experiments have been performed with aqueous KCl solutions in thin growth cells (figure 7.6), evaluating the solute fields by optical interferometry with a Mach-Zehnder interferometer (Nagashima and Furukawa, 1997b,a). The comparison of cellular spacings with linear stability theory, as attempted by Nagashima and Furukawa (2000b), will be discussed below in paragraph 7.2.3. Here first the information that the authors provided about the onset of first cellular instabilities (Nagashima and Furukawa, 2000b,a) is considered. The following comparison is based on liquidus slope and solute diffusivity values for KCl which are 0.77 and 1.2 smaller and larger than the respective

NaCl properties (appendix A.2.11).

Nagashima and Furukawa (2000a) described the solid-liquid interface evolution during freezing of a 30 ‰ KCl solution in a 0.1 mm thin growth cell, where a crystal is pulled at constant velocity through a temperature gradient. In the experiment of interest the crystal was pulled at $V_p = 5 \times 10^{-4}$ cm s^{-1} through the temperature gradient of $G_{cell} = 23$ K cm^{-1}. Although the situation, due to the superimposed temperature gradient, is somewhat artificial, the study is a good documentation of the initial transient. During planar growth the observed interfacial salinity gradient $G_c(t)$ was close to its theoretical interfacial value $C_{int}(1-k)V/D_s$, the difference of ≈ 10 to 15 % eventually being related to spatial resolution limits at the interface. It is also noted that the c-axis direction in the experiments was in the shortest dimension of the thin growth cell and normal to the growth direction.

The observations made by Nagashima and Furukawa (2000a) were as follows. When the crystal pulling was started with velocity V_p, the crystal started growing in the opposite direction. The initial $V \approx 0.15V_p$ was increasing and the interface remained planar until the first interface instability was detected 60 seconds after the onset of freezing, at a growth velocity of $V \approx 1.5 \times 10^{-4}$ cm s^{-1}, and when the concentration at the interface had increased to approximately 39 ‰. For a comparison with the linear stability equations here first the use of $G_b = G_{cell}$ together with the actual growth velocity V, increasing during the experiment, is applied. Inserting these experimental conditions into equation 7.27, one finds that the interface should have been unstable at a concentration of $C_{int}(t) \approx 11.8$ ‰. In principal the interface should have become cellular instantaneously. It appears therefore that the transient equation 7.27 cannot be applied in a situation with bulk concentrations far above the critical value. As noted above, this is not surprising, as the solutal boundary layer is not present instantaneously, contrasting the supercooling in the pure liquid in the experiments by Hardy and Coriell (1969). Constitutional supercooling still has to build up during the first stage of solidification and solute rejection.

It is desirable to formulate a simple criterion for this build-up, instead of performing transient simulations of the system. Considering the classical linear stability theory, and the result that a transient concentration field leads to a similar instability threshold, if C_{int} is approached from below, the following similarity approach is plausible. On the one hand, the necessary condition for instability is that the *instantaneous interfacial gradients* must fulfill $G_{eff}/(mG_c) < S$. On the other hand, one may interpret the term $C_\infty(1-k)/k$ in the steady-state stability equations as a *concentration difference* across the diffusive boundary layer of width D_s/V. Writing equation 7.20 in the form

$$-m\Delta C_{mi}(t) = -m(C_{int}(t) - C_\infty) > \frac{G_{eff}D_s}{V S(\mathcal{A}, k)} \qquad (7.30)$$

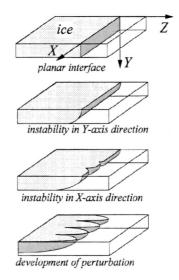

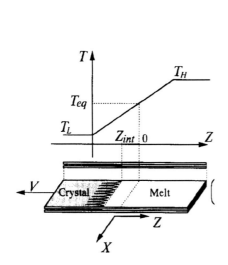

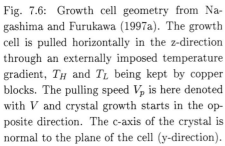

Fig. 7.6: Growth cell geometry from Nagashima and Furukawa (1997a). The growth cell is pulled horizontally in the z-direction through an externally imposed temperature gradient, T_H and T_L being kept by copper blocks. The pulling speed V_p is here denoted with V and crystal growth starts in the opposite direction. The c-axis of the crystal is normal to the plane of the cell (y-direction).

Fig. 7.7: Freezing interface evolution in the horizontal growth cell experiments from Nagashima and Furukawa (2000b), their figure 2. From top to bottom: (i) planar interface, (ii) first instability in the y-direction, (iii) second instability on the edge of the basal plane, (iv) fully developed cellular freezing. The c-axis of the crystal is normal to the plane of the cell and corresponds to the direction of gravity (y-direction).

or

$$-m\Delta C_{mi}(t) > \frac{L_v D_s}{K_i + K_b} \left(1 + \frac{2G_b K_b}{L_v V}\right) \frac{1}{\mathcal{S}(\mathcal{A}, k)} \qquad (7.31)$$

shows the conceptional difference. Now there must be a certain overall constitutional supercooling across the diffusive boundary layer, in addition to the gradient at the interface. This equation may be understood as a more stringent constraint for the onset of cellular growth from the initial transient and should be valid in the case of a quasi-constant growth velocity. For small k and C_∞ it is practically the same as equation 7.27, but for large C_∞ it is very different. It is noteworthy that the criterion 7.30 has been proposed by Bower et al. (1966) as a least interfacial gradient for the stable solidification of alloys.

Interpreting the experiment from Nagashima and Furukawa (2000a) this implies that not the interface concentration should have been critical for the onset of instabilities, but that the difference $C_{int}(t) - C_\infty \approx 12.1$ ‰ is the relevant condition. This prediction is in reasonable agreement with the observed difference $C_{int} - C_\infty \approx 9$ ‰ at the planar-cellular transition. From profiles documented in a parallel publication (Nagashima and Furukawa, 2000b) one finds that the concentration resolution is limited and the observed C_{int} might be in error by 1 to 2 ‰, while observations of G_c seem reproducable to within 20 %. The predicted interface concentrations for the single noted case was 30 % larger than observed, yet the difference may not be significant. It differs in sign from the observations and predictions noted above, where a tendency for a slight overshoot of the stability limit in the transient case was suggestive.

For a further discussion, one may also consider the general condition 7.30 by associating the imposed gradient G_{cell} with the effective G_{eff}. This results in a much larger $\Delta C_{mi} \approx 25$ ‰. The explanation for the discrepancy might be that G_{eff}, as it appears in the stability equations, is linked to the interfacial condition 7.5, which is not fulfilled in the growth cell experiments from Nagashima and Furukawa. This can be seen by computing a time-dependent interfacial temperature gradient $G_{i,v}$ in the ice, not accounting for the local cooling rate, but only for the growth velocity. Assuming $G_b = G_{cell} = 23$ K cm^{-1} and the relevant transitional $V \approx 1.5 \times 10^{-4}$ cm s^{-1}, one obtains $G_{i,v} \approx 8$ K cm^{-1} from the heat flow equation 7.5. This smaller 'effective' interfacial gradient is related to the fact, that in the growth procedure the crystal is initially pulled faster through the temperature gradient (at $V_p = 5 \times 10^{-4}$ cm s^{-1}) than can be compensated by the growth rate V. The result is a transient temperature field that differs from the quasi-steady assumption on which equation 7.31 is based. Hence, using $G_{eff} = G_{cell}$ during the transient stage, is inconsistent with the derivation of the dispersion relation 7.11. Due to this argumentation, it seems more plausible to use V and $G_b = G_{cell}$ in the quasi-steady state equations.

The discussed first instability corresponded to the formation of a wedge in the thinnest (y)-direction, the second frame from above in figure 7.7. On the other hand, Nagashima and Furukawa (2000b) observed a second perturbation normal to the first one (frame 3 in figure 7.7), creating a wavy pattern after 120 s, when the interface concentration difference had reached $\Delta C_{mi} \approx 21$ ‰ and V was $\approx 2.4 \times 10^{-4}$ cm s^{-1}. For these conditions the predictions are $\Delta C_{mi} \approx 8.2$ ‰ for the $G_b = 23$ K cm^{-1} case and $\Delta C_{mi} \approx 15.7$ ‰ when assuming the imposed gradient $G_{eff} = 23$ K cm^{-1} to control the instability. Hence, now it appears that the imposed gradient agrees much better with the observed interface concentration. Although this seems at first glance inconclusive, it can eventually understood as follows. As the first instability took place on a strict planar interface, the coupling of growth velocity with the temperature gradients is very important and $G_b = G_{cell}$ is correct. Once the wedge has formed, the interface velocity is no longer linked to

Planar-cellular transition in forced convection

C_∞ ‰	V, transition, cm/s	V_*, cm/s	ΔS_{int}, ‰	ΔS_{mi} ‰
10	$5.0 \pm 0.5 \times 10^{-4}$	$3.30(4.50) \times 10^{-3}$	1.79 ± 0.21	1.27
14	$3.5 \pm 0.2 \times 10^{-4}$	$3.26(4.50) \times 10^{-3}$	1.68 ± 0.11	1.24
28	$1.3 \pm 0.2 \times 10^{-4}$	$3.19(4.28) \times 10^{-3}$	1.19 ± 0.19	1.16
51.5	$0.7 \pm 0.2 \times 10^{-4}$	$3.21(4.19) \times 10^{-3}$	1.15 ± 0.34	1.02

Tab. 7.1: Observations of the planar cellular-transition by Wintermantel (1972) for different C_∞. At concentrations 28 and 51.5 ‰ there the transition velocity had b to be found by extrapolation. V_* differs slightly from values in brackets estimated by Wintermantel, due to the different molecular transport properties used here. ΔS_{int} is calculated from these three parameters via equation 4.18; ΔS_{mi} is the critical concentration obtained from morphological stability theory.

the planar matching condition. Now the imposed gradient controls the liquid temperature behind the tip of the wedge and $G_{eff} = G_{cell}$ is a reasonable control condition. It seems, of course, likely that nonlinear transient effects like in the simulations by Coriell et al. (1994) will have to be accounted for detailed predictions. Considering equation 7.31, one may expect a modification of the term $2G_bK_b/(L_vV)$, as a function of the growth velocity V, the pulling velocity V_p, and the imposed temperature gradient G_{cell}. For example, for the first perturbation the time scale $1/\Sigma$ for the perturbation development was ≈ 7 s, and may be associated with a length scale $V_p/\Sigma \approx 35\mu m$ and a concentration change of ≈ 3 ‰ near the interface, as an indication of the magnitude of transient effects. In conclusion, although a numerical study seems necessary to interpret this type of growth cell experiments strictly in terms of stability theory, it appears that equation 7.31 predicts the onset of instabilities quiet well. This will be further discussed in the next paragraph in connection with the wave lengths of the initial perturbations.

Forced convection

The dataset has been obtained by Wintermantel (1972) in forced convection experiments, whose results were discussed in sections 4.4.2 and 4.6.1 of chapter 4. In this laboratory setup the solute flux velocity V_*, obtained from an empirically validated equation 4.22, is related via equation 4.18 to $(C_{int} - C_\infty)$. The equations allow the calculation of the concentration difference at the planar cellular transition velocity observed in the experiments. The results are listed in table 7.1, where $(C_{int} - C_\infty)$ is compared to the value predicted by morphological stability theory for the planar-cellular transition. The predicted salinity step, assuming $G_b = 0$ at the transition, is on average a factor 1.24

smaller than observed. If one assumes that the strong stirring may create an interfacial temperature gradient in the liquid similar to the solid, then the observations would be completely consistent with such modified predictions. The study from Wintermantel (1972) therefore strongly supports that equation 7.31 is fundamental to describe the planar-cellular transition.

7.2.3   Wave lengths of initial instabilities

From the review of the studies discussed so far it can be concluded that the *onset of cellular perturbations* can reasonably be predicted from linear morphological stability theory. Concerning the wave lengths at the onset of instabilities, comparably little work has been done. For a discussion of the limited results the following approximations are useful.

Approximate analysis

To study the maxima of equation 7.11, which represent the most rapidly growing wave length, it is useful to make the following approximate analysis (Sekerka, 1965; Coriell et al., 1985). When instability occurs from the transient phase and the wave length of initial perturbations is expected to be given by equation 7.23 and one may rearrange equation 7.24 as

$$\lambda_{mi} = \lambda_{max} = \frac{D_s}{V} \frac{4\pi}{(r^4 - 1)^{1/2}}, \tag{7.32}$$

withe r being given by the cubic equation 7.24. The limiting cases which can be considered are

1. *Near absolute stability.* In this case, setting $\mathcal{A} \simeq 1$ in equation 7.24, one has $r \simeq 1 + (1 - \mathcal{A}^{1/2})k/(k+1)$ and

$$\lambda_{mi} = 2\pi \frac{D_s}{V} \left(\frac{k+1}{k(1 - \mathcal{A}^{1/2})} \right)^{1/2} \tag{7.33}$$

2. *Small $\mathcal{A}$ with $\mathcal{A}^{1/6} \ll 1$.* This implies $r \gg 1$, $r \simeq (4k^2/\mathcal{A})^{1/6}$ and the stability criterion 7.23 takes the approximate form

$$\mathcal{G} = 1 - \left(\frac{27k\mathcal{A}}{4} \right)^{1/3}, \tag{7.34}$$

while the critical wave length is

$$\lambda_{mi} = 2\pi \frac{D_s}{V} \left(\frac{2\mathcal{A}}{k^2} \right)^{1/3}, \tag{7.35}$$

which under diffusive equilibrium becomes

$$\lambda_{mi} = 2\pi 2^{1/3} \left(\frac{D_s}{V}\right)^{2/3} \left(\frac{\Gamma}{mC_\infty(k-1)}\right)^{1/3}. \tag{7.36}$$

3. *Large* $\omega D_s/V \gg 1$. This case does not imply a restriction to the onset of stability but to small wave lengths compared to D_s/V, which allows a simplification of the nominator in the dispersion relation 7.11 and gives λ_Γ. With $\lambda_{max} = 3^{1/2}\lambda_\Gamma$ one finds (Coriell et al., 1985) that

$$\lambda_{max} = 2\pi 3^{1/2} \left(\frac{D_s}{V}\right)^{1/2} \left(\frac{\Gamma k}{mC_\infty(k-1)(1-\mathcal{G})}\right)^{1/2}, \tag{7.37}$$

which for the transient case, replacing C_∞/k by C_{int} and $\mathcal{G}$ by $\mathcal{G}' = \mathcal{G}C_\infty/(C_{int}k)$, can be written as

$$\lambda_{max} = 2\pi 3^{1/2} \left(\frac{D_s}{V}\right)^{1/2} \left(\frac{\Gamma}{mC_{int}(k-1)(1-\mathcal{G}')}\right)^{1/2}. \tag{7.38}$$

The latter approximation was given and discussed by (Coriell et al., 1985). It underestimates the most rapidly growing wave length slightly. The margin of stability λ_Γ is, above threshold, given by a $3^{1/2}$ smaller wave length. Note that in the first two approximations the onset of instabilities implies $\lambda_{mi} = \lambda_{max}$, while in the latter 7.38 these two wave lengths are different

The experiments to be discussed in the following correspond to case (2.), with equation 7.36 of principle interest. It must be noted, that equation 7.23, on which these approximations are based, is valid at the onset of instability, when the critical C_{int} is approached from below, and for the diffusive equilibrium (Sekerka, 1965; Coriell et al., 1985). When $\mathcal{G}' \ll 1$, and instability sets in from the transient stage at much larger C_{int} than critical, one must use the approximation 7.38.

Observations

The first quantitative evaluation of wave lengths at the onset of instability are the already noted experiments with ice cylinders growing into pure supercooled water (Hardy and Coriell, 1969; Coriell and Hardy, 1969). In these studies the observed wave length corresponded to the wave length λ_{max} of most rapidly growing perturbations. The situation is somewhat different from directional solidification, because the supercooling ΔT_{sup} in the liquid is well-defined and needs not to be computed from a solute diffusion model and constitutional supercooling.

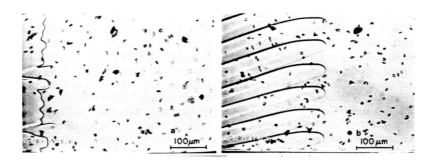

Fig. 7.8: Interface instability evolving during freezing of an aqueous 9 ‰ NaCl suspension of red blood cells freezing at $V = 7.8 \times 10^{-4}$ cm s^{-1} and $G_b \approx 93$ K cm^{-1}. On the left the onset of perturbations is seen while the right picture represents a later stage when spacings were stable. Adapted from Rubinsky and Ikeda (1985).

Some limited graphical documentations during unidirectional solidification of saline solutions can be found in cryomicroscopic studies (Körber and Scheiwe, 1983; Körber et al., 1983; Rubinsky and Ikeda, 1985). Koerber (1988) has reported that the initial perturbations during the freezing of NaMnO$_4$ solutions were comparable to the wave length λ_{max} of maximum growth rate. Although a detailed analysis has not been presented, the cryomicroscopic experiments clearly demonstrated a general feature of perturbation growth: the initial wave length was not stable yet normally increased rapidly by overgrowth of initial perturbations, while an array of deep cells was established. In freezing experiments with either 21.6 ‰ NaMnO$_4$ or 9 ‰ NaCl aqueous solutions the final stable cell spacings were typically a factor of 3 to 4 larger than the initial perturbations (Körber and Scheiwe, 1983; Rubinsky and Ikeda, 1985). An example of the evolution of such an instability is shown in figure 7.8 from Rubinsky and Ikeda (1985). The initial perturbations, seen in the in the upper left figure, are 23 to 25μm, and an increase to 30 to 35μm is already apparent for the few evolving deeper cells. The predicted wave length of maximum growth rate for the conditions noted in the figure, combining equations 7.31, 7.29 and 7.11, is 22μm and in good agreement with observations. It is noted that the approximation 7.38 yields an almost exact λ_{max} only 0.7 % larger than the exact analysis, while equation 7.36 is expected to be inaccurate here: indeed, it gives a considerably larger $\lambda_{max} \approx 56\mu$m.

Growth cell studies by Nagashima and Furukawa

The only systematic experimental study of perturbations at the planar-cellular transition to date was performed by Nagashima and Furukawa (2000b) in the thin growth cell noted above, during freezing of aqueous KCl solutions with an initial concentration of 30 ‰. These authors compared their results to linear stability theory and their relevant observations are reproduced as figure 7.9. To proceed with the above discussion it is important, that the instability documented by them is the second one, the x-axis instability in the growth plane in figure 7.7. It is also noted, that the theoretical equilibrium relation of λ_{max}, calculated by Nagashima and Furukawa (2000b) from equation 7.11 and shown as a solid line, is incorrect, because the used $k = 0.003$ would give an unphysical interface concentration of $C_\infty/k = 10000$ ‰. However, Nagashima and Furukawa (2000b) also presented an analysis of the transient problem, which deserves further discussion. They determined a relation between the observed interfacial salinity gradient G_c and the growth velocity which they inserted it into the dispersion relation 7.11. The wave length λ_{max} of most rapidly growing perturbations was than obtained by assuming $G_{eff} = G_{cell}$, the externally impose gradient being the relevant interface control parameter. This solution is shown as the dashed line in figure 7.9 and agrees reasonably with observations. Although this approach is not predictive, as one needs to observe G_c, it indicates the predictability of λ_{max} from the transient.

To evaluate the data from Nagashima and Furukawa (2000b) the same procedure as noted above was applied, combining equations 7.31, 7.29 and 7.11. Also here two computations were performed. In a first approach, the interfacial gradient in the liquid was assumed as $G_b = G_{cell} = 23$ K cm^{-1} and equation 7.31 was applied, while in the second $G_{eff} = G_{cell} = 23$ K cm^{-1} was assumed. As discussed above, these conditions may be viewed as a control of instabilities by the growth velocity and by the imposed temperature gradient. The instability in question is the second instability in figure 7.9, for which, as outlined above, the interface concentration was in much better agreement with the assumption $G_{eff} = G_{cell}$. Interestingly both wave length predictions, shown as dotted and dash-dotted curves in figure 7.10, are very similar for the growth rate regime in consideration. Only at low growth rates, where the curves have been drawn until the interface concentration becomes eutectic (≈ 196 ‰ for KCl), the difference increases. This is explained by the partial cancellation of the dependencies of λ_{max} on G_{eff} and C_{int}. As the scaling 7.36 indicates, the dependence $\lambda_{max} \sim C_{int}^{1/3}$ is relatively week and only becomes significant at low grow velocities when C_{int} is very different for the two assumptions. In comparison to the observations the wave lengths λ_{max} predicted in this way are 10 to 30 % too low.

Next, as applied by Nagashima and Furukawa (2000b), the observed relation between G_c and V has been inserted into the dispersion relation and the computations have been repeated for the cases $G_b = G_{cell}$ and $G_{eff} = G_{cell}$. For the former case (bold solid curve) this does change very little in the wave length prediction, while for the latter case the agreement appears to be better (bold dashed curve). It is unknown why the calculated wave lengths from Nagashima and Furukawa (2000b), their dashed line in figure 7.9, differ from the present calculations for $G_{eff} = G_{cell}$. Although slightly temperature-dependent thermal properties were used in the present calculations, this cannot account for the difference. Focusing on the present calculations two aspects are important. First, when using a kind of uncoupled and separately observed G_c and G_{eff}, the wavelength changes much stronger upon a small change in G_c, than in a coupled simulation. And second, as this second instability was observed from a wedge-shaped and non-planar interface, as seen in the third frame from the top in figure 7.9, the difference between the selfstanding simulation (dash-dotted curve) and the calculation from the observed G_c is plausible: the former assumes an interfacial gradient $C_{int}(1 - k)V/D_s$, while the actual gradient is smaller, because the interface is no longer planar at the onset of secondary instabilities.

Considering the predictability of the initial perturbations one may say that the self-standing approach would be close to reality, if the concentration gradients would be reduced by 20 to 30%. Several causes may account for the systematic disagreement between predictions and observations. The transient disequilibrium character due to the increasing growth velocity has already been noted, and simulations in the lines of Coriell et al. (1994) are necessary to find out, if such a case may act destabilising, with convective onset at smaller concentrations. The approach to use 7.30 from the transient and with $\mathcal{G} \ll 1$ should also be tested numerically. Eventually the analysis of the experiments from Nagashima and Furukawa (2000b) requires a threedimensional refinement because, as pointed out by the authors and investigated in a later paper (Nagashima and Furukawa, 2000a), also in the thin cell of 100μm thickness gravity effects might be present. Considering the accuracy of observations, it has to be ascertained that the very first perturbations are measured, because already at the early stage a coarsening of initial cell spacings takes place (see figure 7.8 and the discussion in section 7.3 below). Finally, one needs to recall that the quantified instability described here is the secondary instability of the edge of the basal plane. It may therefore be related to a different Γ due to the anisotropy in the ice-water interfacial free energy. A 50 % larger Γ can explain the observed λ from Nagashima and Furukawa (2000b) and would also produce the correct velocity dependence. The correspondingly larger required γ_{sl} is fully consistent with the review of this property in the appendix A.2.10.

Conclusion: Initial morphological instability

While the noted aspects still need to be verified by additional studies, the above discussion of thin growth cell observations reported by Rubinsky and Ikeda (1985) and by Nagashima and Furukawa (2000b) has demonstrated that the wave length of first perturbations on a planar interface, as well as the corresponding interface salinities, can be reasonably predicted. To do so one apparently may use the steady state morphological stability dispersion relation equation 7.11 together with equation 7.30 to predict the onset of cellular instabilities from the transient evolution of the solute fields, assuming the quasi-steady temperature field approximation. The prediction applies to conditions where the transient evolution is by diffusion only.

It has conjectured that, in a situation of increasing growth velocities, like the experiments by Nagashima and Furukawa (2000b), the steady-state temperature assumption is questionable if the interface velocity is far from equilibrium. Some simplistic conclusions were drawn. (i) In the first case of a strictly planar interface, the onset of instability appears to coincide with the assumption, that the externally imposed temperature gradient G_{cell} only appears as a boundary condition on the liquid side of the interface, $G_b = G_{cell}$. (ii) In the second case, when instabilities on a wedge-like interface are observed, it has been suggested that the effective interface gradient it completely controlled by the external forcing, $G_{eff} = G_{cell}$. This may explain the occurrence of two kinds of instabilities in the experiments from Nagashima and Furukawa. The simplistic prediction overestimated $(C_{int}(t) - C_{\infty})$ by eventually 10 to 30 % in the first case and underestimated it by 20 % in the second wegde-shaped case. It must be noted that the applicability of relation 7.30 implicitly assumes solute fields being exponential, like in the steady state derivation. If the differences can be explained by a modified transient solute field, by threedimensional pattern due to the wedge and/or by observational uncertainty, requires further studies.

As the theoretical predictions match the observations quite well, one may look for indications of stability length scales in ice formed from fresh water or dilute solutions in the natural environment. Here again the descriptions of Quincke are of interest (Quincke, 1905b,a), who reported the distance of Tyndall figures observed in warming and internally melting ice, implying a cellular appearance which he called *foam cells*. It is interesting that, associating Quincke's slow freezing experiments in free air with a freezing velocity of 10^{-5} cm s^{-1}, one obtains $\lambda_{max} \approx 4.8$ mm, if the critical $C_{\infty} \approx 1.33$ ‰ NaCl is inserted in equation 7.36, while Quincke described most foam cells of size in the range 5 to 10 mm. For different kinds of very rapid freezing experiments Quincke reported cells 0.1 to 0.3 mm in diameter. Freezing velocities for the rapid growth have not been documented yet, considering the experimental descriptions, likely in the 10^{-2} to 10^{-3} cm s^{-1} range, for which stability theory gives $0.05 < \lambda_{max} \approx 0.22$ mm at the onset of instability. These considerations indicate that, when freezing takes place in the absence of haline convection, morphological stability may also control major solute inclusions in ice formed from fresh water.

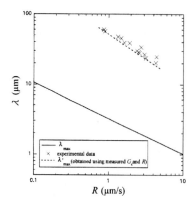

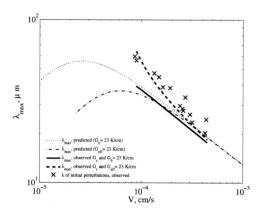

Fig. 7.9: Adapted from Nagashima and Furukawa (2000b). Observed instability wave length at the planar-cellular transition during freezing of aqueous KCl solutions with $G_b = 23$ K cm^{-1} and $C_\infty = 30$ ‰. The dashed line is based on the observed temperature gradient. See the text for a further discussion. The solid line is erroneous due to the linear stability prediction using the k from equilbrium diffusion. See the text for a further discussion.

Fig. 7.10: Predictions of the wave lengths at onset of instability assuming $G_b = 23$ K cm^{-1} (dotted curve) and the extreme case $G_b = G_i$ (dash-dotted curve). Also shown are, as bold solid line, the prediction based on the observed interfacial salinity gradient G_c from Nagashima and Furukawa (2000b) and the range of observations corresponding to figure 7.9 as dashed lines.

7.3   Summary and Outlook: Predictability of stable cell spacings

The present section has summarised the essentials of morphological stability theory for a dilute binary alloy from Mullins and Sekerka (1964) and discussed its consequences for the freezing of dilute NaCl solutions quantitatively. Using molecular properties for NaCl and water it was concluded that, in the absence of a considerable thermal gradient in the liquid, cellular freezing is initiated when the interface concentration exceeds $\approx$ 1.33 ‰ NaCl. This implies that, assuming $k \approx 0.001$ at this critical concentration and diffusive equilibrium, cellular freezing requires $C_\infty > 0.0013$ ‰ NaCl. Interestingly, observations which are consistent with this transition have been traced back to some early work by Quincke (1905b), although a clear experimental proof is still lacking.

Observations which confirm morphological stability theory for water and aqueous solutions have been reviewed. It was emphasised that in most laboratory situations, and for freezing of any natural water with solute transport controlled by diffusion, one has to describe the onset of instability from the transient. However, this appears to be no serious problem, as both the onset of instability and the wave length of initial perturbations can be predicted by only slight modifications of the classical Mullins-Sekerka analysis. Of particular usefulness for an accurate quantitative validation were freezing experiments performed with thin layers of aqueous solutions under the cryomicroscope (Körber and Scheiwe, 1983; Rubinsky and Ikeda, 1985; Koerber, 1988) and in a comparably thin growth cell (Nagashima and Furukawa, 2000b,a). Initial perturbations that appear on a planar interface may be associated with the wave length of most rapidly growing perturbations from linear stability theory.

7.3.1 Spacing versus amplitude

Returning to the goal of the present study, the predictability of cell spacings in sea ice, the observations discussed so far clearly show, that the linear morphological stability theory only predicts the wave length of initial perturbations, while the cell spacings that evolve during further growth are considerably larger. An example for the cryomicroscopic freezing is shown in figure 7.8 and similar graphical presentations are available in other reports of this type of solidification (Körber and Scheiwe, 1983; Koerber, 1988). The spacing increase with amplitude is not specific for aqueous saline solutions, but has been observed for many binary alloys (Trivedi and Somboonsuk, 1985; Coriell et al., 1985; Billi and Trivedi, 1993; Liu and Kirkaldy, 1995). In the reported cryomicroscopic experiments the ratio of initial to stable cell spacings, as seen in figure 7.8, was frequently in the range 3 to 4. However, Rubinsky and Ikeda (1985) reported that it appeared to depend on small uncontrolled differences in the freezing conditions like the thickness of the film and the orientation of the crystal. For the limited observations presented it ranged between 1 and 4 for identical forcing conditions. Coarsening ratios up to 20 have been reported in the above references for other binary solutions. In figure 7.11 the predicted initial λ_{max} for a freezing NaCl solution is shown for the case of negligible temperature gradient in the liquid, along with the sea ice and NaCl ice plate spacings compiled in figure 6.8 of the previous chapter. It is seen that the stable spacings are approximately larger by a factor 10.

Another example micrograph from the cryomicroscope study by Rubinsky and Ikeda (1985) is shown as 7.12. It illustrates that the coarsening of the wave length cell takes place very rapidly. In this figure the smallest amplitude wave lengths take a rather constant value of $\approx 19 \mu$m while the few deeper amplitude cells are spaced by 22μm, slightly but significantly larger. For comparison one may recall figure 7.8, where also in the initial perturbation field some even deeper amplitude cells are visible with a 50 % larger wave length than the small-amplitude perturbations. An quantitative documentation of the cell spacing coarsening for a pivalic acid-ethanol binary has been given by Liu and

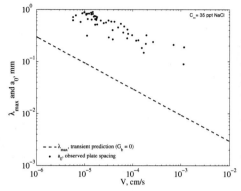

Fig. 7.11: Most rapidly growing wave length λ_{max} on planar interface obtained from thermal steady state linear stability theory. Calculations (dashed line) refer to for aqueous NaCl solutions with $C_\infty = 35$ ‰ and no thermal gradient in the liquid. The sea ice and NaCl ice plate spacings compiled in figure 6.8 are shown as dots.

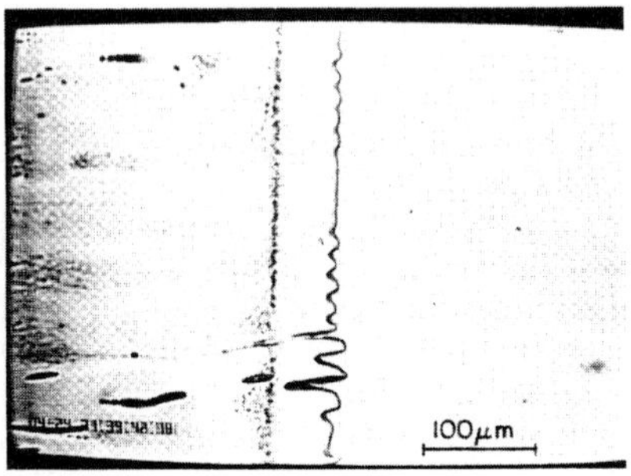

Fig. 7.12: Initial interface instability during freezing of an aqueous 9 ‰ NaCl suspension with $V = 7.8 \times 10^{-4}$ cm s^{-1} and $G_b \approx 155$ K cm^{-1}. The wave lengths of smallest perturbations is ≈ 19 μm and in good agreement with the prediction of 21 μm. Adapted from Rubinsky and Ikeda (1985).

Kirkaldy (1995). From their article figures 7.13 and 7.14 are reproduced. The former shows, how the increasing cell spacing after the onset of instability is aquired by overgrowth of initial cells. The latter, corresponding to a different freezing run, presents the evolution of the average spacing versus the cell amplitude, until a quasi-stable spacing ≈ 20 times the intial values was reached.

It may finally be noted, that the coarsening of cell spacings with increasing cell amplitude, and thus an *under prediction* of the stable wave length by planar stability theory, should not be confused with the *over prediction* of the initial perturbation wave length pointed out by DeCheveigne et al. (1988). The over prediction in the latter study primarily results from the fact, that in most cases considered by the latter authors equation 7.36 is inappropriate. Another frequently mistaken 'theoretical' estimate is due to the assumption of diffusive equilibrium, as pointed out already in connection with $k = 0.003$ applied by Nagashima and Furukawa (2000b) in figure 7.9. If instability sets in from the transient the use of a small equilibrium k implies much too high interface concentrations and, from equation 7.38, an underestimate of the initial perturbation wave length. This likely explains the mismatch between the observed and calculated initial λ_{max} in the frequently cited work by Trivedi and Somboonsuk (1985). Clearly, linear stability theory appears to be much more closer to reality than noted in these early investigations, if one considers the instability onset from the transient.

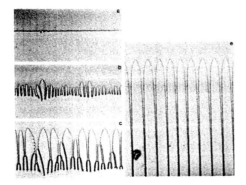

Fig. 7.13: Evolution of a freezing interface of pivalic acid-ethanol in a thin growth cell showing the transition from the initial instability to deep cells: a) 0 s, b) 30 s, c) 60 s, e) 240 s. Adapted from Liu and Kirkaldy (1995).

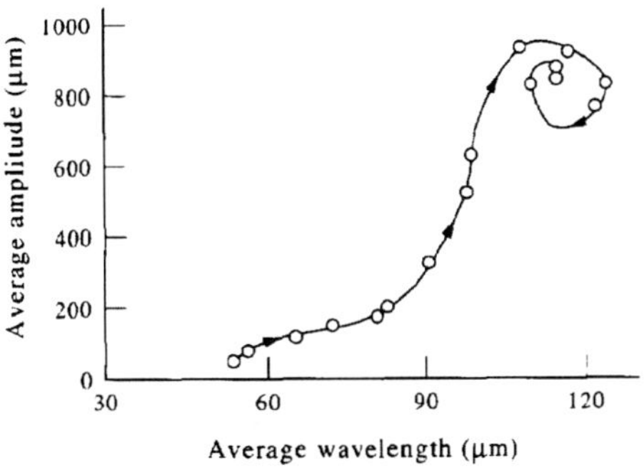

Fig. 7.14: Evolution of the cell spacing of pivalic acid-ethanol in a thin growth cell from initial perturbations to deep cells in the cell spacing-amplitude plane. Adapted from Liu and Kirkaldy (1995).

7.3.2 Theoretical work

Cellular structures

While already Mullins and Sekerka (1964) warned that a detailed agreement of the relationship between cellular spacings and growth velocity may not necessarily be found because the theory was derived for infinitesimal perturbations, it took some time until accurate experimental data became available. In early experiments the conclusion was drawn that a prediction of observed wave lengths required unreasonably large solid-liquid interface energies (Sharp and Hellawell, 1970) and only later the regular difference between initial perturbations and final cell spacings was clearly established experimentally (Somboonsuk et al., 1984; DeCheveigne et al., 1988), with systematic studies leading to more detailed descriptions as in figures 7.13 and 7.14. The first fundamental theoretical work on the transition to and stability of deeper cells had been done by Wollkind and Segel (1970), who extended the linear stability theory to a nonlinear analysis. These authors concluded that only for a certain set of parameters steady cells are possible. In terms of the notation of the present chapter, this range requires a nondimensional wave number $\omega D_s/V$ of order one or less, and thus corresponds to rather long wave lengths, normally not accessible with solid-liquid surface energies and typical freezing conditions of saline solutions. In an early review article, Langer (1980b) confirmed Wollkind and Segels's findings. Alternatively, he found that in a situation with solute diffusion in the solid, the so called symmetric model, at least small amplitude cells can be stable, at a wave length close to the the classical linear instability model, if growth conditions are not far from the threshold $\mathcal{G} \approx 1$. Similar results were obtained by Kerszberg (1983a) for the one-sided solute diffusion (the classical Mullins-Sekerka problem), using a different numerical technique and pointing out the importance of higher order terms in the nonlinear analysis. The question, how the coarsening of cell spacings may take place dynamically was also addressed by Kerszberg (1983b) in a nonlinear simulation. Kerszberg indeed found that close to the threshold the cell spacing rapidly drifted to longer wave lengths and attained a stable value which differed the more from λ_{max} the lower $\mathcal{G}$.

Much effort has been made in later theoretical work on *cellular shapes* to clarify that (i) a band of stable cell spacings is possible and that (ii) this band is restricted to conditions very close to the threshold (Dee and Mathur, 1983; Kerszberg, 1983b; Dombre and Hakim, 1987; Ramprasad et al., 1988; Bennett and Brown, 1989). It was further shown, that also at the planar-cellular transition a nonlinear analysis refines the wave length to a more narrow range than shown in 7.1, and that a conductivity ratio different from one plays an important role when considering the nature of the cellular transition revealed by a nonlinear analysis (Merchant and Davis, 1989). Brattkus and Misbah (1990) found that the predictions may depend on the type of the nonlinear analysis, with experimental validations of the details of the transitions still outstanding. A more recent result could be important in connection with the the freezing of saline solutions. Kopczynski et al. (1996) showed that a relatively small amount of anisotropy can extend the stable band of cell spacings to growth velocities far from the threshold. Some first observational studies on the subject (Nagashima and Furukawa, 1997b; Akamatsu and Faivre, 1998) have been performed. The work from Nagashima and Furukawa on aqueous saline solutions indicated the additional role of the anisotropy of growth kinetics at high growth velocities.

The work noted so far was concerned with the question, if stable cell spacing may exist at all, and which theory may be used to describe them. It seems, however, still not clear, if these regimes are accessible from the onset of stability and how the dynamical evolution in figures 7.13 and 7.14 is explained theoretically. To be summarised in the following chapter, section 8.3.4, the performance of numerical modeling in connection with cell formation is apparently not that good as for dendritic arrays (Hunt and Lu, 1996; Feng et al., 1999). Observations that relate cellular shapes to wave lengths and growth conditions are still sparse but systematics are improving (Georgelin and Pocheau, 1997, 2004).

Cells versus dendrites: natural waters

Cells are normally associated with a low liquid fraction. In the two-dimensional Hele-Shaw limit, to which much of the above work on cellular cells refers, the strict limit is

$v_b = 0.5$ (Bensimon et al., 1986; Kessler et al., 1988). At larger liquid fractions one will normally find dendritic growth. It is useful to illustrate the regimes during natural growth of water by considering the solid fraction $(1 - v_b)$ one wave length or cell spacing a_0 from the freezing interface. When equilibrium solidification is considered the latter is approximately given by $v_b \approx a_0 dT/dz/T_{int}$, where T_{int} is the temperature at the interface where $v_b = 1$. Taking typical values $a_0 \approx 0.05$ cm and $dT/dz \approx 1$ K cm^{-1} for rapid natural freezing, one obtains this solid fraction as ≈ 0.025 for seawater with $T_{int} \approx -2\,°C$. For lake water one must, according to linear stability, at least have $T_{int} < -0.08\,°C$ for an unstable interface[4], and finds solid fractions of order unity. For slower freezing a_0 increases and the classification may still be roughly fulfilled.

The comparison illustrates that the interfacial selection during freezing of seawater should in principle correspond to an array of widely separated dendrites, while lake water freezing likely requires a framework of cellular pattern formation that may be related to the theory of Saffman-Taylor fingering. Deep cells will, however, due to solute conservation principles, not exist in ice formed from lake water even if frozen upwards. The strong anisotropy in surface tension and growth kinetics of ice will likely play an important role for the lake ice system in connection with bubble and impurity entrapment. If it makes a description easier or more complicated, remains still to be shown, in particular when considering that both c-axis and a-axis growth exist in natural lake water, e.g., Shumskii (1955). In seawater the growth rate anisotropy must lead to the plate-like structure, likely close to the interface, but the problem is still 'dendrite-like' in a sense of plate interaction.

Dendritic array evolution

An approach to the problem of dynamical evolution of a dendritic array requires the understanding of how a single dendrite will behave. This is not a simple task, yet the theory that has emerged during the past 50 years will be summarised in the following chapter. Geometric dendritic models may then be matched with the solute and heat balance, similar to the very simple example given for seawater in the preceding paragraph, to find solutions for the spacings. Early approaches (Hunt, 1979; Trivedi, 1984; Kurz and Fisher, 1984) to the problem consider a steady state and will also be summarised in the following chapter. It may be noted in advance that the first basic study on the dynamical evolution of the spacing from its initial Mulllins-Sekerka instability to a dendritic array, was performed by Warren and Langer (1990) on the basis of an improved understanding of individual cells or dendrites. The method has been refined in a later study (Warren and Langer, 1993), being capable to predict the final stable cell spacings in the experiments from Somboonsuk et al. (1984) to within 50%. The reasonable predictabil-

---

[4] This value relates to NaCl as the solute depressing the freezing point.

ity was restricted to large growth velocities. Warren and Langer (1993) noted that this deficit might be attributed to the simple basic concepts of their model which are: (i) the dendrites are not interacting, (ii) dendrite coarsening is modeled in simplified manner by an overgrowth mechanism and (iii) shapes differing from parabolic dendrites are not allowed. Assumption (i) requires the condition $a_0 V/D_s > 1$, which was not fulfilled at low velocities.

Warren and Langer also pointed out that the necessary condition for the predictability of dendrite spacings seems to be the longwave condition $(a_0 V/D_s > 1)$, as otherwise the problem cannot be treated via a model of noninteracting dendrites. For seawater, taking $a_0 \approx 0.05$ cm and $D_s \approx 7 \times 10^{-6}$ cm^2 s^{-1}, this implies growth velocities above 1.4×10^{-4} cm s^{-1}, which is out of the natural growth regime. It is therefore likely that a modified solution to the problem must be sought for cells with $(a_0 V/D_s < 1)$, even if the coarsening appears to be conceptually similar, as seen in the figures 7.13 and 7.13 from Liu and Kirkaldy (1995). Such an approach will be presented in chapter 9.

8. Cellular and dendritic pattern selection in crystal growth

As noted in the previous chapter it is important to consider the interaction between cells to predict the evolution of their spacing. The problem may be introduced by the following aspect of the freezing interface. When a planar interface transforms to cells and dendrites the solution is incorporated interstitially into the evolving solid matrix. If the temperature gradient is imposed in the solid and no supercooling is allowed, the liquid fraction v_b is determined by solute conservation. Assuming a linear liquidus slope m, it is given via the Scheil equation

$$v_b(z) = \left(\frac{S_b(z)}{S_{b0}} \right)^{1/(1-k)} \approx \left(\frac{T_b(z)}{T_{b0}} \right)^{1/(1-k)} \tag{8.1}$$

where T_{b0} and S_{b0} are the melting temperature and corresponding composition at the interface, where $v_b = 1$. $T_b(z)$ and $S_b(z)$ are brine temperature and brine concentration in the ice. For the very small k of saline solutions the exponent can be set to unity. This classical condition after Scheil does not tell anything about the structure of the cellular array normal to the growth direction. It does neither fix the vertical shape of a cell, yet $v_b(T_b(z), T_{b0})$ remains to be solved. If the temperature distribution $T_b(z)$ is not imposed, itself will depend, eventually in some nonlinear manner, on the liquid fraction v_b, and the thermal properties. Some simplified similarity solutions that can be obtained, are discussed in appendix B.

The Scheil equation is a macroscopic continuum equation and ignores all internal microscopic length scales. The problem becomes more complex, when deviations from the Scheil shape $v_b(T_b)$, where T_b only depends on the vertical coordinate z, are allowed, due to horizontal variations in diffusivity, brine concentration or even non-equilibrium interstitial supercooling. These local deviations are largest close to the interface, where brine volume fractions are large, shapes are curved and diffusion of heat and salt is possible in any direction. One is again faced with the problem of dynamical interfacial pattern formation. The main problem in predicting the evolution of deep cells is to match these *local* conditions with the *global* Scheil shape near the interface. Therefore it is essential to understand the physics that determine the shape of a single dendrite growing into a supercooled or constitutionally supercooled solution[1].

The mentioned principle considerations have long been realised and defined two main problems of understanding the process of cellular spacing selection: (i) the shape and velocity of a single isolated dendrite growing freely into a supercooled melt and (ii) the modification of free growth due to the interaction of an array of dendrites. During the past 50 years much work has been done on problem (i) which may be in principle considered as solved. Its history has been reviewed by many authors (Langer, 1980b, 1987, 1989; Glicksman and Marsh, 1993; Trivedi and Kurz, 1994; Caroli and Müller-Krumbhaar, 1995; Boettinger et al., 2000; Glicksman and Lupulescu, 2004). For problem (ii) no concise generally accepted solution exists. In addition to the already noted length scales, the most rapidly growing perturbation λ_{max} and the stable cell spacing a_0, (often noted as the primary spacing λ_1), two other scales are relevant when an array of cells is considered. These are the radius r_{tip} describing the curvature at the tips of the array dendrites and eventually given by the isolated dendrite problem, and the secondary spacing λ_2, a lateral instability along the dendrite stem, which principally requires that the supercooling persists between the dendrites. To understand the physical processes leading to these length scales, it is useful to consider first the case (i) of a dendrite growing freely into a supercooled melt. A more detailed summary of the state-of-the-art may be found in a number of recent overviews (Glicksman and Marsh, 1993; Trivedi and Kurz, 1994; Caroli and Müller-Krumbhaar, 1995; Glicksman and Lupulescu, 2004).

8.1 Free dendritic growth

8.1.1 Ivantsov's transport equation

Ivantsov (1947) first presented a rigorous analysis of the thermal diffusion equation during growth of a dendrite. Assuming an isothermal dendrite, and neglecting the effect of capillarity on the freezing point, he presented an analytic shape-preserving solution for a paraboloid of circular cross section, growing with velocity V into a melt supercooled by

---

[1] In case of directional solidification of salt water, and a negligible temperature gradient in the liquid, freezing is driven by heat flow from a cold boundary and latent heat transported through the solid. A pure melt will under these conditions not develop cellular patterns, as any perturbation is stopped by the latent heat released at the interface. The morphological instability of a directionally solidifying impure melt is, as discussed in the previous chapter, therefore solely related to the principle of constitutional supercooling. Formulations for constitutionally supercooled and supercooled melts are similar but not the same (Mullins and Sekerka, 1963, 1964; Langer, 1980b). These differences also exist for the single dendritic problem (Lipton et al., 1984; Glicksman and Marsh, 1993). Formally the description of free dendritic growth into a pure supercooled melt is simpler and much better validated by experiment. The present summary is largely restricted to this pure case.

$\Delta T_{sup} = T_f - T_\infty$. It is usually presented in terms of a Stefan number Ste as

$$Ste = \frac{L_f}{c_{pw} \Delta T_{sup}}, \tag{8.2}$$

L_f and c_p being latent heat of solidification and specific heat of the liquid (here: water), and a growth Peclet number Pe defined as

$$Pe = \frac{V r_{tip}}{2 \kappa_w}, \tag{8.3}$$

where κ_w is the liquid thermal diffusivity and r_{tip} is the dendritic tip radius. Ivantsov's solution takes the form

$$Ste^{-1} = Pe\, e^{Pe} Ei(Pe), \tag{8.4}$$

where the quantity

$$Ei(x) = \int\limits_{x}^{\infty} \frac{e^{-y}}{y} dy \tag{8.5}$$

is the first exponential integral of Pe. Both the dendrite and the isotherms are paraboloids of revolution. The problem was later generalised to paraboloids of arbitrary elliptic cross-section by Horvay and Cahn (1965). The two-dimensional limit with one infinite elliptic axis is the parabolic cylinder and its growth velocity described by

$$Ste^{-1} = (\pi Pe)^{1/2} e^{Pe} erfc(Pe^{1/2}), \tag{8.6}$$

with the complementary error function

$$erfc(x) = \frac{2}{\sqrt{\pi}} \int\limits_{x}^{\infty} e^{-y^2} dy. \tag{8.7}$$

The axisymmetric shapes corresponding to solutions 8.6 and 8.4 imply that the dendrite and its surface are isothermal at the equilibrium freezing temperature T_f of the pure melt, latent heat only being transported away from the interface by diffusion in the liquid. For a given Stefan number Ste (supercooling ΔT_{sup}, in dimensional units) these equations provide the Peclet number Pe and thus the growth velocity. They are plotted in figure 8.1 and bound the growth of all paraboloids with elliptic cross section. It is of interest for later considerations that equation 8.6 is also the solution to an isothermal solid *planar interface* growing into a supercooled liquid, if $H(t)$ replaces the tip radius r_{tip} in the Peclet number[2]. However, in the absence of capillarity such an isothermal planar interface would be instantaneously unstable.

---

[2] As an analogy, in the classical Neumann-Stefan problem with constant freezing temperature in the liquid (appendix B.2) the error function erf appears, the conventional growth rate η^2 in equation B.11 being equivalent to Pe.

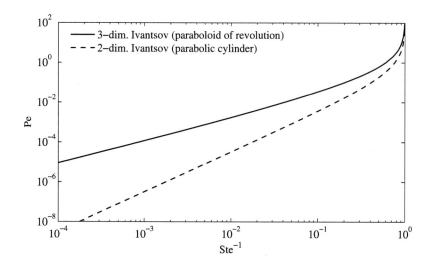

Fig. 8.1: Steady state solution to the diffusion equation of a paraboloid of revolution (Ivantsov, 1947) and a parabolic cylinder (Horvay and Cahn, 1965) growing into a supercooled melt.

Ivantsov's heat transport equations imply that

$$r_{tip} V = const. \tag{8.8}$$

for a given undercooling or Ste. Hence, only the product of V and r_{tip} and the shape of the dendrite are determined by this solution. As observations demonstrate a unique relation between growth velocity and supercooling Ste, there must be a mechanism that selects the tip radius.

8.1.2 Tip radius selection

The first attempt to solve the selection problem was based on the Gibbs-Thomson effect of freezing point depression due to curvature $\mathcal{K}$, as discussed in the previous chapter. The limiting nucleation condition due to curvature

$$\Delta T_\Gamma = \Gamma \mathcal{K} = \frac{\Gamma}{r_\Gamma}, \tag{8.9}$$

may be written as

$$\frac{r_\Gamma}{Ste} > 2l_\Gamma, \tag{8.10}$$

where

$$l_\Gamma = \frac{\Gamma c_{pw}}{L_f} \tag{8.11}$$

is the capillary length. Several approaches have been formulated to incorporate r_Γ into equation 8.4 and all of them yield, in contrast to the classical Ivantsov solution, a radius r_{tip} where the velocity is a maximum, which than has been suggested to be the operating condition of a dendrite. However, when the first simultaneous observations of r_{tip} and V appeared in 1976, this *maximum velocity* hypothesis was found to be incorrect (Langer, 1980b; Glicksman and Marsh, 1993).

Marginal stability conjecture

The first step towards a consistent solution was made by Oldfield (1973), who proposed, on the basis of numerical simulations, a stability argument in the lines of the Mullins-Sekerka stability theory. He suggested that, as a consequence of stabilizing surface tension and destabilising thermal gradients, a dynamically evolving dendrite will produce side branches at a certain distance from the tip, and discovered the limiting law

$$r_{tip}^2 V = const., \tag{8.12}$$

which he suggested as the necessary dynamical condition to select the tip radius in the Ivantsov equation 8.6 for r_{tip}. The problem has been analysed in detail by Langer and Müller-Krumbhaar (1978), who refined the following stability conjecture in terms of morphological stability theory. Identifying the second $\Gamma k/(mC_\infty(k-1))$ in equation 7.38 with l_Γ and setting $\mathcal{G} \approx 0$ (as in the thermal case one is always far from threshold), the marginal stability wave length from linear theory becomes

$$\lambda_\Gamma = 2\pi \left(\frac{2\kappa l_\Gamma}{V} \right)^{1/2}. \tag{8.13}$$

Langer and Müller-Krumbhaar (1978) proposed that r_{tip} will be proportional to λ_Γ and defined $\lambda_\Gamma/(2\pi r_{tip}) = \sigma^{*1/2}$, which leads to

$$r_{tip}^2 = \frac{1}{\sigma^*} \left(\frac{2\alpha l_\Gamma}{V} \right) \tag{8.14}$$

or in a different form using 8.3 to

$$r_{tip} = \frac{l_\Gamma}{\sigma^* Pe} = \left(\frac{2\kappa_w}{\sigma^* V} \right)^{1/2} \left(\frac{\Gamma c_{pw}}{L_f} \right)^{1/2} \tag{8.15}$$

and

$$Pe^2 \sigma^* = \left(\frac{V l_\Gamma}{2\kappa_w} \right) := V_\Gamma \tag{8.16}$$

Equation 8.16 may be used together with the Ivantsov transport equation 8.4 to solve the problem, if σ^* is known, and it has become convenient to present the growth velocity in the nondimensionalised form V_Γ. Langer and Müller-Krumbhaar (1978) obtained numerically $\sigma^* = 0.020 \pm 0.007$ and $\sigma^* = 0.025 \pm 0.007$ for two and three dimensions, respectively, while Oldfield (1973) had found $\sigma^* = 0.02$ in his computations. In a related publication it was shown by Langer et al. (1978) that this scaling law was in reasonable agreement with dendritic growth velocities for ice and succinonitrile. These observations for ice are, including additional work done since then, shown in figure 8.2 to be discussed below. Despite the good agreement it was noted by the authors (Langer et al., 1978; Langer, 1980b), that no reliable first-principles estimate of σ^* exists. Neither was at that time a theory available which could properly describe how an overshooting radius would adjust to it maximum structure without being destroyed. In particular, Langer (1980b) pointed out that the agreement for a highly anisotropic material like ice might be fortuitous.

Microscopic solvability

It took some years until an analytical solution to the problem was found and received its name *microscopic solvability theory* (Langer, 1989; Glicksman and Marsh, 1993; Trivedi and Kurz, 1994; Caroli and Müller-Krumbhaar, 1995). Upon further refinement its main essential was the need of anisotropy in the surface tension to define the selection of the tip radius. In the case of low anisotropy of a few percent an equation of the form

$$\sigma^* = \sigma_0 \alpha_\epsilon^{7/4} \tag{8.17}$$

was found, where $\alpha_\epsilon = (m^2 - 1)\epsilon_m$ is an anisotropy parameter for m-fold anisotropy ϵ_γ of the surface tension, and σ_0 is a constant found to be of order one for both two and three dimensions. The form 8.17 is only approached below $\alpha_\epsilon \approx 0.01$ and becomes more linear at larger anisotropy (Barbieri and Langer, 1989). A comparison of several data sources in principle supports the theory (Muschol et al., 1992; Glicksman and Marsh, 1993). Discrepancies, which remained in the latter comparisons of data and theory for the likely best studied material succinonitrile, have been explained by more recent detailed numerical simulations, which accounted for a deviation from the idealised parabolic tip radius Karma and Rappel (1998). It has also been learned from high quality data of succinonitrile, obtained in space experiments, that, under earth's gravity, natural convection affects the results considerably, being responsible for the differences between theory and observations (Glicksman and Koss, 1994). Several models have been suggested to account for convection and the problem is at present not completely solved (Tennenhouse et al., 1997). A detailed theoretical treatment has recently been published (Xu, 2004).

A further treatment of the selection problem which considers boundary conditions at the root of cells and dendrites being relevant is the 'Interfacial wave theory' formulated by Xu. It has been described as an extension of 'Microscopic solvability' to non-steady oscillating needle solutions in the limit of zero or small surface tension (Xu, 1997; Xu and Yu, 2001). At the time of its formulation it was not generally accepted (Glicksman and Marsh, 1993; Caroli and Müller-Krumbhaar, 1995) and is still seldom cited with a recent discussion given by Saarloos (2003). *Interfacial wave theory* predicts σ^* for succinonitrile reasonably, yet a stringent test of its different behaviour at very low surface tension anisotropy seems to be outstanding, although the available data eventually support its validity: *Interfacial wave theory* predicts $\sigma^* \approx 0.0246$ at very low surface tension anisotropy (Xu and Yu, 2001) and seems in better agreement with observations than *microscopic solvability* (Muschol et al., 1992; Glicksman and Marsh, 1993).

A key question, concerning steady needle solutions and the noted theories, appears to be the relation between side-branching and the dendritic tip shape. Recently it has been suggested that sidebranching conditions quite far from the tip may considerably influence its shape and Peclet number (LaCombe et al., 1999). The detailed effects due to convective influence or a shape change have not been fully separated for the high quality succinonitrile data, but progress is ongoing with numerical phase-field simulations, e.g., Karma et al. (2000); Sekerka (2004b). Although some details are still debatable *microscopic solvability theory* is at present the self-consistent widely accepted theory of dendritic growth selection. The original marginal stability conjecture from Langer and Müller-Krumbhaar (1978) only gives the correct scaling law 8.14. While its agreement with the observed σ^* can be termed fortuitous (Langer, 1980b; Sekerka, 2004a), it may still be viewed as a fundamental finding in the absence of surface tension, pointing to the connection of planar interface stability and the pattern of a parabolic tip (Glicksman and Lupulescu, 2004). The general question appears not, if microscopic solvability theory is correct, but under what conditions a different mechanism can affect the solution for σ^*.

8.1.3 Thermosolutal dendrites

When dendrites grow into a saline solution the problem becomes complicated, because now constitutional supercooling also contributes to the stability characteristics. Under such conditions the dendritic growth velocity becomes strongly dependent on the concentration C_∞, which was first described in some detail for different aqueous solutions including NaCl by Tammann and Büchner (1935). These authors first described that the free growth velocity first increased with concentration, reached a maximum and then decreased monotonically.

A qualitative explanation for this maximum in terms of classical stability theory is, that the much lower solute diffusivity creates a thinner solutal boundary layer and thus implies a smaller stability wave length. However, for large solute concentrations, the freezing point depression through the solute release overcompensates the latter effect.

First quantifications of this effect in terms of the classical stability concept from Langer
and Müller-Krumbhaar (1978) were presented by Karma and Langer (1984), while a
somewhat different simplified model was suggested by Lipton et al. (1984). The principle
simplified treatment in the latter approach is to solve the concentration field according
to Ivantsov's equation 8.4 by defining a solutal Stefan Number

$$Ste_C = \frac{C_{tip}(1-k)}{C_{tip} - C_\infty},\tag{8.18}$$

where C_{tip} is the, a priori unknown, concentration at the tip of the dendrite. The corresponding tip temperature in the thermal Ivantsov equation must then become the liquidus
temperature $T_{tip} = T_f(C_{tip})$ and the coupled system becomes solvable, which has been
done at that time using the marginal stability conjecture (which is quantitatively correct
for the pure substance succinonitrile). The theoretical application has the following main
results:

- The tip velocity increases from infinite dilution to a maximum and then decreases
  monotonically with increasing C_∞.

- The tip radius r_{tip} decreases first rapidly with solute concentration, reaches a minimum and then increases weakly.

- The effective selection parameter σ^* increases almost linearly with solute concentration.

Karma and Langer (1984) showed that the first two features were well predicted, when
compared to observations, but the selection parameter σ^* was not. They proposed, as the
most likely cause, a possible deviation from the idealised tip shape at large concentrations.

Neither could a theoretically more rigorous treatment performed by Ben Amar and
Pelce (1989), nor the account of microconvection near the tip (Li and Beckermann, 2002)
explain the difference in the predicted and observed σ^*. A numerical phase-field simulation has recently indicated that the simplified approaches likely do not apply when
thermal and constitutional supercooling are of the same magnitude, that is, near the
maximum in V (Ramirez and Beckermann, 2005). At present the problem is not solved,
yet it appears that systems with pure substance supercooling or strict constitutional supercooling behave the same and are predictable, while problems arise when these fields
are both dominant.

8.1.4 Summary: Free growth of dendrites

After all, it is useful to summarise the main principles of the theory of free growth of
dendrites into a supercooled melt:

- A dendrite, freely growing into a supercooled melt, selects a single velocity of steady growth and a corresponding radius at its parabolic tip, which are a function of the nondimensional supercooling or Stefan number Ste.

- The functionality on the Stefan number depends on the geometry of the paraboloid and thus in principle on the anisotropy in the growth rate.

- The proportionality factor σ^*, in the scaling $\sigma^* r_{tip} = l_\Gamma / Pe$, fixes the solution and is determined by microscopic solvability theory in dependence on the anisotropy in the surface tension.

- In the absence of anisotropic surface tension no steady needle solutions exist.

From dendrites to cells

The theory of pattern selection of dendrites has many similarities with the problem of viscous fingering in a Hele-Shaw cell (Kessler et al., 1988; Langer, 1989). The latter theory shows that cells propagating into a channel select a width of half the channel width (Saffman and Taylor, 1958). This relative width is increased when surface tension is included and approaches unity when surface tension becomes comparable to viscous drag (McLean and Saffman, 1981; Bensimon et al., 1986; Kessler et al., 1988). Hence, cells grow with solid fractions of order one, while the description of free dendrites requires very low solid fractions. A logical onset of separation between cells and dendrites would be a solid fraction of $1/2$, a value which is decreased when anisotropy in surface tension is included (Molho et al., 1990). It is the transition from cells to dendrites which is difficult to describe, when shapes are changing and sidebranching sets in (Molho et al., 1990; Kurowski et al., 1990; Georgelin and Pocheau, 1997).

8.2   Free growth of ice crystals

8.2.1   Velocity selection

A detailed analysis of the ice-water system in terms of microscopic solvability remains to be done and cannot be provided here, yet some points evident from the available experimental data can be pointed out. Observations of the growth velocity of ice dendrites have recently been extended to -30 °C[3], and these data are shown in figure 8.2 with some older sources (Shibkov et al., 2005). The solid curve in the figure represents the original marginal stability hypothesis from Langer et al. (1978) with the value $\sigma^* = 0.025$. It apparently fits the observations quite well in an intermediate growth regime. This

---

[3] This corresponds to $Ste^{-1} \approx 0.55$, while near the often quoted homogeneous nucleation limit of -37 °C one has $Ste^{-1} \approx 0.85$.

agreement is now known to be very likely fortuitous. To understand, why it happens to match the data in the intermediate regime, while at high and low supercooling large differences emerge, the corresponding morphologies reported by Shibkov et al. (2005) are shown in figure 8.3 are critical.

At undercooling (Ste^{-1} in the present notation) below $\Delta T \approx 0.3$ K, the crystals display a structure with a splitted tip (upper left (a) in figure 8.3) and the slope in the growth law shows a strong change. This morphology has been described earlier (Tirmizi and Gill, 1987; Koo et al., 1991), was found in the growth velocity relationship (Tirmizi and Gill, 1987; Furukawa and Shimada, 1993) and has been associated with natural convection (Tirmizi and Gill, 1987; Koo et al., 1991, 1992). The convection model has however, recently been questioned on the basis of microgravity observations for succinonitrile (Tennenhouse et al., 1997). A different explanation might be a change in the growth kinetics normal to the basal plane at similar supercoolings, first reported by Hillig (1958). Here it suffices to say that the growth law becomes very different at supercoolings lower than ≈ 0.2 to 0.3 K, with some further references given in the following section 8.2.2.

In the intermediate regimes dendrites develop the classical parabolic needle structure with side branches, which becomes sharper with supercooling. The dendrites become branchless at $\Delta T \approx 4$ K (lower left (f) in figure 8.3), and here also the growth law in figure 8.2 changes to smaller velocities. Shibkov et al. (2005) interpreted this change in terms of a kinetic effect, interface kinetics becoming more sluggish. A similar explanation had been suggested many years ago by Pruppacher (1967), from whom also some data points are included in figure 8.2, based on less accurate data and at temperatures down to $\approx -9\,°C$. The regime between -4 and $-9\,°C$ shows rather scattered observations (Shibkov et al., 2005), and a sharp transition seems not to apply. However, below $\approx -9\,°C$, the growth law is essentially different.

Another finding of earlier studies, the observed tendency for plate-like growth in a certain growth velocity regime, needs to be recalled, when morphology is considered. It has now been quantitatively described (see 8.2.2 below), that tip radii at the edge of the basal plane are one to two orders of magnitude larger than normal to it (Vlahakis and Barduhn, 1974; Tirmizi and Gill, 1987; Koo et al., 1992; Furukawa and Shimada, 1993). Shibkov et al. (2005) showed the available data with $\sigma^* = 0.025$ from the marginal stability hypothesis and suggested an improved $\sigma^* \approx 0.0213$ in comparison with tip radius observations from Tirmizi and Gill (1987). This approach is, however, inconsistent, because it does not account for the shift to two-dimensionality in the Ivantsov solution. A detailed discussion remains to be given elsewhere, but it seems most likely that figure 8.2, in connection with earlier discussions by Pruppacher (1967), reflects two modes of growth of ice crystals, one in the c-axis direction and one normal to it, being governed by extremely different anisotropies in the surface energy. For c-axis growth, the axial

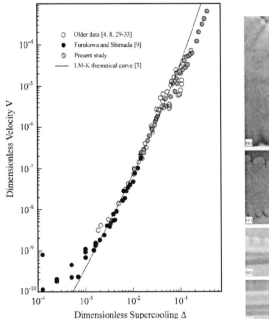

Fig. 8.2: From Shibkov et al. (2005). Plot of dimensionless growth velocity $Vl_{Gamma}/2\kappa_w$ versus dimensionless supercooling (present Ste^{-1} of ice dendrites growing freely in pure supercooled water.

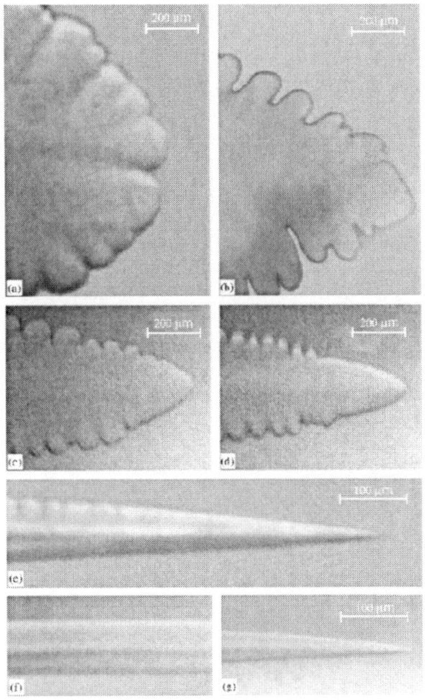

Fig. 8.3: From Shibkov et al. (2005). Morphology of ice dendrites growing at supercooling (a) $\Delta T = 0.3$ K, (b) $\Delta T = 0.5$ K, (c) $\Delta T = 0.7$ K, (d) $\Delta T = 1.1$ K, (e) $\Delta T = 3.8$ K, (e), (f) and (g) $\Delta T = 4.2$ K.

anisotropy $\epsilon_6 \approx 0.002 \pm 0.01$ is rather small (Koo et al., 1991), giving $\alpha_\epsilon \approx 0.07 \pm 0.03$ and $\sigma^* \approx 0.009 \pm 0.005$ from solvability theory (Muschol et al., 1992; Glicksman and Marsh, 1993). This seems to be consistent with the growth law at large supercoolings. For growth normal to the c-axis, the anisotropy is much larger with $\epsilon_2 \approx 0.3$ (Koo et al., 1991) and $\alpha_\epsilon \approx 0.9$. According to solvability theory, this should result in a much larger selection parameter. Extrapolating the simulations from Karma and Rappel (1998), would result in $\sigma^* \approx 5$. For plate-like growth one must then apply a modified Ivantsov model (Horvay and Cahn, 1965) with a solution between the extrema in figure 8.1. This was discussed by Koo et al. (1992), who proposed a different evaluation of σ^* in apparent agreement with observation, yet by matching a convection model which has not been fully validated since then. Calculations, based on an aspect ratio of 28 for the tip radii observed by Koo et al. (1992) and the theory from Horvay and Cahn (1965), indicate that the smaller Pe due to plate-like geometry can closely compensate for the much larger σ^*, and thus explain the *apparent* agreement with $\sigma^* = 0.025$ in the regime where growth is parallel to the basal plane. These considerations indicate the importance of further studies, including a modification of solvability theory results, due to two-fold instead of four-fold anisotropy.

Recalling that there are still some open questions concerning the application of solvability theory (Muschol et al., 1992; Glicksman and Marsh, 1993; Karma and Rappel, 1998; Karma et al., 2000), the potential of the ice-water system to elucidate them is noteworthy. The system can be studied by focusing on the preferred growth rates of ice crystals, with transitions near −4 and −9 °C. The two growth modes both have, consistent with theory, a very different σ^*, indicating that the marginal stability conjecture value 0.025 does not apply in either case, and demonstrating clearly that the previously reported apparent agreement (figure 8.2) is fortuitous. The growth law behaviour at supercoolings less than 0.3 K is more difficult to explain and most probably requires a better understanding of anisotropic growth kinetics and its coupling to morphology and convection.

8.2.2 Tip radii of ice dendrites

The tip radius of the edge of a growing plate-like dendrite is increasingly more difficult to measure at high supercoolings. At $\Delta T_{sup} \approx 1$ K it falls already below ≈ 1 to 1.5 μm (Vlahakis and Barduhn, 1974; Koo et al., 1992; Furukawa and Shimada, 1993). Therefore a direct evaluation of σ_* at high supercoolings has not been possible to date. However, for lower supercoolings observations are available. Of particular influence is the range at supercoolings below 0.3 K, not only because it is the most plausible range at natural cellular interfaces, yet also because microscopic solvability theory apparently has to be modified here due to convection or kinetics (see figure 8.2).

For supercoolings below 0.3 K, the relations between the tip radius and the supercooling for freely growing dendrites observed by several authors (Vlahakis and Barduhn, 1974; Koo et al., 1992; Furukawa and Shimada, 1993) are shown in figure 8.4. It is seen that, at supercoolings of 0.05 to 0.1 K, tip radii of 5 to 10 μm are typical. The difference between the curves is eventually explained by the fact that tip radii at these low supercoolings take some time to stabilise (Koo et al., 1992) and the curve from the latter authors, giving maximum radii after stabilisation, may be the most relevant steady state measure. Their scaling law based on a supercooling range 0.05 to 1 K reads

$$r_{tip} = \frac{1.51}{\Delta T} \tag{8.19}$$

where r_{tip} is given in micrometers. The initial thickness of discs growing in water of such supercoolings has, as one would expect, been found to be about two times as large

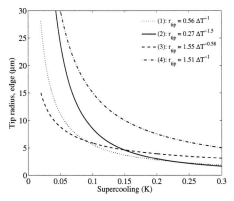

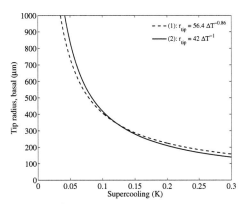

Fig. 8.4: Empirical relations between tip radius of the edge plane and supercooling: (1) - pure water (Vlahakis and Barduhn, 1974); (2) - 1.5 ‰ NaCl (Vlahakis and Barduhn, 1974); (3) - pure water (Furukawa and Shimada, 1993); (4) - pure water (Koo et al., 1992). Relations are based on observations with $\Delta T \geq 0.1$ K, with except of Koo et al. (1992), who obtained data down to 0.035 K.

Fig. 8.5: Empirical relations between tip radius of the basal plane and supercooling: (1) - pure water (Furukawa and Shimada, 1993); 2) - pure water (Koo et al., 1992). Relation (1) is based on observations with $\Delta T \geq 0.1$ K, relation (2) on experiments down to 0.035 K.

(Fujioka and Sekerka, 1974).

Also shown, in figure 8.5, is the basal plane radius r_{bas}, which is much easier to measure. The observations from Koo et al. (1992) and Furukawa and Shimada (1993) agree remarkably well. The best estimate of the ratio of r_{bas} and r_{tip} has been obtained by Koo et al. (1992) to be 28. As noted above, this is relevant when computing the growth velocity via a modified Ivantsov model between equations 8.4 and 8.6, and when interpreting observations in terms of microscopic solvability theory. It will be discussed in later sections that the dependence of the basal plane tip radius is likely to play a role when the bridging of two parallel plates due to a supercooled throughflow is considered.

The slope change in figure 8.2 near $Ste \approx 2 \times 10^{-3}$ or 0.2 to 0.3 K may be illustrated in an empirical form by transforming the dendritic growth velocity laws from (Tirmizi and Gill, 1987)

$$V = 1.87 \times 10^{-2} \Delta T^{2.09} \qquad , \Delta T > 0.2 \ K \tag{8.20}$$

$$V = 3.52 \times 10^{-3} \Delta T^{1.06} \qquad , 0.05 \ K < \Delta T < 0.2 \ K \tag{8.21}$$

with V given in cm/s, by combination with equation 8.19 into

$$r_{tip} = 2.25 \times 10^{-5} V^{-1/2.09} \qquad , \Delta T > 0.2 \ K \tag{8.22}$$

$$r_{tip} \;=\; 7.32 \times 10^{-7} V^{-1/1.06} \qquad , 0.05 \; K < \Delta T < 0.2 \; K \qquad\qquad (8.23)$$

giving the tip radius in centimeters. It is seen that, in the low supercooling regime, the dependence is very different from $r_{tip} \sim V^{-1/2}$ expected from stability theory. Observations from Furukawa and Shimada (1993), giving a slope close to $V^{-1/2}$, refer mainly to the larger supercooling growth regime. The observations from Koo et al. (1992) emphasis the less well understood physics in the regime below a growth velocity of $\approx 6.4 \times 10^{-4}$ cm/s.

8.2.3 Saline aqueous solutions

As summarised above in 8.1.3, it is theoretically expected that the tip radius will decrease upon adding salt to dilute solutions, whereas the growth velocity will increase, both reaching their minimum and maximum at some, not too large, concentration. For aqueous NaCl solutions, the velocity behaviour was first documented by Tammann and Büchner (1935). More recent observations indicate that, for quiescent upward growth, the velocity maximum is realised near $C_\infty \approx 10$ ‰ (Barduhn and Huang, 1987). In an agitated solution it is shifted to somewhat smaller values C_∞, as the earlier observations from Vlahakis and Barduhn (1974) indicate. That this maximum can be completely suppressed when strong stirring is combined with low supercoolings confirms its origin from the theoretical thin boundary layer effect. It may be noted that the unstirred dendritic growth velocities at the maximum were a factor of 3 to 4 larger than for pure solutions (Barduhn and Huang, 1987), in agreement with observations on other solutions of comparable supercooling below 1 K (Ryan, 1969).

No systematic observations of tip radii for saline solutions have been obtained yet. The relation for a 1.5 ‰ NaCl solution obtained by (Vlahakis and Barduhn, 1974) and shown in figure 8.4 indicates no significant difference to pure water at that low concentration. Due to the mentioned difficulties in theoretically predicting σ^* (and hence tip radii) at low supercoolings for pure water, discussed in 8.2, and considering the problems in the theoretical formulations pointed out by Ramirez and Beckermann (2005), no simple solution is expected. However, the scalings for the tip radii of ice growing in pure water (figures 8.4 and figure 8.5), may be expected to apply when only constitutional supercooling exists.

8.3 Scaling laws for cellular and dendritic spacing a_0

8.3.1 Simplified analytical approaches

Having summarised the history of free growth pattern selection of single dendrites it is now possible to understand and evaluate some approaches which have been made to derive the stable spacing of dendritic and cellular arrays. Here first the concentration problem is reintroduced, recalling the similarity between marginal stability Mullins-Sekerka wave length for the thermal case

$$\lambda_\Gamma = 2\pi \left(\frac{2\kappa_w}{V}\right)^{1/2} \left(\frac{\Gamma c_{pw}}{L_f}\right)^{1/2}, \tag{8.24}$$

and contrast it with equation 7.38 (dropping the factor $3^{1/2}$ for the most rapidly growing wave length)

$$\lambda_\Gamma = 2\pi \left(\frac{D_s}{V}\right)^{1/2} \left(\frac{\Gamma}{mC_{int}(k-1)}\right)^{1/2} \left(\frac{1}{(1-\mathcal{G}')}\right)^{1/2}. \tag{8.25}$$

In the limit of small $\mathcal{G}'$, far from equilibrium, the equations are similar with c_{pw}/L_f being replaced by the constitutional supercooling source term $mC_{int}(k-1)$. The numerical difference is the factor two in the diffusion length scale which arises because solute diffusion through the solid is neglected. The thermal case, here with equal solid and liquid conductivities assumed, can easily be extended to arbitrary conductivity ratios, e.g., (Langer, 1980b). In the following only the concentration equation 8.25 will be considered. From comparison to equation 8.15, it is also recalled, that the tip radius of a dendrite is proportional to the marginal stability wave length λ_Γ of a planar interface, the prefactor 2π being replaced by the selection parameter $\sigma^{*-1/2}$.

The problem to be solved is the following. A dendritic or cellular array moves at velocity V through a solution with concentration C_∞ with an effective temperature gradient G_{eff}. It adopts thereby a stable cell spacing a_0. At the root of the cells the brine or liquid volume of the array is given by the Scheil equation 8.1. The tips of the cells may be characterised by a radius of curvature r_{tip} which may, for noninteracting dendrites, evolve the shape of a parabolic needle crystal. Three situations indicate the complexity that arises now: (i) if the dendrites are indeed noninteracting, one must find a condition, how their spacing is limited; (ii) if they are interacting, the dendrite shape and spacing are likely to change and influence each other and one may imagine that for $2r_{tip} \approx a_0$ a kind of planar interface problem is approached, yet with a much larger k; such cells are not parabolic but have, in two dimensions, a straight stem and a circular cap; (iii) rather flat cells with $r_{tip} > a_0$ are another different pattern class found during eutectic freezing. The following discussion mainly focuses on the situation (i) for dendritic growth with $a_0 \gg r_{tip}$.

Matching at the tip

The geometric constraint that emerges from free dendritic growth under purely diffusive conditions is the parabolic nature of a dendrite. If the supercooling is given, then the tip radius r_{tip} of this paraboloid is determined by the interplay between between interfacial physics and heat flow. The challenge to construct a relation between r_{tip} and the cell spacing a_0 has been first undertaken by Hunt (1979) and is illustrated in figures 8.6 and 8.7. It is a two-step procedure and has been adopted by other authors (Trivedi, 1984; Kurz and Fisher, 1984). First, a relation between spacing and tip radius is derived. Then either the selection of the tip radius is defined due to an equation of the form 8.25, or an additional constraint on the thermodynamics of the system is formulated. While the approaches from Hunt, Trivedi and Kurz and Fisher are often noted to be similar (Trivedi, 1984; Spencer and Huppert, 1999; Cadirli et al., 2003), and their parametric form indeed is so, below it will be seen that it is essential to understand their difference.

The simple geometric condition to start with is that at some distance from the interface, where the radius of the dendrite is r_0, the brine volume or liquid fraction is given by the relation

$$v_{b0} = 1 - \left(\frac{8r_0}{\pi a_0}\right)^2 \tag{8.26}$$

for three dimensional cells, while in two dimensions the exponent simply becomes 1. Hunt (1979) now considered the Scheil equilibrium shape shown as the solid cell surface in figure 8.6. He assumed that a tip radius r_{tip} exists that matches the Scheil *slope* dV_b/dz at some arbitrary angle ϕ and fulfills $r_{tip} = r_0/\sin\phi$. He further assumed uniform composition and equilibrium freezing temperature normal to the growth direction and ended up with the equation

$$a_0 = \left(1 - \left(\frac{2r_{tip}\sin\phi}{a_0}\right)^2\right)^{(k-2)} (8\cos\phi)^{1/2} \left(\frac{D_s r_{tip}}{V}\right)^{1/2} \left(\frac{1 - \mathcal{G}'}{\mathcal{G}'}\right)^{1/2}, \tag{8.27}$$

where, in the present notation,

$$\mathcal{G}' = \left(\frac{G_{eff}}{mG_c}\right)_{tip} = \left(\frac{G_{eff}D_s}{mC_{tip}(k-1)V}\right)_{tip} \tag{8.28}$$

is defined at the tip of the cell or dendrite, analogous to equation 7.18 in the Mullins-Sekerka analysis, and relates to the latter via $\mathcal{G}' = \mathcal{G}kC_{tip}/C_\infty$[4]. It is noteworthy that (Hunt, 1979) assumed a hexagonal array and computed the brine volume from $v_{b0} = 1 - (2r_0/a_0)^2$. In all what follows his equations are changed by a factor $\pi/4$ assuming a square array based on equation 8.26. With the further assumption $r_{tip}^2 \ll a_0^2$ one obtains

$$a_0 = (2\pi\cos\phi)^{1/2} \left(\frac{D_s r_{tip}}{V}\right)^{1/2} \left(\frac{1 - \mathcal{G}'}{\mathcal{G}'}\right)^{1/2}, \tag{8.29}$$

---

[4] Comparison with equation 7.38 indicates that the earlier defined C_{int} for a macroscopic interface corresponds to C_{tip}

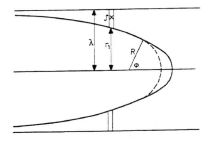

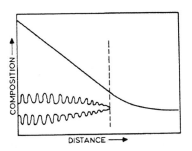

Fig. 8.6: Adapted from Hunt (1979). The Scheil profile (solid curve) is matched by a spherical cap or radius R, r_tip in the present notations. Note that Hunt's cell spacing $\lambda = a_0/2$ and r_1 correspond to r_0 in the present notation.

Fig. 8.7: Adapted from Hunt (1979). The constitutional undercooling at the tip level of dendrites determines their tip radius via free dendritic growth theory.

The essence from equation 8.29 is: It is possible to find a tip radius r_{tip} which gives, at an angle ϕ, the same magnitude and change in the solid fraction $(1 - v_b)$ that an equilibrium Scheil profile would produce. Trivedi (1984) adapted, in a work to be discussed below, the scaling between a_0 and r_{tip} from Hunt.

Kurz and Fisher (1984) made a different geometrical approximation. They assumed an elliptical dendrite, which reaches its eutectic composition at a distance $z_e = (r_{tip}a_0)^{1/2}$ from the tip. This approximation leads, for velocities far from the threshold, to

$$a_0 = \left(\frac{3}{1-k}\right)^{1/2} \left(\frac{D_s r_{tip}}{V}\right)^{1/2} \left(\frac{1-\mathcal{G}}{\mathcal{G}}\right)^{1/2} \tag{8.30}$$

in the present notation. Although this equation appears similar to 8.29 derived by Hunt, the difference between $\mathcal{G}'$ and $\mathcal{G}$ must be pointed out. It follows from the fact, that Kurz and Fisher constrain the tip concentration and shape via the eutectic concentration far from the tip, and not locally, as done by Hunt. Formally, $\mathcal{G}' = \mathcal{G}$ corresponds to the case when diffusive equilibrium freezing is realised near the tip and $C_{tip} = C_\infty/k$, but normally $\mathcal{G}' \ll \mathcal{G}$, and the predicted spacings will consequently be larger, to be discussed below. The important difference is that the Kurz and Fisher approach defines the relation between spacing and tip radius from a condition far away from the cell tips.

Tip radius selection

The tip radius has been predicted in the same way by Trivedi (1984) and Kurz and Fisher (1984), assuming in principle the above discussed marginal stability condition from Langer and Müller-Krumbhaar (1978), and thus setting

$$r_{tip} = \frac{1}{\sigma^{*1/2}} \left(\frac{D_s}{V}\right)^{1/2} \left(\frac{\Gamma}{mC_{tip}(k-1)}\right)^{1/2}. \tag{8.31}$$

This equation is often simplified by assuming $C_{tip} = C_\infty$, which is reasonable at concentrations far above threshold, e.g., Kurz and Fisher (1984). If $(C_{tip} - C_\infty)/C_\infty$ is not close to unity a model for C_{tip} is required and things are more difficult to predict (see paragraph 8.1.3 above). At that time Kurz and Fisher (1984) assumed $\sigma^* = (2\pi)^{-2} \approx 0.0253$, while Trivedi (1984) implicitly applied $\sigma^* = 0.0179$.

A solution for a_0 is now found by combining equations 8.29 and 8.30 with 8.31. Trivedi (1984) used the relation 8.29 between a_0 and r_{tip} from Hunt with 8.31 to predict

$$a_0 = \frac{(2\pi cos\phi)^{1/2}}{\sigma^{*1/4}} \left(\frac{D_s}{V}\right)^{3/4} \left(\frac{\Gamma}{mC_{tip}(k-1)}\right)^{1/4} \left(\frac{1-\mathcal{G}'}{\mathcal{G}'}\right)^{1/2}, \tag{8.32}$$

which here contains, in comparison to the equations given by Trivedi, still the factor $cos\phi$ to be discussed below. Kurz and Fisher (1984) combined 8.30 with 8.31 to

$$a_0 = \frac{3^{1/2}}{(1-k)^{1/2}\sigma^{*1/4}} \left(\frac{D_s}{V}\right)^{3/4} \left(\frac{\Gamma}{mC_\infty(k-1)}\right)^{1/4} \left(\frac{1-\mathcal{G}}{\mathcal{G}}\right)^{1/2}, \tag{8.33}$$

where as mentioned above $\mathcal{G}$ and not G appears.

Hunt (1979) proceeded in a slightly different way. Assuming a minimum undercooling condition at the tip of the cells he derived

$$a_0 = \frac{(2\pi cos\phi)^{1/2}}{1/2^{1/4}} \left(\frac{D_s}{V}\right)^{3/4} \left(\frac{\Gamma}{mC_\infty(k-1)}\right)^{1/4} \left(\frac{1-\mathcal{G}}{\mathcal{G}/k}\right)^{1/2}. \tag{8.34}$$

It can be seen that the latter equation from Hunt now also contains $(1-\mathcal{G})$ in the nominator but $\mathcal{G}/k$ in the denominator, contrasting the Kurz and Fisher scaling. Comparing 8.34 with 8.29, the result is equivalent to a scaling for the tip radius given in an earlier work (Burden and Hunt, 1974)

$$r_{tip} = 2^{1/2} \left(\frac{D_s}{V}\right)^{1/2} \left(\frac{\Gamma}{mC_\infty(k-1)}\right)^{1/2}. \tag{8.35}$$

By comparison with 8.31 one realises that this approach from Hunt implies an effective $\sigma^* = 1/2(C_{tip}/C_\infty)^{1/2}$.

While many aspects of the three equations appear similar, some main differences need to be pointed out. Far from threshold there is normally little difference between C_{tip} and C_∞ and also $(1-\mathcal{G})$ and $(1-\mathcal{G}')$ are both close to one. In this case one main parametric difference between the approaches arises from the factor $k^{-1/2}$ in the scaling 8.33 from Kurz and Fisher, when compared to the equivalent approaches from Hunt and Trivedi. Both Hunt and Trivedi assumed the matching for $cos\phi = 2^{-1/2}$ at the tip, which led to the prefactors of 2.83 for Hunt's model and $2.38/\sigma^{*1/4} \approx 6.51$ in the treatment by Trivedi (1984). Both were here corrected by $\pi^{1/2}/2$, as the liquid fraction was computed for a square array. Furthermore, as there is no conceptual argument, why $cos\phi$ could not take a very small value, these results appear somewhat arbitrary. One might instead suggest to take $cos\phi = 1$ and interpret the result as an upper bound for the spacing. Equation 8.33 due to Kurz and Fisher does not suffer from such a geometrical indeterminacy, as it is based on a condition at the root of the cells. It gives a prefactor $(3/(1 - k))^{1/2}/\sigma^{*1/4}) \approx 4.34/(1 - k)^{1/2}$. While this prefactor lies between the other two scalings, for not too large k the equation from Kurz and Fisher gives the largest spacings, as $k^{-1/2}$ comes in addition.

As discussed above, the values chosen for σ^* by Trivedi and Somboonsuk and Kurz and Fisher were at that time arguable by heuristical arguments or a limited numerical analysis (Langer, 1980b). Meanwhile, with the establishment of microscopic solvability theory, it is clear that the correct value of σ^* needs to be either computed from surface tension anisotropy for the material in question, or determined empirically. The large implied $\sigma^* = 1/2$ in the equation from Hunt corresponds to the *minimum undercooling* principle for array growth (Burden and Hunt, 1974). It gives the smallest possible tip radius controlled by capillarity and thus the smallest possible spacing. While an analogous maximum velocity principle for single dendrites had been found to be incorrect, see the summary above and several reviews (Langer, 1980b; Kurz and Fisher, 1984), such a condition has been theoretically justified for eutectic lamellae (Langer, 1980a) and for cellular solidification in the limit of small nondimensional spacing $a_0 V/D$ (Karma and Pelce, 1990).

8.3.2   Refinement due to observations

The present notation, using $\mathcal{G}'$ from linear stability theory, is different from the notation in some of the original works, but allows a comparison with equation 7.36 and 8.25 for the planar interface close to and far from equilibrium, respectively. Interestingly, $(1 - \mathcal{G}')^{1/2}$ appears in the a_0-scalings in the nominator (equations 8.32 through 8.34), while for the planar interface it is in the denominator (of equation 8.25). While far from equilibrium this is not relevant, it indicates a different behaviour to changes in G'. At first glance, it appears that the exponent $a_0 \sim V^{-3/4}$ in the a_0-scalings is larger than in both stability scalings for the planar surface. However, as normally $\mathcal{G}'$ is not kept constant while changing V, but instead $\mathcal{G}' \sim V^{-1}$, the overall dependence is $a_0 \sim V^{-1/4}$. This is indeed close to the observed slope for many systems (Somboonsuk et al., 1984; Kurz and Fisher, 1984; Bouchard and Kirkaldy, 1997; Cadirli et al., 2003). The same arguments hold for the concentration dependence, also hidden in $\mathcal{G}'$, which is predicted as $a_0 \sim C_\infty^{1/4}$ and also reasonably found in observations (Trivedi and Somboonsuk, 1984; Bouchard and Kirkaldy, 1997; Cadirli et al., 2003).

Several studies performed, also indicate that the predictions closely yield the tip radius, if σ^* is known or predictable (Somboonsuk et al., 1984; Gündüz and Cadirli, 2001; Cadirli et al., 2003), although there are some problems at concentrations where (C_{tip}/C_∞) differs increasingly from 1 and must be predicted (Li and Beckermann, 2002; Ramirez and Beckermann, 2005). For many materials, where σ^* is close to 0.02, both the original scalings from Trivedi (1984) and Kurz and Fisher (1984) are close to reality. However, the proper prediction of r_{tip} involves equation 8.31 with σ^* given by microscopic solvability and *not* equation 8.35 from Hunt, which gives generally too small tip radii.

The largest problem is faced with the predictability of the cell or dendrite spacing a_0, which was already pointed out in the early studies (Kurz and Fisher, 1984; Trivedi and Somboonsuk, 1984) and has meanwhile been found in further experiments (Bouchard and Kirkaldy, 1997; Gündüz and Cadirli, 2001; Cadirli et al., 2003). Kurz and Fisher (1984) reported that their equation 8.33 overpredicted a_0 by a factor of 1.3 to 2, while Hunt's scaling 8.34 was typically a factor 2 to 3 less than observed spacings. Considering later studies (Gündüz and Cadirli, 2001; Cadirli et al., 2003), the overprediction corresponds to a factor of even 2 to 3 and is due to the additional factor $k^{-1/2}$ in the Kurz and Fisher's scaling. This discrepancy is likely linked to the assumption of eutectic composition a distance $z_e = (r_{tip}a_0)^{1/2}$ from the tip made in their approximation. Comparing the scaling 8.32 from Trivedi and Somboonsuk with $\sigma^* = 0.0179$ and $cos\phi = 2^{-1/2}$, either a $\approx$ 30 % underestimate for succinitrile-salol (Cadirli et al., 2003) and succinonitrile-acetone (Trivedi and Somboonsuk, 1984), or a similar overestimate for aluminium-copper alloys (Gündüz and Cadirli, 2001; Spinelli et al., 2004) has been found. The scaling 8.34 from Hunt almost always underestimates the actual spacings by a factor of 1.5 to 2.5.

Scaling compromises

As for many alloys, like Al-Cu, the surface tension anisotropy or σ^* are not well known, the comparison with observations can only be refined for succinonitrile, for which σ^* has been properly determined, e.g, Glicksman and Marsh (1993). It seems then in best agreement with data, and perhaps most plausible, to use an upper bound of equation 8.32 by setting $cos\phi = 1$. One obtains

$$a_0 = \frac{(2\pi)^{1/2}}{\sigma^{*1/4}} \left(\frac{D_s}{V}\right)^{3/4} \left(\frac{\Gamma}{mC_{tip}(k-1)}\right)^{1/4} \left(\frac{1-\mathcal{G}'}{\mathcal{G}'}\right)^{1/2}, \tag{8.36}$$

by applying $C_{tip} = C_\infty$ (but keeping in mind that the case, where C_{tip}/C_∞ differs considerably from unity, requires a correction). The condition $cos\phi = 1$ requires only the tip curvature to be the same as in a Scheil profile and is thus a singular constraint. One might argue, that it is more reasonable to derive a prefactor $(2\pi \int_0^{\pi/2} cos\phi)^{1/2} = 2$, by integrating over the tip half circle, and define the geometric matching condition in an average sense.

A smaller prefactor $3^{1/2}$ can be obtained from the elliptical dendrite approximation by Kurz and Fisher (1984), if simply the factor $k^{-1/2}$ is dropped from their work. However, this is just an ad-hoc change to illustrate how an arbitrary shape can influence the scaling. As noted above, the approximation from Kurz and Fisher is based on the assumption, that the eutectic composition is reached a distance $z_e = (r_{tip}a_0)^{1/2}$ from the tip, letting the liquid fraction become constant. The equivalent matching condition in the model from Hunt is the limit $cos\phi = 0$, when both the slope of the Scheil profile and of the inscribed tip radius become zero (figure 8.6). In this case also the plate spacing a_0 would become zero in the scalings from Hunt. Eventually this bound can be interpreted in a relatively simple manner. If the dynamical selection of the tip radius via σ^* encounters $cos\phi = 0$ within one tip radius from the tip, it can no longer control a_0. In such a geometry one may argue that now indeed the minimum undercooling principle from Burden and Hunt (1974) and Hunt (1979) could become operative. Due to the large tip radius one would have to use equation 8.27. However, as long as $2r_{tip} < a_0$ one may still use equation 8.27 locally at the tip where $cos\phi = 1$, obtain the prefactor $(8\pi^2)^{1/4} \approx 2.98$ and the scaling

$$a_{0,min} = 2.98 \left(\frac{D_s}{V}\right)^{3/4} \left(\frac{\Gamma}{mC_{tip}(k-1)}\right)^{1/4} \left(\frac{1-\mathcal{G}}{\mathcal{G}/k}\right)^{1/2}. \qquad (8.37)$$

The minimum undercooling equation 8.37, once suggested in general (Hunt, 1979), is known to fail for dendrites but may eventually be associated with cellular spacings of large tip radius. It is noteworthy that therein only C_∞ appears while in 8.36 in principle C_{tip} must be known. However, with $C_{tip} \approx C_\infty$ and $\mathcal{G} \ll 1$, the scalings become parametrically identical as in this limit $\mathcal{G}' = \mathcal{G}/k$. The conditions implied by 8.37 closely resemble the case of eutectic freezing, for which it has been shown that the minimum undercooling condition is theoretically justified (Langer, 1980b; Hunt, 2001).

The considerations of the albeit approximate evaluation indicate two main consequences concerning the matching of the Scheil profile 8.26 with parabolic dendrite tips. If the tip radius is selected via σ^* and microscopic solvability theory, the geometric matching still needs to be determined by an ad-hoc condition, yet the resulting variation in the predicted a_0 due to equation equation 8.36 is not too large. A different scaling becomes plausible, if the free selection condition is lacking because the solid fraction becomes constant close to the tip. In this case one can naively argue that the minimum undercooling condition determines a_0 via equation 8.37. This may happen when the eutectic composition is reached close to the tip and/or the tip radius is larger than the cell spacing.

One may imagine that other conditions than eutectic freezing, like convection and kinetic anisotropy may be responsible for such a geometry and thus influence the spacing a_0.

Finally, the role of the term $(1 - \mathcal{G})^{1/2}$ in the scalings from Hunt and Kurz and Fisher, contrasting $(1 - \mathcal{G}')^{1/2}$ from Trivedi needs to be recalled. The former function of $\mathcal{G}$ apparently arises from a matching condition at the root of the cells in the derivations by Hunt and Kurz and Fisher, which appears to be lacking in the local tip approximation by Trivedi (Trivedi, 1984; Trivedi and Somboonsuk, 1985) leading to the latter function of $\mathcal{G}'$. This matching condition becomes certainly relevant for eutectic freezing and increases in importance when $\mathcal{G}$ approaches one, because the relative length of the dendrites decreases and the conditions at the root becomes relevant. Neglecting any such effect in equation 8.36 may then no longer be justified. It will be seen below, that numerical simulations support a form where $(1 - \mathcal{G}')^{1/2}$ in equation 8.36 should be replaced by $(1 - \mathcal{G})^{1/2}$.

8.3.3   Critical discussion: SCN-acetone

Figure 8.8 compares the results of equations 8.36 and 8.37 with the high quality data from Somboonsuk et al. (1984) using the selection parameter $\sigma^* = 0.0195$ for succinonitrile (Glicksman and Marsh, 1993). Observations are shown as circles. The prefactors for the scalings 8.36 and 8.36 then become 6.71 and 2.98 respectively. These are, after all, not very different from the originally suggested values (6.51 and 2.83), but the present values are corrected for a square array, and correspond to upper bounds without arbitrary geometrical assumptions. Both curves sharply drop after a maximum, which is delayed to lower velocities in the minimum undercooling scaling after Hunt (1979), because the latter contains the term $(1 - \mathcal{G})^{1/2}$ in the nominator, while equation 8.36 contains $(1 - \mathcal{G}')^{1/2}$. As already shown by Trivedi (1984) the agreement with the form 8.36 is quite reasonable. That for the two lowest velocity points the deviation is large, can be explained by their proximity to the morphological threshold, when equation 8.36 is not accurate enough. The more accurate calculations in this range were performed by Trivedi (1984) and resulted in a similar agreement as for other points.

For comparison, the numerical simulations by Warren and Langer (1993) noted at the end of the previous chapter are shown in figure 8.9. Notably also Warren and Langer (1993) set the tip radius by the solvability selection criterion, but instead of using an approximate condition of the form 8.29 they simulated the evolution of the tip radius and spacing from its initial planar interface instability by a boundary layer model. This approach appears to closely reproduce the spacing a_0 only at velocities down to $10\mu m/s$. In an earlier work Warren and Langer (1990) had suggested that this failure is likely due to the fact that at this velocity the nondimensional wave number $a_0 V/(2D_s)$ drops below unity, making the assumption of noninteracting dendrites in the tip radius selection questionable. However, while the predicted r_{tip} from Warren and Langer (1990) already

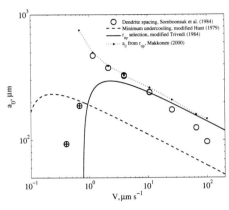

Fig. 8.8: Primary spacing observed by Somboonsuk et al. (1984) for the succinonitrile - 5.5 mol% acetone system (circles) along with several theoretical predictions discussed in the text. The circles denote dendrites, those field with plus signs cells.

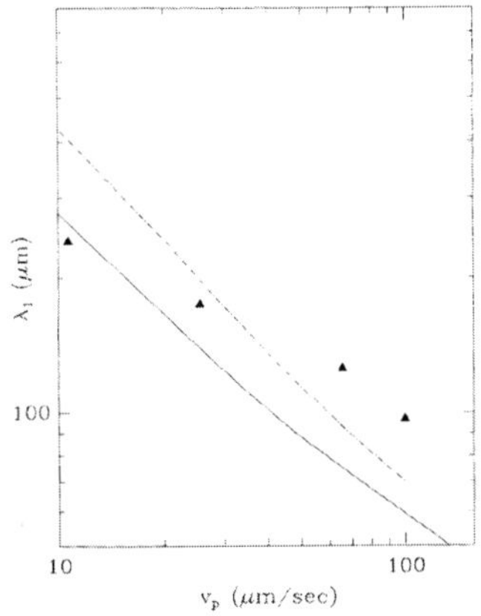

Fig. 8.9: From Warren and Langer (1993). Observations as in figure 8.8, yet for large velocities only, with predictions via a time-dependent boundary layer model.

underestimated the observations by 15 % near 10μm/s, the simplified equation 8.31 performs quite well down to 2μm/s to predict the tip radius within typically 5 %. For some reason the selfstanding simulations of the tip radius and dendrite spacing by Warren and Langer are not clearly superior to the approximate scalings. At high velocities they appear to reproduce the velocity dependence better than the simplified approach, but at low velocities they strongly overestimate the spacings, although they underestimate the tip radius. Some hints may be found in the effect of freezing cell size documented by Somboonsuk et al. (1984): It appears that r_{tip} decreased while a_0 increased, when the height of the cell became less than ≈ 4 times the tip radius, indicating an influence of the walls. In the simulations by Warren and Langer (1990) this constrainment effect is apparently stronger and indicates a more enhanced response of their model to these boundary conditions.

Further approaches

Recently two other approaches to the problem have been proposed. Makkonen (2000) suggested an analysis of the problem in terms of the steady state heat balance of a dendritic field moving through a temperature gradient. He showed that the steady-state heat balance $L_f \partial v_b / \partial z = c_{pw} \partial G_{eff}$ of a stationary dendritic field in a temperature gradient results in the condition

$$a_0 = \left(\frac{\pi r_{tip} L_f}{G_{eff} c_{pw}}\right)^{1/2},$$ (8.38)

with unit undercooling L_f/c_{pw} and imposed temperature gradient $G_{eff} = \partial T/\partial z$, where r_{tip} is the tip radius of paraboloids of revolution spaced by a_0. This is a global condition which in principle states, that the solid fraction within a unit area a_0^2 becomes $(1 - v_b) = r_{tip} G_{eff} c_{pw}/L_f$ one radius from the tip. Interestingly, using the measured tip radii and plotting the resulting 8.38 in figure 8.8, almost exactly matches the observed spacings, but only in the dendritic regime above 1 μm/s. From this agreement, Makkonen (2000) concluded that equation 8.38 is the relevant condition between r_{tip} and a_0. However, Makkonen (2000) only provided observations at a single concentration. The apparent agreement with the data from Somboonsuk is most likely fortuitous and related to the particular solute concentration. When comparing equation 8.36 and 8.38, one needs $2(k - 1)mC_\infty \approx L_f/c_{pw}$ for equality which was almost identically fulfilled in the experiment of figure 8.8. The limitation of equation 8.38 seems likely from the following simple considerations. First, it is valid for a pure melt, and does not consider the Scheil equation change in brine volume with solute concentration. Next, from an observational standpoint, the dependence $a_0^2/r_{tip} \sim C_\infty$, following from 8.36, is supported by experiments (Trivedi and Somboonsuk, 1984; Bouchard and Kirkaldy, 1997; Cadirli et al., 2003) and in conflict with equation 8.38 which predicts no such dependence. While it could be of interest, if Makkonen's scaling can be a limitation for a_0 due to the overall heat budget, it fails clearly at lower concentrations in the study by Trivedi and Somboonsuk (1985), not shown here.

Spencer and Huppert (1999) described the basic results of more complicated simulations of dendritic array interactions (Spencer and Huppert, 1998). The condition between tip radius and cell spacing proposed by these authors from matching the tip and the tail of an Ivantsov dendrite with the Scheil equation is, in the present notation,

$$a_x a_y = (2\pi)^{1/2} \left(\frac{D_s r_{tip}}{V}\right)^{1/2} \left(\frac{1}{\mathcal{G}/k}\right)^{1/2},$$ (8.39)

where now a_x and a_y are the dendrite spacings in the directions normal to the growth axis. The authors proposed this form as a new result which, in the limit $\mathcal{G}/k \ll 1$, determines the relation between spacing and tip radius. However, it can be seen that their equation 8.39 is equivalent to the brine-volume corrected equation 8.29 from Hunt, as in this limit $(1 - \|\mathcal{G}\|)^{1/2} \approx 1$ and $C_{tip} \approx C_\infty$, if one requires $cos\phi = 1$. One sees that the condition emphasised by Spencer and Huppert is essentially the matching of a parabolic radius at the tip to determine the spacing. Spencer and Huppert (1998) addressed this matching problem in detail and performed simulations, where the Scheil shape at the eutectic root of dendrites was matched to the parabolic Ivantsov shape at the tip. They prescribed a_x as the observed twodimensional cell spacing a_0 and a_y as the cell width a_{cell} from Somboonsuk et al. (1984) and simulated the resulting tip radii. Comparing their results also to the simulations from Somboonsuk et al. (1984) they found underestimates of the tip radius by a factor of 2 to 3, which was attributed to the increasing $\mathcal{G}$ no longer being $\ll 1$. However, evaluating the simulations (their tables 1 and 2) does not indicate a correlation between the underprediction and $\mathcal{G}$. Instead, it appears, that the ratio of simulated to observed r_{tip} is proportional to a_{cell}/a_0, a relationship which is close to linear when $a_{cell} < a_0$. In other words: Their model simulations produce a change in r_{tip} induced by changing a_{cell} that actually does not exist in the observations. It appears that, similar to the simulations by Warren and Langer (1990), the confinement effect by walls is strongly overestimated by the model from Spencer and Huppert (1998). The hypothesis proposed by Spencer and Huppert (1998), that a_0 determines r_{tip}, is therefore at least not justified on the basis of their model results.

8.3.4 Numerical Modeling

More recently numerical simulations of the coupled spacing and tip radius selection process have been performed and extended (Hunt and McCartney, 1987; Hunt, 1991; Lu and Hunt, 1992; Hunt and Lu, 1996; Hunt, 2001). The model has been demonstrated to resolve the difference between parabolic dendrites and flat cells. Due to its inability to resolve sidebranching and the competitive growth during spacing adjustment the model has been proposed to yield minimum cell and dendrite spacings. The agreement with observations is similar as for equation 8.36, with deviations of mostly less than 30%, when observed minimum spacings are compared to predictions (Hunt and Lu, 1996; Ding et al., 1997; Feng et al., 1999; Hansen et al., 2002; Spinelli et al., 2004). The scaling laws that were originally given by Hunt and Lu (1996) were physically less ideal than obtained by the intuitive models discussed so far. They also contained a relatively strong dependence on the partioning coefficient k, in contrast to the simple intuitive models. This has been corrected by Hunt (2001), who presented equations for the minimum dendritic spacing in a form that should hold at very small k as well. In the present notation the minimum spacing during *dendritic growth* is given by the characteristic equation

$$a_{0,min} = 5 \left(\frac{D_s}{V} \right)^{2(1+b/2)/3} \left(\frac{\Gamma}{mC_\infty(k-1)} \right)^{(1-b)/3} (1-\mathcal{G})^{1/2} \left(\frac{k}{\mathcal{G}} \right)^{2(1-b)/3} \tag{8.40}$$

with

$$b = 0.3 + 1.9 \, (\mathcal{G}\mathcal{A})^{0.18} \, . \tag{8.41}$$

Interestingly b depends on the stability parameter $\mathcal{A}$ from the Mullins-Sekerka analysis given by equation 7.17. Equation 8.40 differs from the intuitive equation 8.36 in that it indeed implies $(1 - \mathcal{G})^{1/2}$ related to a matching condition from the root of cells and dendrites. It is further of interest that, if the latter term is close to 1, the growth velocity dependence now takes the form

$$a_{0,min} \sim V^{-b}. \tag{8.42}$$

Hence, the exponent $-b$ is at least -0.3 and the exponential dependence always stronger than -0.25 implied by the intuitive analytical models. This is the consequence of array interactions of dendrites (Hunt and Lu, 1996; Hunt, 2001). Its logical heuristic interpretation is that, if there is a connection between plate spacing and tip characteristics of single dendrites, making the prediction of the plate spacing possible, then these scales must also interact by changing the selection and stability criteria. Another important finding due to the model was that the scalings for cellular, non-parabolic solidification were fundamentally different and given as (Hunt and Lu, 1996; Hunt, 2001)

$$a_{0,min} = 8.18k^{-0.335} \left(\frac{D_s}{V}\right)^{0.59} \left(\frac{\Gamma}{mC_\infty(k-1)}\right)^{0.41}. \tag{8.43}$$

It is seen that the spacing during *cellular growth* is strongly dependent on the distribution coefficient k, and that it has a considerably stronger growth velocity dependence. Figure 8.10 shows the numerical predictions from Hunt (2001) both for cells and dendrites in comparison with the above shown observations from Somboonsuk et al. (1984), illustrating the difference to the classical form 8.36 after Trivedi. The agreement for the dendritic points is remarkable, as already pointed out by Hunt and Lu (1996). Figure 8.11 from Ding et al. (1997) demonstrates the performance in the dendritic regime for another set of observations, in comparison to predictions with the boundary layer model from Warren and Langer (1993). The latter also simulates realistic spacings, but the model from Hunt and Lu (1996) predicts both the slope and absolute values better. For the two cellular freezing points in figure 8.10 the agreement is less good. A similar finding was noted in later studies (Ding et al., 1997; Feng et al., 1999) and is also evident from the comparisons made by Hunt and Lu (1996) themselves.

Selection parameter σ^*

It is noteworthy, that the effective tip radius selection parameter σ^* was found to be ≈ 0.015 in two dimensions and ≈ 0.021 in three dimensions, in the absence of surface energy anisotropy and for a pure melt (Hunt, 1991). For strongly interacting dendrites σ^* was found to decrease. The model apparently selects the tip radius independent of microscopic solvability for noninteracting dendrites. The numbers are similar to the classical marginal stability criterion, and comparable to the first simplified numerical studies for single dendrites on the subject, which resulted in respective estimates of

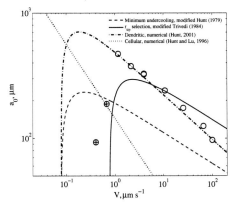

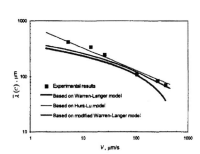

Fig. 8.10: Primary spacing observed by Somboonsuk et al. (1984) for the succinonitrile - 5.5 mol% acetone system (circles), shown with theoretical predictions as in figure 8.8 and the numerical simulation results from Hunt and Lu (1996) for cells and dendrites. The plus signs denote cells, the circles dendrites.

Fig. 8.11: From Ding et al. (1997). Comparison of observations of dendritic spacings of SCN-1.0 wt% acetone with the model predictions from Hunt and Lu (1996) and the boundary layer model results from Warren and Langer (1990, 1993).

≈ 0.020 for two and ≈ 0.020 to 0.025 for three dimensions (Oldfield, 1973; Langer and Müller-Krumbhaar, 1978). For alloy growth σ^* was more variable and Lu and Hunt (1992) reported $0.017 < \sigma^* < 0.025$ for dendritic succinitrile-acetone, yet larger and lower values for cellular spacing limits. Hunt and Lu (1996) also reported a different generalised scaling, when based on the far-field concentration C_∞ instead of C_{tip}. They then reported a larger $\sigma^{*\infty} approx 0.042$ for dendrites. The selection parameter σ^* for cells was strongly dependent on k, being ≈ 0.24 for $k = 0.1$. This indicates that the analytical scalings cannot be applied in this case. As pointed out by Hunt and Lu (1996), these results are not a consequence of an adjustment of tip radii to the spacing, yet reflects that dendritic tips must grow close to marginal stability. However, the selection was found to be influenced by anisotropy in the surface tension. The dependence on surface tension anisotropy was an approximate 10 to 15% decrease in the plate spacing for an anisotropy $\epsilon_4 = 0.02$ (Lu and Hunt, 1992; Hunt and Lu, 1996). According to linear solvability theory, one would expect $\sigma^* \approx 0.08$ for such an anisotropy (Karma and Rappel, 1998), and the mentioned analytical scalings would imply a reduction of a_0 with respect to the zero anisotropy by the factor $(0.021/0.08)^{1/4} \approx 0.72$. The actual decrease in a_0 reported for the model is thus less than implied by equation 8.36.

Tip undercooling ΔT_{tip}

Another important outcome from the numerical studies, that already had emerged during initial applications by Hunt and McCartney (1987), was the result obtained for the undercooling at the tip. In the analytical approaches noted above (Burden and Hunt, 1974; Hunt, 1979), the tip undercooling of cells was, using the present notation, approximated by

$$\Delta T_{tip} = mC_\infty \left(k - 1\right) \left(\frac{r_{tip}V}{D_s} + \frac{\mathcal{G}}{k}\right) + \frac{2\Gamma}{r_{tip}} \qquad (8.44)$$

As mentioned above, this equation may be differentiated with respect to r_{tip} and equated to zero to retrieve the minimum undercooling and the corresponding tip radius given by equation 8.35. The numerical model indicated a slightly different form (Hunt and Lu, 1996) which in the present notation reads

$$\Delta T_{tip} = mC_\infty \left(k - 1\right) \left(\frac{1}{2} + \frac{\mathcal{G}}{k} + \left(1 - \frac{k}{2}\right) \left(\frac{\mathcal{A}}{k^2}\right)^{1/2}\right). \qquad (8.45)$$

This equation shows, that the undercooling at the tips of cells is independent of the tip radius and not controlled by the latter. For growth far from equilibrium, when $\mathcal{G}/k \ll 1$, but the system is still far from absolute stability $\mathcal{A} = 1$, the cellular tip undercooling has a minimum $\approx mC_\infty(k-1)/2$, which is half of the earlier intuitive analytical assumptions.

The undercooling at the tips of dendrites in the numerical model could, according to Hunt and Lu (1996), be described as

$$\Delta T_{tip} = mC_\infty \left(k - 1\right) \left(\frac{\mathcal{G}}{k} + \left(\frac{\mathcal{A}}{k^3}\right)^{1/3}\right), \qquad (8.46)$$

and thus is, at low to intermediate growth velocities with $\mathcal{A} \ll 1$, less than than the undercooling during cellular growth. Again, the parameter $\mathcal{A}$ provides a contribution that becomes only important near absolute stability.

The differences between dendritic and cellular growth regimes were discussed to some degree by Hunt and Lu (1996). The leading term for the dendritic tip undercooling under moderate growth velocities is simply $\Delta T_{tip} \approx D_s G_{eff}/V$. It was first proposed by Bower et al. (1966) as an interfacial condition for the directional solidification of arrays and has since then been confirmed by observations (Burden and Hunt, 1974; Hunt and Lu, 1996; Pocheau and Georgelin, 1999). It is recalled that this is also the stability condition 7.30 for the onset of cellular growth from a planar surface suggested in the previous chapter and validated by observations of freezing saline solutions. The undercooling, or in the purely solutal case constitutional supercooling, therefore will not drop below the value at which the planar interface has become unstable, but it will neither be much larger.

8.3.5 History dependence

Hunt and Lu (1996) pointed out that their model can only predict the minimum spacing under steady conditions and not the generally observed range of spacings and the adjustment during increasing and decreasing growth velocities. While cells decrease their spacing by tip splitting, dendrites are generally observed to do so by transformation of secondary and tertiary arms into a new dendrite, and these processes are not resolved. Also for cells a maximum spacing two times the minimum spacing due to tip splitting is arguable, the picture is less clear for dendrites. Meanwhile it has been shown by a number of experiments that the allowable range of spacings is strongly history-dependent (Bechhoefer and Libchaber, 1987; Huang et al., 1993; Ding et al., 1997; Feng et al., 1999; Ma, 2002). The main conclusions from the latter studies are

- Upper and lower bounds of allowable spacings are rather sharp and differ by a factor ≈ 3.5 for unsteady growth compared to a factor 2 for steady growth.

- With *increasing* growth velocities, spacings may remain stable upon an increase by a factor of up to 10. Stable average spacings are larger than during constant growth velocity with a difference of up to a factor 2.

- With *decreasing* growth velocities, adjustment follows after a decrease in V by a factor 2 and the average spacing is close to the constant velocity spacing.

History-dependent studies also support that the dependence between tip radius and spacing is not unique. As demonstrated clearly by Bechhoefer and Libchaber (1987), the tip radius responds quite rapidly to an even small velocity change, while the spacing adjustment may require a considerable change.

8.3.6 Concluding remarks on spacing selection

Data for the succinonitrile-acetone system support that equation 8.36, slightly modified after Hunt (1979) and Trivedi (1984), reasonably predicts the steady state spacing, as shown in figure 8.8. Quite good quantitative agreement with Trivedi's scaling similar to 8.36 has been documented (Gündüz and Cadirli, 2001; Cadirli et al., 2003; Rocha et al., 2003; Spinelli et al., 2004), with quantitative differences of normally less than 30 %, and in most cases attributable to the less well known σ^* for other alloys. In most noted studies there is a tendency for a_0 to fall below the predictions at highest growth velocities, as also in figure 8.8. Exponents for $a_0 \sim V^e$ vary typically between -0.2 and -0.5 (Bouchard and Kirkaldy, 1997; Gündüz and Cadirli, 2001; Cadirli et al., 2003). Over smaller velocity regimes this can be explained by scatter, but on average the slope exceeds the value of -0.25 and is more near -0.35 to -0.4. A general scaling $a_0 \sim V^{-1/3} C_\infty^{1/6}$ has been suggested by Kirkaldy and coworkers (Liu and Kirkaldy, 1995; Bouchard and

Kirkaldy, 1997), based on a different concept than the tip radius selection. However, the scaling is not analytically derived, requires calibration constants and results in a too small concentration dependence (Quaresma et al., 2000; Rocha et al., 2003; Spinelli et al., 2004).

The numerical model from Hunt and Lu predicts the observations in figure 8.8 with remarkable accuracy. The model shows that the relations between a_0, r_{tip}, V and C_∞ are more complicated than in the analytical expressions and that the assumption of growth at the minimum tip undercooling is incorrect, although it might give reasonable cell spacings. While the numerical model produces somewhat different relationships than the analytical simplifying approaches, the parametric form of the relations is similar. The practically most important outcomes are the stronger $a_0 - V$-dependence and the finding that the minimum undercooling assumption is incorrect both for cells and dendrites. This had been pointed out by Hunt and McCartney (1987) from initial numerical studies and comparison with the approaches from Trivedi (1984) and Burden and Hunt (1974). However, although the approach from Trivedi (1984) was shown to be incorrect in terms of the interfacial undercooling (Hunt and McCartney, 1987), it seems to give reasonable dendrite spacings. This can be expected for situations when the difference between C_∞ and C_{tip}, the latter being used in the Trivedi scaling, is small. The reason seems to be that, while the minimum undercooling approach is not justified, the relation 8.29 is reasonable for certain conditions. Furthermore, considering cellular growth, equation 8.37, based on the incorrect minimum undercooling, gives reasonable estimates of cell spacings, because the selected wave length is close to this condition (Lu and Hunt, 1992; Hunt and Lu, 1996).

It is emphasised here that the scaling 8.40 from the numerical model contains the factor $(1-\mathcal{G})^{1/2}$ as in the minimum undercooling relation 8.36 and in contrast to $(1-\mathcal{G}')^{1/2}$ in the scaling 8.36 after Trivedi. As noted above, the factor $(1-\mathcal{G})^{1/2}$ may be in principal understood to reflect a matching condition at the root of the cell that is missing in the scaling from Trivedi. In figure 8.10 it is seen that this condition delays the maximum in a_0 to lower growth velocities. That a larger velocity is observed is not necessarily in conflict with the predictions, because the transition from dendrites to cells is more complex than the simple approach can capture. It seems that the transition to cellular growth is better resembled by relation 8.36 from Trivedi which, however, is otherwise less realistic. It may be recalled in this connection that the influence of the sample thickness on the results is neither fully understood. In other simulations (Warren and Langer, 1993; Spencer and Huppert, 1998) a much stronger response of r_{tip} to the sample thickness, than observed by Somboonsuk et al. (1984), was predicted. Implications of sample thickness for the dendrite-cellular transition and the spacing for these observations were emphasised by Billia et al. (1996).

The numerical model shows that a selection parameter of $\sigma^* \approx 0.02$ for the tip radius emerges also in the lack of surface tension anisotropy (Hunt, 1991; Lu and Hunt, 1992). This behaviour, and the decrease in σ^* for large interfacial undercoolings appears to be similar as proposed by the 'Interfacial wave theory' (Xu and Yu, 2001). It indicates the possible relevance of the latter, which incorporates conditions at the finite root of cells, for a better understanding of the interaction between cell spacing and tip shapes in arrays. While scalings 8.40 (numerical) and 8.36 (analytical) appear to be reasonably verified by observations of dendrites, the situation is worse for cells, when considering the numerical equation 8.43 from Hunt and Lu (1996), or eventually the minimum undercooling relation 8.37. One reason is, that the transition from dendrites to cells, accompanied by a decrease in wave length (as seen, e.g., in figure 8.10), is not only related to a principle shape change as incorporated by the numerical model, but also to the disappearance of sidebranches. The latter is more difficult to describe in a simple analytical manner (Molho et al., 1990; Kurowski et al., 1990; Georgelin and Pocheau, 1997). Regarding cellular growth it is evident from both observations (Georgelin and Pocheau, 2004) and the numerical model (Hunt and Lu, 1996) that the selection of r_{tip} is no longer in a simple manner related to solvability or marginal stability theory. As found in the numerical model, σ^* then is strongly dependent on the solute distribution coefficient k, which plays only a minor role during dendritic growth (Hunt and Lu, 1996).

8.4 Freezing of saline solutions: Validation of theories

Having summarised the main physics behind the scalings for the cellular spacings the relations can now be compared to the freezing of saline solutions. However, to do so it is important to ensure that diffusion is the controlling transport mechanism. This means that either upward freezing experiments or very rapid downward freezing experiments can be used for comparison. A rough estimate of the downward freezing velocity, where the influence of convection might be negligible, can be obtained from the limit $k_{eff} \approx 1$. Considering the freezing runs from Weeks and Lofgren (1967) discussed in an earlier section, this seems to happen at $\approx 5 \times 10^{-4}$ cm s^{-1}. It seems therefore not feasible to compare the observations of a_0, compiled in chapter 6 for natural growth velocities, to the scalings. However, some laboratory data are available at high growth velocities and upward freezing, and these will be analysed first, before an eventual extension to the regime influenced by convection can be made.

8.4.1 Downward freezing: Laboratory data from Lofgren and Weeks (1969a)

The observations from Lofgren and Weeks (1969a) have already been discussed in chapter 6 where their original plots of the data distribution were shown in figures 6.1 and 6.1. At certain salinities these authors performed downward freezing runs with a cold boundary of $-70\,°C$, during which very rapid freezing with almost complete solute entrapment was realised. The corresponding observations of a_0 are shown versus the growth velocity in figures 8.12 through 8.15 for four solution salinities and compared with the scaling laws 8.40 from the numerical model and the modified analytical scaling 8.36 after Trivedi. The lower bound 8.37 based on the the minimum undercooling principle after Hunt (1979), eventually valid for cellular freezing, is also shown.

Figure 8.12 shows the observations at the lowest salinity of $C_\infty \approx 1\,‰$. The agreement with the numerical model and the Trivedi prediction, which lie very close to each other, is convincing. Absolute values of a_0 deviate by less than 20% from the observations. The slope appears to be reasonably reproduced although a logarithmic fit results in $a_0 \sim V^{-0.49}$ compared to -0.75 from the predictions. This is mainly associated with one datapoint from the very initial freezing, a position within the sample just two millimeters from the cold boundary. The deviation of the latter may be related to initially imperfect orientation of crystals, or uncertainties in the growth velocity.

The comparison at the next higher salinity $C_\infty \approx 5\,‰$ shows a similar result (figure 8.13). The data points indicate a change in the slope near in the range 3 to $\approx 4 \times 10^{-4}$ cm s^{-1}, now indicating a convective velocity threshold as suggestive from the corresponding salinity entrapment observations from Weeks and Lofgren (1967). Above this velocity the observations are approximately 5 to 10 % below the predictions of the numerical model. The observed power law dependence $a_0 \sim V^{-0.53}$ is also here weaker than $b \approx -3/4$ implied by the model and analytical expressions, yet the data scatter indicates that the difference is not highly significant.

The agreement is considerably less for a salinity of $C_\infty \approx 30\,‰$ (figure 8.14). Again a sharp change in the slope may be seen at a velocity of 3 to $\approx 4 \times 10^{-4}$ cm s^{-1}. Considering only the high velocity data the observed a_0 is almost a factor 2 below the numerical model scaling. The discrepancy is largest at the highest salinity $C_\infty \approx 100\,‰$ in figure 8.15. Also these observations indicate that the $a_0 - V$ dependence is similar to the predictions but the absolute values are considerably smaller. At low growth velocities the slope is again different. The transition velocity seems to be similar for the salinities 5, 30 and 100 ‰. The magnitude of the predicted plate spacing in the presumed diffusive regime deviates the more from the observations the larger the salinity.

Tho illustrate the salinity dependence of the overprediction by the model the high growth velocity data for each salinity have been normalised to a single velocity $V_n = 10^{-3}$ cm s^{-1} and are shown in figure 8.16. The dependence is similar to figure 6.2 from Lofgren and Weeks (1969a) where all observations are shown. The spread at each salinity indicates the scatter in the data and deviations from a single power law. It is noted that the initial velocities for the runs at different salinities were different, at least due to the polynomial fits by Lofgren and Weeks (1969a), which may explain some of the deviations

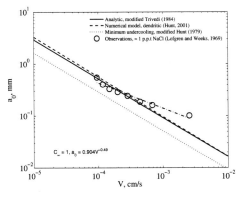

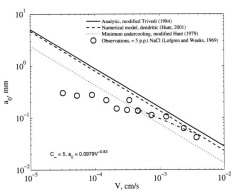

Fig. 8.12: Plate spacing during downward freezing of an 1‰ NaCl solution Lofgren and Weeks (1969a) along with the scaling laws 8.40, 8.36 and 8.37.

Fig. 8.13: As in figure 8.12 yet for a solution salinity of $C_\infty \approx 5$ ‰ NaCl.

in terms of a history dependence. However, it clearly emerges that the data do not reveal a clear concentrations dependence on a_0. If at all, the plate spacings are larger at low concentrations, yet the concentration dependence implied by the dendritic models is not seen in the observations. In additions, in figure 8.16 also equation 8.43 given by Hunt and Lu (1996) and Hunt (2001) for cellular growth has been plotted. It is noteworthy that, although it considerably overestimates the spacing at low concentrations, it yields an opposite concentration dependence.

The findings are further corroborated by figure 8.17 which shows simulations at and observations interpolated to $V_n = 10^{-4}$ cm s^{-1}. This is no longer a strict diffusive regime and predicted spacings are much larger. Due to the slower growth, the data are less scattered and indicate a decrease in a_0 with concentration, which is similar to the prediction for cells at low concentrations by the numerical model.

8.4.2 Upward freezing

Upward freezing experiments of saline solutions with well documented growth conditions and microstructural observations have been documented by Rohatgi and Adams (1967a) for KCl and Milosevic-Kvajic et al. (1973) for NaCl. Unfortunately both teams have presented their results in terms of the 'local solidification time' t_f, the time between passage of freezing and eutectic temperatures, which was either measured or modelled. Although such a presentation gave a good data collapse according to the relation $a_0 \sim t_f^{1/2}$, the approach may only be termed *diagnostic*. It does neither predict nor model the selection of the plate spacings. This was already discussed in section 6.2.2, where it was

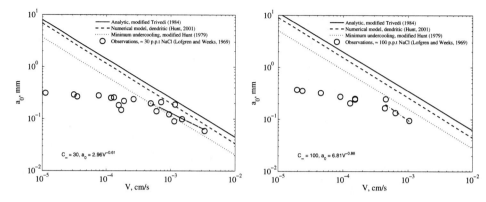

Fig. 8.14: As in figure 8.12 yet for a solution salinity of $C_\infty \approx 30$ ‰NaCl.

Fig. 8.15: As in figure 8.12 yet for a solution salinity of $C_\infty \approx 100$ ‰NaCl.

pointed out that the proposed 'lateral freezing model' from Rohatgi and Adams (1967b) is rather a solute budget and, in the presented form, has no predictive capability for the relationship between a_0 and V. However, both studies have tabulated some selected data from which the freezing velocity can be determined and these are discussed in the following.

Results from the study by Milosevic-Kvajic et al. (1973) are shown in figure 8.18 for an 10 ‰ NaCl aqueous solution. A temperature gradient of 10 K cm^{-1} in the liquid was reported by the authors, which leads, as seen in earlier figures, to a non-constant exponent. Although the velocities relate to the passage of the 0 °C isotherm, the overestimate is very likely only a few percent. It is seen that the agreement with the predictions is reasonable. Also here the overall velocity dependence appears to be weaker than predicted by the numerical model. As the temperature gradient in the liquid was only approximately given and may have been a function of time, thermal convection and the width of a solutal boundary layer, a detailed discussion of the deviation between the predicted and observed plate spacings is not attempted here. The main features appear similar as for the downward freezing experiments from Lofgren and Weeks (1969a), except that a convective threshold velocity for haline convection does not separate the data. The observations relate to three relatively thin 2 to 4 cm ice cores and one may suspect also here initial adjustment effects and inclined crystals which produce apparently larger plate spacings. Therefore it is believed that the results may be interpreted as a slight overprediction of a_0 by the models, consistent with figure 8.16 at this salinity.

Milosevic-Kvajic et al. (1973) have reported on, but not tabulated, freezing experiments at lower concentrations of 1 and 0.1 ‰. In the latter case the reported plate spacings were found to be approximately 30 % smaller. For the given temperature gra-

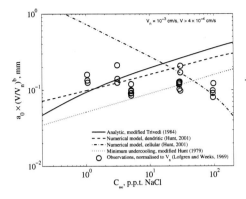

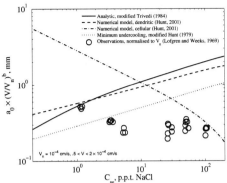

Fig. 8.16: High velocity observations of a_0 from figure 8.12 through 8.15 normalised to $V_n = 10^{-3}$ cm s^{-1} using the individual logarithmic fits for each salinity. The observations do not reveal a salinity dependence.

Fig. 8.17: Medium $0.5 < V < 2 \times 10^{-4}$ cm s^{-1} velocity observations of a_0 from figure 8.12 through 8.15 normalised to $V_n = 10^{-4}$ cm s^{-1}. The observations indicate a salinity dependence reflecting cellular growth.

dient the predictions yield a plate spacing which is smaller by a factor of three (not shown) and hence again indicate that the models yield a concentration dependence that is not revealed by the observations. Milosevic-Kvajic et al. (1973) note that the liquid temperature gradient was related to thermal convection during upward freezing, when the thermal stratification at low salinities is unstable. This has been verified in a parallel study (Kvajic et al., 1971). Although not reported by the authors, it is likely that temperature gradients and thermal convection conditions have been different for the freezing runs at 10 and 0.1 ‰, at least close to the interface. One may, for example, interpret the ≈ 20 % smaller growth rate during freezing of pure water or $C_\infty = 0.1$ ‰, compared to the $C_\infty = 10$ ‰ run, in terms of a larger effective temperature gradient in the liquid. This is consistent with a two times larger difference between freezing and maximum density temperatures for the pure and dilute cases. The thermal gradient increases $\mathcal{G}$ and thus decreases a_0. The observed decrease is roughly consistent with an ad-hoc estimate. For a quantitative comparison, a study of this double-diffusive system with unstable thermal and stable haline stratification is required (Turner, 1973). Additional complications may arise due to the confinement and wall effects in the freezing cylinder. It seems, from the approximate discussion, that also the experiments from Milosevic-Kvajic et al. (1973) imply plate spacings which are almost independent of concentration, contrary to the model predictions.

Another growth series has been tabulated by Rohatgi and Adams (1967a) for the

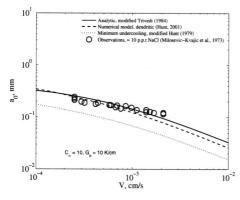
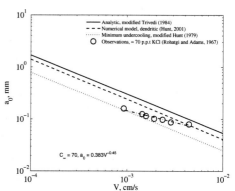

Fig. 8.18: Plate spacing a_0 during the upward freezing of a 10 ‰ NaCl aqueous solution in a cylinder with an interfacial temperature gradient ≈ 10 K cm^{-1} in the liquid (Milosevic-Kvajic et al., 1973) compared to model predictions. The observations refer to sectioned ice cores from three freezing runs.

Fig. 8.19: Plate spacing a_0 during the upward freezing of a 70 ‰ KCl aqueous solution (Rohatgi and Adams, 1967a) to model predictions.

upward freezing of an ≈ 70 ‰ KCl (1 N) aqueous solution (From their table I growth velocities were determined and related to plate spacings from their table III). The comparison with predictions, where the different m and D_s for KCl were accounted for, are shown in figure 8.19. Also here, a weaker $a_0 - V$-dependence than predicted is observed, and the overestimate by the predictions is similar as one may guess from figure 8.16, at a similar supercooling mC_∞. On the other hand, Rohatgi and Adams (1967b) did, in a parallel publication, report experiments at ≈ 5 and 10 times lower concentrations with smaller a_0 (their figure 10), and a concentration dependence which is comparable to the model predictions. However, observations were not compared at the same growth velocities but at distances of 1 to 3 cm from a -70 °C chilled surface. A possible explanation is that, initial growth velocities might be different due to adjustment, convection and supercooling effects. A closer look shows, that the initial values of a_0 for the low concentration runs at 0.1 and 0.2 N KCl are much smaller than predicted by the numerical model. They are close to the minimum undercooling prediction. This indicates initial eutectic freezing for the dilute runs, and one may suspect that such a spacing could eventually survive and thus can explain the salinity dependence.

Another set of experiments, for which Rohatgi and Adams (1967b) reported an increase in a_0 with solute concentration, was the freezing of droplets falling through a -20 °C cold organic liquid. Again, freezing velocities were not given, yet assuming these were similar, the observed concentration dependence is very close to $a_0 \sim C_\infty^{1/4}$ from the models. This is, however, a situation which differs from a free freezing interface selection problem. High concentration droplets likely freeze slower, because the effective temperature forcing is less. Droplet freezing may also be much more affected by expansion phenomena. Although it seems that the freezing of droplets follows the predicted concentration dependence, the overall interpretation of the particular freezing experiments from Rohatgi and Adams is less clear. The two applied methods may not produce comparable stable conditions at different concentrations. While non-equilibrium and eutectic freezing are plausible for the cold surface configuration, one may expect confinement, non-diffusive solute transport, and freezing rate differences to be relevant for the falling droplets.

8.4.3   Discussion: concentration dependence of a_0

The observations indicate that the analytically and numerically estimate scaling laws 8.40 and 8.36 yield qualitatively correct estimates of the sea ice plate spacing in the diffusive growth regime at low concentrations. The concentration dependence of roughly $a_0 \sim C_\infty^{1/4}$ implied by the theories is, however not revealed by the observations from Lofgren and Weeks (1969a) and, considering the specific growth conditions under thermal convection, neither supported by the study from Milosevic-Kvajic et al. (1973). It has further been argued that the concentration dependence found in the experiments by Rohatgi and Adams (1967b) may not be based on equivalent growth conditions, because the rather small spacings for the low concentration experiments indicate eutectic freezing. The exponential $a_0 - V$-dependence revealed by the observations is weaker than the models predict.

It is unlikely that the lack in the concentration dependence can be related to differences in haline convection, as all freezing runs at different concentrations indicate a similar convective threshold velocity of $\approx 5 \times 10^{-4}$ cm/s. One may therefore ask, if the generally observed plate-like freezing of saline solutions can be modeled by a dendritic model approach and point to the difficulties when interpreting the observations from figure 8.2 by a general growth law. As discussed in connection with the latter figure in section 8.2.1 above, it is the high anisotropy in the surface tension of ice which would imply a much larger σ^*, most likely near 5 (Karma and Rappel, 1998) and thus a factor 250 above the marginal stability values, when growth normal to the c-axis is considered. One may question, if the analytical equation 8.36 can be applied at all in this case. On the other hand, it is apparent from figure 8.2 that the classical marginal stability approach with $\sigma^* \approx 0.025$, despite of not accounting for the correct solvability solution and the plate-like growth, predicts the growth rates of ice dendrites reasonably. One may then further argue that the procedure leading to the analytic equation 8.36 via matching conditions of the solid fraction should be applicable to the problem of plate-like growth of ice dendrites. As mentioned above in the numerical model the plate spacing's response to changes in anisotropy was much weaker than implied by the analytic equation (Lu and Hunt, 1992; Hunt and Lu, 1996), which would also favour the applicability.

To clarify the questions numerical simulations at high anisotropy would be needed. A simple explanation for the good quantitative agreement could be that the plate structure of sea ice forms behind a dendritic tip array, as indicated by the instabilities on the edge of the basal plane in figure 7.7 from Nagashima and Furukawa (2000b). This is also supported by observations from Harrison and Tiller (1963), who reported on the transition from knife-edged to scallop-edged cells with increasing constitutional supercooling. Moreover, in their optical observations, c-axis dendrites existed besides basal plane dendrites, with no significant difference in the spacing between them. Whatever may explain the agreement between the three-dimensional predictive scalings and the plate spacing observations for saline solutions, it is not obvious how the concentration dependence of $a_0 \sim C_\infty^{1/4}$ could disappear due to the differences. Reconsidering observations of other alloys, an exponent of $1/4$ has been sometimes reported and confirmed (Cadirli et al., 2003; Bouchard and Kirkaldy, 1997), but there appear to be many observations where it is much weaker or even nonexistent (Bouchard and Kirkaldy, 1997; Rocha et al., 2003; Spinelli et al., 2004). In the succinonitrile-acetone system discussed above the exponent was approximately 0.1 to 0.15 for the dendritic regime (Trivedi and Somboonsuk, 1984).

It seems to be safe to conclude that the predictive equations 8.40 and 8.36 overestimate the dependence of the plate spacing on both growth velocity and concentration. The overestimate is implied by equations derived from tentative analytical approaches (Hunt, 1979; Kurz and Fisher, 1984; Trivedi, 1984), a mathematically more sophisticated detailed shape matching formulation (Spencer and Huppert, 1998) and numerical models (Warren and Langer, 1990; Hunt and Lu, 1996) which all give similar scalings. While in many cases the velocity dependence of the spacing has been properly predicted for other alloys, the overprediction of the concentration dependence seems not to be specific for saline solutions yet is frequently observed. This indicates a deficiency in the dendritic models.

It is interesting that equation 8.43 given by Hunt and Lu (1996) and Hunt (2001) for cellular growth yields an opposite concentration dependence, and that this dependence is close to the observed one at lower growth velocity and low concentration (figure 8.17), when cells are most likely to occur. In a quantitative sense its applicability to the low $k = 0.001$ for non-dilute aqueous NaCl solutions (see chapter 4) is uncertain, because Hunt and Lu (1996) only used a fifty times larger minimum k in its derivation. However, the relationship between a_0 and V should be the same. Cellular growth apparently implies a mechanism which leads to an opposite concentration-dependence than dendritic growth. This behaviour indicates that the tip radius, which decreases with concentration, controls the cellular spacing in a different way than assumed in the analytical relation 8.29. The weaker growth velocity dependence in the cellular scaling seems also to agree better with the ice growth observations, indicating that these mechanisms may be operative in sea ice. It is noteworthy that a dependence $a_0 \sim V^{-1/2}$ was found by Kessler and Levine (1989) in a numerical study. These authors suggested a path to the understanding via the solvability condition $r_{tip}^2 V = const.$, which would produce the observed dependence if cells are close to spherical and $a_0 \approx 2r_{tip}$. Furthermore do recent numerical phase-field simulations of mostly cellular shapes do also support a growth law $a_0 \sim V^{-1/2}$ (Boettinger and Warren, 1999; Lan and Chang, 2003; Takaki et al., 2005).

In summary, one may suggest that the above considerations of different concentration dependencies indicate some mixed mode between dendritic and cellular pattern for sea ice. While in figure 8.16 only the high growth velocity data were considered, the original figure 6.2 from Lofgren and Weeks (1969a) shows a similar behaviour also at lower growth velocities with convection present. Observations for the velocity regime $0.5 < V < 2 \times 10^{-4}$ cm/s are shown in figure 8.17, emphasising the decrease in plate spacing at low concentrations and lower growth velocities, when cellular pattern are most plausible.

8.5 Summary and outlook

It has been the aim of the present chapter to provide an overview of attempts to predict the dendrite or plate spacing during directional solidification. To do so the history of the problem of free growth of a dendrite has been shortly sketched. Its presently widely accepted solution is the *microscopic solvability theory*, which determines uniquely the tip radius and growth rate a parabolic needle crystal will select under a given supercooling.

The theory has been illustrated in connection with observations of the pattern and growth velocity of ice dendrites. As ice is a highly anisotropic material, both in terms of solid-liquid surface energy and growth kinetics, the transition between different growth regimes is complex, yet can be principally understood in terms of solvability theory. It is important to realise the different tip radii of ice dendrites as stability length scales which emerge as a consequence of the large anisotropy in the surface tension.

Next, theories of plate selections were reviewed and discussed. The solution of the free dendrite growth problem justifies assumptions about the geometry at the solid-liquid interface, being an array of tips of parabolic needles. Matching their geometry with the Scheil-equation of equilibrium solute redistribution results in an equation between the spacing and the tip radius (Hunt, 1979; Kurz and Fisher, 1984), which then may be supplemented by a stability condition for the latter Trivedi (1984). Other matching principles arrive at the same proportionality $a_0 \sim V^{-1/4} G_{eff}^{-1/2}$ which, if G_{eff} is simply proportional to V, yields $a_0 \sim V^{-3/4}$. A slightly stronger $a_0 - V$-dependence is suggested by different time-dependent models (Warren and Langer, 1990; Lu and Hunt, 1992; Hunt and Lu, 1996).

It has been shown that the analytical and numerical scaling laws for *dendritic spac-*

ings allows a reasonable quantitative prediction of the plate spacing during freezing of aqueous saline solutions at not too large salinities. The observed $a_0 - V$-dependence is, however, clearly weaker than predicted by any of the relations. Neither is a significant concentration dependence found in the observations. The comparison has been made in a growth velocity regime where presumably the influence of rigorous global haline convection is small, although local effects can not be excluded. The conclusions seem also to apply to upward freezing experiments from Milosevic-Kvajic et al. (1973), if plausible modifications due to thermal convection are accounted for. The comparison indicates a deficiency in the predictions of dendritic spacings by analytical and numerical models that is not unique for sea ice, but found for other alloys. Alternative semi-empirical scaling laws, suggested by Kirkaldy and coworkers and yielding a velocity exponent of -0.5 to -0.66 (Liu and Kirkaldy, 1995; Bouchard and Kirkaldy, 1997), do not solve the concentration problem.

Cellular spacings follow, according to the numerical model from Hunt and Lu (1996) and recent phase field simulations (Boettinger and Warren, 1999; Lan and Chang, 2003; Takaki et al., 2005), a weaker velocity dependence. The fact that also the concentration dependence in the numerical model is the opposite to dendritic freezing, indicates that the plate spacing selection during freezing of saline solutions is governed by a combination of these cases. Although the data in figure 8.16 must be rated as limited, because they are likely influenced by convection at low growth velocities, the concentration dependence of a_0 at low concentrations seems to agree with the numerical predictions for cellular growth.

The problem of plate spacing prediction apparently is not solved in a concise manner for quasi-isotropic binaries, because the overestimate in the concentration dependence is also found in other systems. For the ice-water system not only the large anisotropies in growth kinetics and solid-liquid interfacial energy add to this indeterminacy, but, at natural growth velocities, also the effects of haline convection. All mechanisms may critically influence the crossover between cellular and dendritic shapes, and thereby change the spacing dependence on growth velocity and concentration. Phase field simulations for widely separated dendrites appear not realistic at the moment, and more observations appear necessary to verify heuristic transition criteria for saline solutions. A very simple approach is now proposed in the next chapter.

9. Cellular spacing and morphological stability in saline water

Considering the prediction of spacings for dendritic arrays, the scalings given by Hunt (2001) on the basis of numerical model simulations are at present those which agree best with observations. However, as discussed in the previous chapter, the concentration and velocity dependence of a_0 for saline solutions is not correctly predicted.

Discrepancies between theoretically predicted and observed spacings, in particular in the velocity and concentration dependencies, are not unique for saline solutions but found for other systems as well Trivedi and Somboonsuk (1985); Billia and Trivedi (1993); Hunt and Lu (1996); Bouchard and Kirkaldy (1997). The problem is also met with alternative semi-empirical relations (Billia and Trivedi, 1993; Bouchard and Kirkaldy, 1997). In other systems the disagreement has sometimes been explained in terms of history-dependent spacing adjustment (Hunt and Lu, 1996; Huang et al., 1993). However, in any case the numerical predictions are expected to provide a lower bound on the spacings (Hunt and Lu, 1996; Huang et al., 1993), which is not found for the case of saline ice. Hence, the Lu-Hunt model appears only to be a first step to a concise theory.

Here an approach to the problem is made that largely ignores the geometrical constraints of the selection process. It is very similar to a criterion proposed by Bower, Brody and Flemings, who once suggested the condition

$$-m(C_{int} - C_\infty) = \frac{DG_{eff}}{V}, \tag{9.1}$$

for the equilibrium growth of a cellular interface on the basis of observations (Bower et al., 1966). It has been confirmed by recent observations identifying C_{int} with the concentration C_{tip} at the tip of cells (Pocheau and Georgelin, 2003). In the notation of morphological stability theory the condition corresponds to $\mathcal{G} = 1$, if C_∞/k at equilibrium is replaced by C_{int} in equation 7.18. It is equivalent to equation 7.30 with $\mathcal{S} = 1$, which has been verified in section 7.2.2 as the condition for the onset of instability of a planar interface in thin growth cells. The criterion 9.1 may thus be interpreted as the *marginal stability* of a *macroscopic interface* and it motivates the following simple solution to the plate spacing problem.

9.1   Marginal array stability: Basic solution

Treating the marginal stability of the macroscopic interface in the same manner as the planar case, the Mullins-Sekerka theory from section (2) is applied *macroscopically* with two principal modifications.

9.1.1   Effective interfacial conductivity

First, one has to use an effective thermal conductivity, with the solid-liquid mixed phase conduction replacing the pure ice conductivity, and an effective latent heat of fusion for the mixed phase. This cannot simply be done by introducing a parallel heat conduction model, as it actually involves the solution of the thermal diffusion field at an interface where the dendrites are thickening laterally. An approximate analysis of the problem and its observational validation has been given in the appendix B.6. The results are shown in figure B.10, which indicates that the problem can be treated in a macroscopic sense, by using the latent heat of fusion of pure ice in connection with an enlarged effective thermal conductivity

$$K_i' = \frac{K_i}{r_{tz}},\tag{9.2}$$

where r_{tz} is ≈ 0.56 for $C_\infty = 35$ ‰ NaCl. Figure B.10 shows that the factor r_{tz} is relatively insensible to the solute concentration in the range 10 and 100 ‰. On the other hand, one may argue that the temperature gradient under certain conditions can sustain its steepening near the interface as proposed in figure B.6 of the appendix. If this is the case, one would simply have to use the conductivity K_i for pure ice in the computations. This condition results in ≈ 30 % larger predictions of a_0 at moderate salinities, while the approaches do not differ for dilute solutions where $r_{tz} \approx 1$. The conditions that may favour the too mentioned cases will be discussed below.

9.1.2   Maximisation of k

The solution to the problem is obtained by realising the second difference between a planar and macroscopic interface: The distribution coefficient k is no longer a material property. It is now a free parameter that can be maximised as follows. For completeness the approximate equations for the marginal stability of a planar interface (Coriell et al., 1985) given in section 7.1.4 are repeated here: To study the extrema of the morphological stability dispersion relation 7.11 the surface tension parameter

$$\mathcal{A} = \frac{k^2 \Gamma V}{mC_\infty(k-1)D_s} := \mathcal{A}_0 \frac{k^2}{k-1},\tag{9.3}$$

and the constitutional supercooling parameter

$$\mathcal{G} = \frac{k}{mC_\infty(k-1)}\left(\frac{L_vD}{K_i'+K_b}\right)\left(1 + \frac{2G_bK_b}{L_vV}\right) := \mathcal{G}_0\frac{k}{k-1}, \tag{9.4}$$

are relevant for the planar case. Here the corresponding $\mathcal{A}_0$ and $\mathcal{G}_0$ are defined in addition. They do not depend on k.

As in section 7.1.4 for the planar interface, the variable

$$r = \left(1 + \left(\frac{2D_s\omega}{V}\right)^2\right)^{1/4} \tag{9.5}$$

is obtained from the cubic equation

$$r^3 + (2k-1)r - \frac{2k}{\mathcal{A}^{1/2}} = 0, \tag{9.6}$$

and the relationship between k, $\mathcal{G}$ and $\mathcal{A}$ at onset of instability is

$$\mathcal{G} = 1 - \frac{3}{2}\mathcal{A}^{1/2}r + \frac{\mathcal{A}}{4k}\left(1 - (1-2k)r^2\right), \tag{9.7}$$

where r is the one real cubic root greater the unity from the cubic equation 9.6.

The maximum k is obtained from equations 9.3 through 9.7. For given thermodynamic system properties it depends on $\mathcal{A}_0$, $\mathcal{G}_0$ and V. However, a useful approximation for conditions considerably far from absolute stability ($\mathcal{A} \approx 1$) given by Coriell et al. (1985) was already mentioned as equation 7.24. Now, in terms of $\mathcal{A}_0$ and $\mathcal{G}_0$ one may write the latter as

$$\mathcal{G}_0\frac{k}{k-1} = 1 - \left(\frac{27k^3\mathcal{A}_0}{4(k-1)}\right)^{1/3}. \tag{9.8}$$

which is a fifth-order polynom to be solved for k. If one restricts the analysis to $V < 10^{-3}$ cm s^{-1}, this equation gives a maximum overestimate of the exact $(1-k)$ by ≈ 6 % at large concentrations near 200 ‰NaCl. As will be seen next, this only implies an underestimate of the wave length by 2 %. The approximation 9.8 is thus sufficient for applications to natural growth conditions of sea ice (and laboratory observations discussed here).

The wavelength λ_{mi} corresponding to the maximum k is then readily obtained from equation 7.35 and recalled as

$$\lambda_{mi} = 2\pi 2^{1/3}\left(\frac{D_s}{V}\right)^{2/3}\left(\frac{\Gamma}{mC_\infty(k-1)}\right)^{1/3}. \tag{9.9}$$

The essence of the solution is the dependence $\lambda_{mi} \sim (1-k)^{-1/3}$ with k depending on C_∞.

9.2 Observations

9.2.1 Validation of array stability: k

The marginal stability or array stability condition corresponds to the maximum possible k for the macroscopic interface and thereby to the minimum $(C_{int} - C_\infty)$. It may then be interpreted as a minimum undercooling condition. Disregarding the correction corrected due to the capillary term implied by $\mathcal{A}$, which is small under natural growth conditions, the condition corresponds to the conjecture from (Bower et al., 1966). The important extension of the present approach is, that the condition is formally related to an effective k at the macroscopic interface. This value of k in turn allows the prediction of the most rapidly growing wave length λ_{mi} from morphological stability theory. Hence, in contrast to the approaches from the previous chapter, this condition does not imply any assumption about the geometry of the tip of the cells. It does, however, yield the spacing of the cells.

As already mentioned, there are confirmations of the minimum undercooling condition for metallic alloys, yet these are to date relatively sparse, as they require detailed temperature and composition observations (Bower et al., 1966; Pocheau and Georgelin, 2003). Observations for NaCl solutions are available from a study by Terwilliger and Dizio (1970). These authors used micro-conductance probes and obtained reasonable resolution of the solutal boundary layer during upward freezing. From these concentration profiles and the final salinity in the ice, which was close to $k_{eff} = 1$, Terwilliger and Dizio obtained k at different salinities. These values are shown as circles in figure 9.1. The freezing experiments were performed with two types of coolants which gave slightly different values of k for the thermal forcing conditions. Over the range 4 to 40×10^{-4} cm s^{-1} the predicted k is not very sensitive to the growth velocity and predictions are shown for 10^{-3} cm s^{-1}. Freezing was performed in containers filled with solutions approximately 20 K above the freezing point. This implies that one has to use equation (B.6) to compute the temperature gradient in the liquid, to be inserted in equation 9.4 above. For the two sets of thermal forcing documented by Terwilliger and Dizio (1970) the temperature gradients in the liquid are approximately 50 to 100 % of the gradients in the solid. The predictions are shown for $\Delta T_i / \Delta T_b = 2$ when $G_i \approx G_b$.

The agreement between theory and observations is convincing and does only change slightly if $\Delta T_i / \Delta T_b \approx 5$ for the stronger thermal forcing conditions is used in the theory (not shown). It appears that the approach $K_i' = K_i / r_{tz}$, applying an effective interfacial conductivity, agrees slightly better with the observed k. If, as discussed by Terwilliger and Dizio (1970) some thermal convection was present, the solid curve would be closer to the observations. It is notable that the lower k values at each concentration represent the weaker thermal forcing in these experiments, and that the slight difference in k

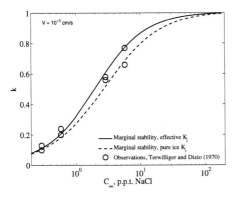

Fig. 9.1: Interfacial distribution coefficient k during upward cellular freezing of aqueous NaCl solutions (Terwilliger and Dizio, 1970). The predictions are from the present marginal stability theory with 10^{-3} cm s^{-1} and a temperature drop ratio $\Delta T_i/\Delta T_b = 2$ for ice and brine. Dashed and solid curves resemble predictions based on pure ice and effective thermal conductivity of the system.

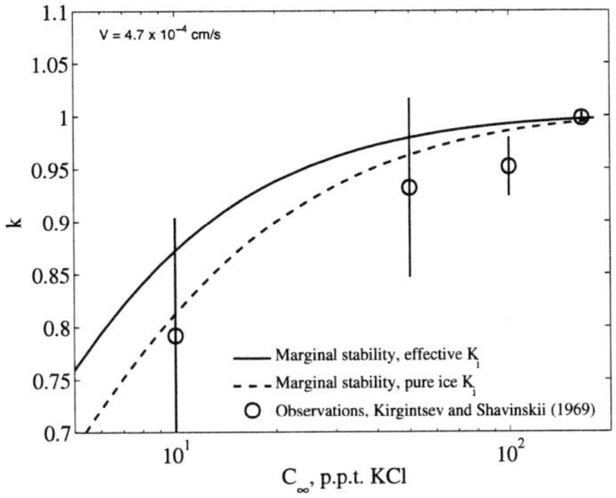

Fig. 9.2: Interfacial distribution coefficient k during upward cellular freezing of aqueous KCl solutions Kirgintsev and Shavinskii (1969a) at $V = 4.7 \times 10^{-3}$ cm s^{-1} and a temperature gradient $G_b \approx 20$ K cm^{-1} in the brine. Dashed and solid curves resemble predictions based on pure ice and effective thermal conductivity of the system.

is also predicted by theory. Although the observations seem to support the effective conductivity approach, the comparison is based on only a few observations and should not be overrated, however.

A second set of observations which is suited to validate the present approach has been published for the upward freezing of aqueous KCl solutions by Kirgintsev and Shavinskii (1969a). These authors performed experiments for a much larger concentration range at a growth velocity of $V = 4.7 \times 10^{-3}$ cm s^{-1}. Here the temperature gradient in the brine was documented as $G_b \approx 20$ K cm^{-1}. The evaluated k shown in figure 9.2 is not based on observations of the solute diffusion boundary layer, yet on sectioning of the completely frozen bottles. Here, an average over k during freezing of the first 4 tenth of the bottles is shown. Although the standard deviations, obtained from 5 samples at each salinity, are rather large, these observations in principle confirm the theory also in the high concentration regime. A possible explanation for the overestimate in k may be the slower freezing compared to the experiments from Terwilliger and Dizio. Slower freezing enhances the relative influence of thermal convection, increasing G_b and decreasing k.

It can be concluded that the observations from Kirgintsev and Shavinskii (1969a) and Terwilliger and Dizio (1970) reasonably confirm the marginal stability hypothesis in

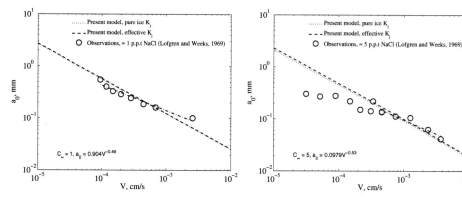

Fig. 9.3: Plate spacing during downward freezing of an ≈ 1 ‰ NaCl solution Lofgren and Weeks (1969a). The present marginal stability predictions are shown for comparison.

Fig. 9.4: As in figure 9.3 yet for a solution salinity of $C_\infty \approx 5$ ‰ NaCl.

terms of k over a wide range of salinities.

9.2.2 Plate spacings

Downward growth

In figures 9.3 through 9.6 the present theoretical predictions of λ_{mi} from equation 9.9 are compared with the plate spacings observed by Lofgren and Weeks (1969a). At high growth velocity the agreement appears convincing for any salinity. Both the absolute values and the slope in the $a_0 - V$ dependence are closely predicted. The predictions slightly underestimate the observed spacings, on average by 10 to 20 %. The figures should be compared with figures 8.12 through 8.15, where the agreement with several discussed models was shown to be much worse at high salinities. The discussed models only yielded a reasonable quantitative prediction of a_0 at the lowest salinity 1 ‰ NaCl (figure 8.12). Also here the present approach is more convincing in terms of the $a_0 - V$ slope.

It is important to note that the effective macroscopic K_i' (dashed curves) matches the observations better in any case than thee pure ice K_i. This parallels the comparison of the model predictions with the observed k from Terwilliger and Dizio (1970). It demonstrates the small, but significant effect of a consistent heat transfer model at the tip of dendrites (appendix B.6).

As mentioned in the previous chapter, in the comparison with other models the convective threshold velocity had appeared to be near $\approx 5 \times 10^{-4}$ cm/s. The present approach

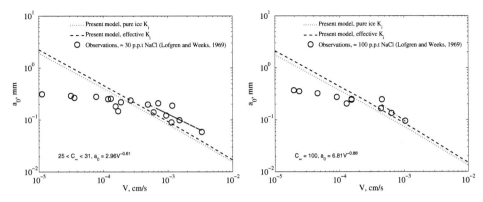

Fig. 9.5: As in figure 9.3 yet for a solution salinity of $C_\infty \approx 30$ ‰ NaCl.

Fig. 9.6: As in figure 9.3 yet for a solution salinity of $C_\infty \approx 100$ ‰ NaCl.

indicates, for all freezing runs, a transition in the $a_0 - V$ slope near $V = 2 \times 10^{-3}$ cm s^{-1}. This again points to the importance of haline convection in the whole natural growth regime. However, as pointed out in an earlier section 6.1.1, a discussion of this problem on the basis of the dataset from Lofgren and Weeks (1969a) is problematic, as the freezing at low growth velocities was very likely affected by thermal convection due to sidewall heating. An interpretation is delayed here to the following chapter 10.

Upward growth

The observations of a_0 in upward growing NaCl ice, tabulated by Milosevic-Kvajic et al. (1973), are compared to the predictions in figure 9.7. In these experiments an average temperature gradient of ≈ 10 K cm^{-1} in the liquid has been reported. This gradient resembles the effect of thermal convection. At low growth velocities the agreement is good, yet the predicted $a_0 - V$-dependence appears to be different at large growth velocities. It is worth a note that the observations appear as groups with a slope comparable to the predicted one. It must also be noted that the growth velocity was not observed, yet obtained from the passage of the 0 °C isotherm.

In figure 9.8 the results from the high concentration aqueous KCl freezing run tabulated by (Rohatgi and Adams, 1967a) are shown. Here the dependence of a_0 on V is closer to the predictions, yet the theoretical values are a factor of 1.5 to 2 too low. This cannot be explained by the different molecular properties of aqueous KCl, because these have already approximately been included. The freezing procedure from Rohatgi and Adams (1967b,a) was, however, different from other methods. The solutions were poured into a tube fitted on a -70 °C cold chill. The reproducibility of this approach is unclear, when comparing the plate spacings from a later work (Rohatgi et al., 1974), which

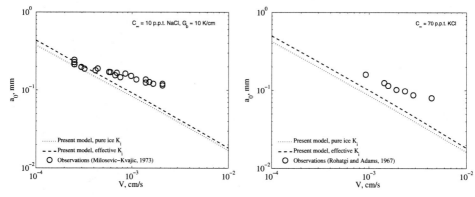

Fig. 9.7: Plate spacing a_0 during the upward freezing of a 10 ‰ NaCl aqueous solution in a cylinder with an interfacial temperature gradient ≈ 10 K cm^{-1} in the liquid (Milosevic-Kvajic et al., 1973). The present marginal stability predictions are shown for comparison.

Fig. 9.8: Plate spacing a_0 during the upward freezing of a 70 ‰ KCl aqueous solution (Rohatgi and Adams, 1967a) The present marginal stability predictions are shown for comparison.

were almost a factor of 2 smaller. As growth velocities were not reported in general, no other observations than those in figure 9.8 can be compared to the present theory. As mentioned earlier, this method resulted in a concentration dependence of a_0 which was very different from the one observed by Lofgren and Weeks (1969a). This concentration dependence is also different from the present predictions.

Although it seems that the present theory underpredicts the plate spacing for both upward growth experiments considered, while it overpredicts the dependence of a_0 on V, the situation is not that clear. It is unknown, to what degree the growth conditions of the two upward freezing experiments resemble the case of directional solidification with well controlled boundary conditions. Several aspects and possible biases have already been discussed in section 8.4.1 of the previous chapter. The most important one, with respect to the latter two studies, is thought to be the crystal inclination during early growth.

The shown observations from the studies by Milosevic-Kvajic et al. (1973) and Rohatgi and Adams (1967b) refer to 2 or 3 centimeters of rapid ice growth. During this initial rapid growth the structure may be different and often still has to acquire an equilibrium. Most important, it has been found that the lamellar plates in sea ice may be inclined by typically 50 to 60° within the first one or two centimeters (Fukutomi et al., 1952; Weeks and Assur, 1964). This implies an overestimate of the intrinsic plate spacing by a factor of 1.5 to 2, if sections normal to the cooling interface are analysed. It could explain the difference between the predictions and the observations in figures 9.7 and

9.8. An inclination correction, as recommended by Weeks and Assur (1964), has neither been reported by Lofgren and Weeks (1969a). The observations from Weeks and Assur indicate a decrease in the bias at larger distances from the surface to $\approx$ 20 % at 3 cm and less than 10 % at 5 cm distance. In the study from Lofgren and Weeks (1969a) the analysed ice cores were more than 10 cm thick and only the first one or two (highest velocity) points might have been severely affected by this inclination. To what degree downward freezing favours a more rapid inclination adjustment than upward freezing is unknown. Effects of micro-convection may play a role in this respect.

In summary, there is a slight underprediction of plate observed spacings from the studies of Rohatgi and Adams (1967a) and Milosevic-Kvajic et al. (1973). The worse agreement in figures 9.8 and 9.7 is, due to the different ice growth methods and implied unmodeled physics, not necessarily in contradiction with the results in figures 9.3 through 9.6.

9.3 Concentration dependence

The largest discrepancy of the models discussed in the previous chapter was their deficiency to predict the observed concentration dependence of the plate spacing (figures 8.16 and 8.17). From figures 9.3 through 9.6 it is already clear that the present approach gives reasonable predictions at concentrations from 1 to 100 ‰ NaCl. In figure 9.9 the theoretical predicted λ_{mi} from equation 9.9 is compared with the plate spacings from Lofgren and Weeks (1969a), interpolated to $V_n = 10^{-3}$ cm s^{-1}. The bars indicate the standard variation from the interpolation and may be interpreted as the uncertainty at each concentration. It is seen that the present approach predicts the concentration dependence of a_0 in a reasonable manner.

Also here it seems that, using the effective K_i', the basic aspects of the observed concentration dependence are better resembled by the predictions than for the pure ice K_i. The K_i' approach shows indications of an intermediate minimum in a_0. The observations in figure 9.9 are, however, too uncertain for a detailed validation.

Figure 9.10 makes the same comparison at a lower growth velocity of $V_n = 10^{-4}$ cm s^{-1}, where convection is expected. As already mentioned in the previous section in comparison with all other models, the observed plate spacings are now considerably lower. It is also noted, that convection apparently leads to an enhancement of the minimum in a_0 at intermediate concentrations, which is only slightly indicated in the purely diffusive model predictions. As mentioned, an account of convection is necessary and will be presented in the following chapter. The following discussion considers the generally expected concentration dependence at high growth velocities, in the absence of convection.

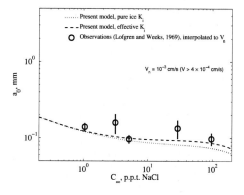
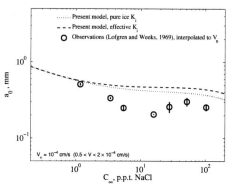

Fig. 9.9: High velocity observations of a_0 from figure 8.12 through 8.15 normalised to $V_n = 10^{-3}$ cm s^{-1} using the individual logarithmic fits for each salinity. The present marginal stability predictions are shown for comparison.

Fig. 9.10: Medium $0.5 < V < 2 \times 10^{-4}$ cm s^{-1} velocity observations of a_0 from figure 8.12 through 8.15 normalised to $V_n = 10^{-4}$ cm s^{-1}. The present marginal stability predictions are shown for comparison.

9.3.1 Basic aspects

Unfortunately, there are little studies on the concentration dependence of a_0 during the freezing of saline solutions. The few observations available are discussed next. The following discussion also compares the conditions that may favour the different concentration dependencies mentioned so far.

- *Tip radius control of a_0.* It is first recalled that the analytical and numerical simplified approaches described in section 8.3 imply a concentration dependence

$$a_0 \sim C_\infty^{1/4}. \tag{9.10}$$

  This dependence is expected when the tip radius of dendrites constrains the cell spacing. While the observed dependence in most systems is weaker, with an exponent between 0.1 and 0.2, some systems yield this exponent quite closely (Cadirli et al., 2003). On the other hand, when cellular, non-dendritic freezing is considered, numerical models predict an inverse dependence with an even steeper slope.

- *Classical marginal stability.* The application of equation 7.38 based on the assumption of a constant interfacial k as a material property may be termed *classical* morphological stability theory. It predicts for the initial instability

$$\lambda_{max} \sim C_\infty^{-1/2}. \tag{9.11}$$

As mentioned earlier, this approach is only applicable to the initial perturbations and can hardly be expected as a general result for the final spacings of deep cells. It is the approximate dependence predicted numerically for the spacing of flat cells (Hunt and Lu, 1996; Hunt, 2001). The initial spacing is linked to the tip radius selection in dendritic growth, as discussed in the previous chapter (section 8.1), and is the observed scaling for r_{tip}. So far, no systematic observations of the salinity dependence of r_{tip} during freezing of saline solutions have been obtained.

- *Cellular morphological stability theory.* The present approach assumes that k adjusts to C_∞ and always keeps the marginal stability condition at a macroscopic interface. If the small stabilising effect of surface energy is neglected by assuming $\mathcal{G} = 1$, then equations 9.4 and 9.9 imply that a_0 decreases with C_∞ according to

$$a_0 \sim \left(\frac{1}{\Delta C_{mi}} + \frac{1}{C_\infty} \right)^{1/3}, \tag{9.12}$$

with $\Delta C_{mi} = \mathcal{G}_0 C_\infty$ being the critical concentration step (largely independent of C_∞). As the latter is, under absence of strong liquid temperature gradients, near ≈ 1.3 ‰, a plateau with little further decrease in a_0 is reached already at a concentration of 10 ‰.

9.3.2 Observations

The concentration dependence in the unidirectional growth experiments by Lofgren and Weeks (1969a) seems mainly governed by the scaling 9.12 and confirms the proposed simplified cellular morphological stability theory. Perovich and Grenfell (1981) also reported a negligible increase in a_0 for freezing of NaCl solutions at 16 and 31 ‰.

In the brackish water regime of naturally growing ice equation 9.12 predicts, owing to $C_\infty > \Delta C_{mi}$ as a requirement for cellular growth, at most a factor $2^{1/3}$ larger cell spacings. An additional effect comes from the increase of solute diffusivity D_s with freezing temperature, giving an expected maximum increase in a_0 by ≈ 40 % with respect to seawater concentrations. This is consistent with figure 9.9. Some observations of plate spacings exist for salinities of brackish water. Gow et al. (1992) reported a plate spacing of $a_0 \approx 0.7$ mm for 11 cm thick ice, which grew over the course of a day from brackish water of ≈ 3.7 % salinity in the Bay of Bothnia. Comparably rapid growth of ice from normal seawater results in spacings of 0.4 to 0.5 mm (figure 6.7), roughly confirming the theoretically expected difference. The lack in microstructural observations of brackish ice is likely related to the smaller brine volumes and less distinct appearance of brine layers. Detailed salinity observations of Baltic Sea ice performed recently (Granskog et al., 2004b) indicate that, in case of columnar growth, the same entrapment mechanims as for normal seawater are operative.

Possible explanations for the slight *increase* in a_0 with C_∞ over the brackish concentration range 0.1 to 10 ‰ observed by Milosevic-Kvajic et al. (1973) have already been discussed in section 8.4.2 in terms of variations in thermal convection. There are, however, unresolved discrepancies between the observations from Lofgren and Weeks and the study by Rohatgi and Adams (1967b). The latter authors performed freezing experiments with NaCl solution droplets falling through a cold organic liquid. The dependence from equation 9.10 implies that an increase of C_∞ by a factor of 10 should increase a_0 by a factor of 1.78, which is very close to what has been observed by Rohatgi and Adams. Hence, droplet freezing apparently produces the behaviour which is predicted numerically and analytically on the basis of geometrical constraints due to the tip radius. Besides the freezing of droplets Rohatgi and Adams (1967b) also obtained observations of plate spacing acquired during directional solidification from a cold interface. These observations have not been compared in detail in terms of solute concentration. However, from the figures in the noted study and a later work (Rohatgi et al., 1974) it appears that, if at all, only a weak concentration dependence is present. The unidirectional freezing system from Rohatgi and Adams thus does *not* show the classical dependence $a_0 \sim C_\infty^{1/3}$.

The inward freezing of droplets is a different system compared to a freely evolving cellular interface. This geometry likely does not favour the generally observed overgrowth and coarsening of cells. A simple explanation is that inward freezing is accompanied by confinement opposing the coarsening effect. The process may also be influenced by expansion constraints if the crystals growth inwards from an initially planar sphere. However, it is not immediately apparent, why only this mode of freezing should be controlled by the tip radius geometry. The question requires further investigations. Before the low growth velocity data from Lofgren and Weeks, can be considered in this respect (figure 9.10), the effects of convection need to be quantified.

9.3.3   Scaling transition. First principles

The following simple concept is here suggested for further studies. The brine volume v_{int} in the solid close to the interface is considered. For a cellular interface of any unspecified morphology it is by definition

$$k = \frac{C_i}{C_{int}}, \tag{9.13}$$

where C_{int} is the brine salinity immediately at the interface. Assuming a solid dendritic fraction $1 - v_{int}$ on the solid side of the interface, one has

$$C_i = \frac{C_{int}\rho_b v_{int}}{\rho_b v_{int} + \rho_i(1 - v_{int})}, \tag{9.14}$$

where ρ_b and ρ_i are the densities of brine and pure ice. Then

$$v_{int} = \frac{k}{k + \frac{\rho_b}{\rho_i}(1-k)} \approx k \tag{9.15}$$

shows that, neglecting a small density effect, the interfacial k may be interpreted as the brine or liquid volume v_{int} at the dendritic level. The solid fraction at the interface is correspondingly $(1 - v_{int}) \approx (1 - k)$. As noted in section 5.2, these considerations refer to the brine volume on the solid side of a dendritic operating interface (v_{int-}). The brine volume on the liquid side is, by definition, $v_{int+} = 1$.

The interface described has either to be imagined as diffuse or as a jump-transition in k and v_{int}. To proceed in this way, the following approach considers an interface geometry of parallel plates with tip radius r_{tip}, and defines the diffuse interface as the width of one tip radius r_{tip}. This is feasible from the basic principles of free dendritic growth discussed in the previous chapter. The geometrical relation

$$a_0(1 - v_{int}) \approx \frac{\pi}{4}2r_{tip} \tag{9.16}$$

gives the average solid fraction for this two-dimensional diffuse interface. One may then suggest

$$a_0(1 - k) > \frac{\pi}{4}2r_{tip} \tag{9.17}$$

as a condition for the solid fraction at the interface. The tip radius may be computed from the local selection condition 8.19, assuming that it adjusts to the minimum interfacial undercooling. It can then be argued that, if solid fractions given by $a_0(1-k)$ drop below the solid fraction imposed by the tip radius at marginal stability, then the latter will control the interfacial undercooling by lowering it. As long equation 9.17 is fulfilled, the tips may be sharper, with some spongy growth and decay, yet as they do not control the overall interfacial heat transfer, these sharp tips do not influence the marginal stability condition for the macroscopic interface. In the latter case the scaling 9.12 is expected. In the former case the wider tip morphology can control the macroscopic undercooling and one may expect to some degree the dependence 9.10.

These considerations indicate qualitatively a possible transition between the regimes 9.12 and 9.10. The problems in evaluating expression 9.17 quantitatively are that (i) only a few observations of tip radii of solutal dendrites do exist (section 8.2.2), (ii) tip radii for pure ice dendrites are apparently time-dependent (Koo et al., 1992) and (iii) the treatment of the interface as diffuse with maximised distribution coefficient k is heuristic. Assuming that for a given constitutional supercooling the pure water relations for the tip radius in figure 8.4 can also be applied for saline solutions, then the critical supercooling of ≈ 0.07 K for the NaCl system (at moderate concentrations) implies approximately $7 < r_{tip} < 22$ μm. For $C_\infty \approx 10$ ‰ and $V \approx 5 \times 10^{-4}$ cm s^{-1}, equation 9.17 then predicts a transitional $110 < a_0 < 346$ μm. This is not inconsistent with a slope transition near $a_0 \approx 200$ μm in figure 9.7. However, in figures 9.3 through 9.6, for the downward growth, such a transition to a *weaker* $a_0 - V$-slope with *increasing* growth velocity is not evident. One possible explanation is the difference between upward and downward growth: from simulations it is known that local microconvection sharpens the tip (Koo et al., 1992; Li and Beckermann, 2002).

As mentioned, tip radius observations for NaCl solutions are lacking, mainly due to resolution problems. A further analysis might make use of indirect observations. Detailed data on the dependence of the free dendritic growth velocity on concentration are available (Barduhn and Huang, 1987) and indirect observations of interfacial undercooling have also been reported (Wintermantel, 1972). Such information may be combined with models of tip radii, e.g., Li and Beckermann (2002), to obtain a better understanding of expected thresholds in dependence on growth velocity and concentration.

The considerations made here should be considered as a qualitative argument and possible explanation, why different scalings of a_0 could be expected. They demonstrate the limitations of the present prediction which neglects the morphology and selection principles of single dendrites. Nevertheless, it appears that the present model and its scaling 9.12 holds during downward growth of saline ice. The detailed behaviour due to transitions in cell tip morphology has still to be demonstrated and observations are essentially needed.

9.4   Other factors influencing morphological stability

The above discussion has considered two major factors that are expected to influence the stable plate spacing during cellular freezing: *natural convection* and *tip geometry of dendrites*. In the corresponding planar approach the following other factors may be relevant.

Interface kinetics

Very generally, one may say that the molecular scale of ice is of the order of 10^{-3} μm, and most microscopic physical processes, for example interface steps and roughness, quasi-liquid films or electrical responses of the ice surface, are operative at scales below 1 μm (Drost-Hansen, 1967; Fletcher, 1970; Hobbs, 1974; Petrenko and Whitworth, 1999). A crude velocity bound of the influence on the diffusional scale controlling morphological stability can be obtained by dividing the salt diffusion coefficient D_s by the noted length scales. This gives a lower bound on V of the order of 10^{-2} cm/s, a value much larger than natural growth rates of ice. Hardy and Coriell (1973) found, however, considerable modifications of interface kinetics when solute was present. The behaviour was complex and depended on the ph-values and concentrations of the solutions. In their study, maximum linear kinetic coefficients, observed at high ph, would have corresponded to a kinetic undercooling of 0.005 K at $V \approx 10^{-4}$ cm/s. Hence, as this is 6 to 7 % of the critical constitutional supercoling, interface kinetics may barely effect the natural growth regime, but would be relevant at higher growth velocities.

A macroscopic paradigm of interface kinetics seems to be more critical. On the one hand, the present macroscopic application of morphological stability to a cellular array may be viewed as an analogy to interface kinetic problems. The relevant critical length scale is now the diameter at the tip of the dendrites. Interface kinetics may thus be associated with the geometrical and dendritic selection principles noted above. With $7 < r_{tip} < 22\mu$m at marginal stability, the critical velocity scale $D_s/(2r_{tip}$ becomes larger than 10^{-3} cm/s, which may provide an idea of the importance of these effects and the applicability of the present non-geometric approach. On the other hand, the effect of interface kinetics can be interpreted as an increase in the effective surface energy (Hardy and Coriell, 1973), and thus effect the results due to the dependence $a_0 \sim \Gamma^{1/3}$. This indicates the potential influence of impurities.

Surface tension anisotropy

The effect of anisotropy of surface tension and interface kinetics on morphological stability has been analyzed by Coriell and Sekerka (1976a) and extended from a quasi-static to a time-dependent analysis by Coriell et al. (1994). It was found that, in the kinetically controlled regime of large velocities, anisotropy in interface kinetics can stabilize a planar surface. The theory was compared by Nagashima and Furukawa (1997a) to experimental data from the growth of a 3 % KCl solution in a thin growth cell. The experiment was devised to analyze the tilting and tip splitting behaviour of cells and dendrites and the theoretical stabilizing effect of anisotropy kinetics was confirmed. The kinetic anisotropy of supercooling was found to be 0.005 K at a velocity of 7×10^{-4} cm/s, dropping rapidly at lower velocities. An effect on natural sea ice growth may be neglected.

The bulk effect of the surface tension in the linear problem is to appear as the effective capillary length in the equations for the stability wave length, which then scales as $\lambda_{mi} \sim \Gamma^{1/3}$. If, as discussed in the appendix A.2.10, γ_{sl} for a prismatic surface two times its value for the basal surface, then the initial λ_{mi} should be 26 % larger for an instability that exposes the prismatic plane. Kvajic and Brajovic (1971) performed a comparative study of growth normal and parallel to the c-axis. After the planar-cellular transition the array of crystals growing normal to the c-axis contained more solute. This appears to be consistent with a smaller Γ of the exposed planar interfaces. Also the dendritic pattern shown by Harrison and Tiller (1963) indicate a slightly larger cell spacing for growth normal to the c-axis. In the present calculations an average value $\gamma_{sl} \approx 30$ mJ m^{-2} was used. If the suggested basal plane value $\gamma_{sl} \approx 20$ mJ m^{-2} is applied, all present predictions of the plate spacing would decrease by ≈ 13 %. The presently used value appears to be justified, as it has been derived from morphological stability theory with perturbations normal to the c-axis (Coriell et al., 1971; Hardy and Coriell, 1973). Again, a slight modification may result from impurity effects which are not well understood (Hardy and Coriell, 1973).

Soret diffusion

Coriell and MacFadden (2002) have reviewed some recent applications of morphological stability. One interesting feature they point out is the Soret-effect, the temperature-dependence of the solute diffusion coefficient, which can be applied to the case of seawater. For a negligible temperature gradient in the liquid their analysis implies the condition

$$k \gg \frac{(\frac{\partial D_s}{\partial T} L \rho_i)}{2(K_i + K_w)} \qquad (9.18)$$

to ensure that Soret-diffusion does not influence morphological stability considerably. Considering the diffusion-coefficient along the liquidus, derived in the appendix A.3, its derivative $\partial D_s / \partial T_f$ is found to be 0.027×10^{-5} $cm^2 s^{-1} K^{-1}$ at a concentration of ≈ 35 ‰. It is 0.2×10^{-5} $cm^2 s^{-1} K^{-1}$ at a typical temperature of -0.08 °C, expected for the onset of morphological instability in lake water. For seawater and a cellular interface with $k \approx 1$ the effect can be neglected. For lake water this gives a value of ≈ 0.011 for the right hand side. As this is an order of magnitude larger than the planar distribution coefficient, one expects that Soret-diffusion may play a role concerning the onset of instabilities during the planar growth of lake ice. This problem requires further studies.

Electromagnetic effects

Studies on the influence of electric currents on morphological stability have been summarized by Coriell and McFadden (1993) and some later work by Coriell and MacFadden (2002). A magnetic field may also influence convective stability, as known from the classical work by Chandrasekhar (1961). Under natural growth conditions of sea ice, none of these effects is expected to be relevant.

Expansion

All calculations in the present work have been done by neglecting the density change upon freezing. In laboratory experiments with a fixed solid upper surface the principle effect of the latter would be a rescaling of the length scale D_s / V to $\rho_b / \rho_i D_s / V$ (Terwilliger and Dizio, 1970; Hurle et al., 1982; Caroli et al., 1985b; Davis, 1993). For the present problem of cellular growth one needs to introduce an effective density ρ_i' of the porous ice. The relative magnitude of the rescaling will then roughly scale with the mean solid fraction of the ice and the brine density. Interestingly, if an average is calculated from equation B.31 with a boundary temperature of $-20\,°C$ (and concentration 224 ‰ NaCl), it turns out that the rescaling factor ρ_i'/ρ_b is between 0.91 and 0.92 for any salinity C_∞ between 1 and 100 ‰.

Two main effects of expansion on morphological stability of a planar surface are found in an analysis of the problem (Caroli et al., 1985b; Davis, 1993). On the one hand, the concentration gradient at the interface decreases by ρ_b/ρ_i' and a larger concentration step is necessary for instability. On the other hand, the non-slip condition at the interface implies a curvature of streamlines, which for expansion leads to a lateral flow from the crests to the troughs. This increases the solute gradient at the crests and opposes the first effect. According to Caroli et al. (1985b), the first effect dominates at low velocities while the second becomes relevant at high velocities.

The influence of the rescaling in the present simulations has been simply investigated by increasing D_s by 10 %. At $V = 10^{-2}$ cm/s this implies an increase in λ_{mi} by ≈ 6 %, which decreases to ≈ 4 % at $V = 10^{-4}$ cm/s. This is the underprediction that may be expected for laboratory observations as obtained by Lofgren and Weeks. In the field such an effect is not expected, because the floating ice compensates the density change by its freeboard. Another effect of the rescaling is a slight shift in the onset of convection under laboratory conditions. It will be discussed in the next chapter.

Solute diffusion

By applying the equations of planar classical morphological stability theory, solute diffusion has been implicitly assumed as one-sided, i.e. restricted to the liquid side of the interface. This may be criticised. One may argue, that in the cellular application a lumped bulk concentration gradient at the interface should be used for the solute diffusivity. For saline solutions the effective K_i' introduced for the ice side of the interface is a factor ≈ 7 to 8 larger than K_b in the liquid. Hence, in the present formulation this implies a decrease in the lumped concentration gradient by 6 to 7 %. However, the length scale D_s/V in the boundary layer is not altered. Keeping the maximisation principle of k, the overall effect of changing the lumped gradient is very small. The maximisation of k at the interface, as it overrides the concentration dependence of a_0, may also be thought to neutralise the difference due to the lumped gradient.

Summary

Among the noted factors that may effect the plate spacing during planar morphological instability only the rescaling of D_s/V by expansion ρ_b/ρ_i', can be expected to provide a small systematic increase in a_0 of $\approx 5\ \%$. As mentioned in section 9.2.2 above, a much stronger bias expected at high growth velocities may be the crystal inclination. For most of the observations from Lofgren and Weeks, however, the corresponding effect is likely less 10 %. Hence besides the tip geometry of dendrites, which is not included in the present model, a proper treatment of natural haline convection is the most important task.

9.5   Discussion: a_0 at seawater concentrations

The results of the present chapter are summarised as follows. A marginal stability condition for a cellular interface has been used to obtain a solution for the cell spacing in terms of morphological stability theory. The approach involves essentially the same equations as for the Mullins-Sekerka analysis of the planar interface outlined in chapter 7, with two modifications. The first is the use of an effective interfacial conductivity K_i' for the cellular ice which can be computed according to a heat flux model in appendix B.6. The second modification is that the interfacial k is no longer a material parameter. The marginal stability condition corresponds to its maximum.

Although the maximisation of k is a simple concept it has not been suggested in earlier work. The predictions of the plate spacings obtained in this way agree much better with observations from Lofgren and Weeks (1969a) than any of the dendritic growth models discussed in detail in the previous chapter. The comparison has shown that the application of the interfacial conductivity K_i' significantly improves the agreement of k and a_0 with observations. Another outcome of the present comparison is, that at large enough concentrations the concentration dependence is to a certain degree affected by the temperature dependence of molecular properties like m and D_s (figure 9.9).

It is an important validation aspect of the present model, that the observed salinity dependence during downward freezing of salt water is correctly predicted, contrasting the failure of other models in this respect. Some open questions, concerning the different salinity dependencies observed by Rohatgi and Adams (1967b) for droplets frozen in a cold organic liquid, have remained. The freezing conditions in this system appear to create a mode of spacing selection which better resembles the dendritic growth models discussed in the previous chapter 8. The somewhat worse agreement between theory and observations for some other studies may also be related to uncontrolled boundary conditions and non-steady state during very rapid growth.

Due to the limited observations it is unknown, to what degree the tip geometry influences a_0 and its concentration dependence during natural sea ice growth. However, much more severe is the impact of natural haline convection on the plate spacing. In figure 9.11 the laboratory observations from Lofgren and Weeks are shown for all runs of natural seawater salinities (25 to 31 ‰ NaCl). Also shown are the plate spacings

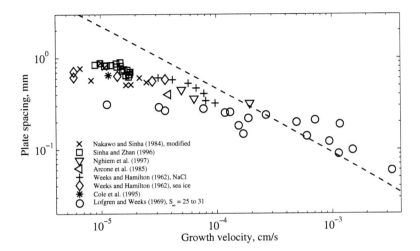

Fig. 9.11: Compilation of plate spacings observed in natural sea ice from chapter 6. The circles are the laboratory observations from Lofgren and Weeks (1969a) at comparable NaCl concentrations. The present model prediction is shown as a dashed line.

observed in natural sea ice, compiled in chapter 6. In this figure only the predictions for the effective interfacial conductivity K_i' are shown. The convective threshold velocity of $\approx 2 \times 10^{-4}$ cm s^{-1} indicated by most freezing runs from Lofgren and Weeks (1969a) is also supported by the natural sea ice observations. Unfortunately, available natural sea ice data are very sparse in the range 1 to 2×10^{-4} cm s^{-1}. As mentioned earlier, there is a serious discrepancy between *all* natural plate spacing observations and the values obtained in the laboratory study by Lofgren and Weeks (1969a). It is likely attributable to thermal convection effects, as they have been quantified in chapter 5 for the comparable freezing experiments by Cox and Weeks (1975).

The lack in confident observations hampers a detailed interpretation of figure 9.11 in terms of the convective transition and the intermediate velocity dependence of a_0. However, it is apparent that the effect of natural convection on a_0 increases with decreasing growth velocity. Near 10^{-5} cm s^{-1} the theory overpredicts the observed plate spacings by a factor of three to four.

10. Morphological stability and compositional convection

In the the previous chapter is was shown, that a cellular modification of morphological stability theory can be used to predict the plate spacing of ice grown unidirectionally from saline solutions at high growth velocities. Below $\approx 2 \times 10^{-4}$ cm s^{-1} the observations start to deviate considerably from the predictions. Hence, the theory needs to be modified for the complete natural growth regime. The present chapter formulates the necessary theoretical extension of morphological stability theory to account for free haline natural convection.

Based on their laboratory data it was anticipated by Lofgren and Weeks (1969a) that solutal convection could have played a role to reduce the plate spacing. Lofgren and Weeks (1969a) suggested that a minimum in the plate spacing near a salinity of 25‰ may eventually be attributed to the salinity, where the temperature of maximum density drops below the melting point. It needs to be pointed out that the threshold of haline convection from growing sea ice has little to do with this limit. The brine release during ice growth produces, in general, density gradients which are orders of magnitude larger than thermal contraction below the temperature of density maximum.

Very basically the discussion is concerned with the following problem: morphological stability theory predicts a certain constitutional supercooling or salinity difference between the interface and the water below. When the boundary layer through which the supercooling occurs exceeds a certain limit it will become unstable. This increases the solute gradient and decreases the thickness of the solutal boundary layer. The consequence is a decreases in the morphological stability wave length which depends on D_s/V.

10.1   Hydrodynamic stability and the Rayleigh-Bénard problem

The free convection problem is related to the classical hydrodynamic Rayleigh-Bénard stability problem (Chandrasekhar, 1961; Nield, 1967; Gershuni and Zhukovitskii, 1976). Rayleigh (1916) first considered the stability of a fluid confined between two plates of distance H_*. The nondimensional number that controls the onset of convection in such a system is the Rayleigh number Ra

$$Ra = \frac{\beta \Delta C g H_*^3}{\nu D_s}, \tag{10.1}$$

with imposed concentration difference ΔC, haline contraction coefficient β, solute diffu-sivity D_s, kinematic diffusivity ν of the fluid and gravity acceleration g. The magnitude of Ra, at which convective fluid motion sets in from a state at rest, depends on the bound-ary conditions. Ideal boundaries are characterised by the no-slip or free-slip conditions for the velocity and the constant temperature or constant heat flux thermal boundary conditions. Theoretical values of all ten possible combinations have been summarised by Nield (1967). Lord Rayleigh (1916) presented the first theoretical solution to the problem for the case of two free and two constant temperature boundary conditions, for which he obtained the critical $Ra_c = 27/4\pi^4$. The corresponding value for two rigid boundaries has been later established as $Ra_c = 1707.762$ (Reid and Harris, 1958).

10.1.1 Convection at large Rayleigh Numbers

When the Rayleigh Number is increased beyond the convective threshold the convec-tion becomes first threedimensional and, at even higher Ra, intermittent and turbulent (Malkus, 1954; Howard, 1966; Krishnamurti, 1970; Busse, 1981). The turbulent heat, or here the salt flux F_S, due to the convective motion, may be described in form of a Nusselt Number

$$Nu = c_{Nu2}Ra^b. \tag{10.2}$$

$Nu = F_S/F_0$ defines the increase over the molecular heat or solute flux F_0 of fluid layer of thickness H_* with a linear gradient. For the classical two-plate system the convective salt flux is thus $F_S = NuD_s\Delta C/H_*$. The coefficients c_{Nu2} and b are variable, and in particular they change at the mentioned transitions. They, and the transitions themselves, depend on the Prandtl number, the boundary conditions, and the systems's geometry which may involve sidewall heating and aspect ratio confinement. The corresponding scaling laws are complex, yet they approach the exponent $b \approx 1/3$ in the limit of infinite Rayleigh Number, while the factor c_{Nu2} has been found to depend slightly on the Prandtl number (Denton and Wood, 1979; Busse, 1981; Siggia, 1994; Grossmann and Lohse, 2000; Ahlers et al., 2002).

10.2 Stability of a semiinfinite layer

The case of haline convection due to the undidirectional freezing of sea ice is different from the classical Rayleigh-Benard problem. Due to the solute flux at the advancing interface a diffusive boundary layer starts growing and the interface becomes unstable in a transient manner. The free convection takes place in form of intermittent turbulent*plumes* or *thermals*. The problem is comparable to confined convection at infinite Ra (Grigull et al., 1963; Howard, 1966; Sparrow et al., 1970; Turner, 1973). This is particularly true in the case of small solute diffusivity, when the quasi-semiinfinite approximation with $b \sim 1/3$ is closely fulfilled in laboratory experiments of finite depths. Notably this is often not the case for thermal convection, when laboratory Rayleigh Numbers are not large enough, and also sidewall heating can be rather critical.

To understand the parameterisation of free convection into a semiinfinite layer, it is useful to consider a system with a given Rayleigh number far from the onset of convection. The turbulent heat or mass flux F_S due to the thermals is generally described in form of a Nusselt Number

$$Nu = c_{Nu} Ra^{1/3}. \tag{10.3}$$

In contrast to equation 10.2 for the two-sided case, the exponent b is now fixed, and c_{Nu} takes a different value[1]. Also, with the molecular flux $F_0 = D_s \Delta C / H_*$, one finds, as $Ra \sim H_*^3$, that the turbulent flux becomes independent of depth H_*. It may be written as, e.g., Turner (1973):

$$F_S = Nu F_0 = c_{Nu} \left(\frac{\beta g D_s^2}{\nu} \right)^{1/3} \Delta C^{4/3}. \tag{10.4}$$

For c_{Nu} values between ≈ 0.1 (Howard, 1966) and ≈ 0.14 (Grigull et al., 1963) were once suggested, based on an evaluation of earlier data sources. These were summarised by Turner (1973). The following discussion sketches the more recent work on the topic, with emphasis on the Prandtl number dependence of the flux.

10.2.1   Theoretical studies

In later studies it was shown experimentally (Onat and Grigull, 1970) and theoretically (Foster, 1968a; Mahler and Schlechter, 1970; Gresho and Sani, 1971) that the convective onset during cooling or heating of a semiinfinite layer depends on the Prandtl ($Pr = \kappa/\nu$) or Schmidt ($Sc = D_s/\nu$) number. This contrasts the onset of convection in the classical problem and is explained by nonlinear gradients that evolve due to the diffusive propagation of density anomalies from one boundary.

It has become convenient to define a critical Rayleigh Number for the time-dependent problem in terms of the diffusional distance $H_* = (D_s t_c)^{1/2}$, where t_c is the critical time at

---

[1] At infinite Ra and constant temperature boundaries one may argue that $c_{Nu} = 16^{1/3} c_{Nu2}$, to be discussed below

the onset of convection. The onset of convection may be predicted from linear perturbation theory, yet this requires some assumptions about the amplification of perturbations (Foster, 1968b; Mahler and Schlechter, 1970; Gresho and Sani, 1971). For the case of a solid boundary with constant temperature or salinity the authors of the latter studies suggested a range of $170 < Ra_c < 450$ for $Pr \approx 7 - 10$ and $Ra_c \approx 130 - 450$ in case of an infinite Prandtl/Schmidt number to be most realistic. Jhaveri and Homsy (1982) devised a stochastic method to solve the problem of indeterminate amplification and arrived at $Ra_c \approx 350$ for water ($Pr = 7$) and a conducting boundary. A general result of the noted studies is, that the results become independent of Prandtl or Schmidt Number once the latter exceeds a value of approximately 200. Hence, for haline convection problems during freezing of aqueous NaCl solutions with $Sc \approx 3000$, the infinite Schmidt/Prandtl number results should apply. A more recent study by Bassom and Blennerhassett (2002) indicates the complexity of the time-dependent problem, in particular the difficulties to obtain observations for validation. In the following the basic results from linear theory for different boundary conditions are sketched.

Constant salinity step

Foster (1968b) finally expressed the critical Ra_c for the transient problem of convection in terms of the critical time t_c and horizontal wavelength λ_c at the onset of convection. For $Pr = \infty$, and a salinity step change at a rigid boundary, he suggested

$$t_c = 59 \left(\frac{\nu^2}{D_s \beta^2 g^2 \Delta C^2} \right)^{1/3} \tag{10.5}$$

$$\lambda_c = 51 \left(\frac{D_s \nu}{\beta g \Delta C} \right)^{1/3}. \tag{10.6}$$

With respect to the above discussion, using $H_* = (D_s t_c)^{1/2}$ in equation 10.1 for Ra_c, equation 10.5 implies a critical $Ra_c = 59^{3/2} \approx 453$.

Constant salt flux F_S

For a constant salt flux $F_S{}^2$ (mass salt per unit area and unit time) imposed at a rigid boundary Foster (1968b) gave alternatively

$$t_{cq} = 27 \left(\frac{\nu \rho_\infty}{g \beta F_S} \right)^{1/2} \tag{10.7}$$

$$\lambda_{cq} = 36 \left(\frac{D_s^2 \nu \rho_\infty}{g \beta F_S} \right)^{1/4} \tag{10.8}$$

---

[2] The use of $F_S = V \rho_i S_{int}$, where S_{int} is the interface salinity, implies that all salt is rejected from the ice. This is reasonable as long the growth is planar.

To relate equation 10.7 in a comparable manner to Ra_c and $(D_s t)^{1/2}$, as was done with equation 10.5 for the conducting boundary, one has to take into account that the interface salinity and profile evolve differently for a constant heat flux (Carslaw and Jaeger, 1959; Foster, 1968a,b). In the latter case the salinity anomaly at the interface evolves as (Carslaw and Jaeger, 1959, p. 75)

$$C_{int}(t) - C_\infty = \frac{2F_S}{\sqrt{\pi}\rho_\infty} \left(\frac{t}{D_s}\right)^{1/2}. \tag{10.9}$$

As $Ra_{cq} \sim t_{cq}^2$ for the constant heat flux, one may compute $Ra_{cq} = 27^2\sqrt{\pi}/2 \approx 646$. This value is, counter-intuitively, larger than for the conducting boundary. The explanation is the different evolution of the boundary layers with time leading to different temperature scales t_c and t_{cq}. Expressing the insulating boundary Rayleigh number in terms of $(D_s t_c)$ instead of $(D_s t_{cq})$, one would obtain an equivalent estimate $Ra_c \approx 27^{3/2}\sqrt{\pi}/2 \approx 124$. This value is of some relevance in the following discussion, as most authors have evaluated their observations in terms of t_c and ΔC_∞ or ΔT_∞.

From equations 10.7 and 10.9 one now obtains

$$C_{int}(t_{cq}) - C_\infty = 27^{1/2}\frac{2}{\sqrt{\pi}} \left(\frac{F_S^3 \nu}{\rho_\infty^3 g\beta D_s^2}\right)^{1/4} \tag{10.10}$$

for the concentration excess at the interface at onset of convection.

Rigid versus free boundary conditions

The impact of different slip boundary conditions on Ra_c has been analysed by Foster (1968a). According to Fosters' calculations at $Pr = 10$, the rigid conducting boundary requires an approximately 1.5 to 1.8 larger Ra_c than the free conducting boundary, depending on assumptions about the growth of perturbations. At $Pr = \infty$ the factor is somewhat larger. For plausible ranges of the perturbation growth parameter (that in the end prove in agreement with observations) the factor was found near 2.2 ± 0.2 for conducting and insulating boundaries. Mahler and Schlechter (1970) derived similar theoretical differences. Observations for a conducting boundary at large Pr yielded a factor in the range ≈ 1.5 (Blair and Quinn, 1969) to ≈ 2.6 (Mahler and Schlechter, 1970).

Nonlinear analysis

In a later study Foster (1972) has included nonlinear terms in his analysis. The modified scalings were given at different Pr for a free, insulating (constant salt flux) boundary. At infinite Pr the derived critical time was a factor $14/27$ smaller, and the wavelength a factor $4/3$ larger than in the linear study from which equations 10.7 and 10.8 were derived. Due to the scaling $Ra_c \sim t_{cq}^2$ for the insulating boundary. this implies a $(14/27)^2 \approx 0.27$

smaller Rayleigh Number. Attributing a reduction factor of 2.2 to the free boundary condition, this indicates a critical Ra_c from the nonlinear approach, which is 60 % of the value suggested due to the linear model.

Because the nonlinear analysis is a more accurate description of the system, the corrected prefactors in equations 10.7 and 10.8 for critical time and wave length become ≈ 21 and 30, respectively. Evaluating the Rayleigh number for the insulating boundary from t_c and ΔC (equations 10.5 and 10.9) gives $Ra_c \approx 21^{3/2}\sqrt{\pi}/2 \approx 85$. For a lower $Pr \approx 7$, if the same ratio between linear and nonlinear Ra_c can be assumed, the linear simulations from Foster (1968a, 1972) then would imply a slightly larger $Ra_c \approx 120$,

for a conducting boundary no similar calculations have been presented. If the same reduction due to nonlinearity also holds in this case, the critical $Ra_c = 59^{3/2} \approx 453$ for $Pr \approx \infty$ would reduce to $approx 270$ in the nonlinear theory. Bassom and Blennerhassett (2002) pointed out the difficulties to perform a nonlinear analysis for the conducting boundary.

10.2.2 Experiments and interpretations

Observations: two convection stages

Onat and Grigull (1970) found, in horizontal plate heating experiments with different liquids, $Ra_c \approx 78$ for high Prandtl Number experiments, and $Ra_c \approx 125$ for water ($Pr = 7$). The experiments by Onat and Grigull likely represent insulating boundary conditions and the occurence of the first perturbations is in good agreement with Foster's nonlinear analysis, if Ra_c is expressed in terms of t_c. The results were obtained from the first deviation of the interface cooling from a $t^{1/2}$ behaviour. However, the time series indicate that a strong perturbation of the temperature field does not occur before $Ra_c \approx 200$ for large Pr and $Ra_c \approx 300$ for water.

A two stage development of disturbances has been described by Mollendorf et al. (1984) in experiments at $Pr = 7$. These authors reported initial perturbations, visible on a Schlieren photograph, near $Ra_c \approx 170$ and measurable interface temperature changes near $Ra_c \approx 320$. In addition, they evaluated $Ra_c \approx 460$ from the optical (Schlieren) thickness of the boundary layer at the onset of convection. In the following discussion it will prove useful to distinguish between these two critical Rayleigh numbers:

- Ra_c – Onset of convective motion in the boundary layer.

- Ra_{c2} – Appearance of first temperature or concentration perturbations at the interface.

Goldstein and Volino (1995) also observed two types of instabilities. Interpreting their results in terms of t_c defined above, one needs to divide their documented Rayleigh numbers by $pi^{3/2}$. The first instability was identified at $Ra_c \approx 228$ due to small visible velocity perturbations at the interface. This is the recommended value by the authors, who attributed the very first $Ra_c \approx 137$ to defects in the heating plate. The second relevant $Ra_{c2} \approx 480$ was associated with the onset of temperature changes at the interface. All following observations had a standard deviation of ≈ 30 %.

The mentioned values obtained by Mollendorf et al. (1984) and Goldstein and Volino (1995) corresponded to Pr of water and constant heat flux boundary condition. For large Pr of order 10^3 to 10^4, Davenport and King (1974) have reported $Ra_{c2} \approx 320$ to 360.

Analogy to the classical problem

Hence, experiments indicate that the very first interface perturbations appear for water ($Pr \approx 7$) at a Rayleigh number of the order of $Ra_c \approx 125$ to 170. It is worth a note, that the much larger critical $Ra_c \approx 1100$ has been frequently suggested in analogy to the linear problem (Howard, 1966; Sparrow et al., 1970; Goldstein and Volino, 1995). This is not consistent in a heuristic sense. The latter is the critical Rayleigh number for a system with one free and one rigid conducting and two conducting boundary conditions (Reid and Harris, 1958; Nield, 1967). The situations discussed by the noted authors correspond, however, to a constant heat flux at the heated, rigid boundary, with the second boundary being situated somewhere in the liquid. If this second free boundary, which is advancing in the one-sided time-dependent case, is also viewed as insulating, then the appropriate critical Ra_c would be 320 (Nield, 1967); if it is viewed as having a mixed boundary condition, Ra_c would be larger. For example, for a Biot number of order one, which seems a plausible ad-hoc choice, it would be ≈ 514 (Sparrow et al., 1964). These numbers in fact compare to the second observed instability Ra_{c2} in the observations.

The first instability is apparently not yet associated with a large heat flux increase. It may be compared to the classical two-sided problem, if one simply splits the latter into two layers (Gershuni and Zhukovitskii, 1976). From half the layer and half the temperature differences, a kind of critical Rayleigh Number for the insulating rigid-rigid case would be $Ra_{c1/2} \approx 720/2^4 = 45$. The critical Ra range 45 to 514, from considerations based on classical theory, may thus be associated to some degree with the different observed transitional Rayleigh numbers. A value of $Ra_c \approx 40$ is also suggestive from a slightly nonlinear analysis of the problem by Bassom and Blennerhassett (2002). These authors pointed out that the intrinsic onset of instabilities might remain undetected in observations, because early convection (i) does not effect the slope in a time versus ΔT diagram, and (ii) sets in at very large wavelengths. In the frame of this interpretation the onset of convection might be be identified with the first observations of time-dependent flow. In this context it is further worth a note, that for the classical problem two transitions are observed: For $Pr \approx 7$ these are (i) a transition from two to threedimensional rolls at $Ra/Ra_c \approx 12$ and (ii) a transition to time dependence near $Ra/Ra_c \approx 18$ (Krishnamurti, 1970; Busse, 1981). These distinct flow-related regime transitions in the classical problem appear also to take place during one-sided heating, with simplistic intuitive estimates apparently being quite realistic.

The analogy can be extended to different boundary conditions from Nield (1967). For the conducting boundary a first critical transition may be expected, again from the division of the two-sided system, near $Ra_{c1/2} \approx 1707.8/2^4 \approx 107$ for the rigid case, and near $Ra_{c1/2} \approx (27/4\pi^4)/2^4 \approx 41$ for free boundaries. The smallest $Ra_{c1/2} \approx 120/2^4 = 7.5$ follows from two free insulating boundaries. The comparison provides a very basic path to the understanding of two kinds of observed convective instabilities: A first one, which may be approximated by splitting the two-sided stability problem, and a second one, approximately corresponding to the unsplitted two-sided problem with appropriate boundary conditions. The numbers cannot be expected to be exact solutions to the problem, as the transient basic profile is, in contrast to the classical case, nonlinear. Solutions for the change in the critical Ra_c due to a nonlinear profile have been obtained by Currie (1967) and Nield (1975) by approximating the basic temperature profile for the two-sided problem with equal boundary conditions. Nields's analysis for the insulating boundary condition yielded $720/1.198 \approx 601.1$, while Curry's result for the conducting boundaries was a reduction of $1707.8/1.274 = 1340$. Nield (1975) noted that for an insulating boundary with a step increase the reduced critical Ra is even smaller, $720 \times (8/15) = 384$. In all the mentioned nonlinear modifications, the minima correspond to the application of buoyancy forces more than half the plate distance away from the rigid boundary. This means that the intuitive approach to split the classical two-sided system, in order to estimate Ra_c for one boundary layer, is not fully justified. However, these numbers may still be meaningful as lower bounds, considering a remote triggering of the instability. The corresponding semiinfinte numbers then become $Ra_{c1/2} \approx 601.1/2^4 \approx 37.6$, $1340/2^4 \approx 83.8$ and $384/2^4 = 24$.

The suggested number $Ra_{c1/2}$ may thus, in the above sense of splitting the classical problem, be viewed as corresponding to the first semiinfinite instability Ra_c, while Ra_{c2} may be identified with the two-sided problem with one free and one rigid boundary. Also for the latter case Nield (1975) presented, for two insulating boundaries, some results relevant for the modification due to a nonlinear gradient. For a piecewise linear profile that approximates a flux applied at the rigid boundary, he proposes a reduction of the classical critical Ra_c to $320/1.094 = 292.5$. The minimum Ra_c results from applying a step function somewhere within the layer and is $320/1.733 = 184.6$. The former is likely the appropriate condition for sea ice, although the latter may be of partial interest under certain freezing conditions with a morphologically unstable spongy, intermittently proceeding ice surface. The classical problem does not involve a Prandtl number dependence of the onset of instabilities. However, the value 292.5 seems to resemble the flux-related Ra_{c2} for infinite Pr from other theoretical and experimental studies quite reasonably.

Best estimates of the flux-related Ra_{c2}

To be discussed in the next paragraph, even if velocity fluctuations occur first at Ra_c, one may argue that it is the timescale related to first temperature (or concentration) changes at the interface, which is most appropriate to describe the turbulent flux in a quasi-steady state. The experimental data may then be summarised, for a rigid insulating boundary: $320 < Ra_{c2} < 480$ appears as the most reasonable range in terms of a critical flux Rayleigh number for water ($Pr \approx 7$), while the onset of first convective motion was observed for $125 < Ra_c < 170$. At infinite Pr or Sc, corresponding to the case of salt diffusion during freezing, simulations (Foster, 1968a; Mahler and Schlechter, 1970; Gresho and Sani, 1971) and observations (Onat and Grigull, 1970) indicate Ra_{c2} smaller by a factor ≈ 1.5. Hence, $200 < Ra_{c2} < 320$ is expected for the flux-related Ra_c at infinite Pr, while, $80 < Ra_c < 110$ relates to the onset of convection. The latter numbers are fully consistent with the nonlinear simulations by Foster (1972).

The critical Rayleigh number estimated from Foster's work for the conducting boundary is slightly larger, with a value of $Ra_c approx 270$ at infinite Pr. The observations from Blair and Quinn (1969) which resemble these boundary conditions are consistent, with $Ra_{c2} \approx 300$. The lower $Ra_{c2} \approx 110$ obtained by Mahler and Schlechter (1970) corresponds, according to these authors, to experiments closer to free boundary conditions. While observations and theory appear also consistent for the conducting boundary, there is a difference. $Ra_{c2} \approx 300$ reported by Blair and Quinn (1969) corresponds to the onset of heat flux perturbations and thus to the secondary Ra_{c2} found in the insulating case. These authors also used Schlieren photographs for which they observed the same Ra_c due to interface motion. They reported, however, also a largescale motion on the length scale of the cell that occurred earlier, but whose origin remained unexplained. So far no strict observations of two stages of Ra_c and Ra_{c2} have been reported for a conducting boundary. However, this is due to the fact that most experiments refer to insulating heating conditions. The stability diagram of wave lengths is different for the insulating and conducting cases (Bassom and Blennerhassett, 2002). It bears some similarities to the classical problem, where the convection with insulating boundaries sets in at infinite wave length, while for conducting boundaries the wave length is of the order of the layer thickness (Nield, 1967). More work is needed on the question if, for the conducting boundary, the onset of motion and surface fluxes is more simultaneous than for the insulating case.

10.2.3 Nusselt Number

The connection between a critical Ra_c and the Nusselt number Nu may be understood by introducing a heuristic approach to turbulent convection first suggested by Howard (1966) and emphasised by Foster (1969, 1972). Equation 10.3 may be written in the form

$$Nu = c_{Nu}Ra^{1/3} = \left(\frac{Ra}{Ra_{c*}}\right)^{1/3}, \tag{10.11}$$

reflecting the following idea. Howard assumed, that in the steady state the turbulent convection will operate intermittently at the critical Rayleigh Number Ra_{c*}. For the latter he suggested a value of 1000, at a time when little experimental and theoretical work had been done. Intuitively it seems appropriate to associated Ra_{c*} with Ra_{c2} for the change in the interface temperature. Assuming $480 > Ra_{c2} > 320$ and $320 > Ra_{c2} > 210$ implies $0.128 < c_{Nu} < 0.146$ and $0.146 < c_{Nu} < 0.168$ for water and infinite Pr, respectively. It is noted that relation 10.11 assumes that the breakdown phase of plumes is much shorter than they conductive build-up phase. Nonlinear simulations by (Foster, 1971) indicate that the conductive build up time is approximately 60 to 70 % of the fluctuation period. The ratio of second and third critical Rayleigh number stages from the studies in the previous paragraph may indicate, due to $Ra \sim t^2$, a similar conduction phase of 65 to 70 %. Hence, choosing the Rayleigh number which resembles the interface temperature perturbations, appears appropriate in the scaling 10.11 for the Nusselt number.

For the insulating boundary the picture appears quite clear from simulations and observations. It appears that Ra_{c2}/Ra_c is in the range 2 to 2.5. Convective motion sets in at approximately half the Rayleigh number in equation 10.11 which resembles the heat or solute flux and overall period of release of a plume. The corresponding timescales scale approximately as $2:3$. The Nu-scaling for a conducting boundary, equivalent to 10.11, would likely fall in the noted ranges, but the dynamics may be different. Equation 10.11 can be compared to the following observations of fluxes.

Water. $Pr \approx 7$.

Katsaros et al. (1977) obtained $c_{Nu} = 0.156$ through an experiment with evaporative cooling for $Pr = 7$ and free boundary conditions. Reducing it by a factor $2.2^{1/3}$, one obtains $c_{Nu} \approx 0.120$ for a rigid boundary, in good agreement with c_{Nu} estimated from Ra_{c2} above. From fluctuation observations Katsaros et al. (1977) considered that their boundary condition was likely intermediate between a constant temperature and a constant heat flux. For two-sided heat fluxes the most probable prefactor has been suggested to be $0.05 < c_{Nu2} < 0.055$ for water (Denton and Wood, 1979). For comparison with the present case, assuming two noninteracting boundary layers, the conversion factor $16^{1/3}$ applies and leads to a similar range $0.126 < c_{Nu} < 0.139$.

Large Pr**.**

The relation $Nu = 0.19 Ra^{0.329}$ for infinite Pr was derived from mass transfer at electrodes by Fenech and Tobias (1960), from which a slightly reduced $c_{Nu} \approx 0.172$ is obtained by correcting the exponent to $1/3$, accounting for the typical Rayleigh number range in the experiments. The technique is proposed to yield a conducting boundary. Goldstein et al. (1990) used the same electrochemical transfer technique to determine the Nusselt number for large Sc in the two-sided configuration and found $Nu = c_{nu2} Ra^{1/3}$ with $c_{nu2} = 0.0659$. Assuming also here, that $3 \times 10^9 < Ra < 5 \times 10^{12}$ was large enough to create two non-interacting convection regimes, the factor $16^{1/3}$ converts this to the one-sided $c_{Nu} \approx 0.166$. Selman and Tobias (1978) have reviewed several studies with this technique, the results being in the range $0.15 < c_{Nu} < 0.19$. Other observations and theoretical approaches to estimate upper limits of Nu for infinite Pr have been summarised by Goldstein et al. (1990).

Chan (1971) has obtained $c_{Nu} \approx 0.152$ from a theory of optimum turbulence, in the limit of infinite Pr, two rigid boundaries and heating from one side. More recently, Vitanov (2000) has solved this problem for one and two free boundaries. At large Ra his bounds correspond to $c_{Nu} \approx 0.167$ (free-rigid) and $c_{Nu} \approx 0.325$ (free-free). An earlier estimate derived by Herring (1963) for two free boundaries is $c_{Nu} \approx 0.31$. For the rigid-rigid case (Canuto and Goldman, 1985) proposed $c_{Nu} \approx 0.061$ on the basis of a turbulence model. Again, the difference to the aforementioned studies appears to relate to the fact, that the latter authors relate their scaling to global transport of a box with two boundary layers, while the former studies focus on the heat transport at one boundary. The conversion $0.061 \times 16^{1/3} \approx 0.154$ brings the results in agreement. The prediction for $Pr \approx 7$ from Canuto and Goldman (1985) was $c_{Nu} \approx 0.11$ and is consistently lower.

The agreement of the turbulence scalings with the simplified approach $c_{nu} \approx Ra_{c2}^{1/3}$ depends, of course, on the choice of the length scale to evaluate the critical Rayleigh Number. The consistency implies that, in terms of the heuristic view of intermittent plume convection, a Rayleigh Number based on the length scales $(D_s t_c)^{1/2}$ or $(\kappa t_c)^{1/2}$ reflects the turbulent heat or mass fluxes in the quasi-steady turbulent state.

In the present work the proportionality $c_{Nu} \approx 0.15$, implying $Ra_{c2} \approx 296$, is adopted at infinite Schmidt Number Sc. It should be valid for a rigid boundary that is either conducting or insulating. This condition will now be implemented as a boundary condition into morphological stability theory.

10.3   Incorporation of convection into morphological stability

10.3.1   Previous work on planar interface stability convection

In previous studies the impact of convection on the onset of morphological instabilities for a planar interface has been considered (Coriell et al., 1981; Hurle et al., 1982, 1983; Caroli et al., 1985a; Jenkins, 1990). As in the classical Mullins-Sekerka analysis the authors considered the onset of instabilities from a steady state concentration profile, the interface moving at constant velocity V. It was first shown by Hurle et al. (1982), that the critical Rayleigh number for the onset of compositional convection, if based on the length scale D_s/V, depends on the Schmidt number Sc and the segregation coefficient k. In the limit of $k \to 0$, the situation equivalent to an insulating boundary, they found $Ra_c = 2(1 + Sc^{-1})$. For larger k in the range 0.3 to 1 and Schmidt Numbers of 10 to 81 the critical Ra_c took values between 10 and 13 (Hurle et al., 1982, 1983; Caroli et al., 1985a), a situation equivalent to approaching a conducting boundary condition. Smith (1988) has extended the problem to solidification of a pure melt and found a weak effect of the ratio of thermal conductivities in the solid and liquid. His conversion of the values to classical Rayleigh numbers by a constant factor were, however, arbitrary.

An alternative approach, with similar results and an account for free boundary conditions, has been formulated by Hwang and Choi (1996). Also in the frozen time analysis of two convective modes in a porous medium (to be discussed in chapter 11) and the liquid underneath, $Ra_c \approx 10$ has been found (Worster, 1992). These Rayleigh numbers are apparently an order of magnitude smaller than obtained from observations, and from the above discussed time-dependent nonlinear analysis of the onset of convection based on the length scale $(D_s t)^{1/2}$.

Interestingly, $Ra_c = 2$ has also been obtained by Gresho and Sani (1971) and Currie (1967) in a frozen-time analysis of convective stability based on $(D_s t)^{1/2}$, for any Prandtl number, yet for a conducting boundary. The latter should therefore be contrasted with $10 < Ra_c < 13$ from the mentioned solidification studies. For a conducting boundary condition the interfacial concentration gradient is (Carslaw and Jaeger, 1959, p. 75) given by $G_c = \Delta C/(\pi D_s t)^{1/2}$. Comparison with the steady-state interfacial solute gradient implies $D_s/V = (\pi D_s t)^{1/2}$. One therefore expects Ra_c to be smaller by $\pi^{3/2}$, when considering a frozen-time analysis based on $(D_s t)^{1/2}$ compared to D_s/V. In this context the studies appear consistent, even if the Prandtl number dependence is not included in both approaches. It is further important to note, that Gresho and Sani compared these results with the two orders of magnitudes larger Ra_c obtained from a transient analysis of the problem. They pointed out, that the frozen-time analysis is not applicable when a conducting (step-change) boundary condition is considered.

It seems therefore legal to ask, if the frozen-time solution $Ra_c = 2(1 + Sc^{-1})$ from

Hurle et al. (1982) is relevant when considering the onset of convection from the transient. A study by Tait and Jaupard (1989), who reported $1 < Ra_c < 8$ during solidification of the binary alloy NH_4Cl-H_2O , has been suggested to agree approximately with these low values of Ra_c, in particular when a modification due to free boundary conditions is allowed (Hwang and Choi, 1996). In contrast, the limited observations from McCay et al. (1989), based on the actual observed boundary layer thickness, indicate a 2 orders of magnitude larger critical Ra_c. The growth was more rapid in the experiments from Tait and Jaupard (1989) and the velocity was decreasing, which may limit the applicability of steady state scalings and could have yielded smaller critical times (equation 10.7). The discrepancy may also have to do with an indirect determination of the scale D_s/V by Tait and Jaupard (1989) and Hwang and Choi (1996), not taking into account the interface relation $D_s/V = (\pi D_s t)^{1/2}$ that might be relevant for a conducting interface.

As all noted approaches are very approximate one must conclude, that none of them can be viewed as a proper validation of the critical Ra_c for convective onset during directional solidification. For the present application, two aspects of the mentioned studies are thought to be most relevant:

- It was the main conclusion from the morphological stability studies (Coriell et al., 1981; Hurle et al., 1982; Caroli et al., 1985a; Jenkins, 1990), that the convective and morphological instability modes do not interact severely under most natural conditions. The reason is that convection sets in at considerably larger wave length than morphological stability. In fact, this condition is also fulfilled for sea ice: from equation 10.13 one obtains, using $\Delta C \approx 1.3$ ‰, a convective wave length of 2.2 mm, while the plate spacing at the convective onset, to be discussed below, is $a_0 \approx 0.3$ mm.

- A final conclusion about the experimental and theoretical results for the critical Ra_c must remain open. There are indications that the frozen-time analysis, yielding Ra_c of order 10, does not apply at all to the intermittent problem. But even if the onset of first instabilities occurs at such a low Rayleigh number, a considerable change in the solute flux and $Nu > 1$ likely requires a larger $Ra_c \approx 300$.

10.3.2 Convective onset: The length scale D_s/V

As indicated in the previous paragraph, it is unclear, if the natural length scale D_s/V is appropriate to compute Ra from equation 10.1. It is clear that, in the present problem, the length scale D_s/V resembles $G_c/\Delta C_{mi}$, the ratio of interfacial solute gradient and the marginal stability constitutional supercooling. However, eventually the scale D_s/V needs to be corrected with respect to $(D_s t)^{1/2}$, to which the above discussion of transient convection problems applies. As mentioned, the transient and steady state interfacial gradients imply $D_s/V = (\pi D_s t)^{1/2}$ for the conducting interface and, due to equation 10.9, $D_s/V = 2(D_s t/\pi)^{1/2}$ when comparing interface gradients for a constant heat flux. In addition to the flux boundary conditions, also the validity of the rigid boundary assumption has to be addressed critically. One might, for example, suggest that a partially free boundary condition is appropriate for a cellular interface, which would reduce Ra_c by ≈ 2 to 3. If one assumes that the intrinsic sea ice boundary is likely between conducting and insulating, which means that the length scale $H_* \approx D_s/V$ should be decreased by a factor between $\sqrt{\pi}$ and $2/\sqrt{\pi}$, with respect to $(D_s t)^{1/2}$. These two modifications would eventually cancel in equation 10.11.

Hence, a rescaling of D_s/V depends on the boundary conditions at the interface. Within the present phenomenological approach the related questions cannot be answered yet. An argument that can be forwarded is that, as the Nusselt number from equation 10.11 basically relates to the increase in the interfacial solute flux, the scale D_s/V should be appropriate, if primarily this flux is to be computed and not the details of the boundary layer. From this point of view it is most important to apply the correct Nusselt number. However, $c_{Nu} \approx 0.15$ is reasonably backed by observations and implies a critical Rayleigh number $Ra_{c2} \approx 0.15^{-3} \approx 296$ for the onset of strong convective salt fluxes. It will be seen below, that this model with parameters c_{Nu} and length scale D_s/V is consistent with observations at low growth velocities. Nevertheless, the detailed relation between c_{Nu}, D_s/V and possibly different transitions must be addressed in future studies.

10.3.3 Wave length, critical time and critical growth velocity

Refined predictions

The constants in equations 10.5 and 10.6 given by Foster (1968b) need now to be changed to be consistent with the assumed critical $c_{Nu} \approx 0.15$ and $Ra_c = 296$. The tentatively modified equations are

$$t_c = 44 \left(\frac{\nu}{D_s \beta^2 g^2 \Delta C^2} \right)^{1/3} \tag{10.12}$$

for the critical time and

$$\lambda_c = 44 \left(\frac{D_s \nu}{\beta g \Delta C} \right)^{1/3}. \tag{10.13}$$

for the wave length at the onset of instability. The proportionality for t_c follows simply from $Ra_{c2}^{2/3} = 296^{2/3} \approx 44$. The factor 44 for the wavelength λ_c is based on $\lambda_c \approx \frac{2\pi}{c_*}(D_s t_c)^{1/2}$, where $c_* \approx 0.95$ was retained from the original linear relations given by Foster (1968b). The latter condition implies a wave length to boundary layer width ratio of $\frac{2\pi}{c_*} \approx 6.6$. Foster (1972) pointed out the approximative character of these equations

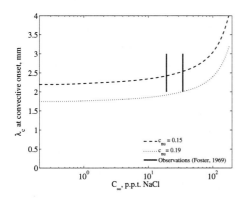 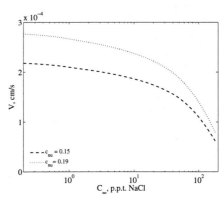

Fig. 10.1: Wave length at the onset of convection due to equation 10.13, modified after
Foster (1968b). The observations are approximate values reported by Foster (1969).

Fig. 10.2: Critical growth velocity at the onset of convection in dependence on the solute
concentration, due to equation 10.14.

when estimating wavelength and critical time. In Foster's linear predictions for free
boundary conditions the critical time is smaller by a factor of ≈ 1.7. The wave length
in equation 10.13 would decrease by a smaller factor of 1.3, provided that $c*$ does not
change. This is not the case. Actually, $c* \approx 0.5$ for free boundaries (Foster, 1968a, 1972).
Hence, the wave length will be almost the same.

Within the above framework the critical ice growth velocity for the onset of haline
convection can be computed, if a mechanism that controls the interface salinity is given.
This mechanism is assumed to be the marginal stability growth of the cellular array.
From equation 10.1 with $H_* = D_s/V$ and equation 10.11 and $Ra_{c2} = c_{nu}^{-3}$ becomes

$$V_{cr} = \left(\frac{\Delta C_{mi} \beta g D_s^2}{\nu Ra_c} \right)^{1/3}, \tag{10.14}$$

where ΔC_{mi} is the difference between interface and far-field concentration due to the
marginal equilibrium from morphological stability theory.

Convective wavelength

The theoretical predictions of λ_c and V_{cr} are shown in figures 10.1 and 10.2 for two values
of c_{nu}. It appears that the approximate wave lengths documented by Foster (1969)
from experiments for a limited salinity range, shown in figure 10.1, agree reasonably
with $c_{nu} = 0.15$. Foster (1968b) also noted unpublished laboratory experiments from
Coachman, which are at the upper limit of his observations. Of course, more observations
are necessary. Figure 10.1 shows that such observations can improve the quantification

of the constants in the scalings. As mentioned above, the actual wave length may differ from equation 10.13 due to effects of nonlinearity. If, however, a free boundary condition would be the more appropriate one, then the flux coefficient c_{nu} would be increased, but the wave length would likely be unaffected. Therefore figure 10.1 must be understood to refer to a rigid boundary condition.

Critical growth velocity

The critical growth velocity for the onset of convection is, for natural solution salinities, typically 2×10^{-4} cm/s. This value agrees reasonably with the slope change observed in figures 9.3 through 9.6. From equation 10.14 it is seen, that the theoretical concentration dependence of V_{cr} follows mainly from the temperature dependence of the solute diffusivity. A closer look on the observations from Lofgren and Weeks (1969a) is given in figure 10.3. Here the observed plate spacings have been normalised by the model predictions (without convection), and are shown in relation to the observed growth velocity, normalised by the predicted V_{cr}. For perfect agreement the data should scatter around 1 and drop below this value for $V/V_{cr} < 1$ due to the onset of haline convection. This is apparently not the case. As mentioned earlier, the model slightly underpredicts the plate spacings at high growth velocities, which may have several reasons:

- Crystal inclination is strongest during very early growth. In the figure three observations from levels closer then < 3 mm to the cooling interface are denoted as dots and two of them support this view. According to observations from Weeks and Assur (1964) the inclination bias is expected to decrease to less than 10 % at 4 to 5 cm from the cooling interface. As this level corresponds tyically to $V_{cr}/V \approx 1$ to 4 in figure 10.3, it is difficult to validate a convection threshold in this regime. The slight trend at high growth velocities is, however, not unreasonable in this context.

- Growth velocity control has been problematic in some freezing runs. An example are the connected squares for a run at ≈ 30 ‰, where large deviations are seen. The overall scatter appears to be of the order ± 20 % in a_0.

- The laboratory study likely suffers from an increasing influence of thermal convection. According to figure 10.9, to be discussed below, the latter effect may have become relevant at velocities above the haline convective transition.

Due to these aspects a definite conclusion and validation of c_{nu}, based on the observations from Lofgren and Weeks (1969a), is not feasible, yet an approximate interpretation can be attempted. Considering the first slope change near $V/V_{cr} \approx 2$ to 2.5, this implies Ra in the range 20 to 40. Recalling the above discussion in paragraph 10.2.2, it is not unreasonable that first convective effects appear at such low Ra. In the lines of two

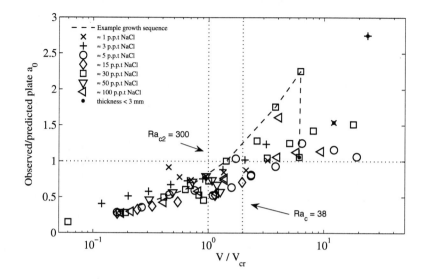

Fig. 10.3: Observed plate spacings from Lofgren and Weeks (1969a) normalised by the present model predictions (without convection), versus the growth velocity, normalised by the critical velocity V_{cr} at the onset of convection. Dots denote three observations very close to the freezing interface, most likely biased by inclination. The connected squares for a run at 30 ‰ indicate uncertainties in growth velocity and plate spacing determinations.

different stages of convective onset, effects of fluid flow may be expected at the lower Ra_c, even if it is not associated with a larger c_{nu} (based on the flux-related Ra_{c2}) at later stages. The transition to turbulent convection is already a complicated process related to different critical Rayleigh numbers. It is therefore likely that nonlinear interactions between morphological, haline and thermal convective modes and boundary conditions complicate the picture, in particular near the threshold. For the planar surface a yet unexplained regime of an oscillatory coupled mode of morphology and convection has been predicted near the convective transition (Coriell et al., 1981; Hurle et al., 1982; Caroli et al., 1985a; Jenkins, 1990). An analysis of these coupled problems for the cellular interface is likely very difficult.

It is further interesting that the local minima in figures 9.3 through 9.6 all merge near $V/V_{cr} \approx 1$ in figure 10.3. This might be interpreted as the secondary flux-related convective transition. Hence, what appears as local scatter in each single salinity plot, may well have to do with the onset of a rigorous mode of haline convection at $Ra_{c2} \approx 300$. Figure 10.3 shows no clear transition near $Ra \approx 10$ or the corresponding $V/V_{cr} \approx 3$.

Discussion

Due to the scatter of the observations and the noted systematic problems, no refinement of a critical Rayleigh number or c_{Nu} is possible here. Nevertheless, it can be said that, on the basis of the limited available data, there is reasonable agreement with the modified predictions of convective wave length from Foster. The predicted critical velocities of convective transitions, when coupled to the present cellular marginal stability criterion, are reasonably resembled by observations. That the behaviour near the transition may not accurately be described by the equations is, due to the above discussion, not unexpected. As will be seen in the following section 10.4, the convective approach is capable in predicting the plate spacings quite accurately at natural growth velocities, which are typically more than a factor of 3 below the threshold. Nevertheless, the question of the stress boundary condition at the interface requires further investigation. As mentioned above, c_{Nu} can be as large as 0.32 for a slip-free boundary. This would be consistent with the early transition in figure 10.3. On the other hand, a boundary disturbed by intermittently growing dendrites will likely not be effectively free. These considerations indicate the difficulties to obtain even a useful heuristic approach. It is possible, that transitional boundary conditions depend on changing fluid flow, thermal convection and morphology of the interface.

Some aspects of earlier work on this subject are noteworthy. When Foster (1969) analysed his experiments on sea ice freezing, he applied equation 10.8 for a constant solute flux. The latter was calculated on the basis of bulk ice salinities. This equation predicts a much stronger concentration dependence of λ_c than observed. It seems that the constant salt flux model, in particular when based on bulk salinities, does not resemble the two essential physical mechanisms, which are solute redistribution at a cellular interface and brine fluxes from the interior of the ice. Therefore it only should be applied to planar freezing. On the other hand, the growth can be expected to have started planar. The first onset time of convection is therefore likely related to equation 10.7 with a prefactor 21 from the nonlinear analysis. For $C_\infty = 24.7$ ‰ and $V = 1.1 \times 10^{-4}$, the conditions in one of Foster's experiments, one obtains $t_c \approx 68$ s. This should be a slight upper estimate, as the actual interface concentration likely was larger than $C_\infty = 24.7$. It is close to the observed critical time of ≈ 60 s documented by Foster. Foster, using the salt flux based on bulk ice salinities in his calculation, obtained a 4 times too large critical time. Even if the first convection onset might have occurred after the onset of cellular freezing, it is likely that the first critical time is related to the early planar growth.

Once the cellular interface is established, the critical time will likely appear as the periodicity of the convection process given by equation 10.12. For typical sea ice growth conditions the period is expected to lie close to 2 minutes. No such observations have been reported by Foster. Sparrow et al. (1970) have reported the period of such temperature fluctuations above a heated rigid water interface. They computed a Rayleigh number $70 < Ra_\tau < 100$ on the basis of the fluctuation period τ in the convecting state, which is close to the expected critical Ra_c. This indicates the potential of frequency analysis of near-interface temperature and salinity for validation of the present theoretical approach. While these values approximately agree with the first Ra_c for water, observations have been obtained relatively far away from the interface.

The problem of wave length prediction based on Fosters's equations was also discussed by Weber (1977) for saline ice growing with a planar interface. Weber evaluated equations 10.5 and 10.6 in connection with the interface salinity obtained from a coupled growth-diffusion model. Also in this connection the equations 10.7 and 10.8 based on a quasi-constant solute flux appear more appropriate. The planar freezing involves different scalings and is related to other aspects of practical relevance. It may be noted that the planar equations enable the prediction of the water salinity, at which convection precedes the onset of morphological instabilities. This limit can be associated with the planar-cellular transition in natural waters and is discussed in chapter 12.

10.3.4 Interface solute gradient G_c

The fundamental change in a system that starts to convect from a boundary is that the solute (or temperature) gradient is steepened. As at the interface the transport is by diffusion only, the interface gradient G_c must increase by the Nusselt number Nu.

Nusselt number

If the critical velocity V_{cr} for the onset of convection is known, the Nusselt number follows from equation 10.11 as

$$Nu = \frac{V_{cr}}{V} \qquad (10.15)$$

for $V < V_{cr}$, while $Nu = 1$ for $V > V_{cr}$. In this equation Nu can increase to infinity if V becomes very small. Nu, however, is limited by the available solute at the interface. This upper limit can be expressed by the equation

$$Nu = \frac{1 - k_{eff}}{1 - k}, \qquad (10.16)$$

where k is the interfacial distribution coefficient without convection and k_{eff} its reduced value due to convection. As k_{eff} cannot be smaller than 0, the Nusselt number cannot exceed $1/(1 - k)$. This model for Nu will be called the 'sharp model', as it implies an abrupt change in the $Nu - V$ relation once the upper bound is reached. As will be discussed below, equation 10.16 should represent the upper limit for Nu, because also at very low growth velocities there must be some k_{eff} to retain the cellular structure.

While equations 10.15 and 10.16 strictly account for the convective influence, a slight modification is introduced here to account intuitively for the changing interface conditions during increasing Nu. Instead of equation 10.15, the scaling

$$Nu \approx k_{eff}\frac{V_{cr}}{V} + (1-k) \tag{10.17}$$

is suggested. It accounts for the fact that the intermittent convective solute flux, and hence the Nusselt number, is expected to depend on the solute source within the ice interfacial regime. The main argument is that k_{eff} corresponds to the brine volume fraction at the interface. If this fraction drops due to the increasing Nusselt number, the constant salinity step condition, the solute gradient near the interface and the boundary layer will be weakened. Equation 10.17 is a simple approach to describe the self-limiting character of the convection. The term $(1-k)$ is added to ensure $Nu = 1$ at $V_{cr}/V = 1$. Eliminating k_{eff} from equations 10.16 and 10.17 yields

$$Nu = \frac{\left(\frac{V_{cr}}{V} + 1 - k\right)}{\left(\frac{V_{cr}}{V}(1-k)+1\right)} \tag{10.18}$$

for V_{cr}/V. For small V also here the limit $Nu \approx (1-k)^{-1}$ is approached.

In figure 10.4 the suggested self-sustained Nusselt number is compared to the simple approach $Nu = V_{cr}/V$ for $C_\infty = 35$ ‰ NaCl, resembling the seawater system. The difference between the parameterisations is largest near the threshold from equation 10.16. The slight increase in Nu at low velocities is a consequence of decreasing k. The more smooth behaviour of equation 10.18 will be discussed below. It appears to agree better with observations, yet is, of course, a tentative approach to approximate a rather complex interfacial problem.

Effective G_c'

If ones assumes a fluctuating ice-water interface that persistently adjusts to marginal equilibrium by dendritic or cellular growth, one expects that the Nusselt number in the liquid only partially controls the interfacial gradient at which the dendrites operate. The bulk gradient on the ice side of the interface will also play a role. This is accounted for as follows. Having obtained Nu in dependence on the growth velocity (and concentration), the modification of the interfacial concentration gradient is simply formulated as

$$G_c' = G_c r_{Gc}, \tag{10.19}$$

where G_c is the liquid solute gradient in the absence of convection and

$$r_{Gc} = \frac{\left(Nu + 1 + \frac{1}{n_k}\right)}{\left(2 + \frac{1}{n_k}\right)}, \tag{10.20}$$

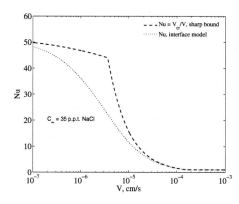
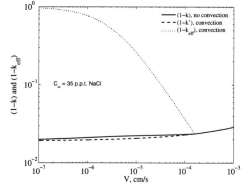

Fig. 10.4: Comparison of expressions 10.15 and 10.18 for the Nusselt number, simulated for $C_\infty = 35$ ‰. Note the transition to $Nu > 1$ at $V_{cr} \approx 1.6 \times 10^{-4}$ cm/s.

Fig. 10.5: Interfacial solute distribution without (k) and with (k') convection for $C_\infty = 35$ ‰. The effective k_{eff}, distributed into the ice, decreases considerably once turbulent convection begins.

where $n_k = K_i'/K_b$ is the ratio of effective conductivities in the ice and brine. The factor r_{Gc} accounts for (i) the salinity gradient in the liquid increasing with the Nusselt number, (ii) the liquidus salinity gradient in the ice and (iii) a slight difference between the corresponding liquidus temperature gradients on both sides of the interface. This computation of a lumped interfacial concentration gradient assumes that the solute gradient on the ice side is not severely modified by the intermittent convection ahead of the interface. The effective G_c' is simply taken as the average on both sides of the cellular interface [3]

Modified wave length λ_{mi}' and rescaled D_s/V

From chapters 7 and 9 it is clear that, according to the difficulties to obtain a solution for the plate spacing in the absence of convection, an analysis of the problem with nonlinear turbulent convection would be very complicated. The possible incorporation of simplistic boundary layers imposed by stirring has been discussed earlier (Hurle, 1969; Coriell and Sekerka, 1976b). The modification due to stirring depends very much on simplifications made about the boundary flow structure. Observational validations are to date lacking in any case.

The present approach to the problem is very simple. The only modification to the

---

[3] One might, at this point, suggest that the interface gradient should in general be calculated as an average on both sides of the interface, if the latter is fluctuating. This would change G_c by the factor $(1 + 0.5 K_b/K_i')$ and imply a decrease in the predicted a_0 by less than 4 % for seawater. To be discussed below, other implications of interface fluctuation will likely be more relevant.

stability equations is to rescale the diffusion length D/V by r_{Gc}^{-1}. The system is then solved as in section 9.1.2. First, an initial maximum k' is obtained from equation 9.8 by using an initial non-convective k-estimate and $D_s/(Vr_{Gc})$. Reiterating gives the final k' very rapidly, within 2 or 3 iterations. Finally, the modified wave length is obtained from

$$\lambda'_{mi} = 2\pi 2^{1/3} \left(\frac{D_s}{r_{Gc}V}\right)^{2/3} \left(\frac{\Gamma}{mC_\infty(k'-1)}\right)^{1/3}, \qquad (10.21)$$

which is essentially the same as equation 10.21, yet with k replaced by k', and a new factor $r_{Gc}^{(-2/3)}$ from the rescaling of the diffusion length D_s/V. The rescaling of D_s/V is principally comparable to other analyses for a planar interface (Hurle, 1969; Coriell and Sekerka, 1976b; Caroli et al., 1985b). Equation 10.21 is the consequent modification of the present approach of cellular marginal stability.

Figure 10.5 illustrates the basic effect of convection in terms of solute redistribution for the case $C_\infty = 35$ ‰. Above $V \approx 1.6 \times 10^{-4}$ cm/s the interfacial k and is not effected by convection and corresponds to the effective k_{eff} in the ice far from the interface. Convection changes the value of k, at which the dendritic interface operates, only slightly to k'. The effective solute distribution k_{eff} is, however, decreased considerably by the turbulent salt transport near the interface. The difference between k' and k_{eff} needs to be pointed out. The latter is the fraction of solute that remains in the ice. The former is an optimisation criterion determined by marginal stability. It defines the interface level, where the solute is still there. As mentioned earlier, in this simple approach the morphology of the interface is not considered.

The modification of the system due to convection is summarised as follows. The basic assumption is that the interfacial k reflects the marginal morphological stability state and a salinity step ΔS_{mi}, which it also maintains in the presence of convection. At some critical velocity V_{cr}, based on the Rayleigh number Ra of the diffusive boundary layer D_s/V, convection sets in and enhances the solute flux. The principle effect on the marginal stability state is to rescale the solutal boundary layer to D_s/V_{cr}. This would, due to equation 9.9, imply a constant plate spacing. However, increasing the Nusselt number limits the salt being available from the skeletal layer. In addition it seems appropriate, due to the intermittent and fluctuating nature of the interface, to introduce a lumped solute gradient via equation 10.19. Equation 10.21, the principal result of the present analysis, describes the modification of the plate spacing model due to these effects of convection. These predictions of λ'_{mi} will now be compared to the already discussed laboratory and field observations of plate spacings.

Aspects of expansion

All calculations here have been done by neglecting the density change upon freezing. As discussed in section 9.4, the principle effect of the latter would be a rescaling of the length scale D_s/V to $\rho_b/\rho_i' D_s/V$, if a solution is frozen from a fixed solid cooling surface. The effective density ρ_i' refers to porous ice. The rescaling effect thus scales with the mean solid fraction of the ice and the brine density. Calculating it from equation B.31 with a boundary temperature of $-20\ °C$, the rescaling ρ_i'/ρ_b was found to have a value between 0.91 and 0.92, for any salinity between 1 and 100 ‰. The dimensional critical velocity thus increases by $\approx 10\ \%$ and the transition Rayleigh numbers indicated in figure 10.3 need to be increased by 30 %. Apparently this difference is not resolvable by the scattered observations.

The expansion effect without convection has, for the growth velocities in the experiments from Lofgren and Weeks, been estimated to imply an increase in the plate spacing by 4 to 6 %. When convection sets in, it therefore does so at a higher V_{cr}. It turns out that the increased Nusselt number closely cancels the latter effect in the growth velocity range $10^{-5} < V < 10^{-4}$ cm/s. In laboratory experiments it thus only implies a slight overestimate at large growth velocities. For natural sea ice in the field, the effect is not expected, as the rising freeboard compensates for the expansion.

10.4   Plate spacings with convection

10.4.1   Laboratory observations

In figures 10.6 and 10.7 the convectively modified predicted plate spacings according to equation 10.21 are compared with observations from Lofgren and Weeks (1969a) at salinities of ≈ 5 and 100 ‰ NaCl. The following aspects apply also to other salinities in figures 9.3 through 9.6 from the previous chapter. The account of convection leads to smaller plate spacings than predicted by the purely diffusive model. The predicted change in the $a_0 - V$-slope agrees with the observations. The predictions including the convection model are, however, still too large for 5 and 100 ‰. A rather large increase of the Nusselt number parameter from $c_{nu} = 0.15$ to 0.30 is necessary to match the observations at a salinity of 100 ‰ NaCl, while the predicted a_0 at 5 ‰ still remains too large. Recalling the above discussion, this c_{nu} would be only realistic for free boundary conditions.

Figure 10.8 compares the laboratory observations from Lofgren and Weeks at the lowest salinity of 1 ‰ NaCl with the predictions. Due to the low solute content also the convective forcing is weak. It is impossible to give preference to a value of c_{Nu}, yet a slight overprediction of a_0 is also seen in these data. The observations apparently require c_{Nu} in excess of 0.30 to predict the correct plate spacings. Such a large Nusselt number cannot be created by haline convection alone, if the interface is rigid. Although it is possible that the reduced Nusselt number model from figure 10.4 does not correspond to reality and the sharp formulation 10.15 is more reliable, this cannot explain the observations at

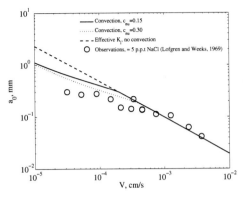
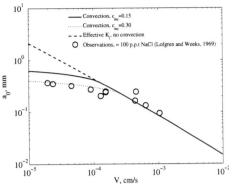

Fig. 10.6: Plate spacing during downward freezing of an ≈ 5 ‰ NaCl solution Lofgren and Weeks (1969a). The present predictions are shown without convection and with convection for two different Nusselt number scalings c_{nu}.

Fig. 10.7: Predictions versus observations as in figure 10.6, yet for a solution salinity of $C_\infty \approx 100$ ‰ NaCl.

1 ‰ NaCl, as both approaches almost coincide in the velocity regime in question. These considerations indicate a different cause for the deviation: Thermal convection or heat fluxes through sidewalls.

Thermal convection

The possible influence of thermal convection is demonstrated on the basis of figure 10.9. In this figure the actual growth velocity documented by Lofgren and Weeks (1969a) has been normalised by the velocity scale $\Delta T_i K_i / (H_i L_v)$, based on the temperature difference ΔT_i between liquidus temperature and cold boundary (-20 or -70 °C), the thickness H_i, and the pure ice properties L_v and K_i. In saline ice one expects $K_i' > K_i$, but as the temperature profile will be also curved (appendix B.6), the normalised velocity ratio should be of order one. Neglecting some uncertain values, which correspond to the onset of freezing, this is reasonably fulfilled at large growth velocities, in particular for the runs at -70 °C. However, at growth velocities below 10^{-4} cm/s the ratio rises to large values. As discussed in some detail in section E.4, for a similar experimental setup in the experiments from Cox and Weeks (1975), this indicates a strong effect of thermal convection. In the latter study temperature profiles were available and allowed the evaluation in figure E.5, showing that thermal convection steepens the temperature gradient within the skeletal layer. It consequently should increases the interfacial solute gradient at marginal stability. The plate spacing is expected to decrease. In the experiments from Cox and Weeks (1975) the effect was already relevant near 10^{-4} cm/s. The observations in figure 10.9 are from many different runs at different salinities and indicate considerable scatter. Thermal convection likely depends on two factors: (i) superheat in the tank can imply double-diffusive fluxes, (ii) sidewall and bottom heating through the insulation can create a convective circulation. Due to the large range of salinities used, both effects may have been present.

Weeks and Cox (1974) have noted that they used to adjust both the freezing temperature and the thermal insulation during the experiments, in order to ascertain unidirectional growth. Adjustment was performed to acquire a slightly smaller ice thickness in the center of the tank, from where the samples were taken. The main result in figure 10.9 is an increase in the effect of thermal convection with decreasing velocity. However, many of the runs are indicative of some drift in the normalised velocity at higher growth velocities, eventually reaching a minimum in the range 1 to 2×10^{-4} cm/s. This may reflect both the adjustment procedure and/or the onset of rigorous haline convection near $Ra \approx 300$, in connection with increasing thermal convection. As the mechanisms all appear simultaneously, it is not possible to discriminate them and give an ad-hoc quantification. Theoretical modeling is necessary.

The heat fluxes associated with haline convection alone can be expected to have little impact on the freezing interface, if the water body is at the freezing point. This follows from the large thermal to solutal diffusivity ratio and holds even if Nusselt numbers of the order 50 are realised. However, if there are heat sources in the water tank, (residual heat or imperfectly insulated walls), then the onset of haline convection might be associated with considerable heat transport towards the freezing interface. It is possible that the first critical velocity, apparent from figure 10.3 and likely corresponding to $Ra_c \approx 30$ to 40, is a sign of interfacial heat fluxes triggered by the first mode of, normally less effective, haline convection. In view of the very different appearance of field observations, to be discussed in the final section 10.5, this appears as a plausible aspect that deserves further studies. Here it suffices to say that figure 10.9, in connection with the discussion from chapter 5 and what follows, strongly supports the idea that thermal convection in the experiments from Lofgren and Weeks (1969a) has influenced the plate spacings in a manner different from natural growth conditions.

Concentration dependence of a_0

At a velocity of 10^{-3} cm/s no effect due to haline convection is expected and the predictions in figure 9.9 are retained. Figure 10.10 now shows the influence of convection at $V_n = 10^{-4}$ cm s^{-1}, that is slightly above the critical velocity. That the quantitative agreement with observations is improved at any salinity has become clear from earlier figures. An important new result of figure 10.10 is that the convection model applied

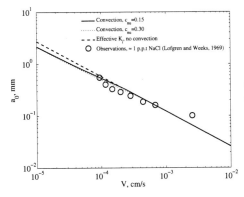
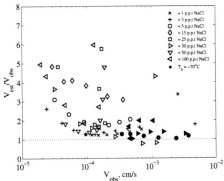

Fig. 10.8: Predictions versus observations as in figure 10.6, yet for the lowest solution salinity of $C_\infty \approx 1$ ‰ NaCl in the experiments from Lofgren and Weeks (1969a). In this runs the effect of solutal convection is smallest.

Fig. 10.9: Growth velocity in the experiments from Lofgren and Weeks (1969a), normalised by the velocity $\Delta T_i K_i / (H_i L_v)$ of pure ice under a linear temperature gradient. Each symbol represents a single run with decreasing growth velocity. Dots denote runs with a boundary temperature of -70 °C, others are for -20 °C.

here realistically reproduces the observed concentration dependence in the plate spacing. This result is further corroborated by figure 10.11 for a lower growth velocity.

In both figures the simulations are shown for the standard parametrisation $c_{Nu} = 0.15$ and the two times larger c_{Nu}. As noted earlier, the larger c_{Nu} gives better quantitative agreement, yet is physically less reliable. An interesting point in figures 10.10 and 10.11 is that $c_{Nu} = 0.15$ correctly predicts the concentration of the minimum in a_0, while $c_{Nu} = 0.30$ does not. This may be taken as a further indicator, that it is not the present haline convection approach which needs to corrected, but that rather the process of thermal convection is present in the laboratory observations. It may finally be noted that the concentration of the minimum in the plate spacing increases with decreasing growth velocity. This aspect had already been noted by Lofgren and Weeks (1969a). However, it does not, as speculated by these authors, have to do with the salinity of ≈ 23 ‰, below which an NaCl solution expands below the freezing point. Rather it follows from a complicated interplay of interface stability, molecular diffusivity and turbulent haline convection.

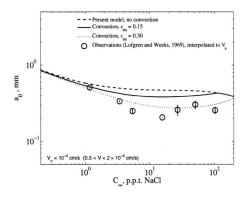
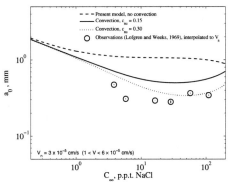

Fig. 10.10: Observations of a_0 interpolated to $V_n = 10^{-4}$ cm s^{-1} as in figure 9.9. The present predictions are shown without convection and with convection for two different Nusselt number scalings c_{nu}.

Fig. 10.11: Predictions versus observations as in figure 10.10, yet for a lower growth velocity of $V_n = 3 \times 10^{-5}$ cm s^{-1}.

Natural brackish water

Statistics of microstructural features of natural ice grown from brackish water seem to be completely lacking, and also the plate spacing has seldom been reported, even for a region of intense sea ice research like the Baltic Sea. This may have to do with the fact that, k_{eff} being similar to normal sea ice (Fransson et al., 1990; Granskog et al., 2004a), the ice salinity and brine liquid fraction is one order of magnitude smaller and brine layers are less clearly visible in brackish ice. Gow et al. (1992) determined plate spacings of 0.7 to 0.8 mm from thin sections at only two sites during a larger field program in the Bay of Bothnia. They reported a plate spacing of $a_0 \approx 0.7$ mm for 11 cm thick ice, which had grown overnight from brackish water of ≈ 3.7 %. Comparably rapid growth of ice from normal seawater results in spacings of 0.4 to 0.5 mm (figure 6.7). The predictions in figures 10.10 and 10.11, which likely resemble the growth velocity regime of thin ice of that kind, are consistent with this difference in a_0 between brackish and sea water. Future studies need to explore this.

10.5   Concluson: Sea ice field observations

Before the field observations of sea ice plate spacings compiled in section 6.1 are compared to the model predictions, it is useful to summarise the results of the present chapter.

Haline convection in a thin boundary layer below growing sea ice has been discussed. This has been done on the basis of several studies by Foster (1968a,b, 1969, 1971, 1972),

who once derived scaling laws for the critical time and wave length at the onset of
turbulent convection for a seminfinite layer. While the nonlinear dynamics of instability
evolution for the semi-infinite fluid layer are not yet fully understood, the discussion of
several ideas indicates that weak motion may be expected near $Ra_c \approx 40$, while the flux-
related Rayleigh number that applies to the present problem should be near $Ra_{c2} \approx 300$.
Based on these values, Foster's parametrisation for a constant concentration boundary
condition has been slightly modified, arguing that $\Delta C_{mi} = C_{int} - C_\infty$ is dictated by the
morphological marginal stability condition at the interface. This coupling of marginal
morphological stability with a turbulent convective boundary model predicts the critical
wave length of the convective eddies in the boundary layer in agreement with some limited
observations. It also yields a critical velocity for the onset of convection and a Nusselt
number in dependence on growth velocity. The latter in turn can be used to predict how
the effective interface solute gradient is influenced by convection and how this effects the
marginal stability wave length of the interface and thus: the plate spacing.

Most features of the observed concentration dependence of the plate spacing are
reasonably predicted by the convective modification. The comparison of predicted plate
spacings with laboratory data from Lofgren and Weeks (1969a) indicates a slight over
prediction of a_0 at growth velocities above the onset of convection. It also shows that
an unrealistically large turbulent convection parameter $c_{Nu} \approx 0.30$ must be used in
equation 10.11 to reach proper agreement with the observations. It has been argued
that these discrepancies could be the consequence of caveats of the laboratory set-up and
non-equilibrium freezing due to the following aspects:

- In the laboratory, where ice is frozen downward from a rigid surface, expansion
  increases the boundary layer, and the onset of convection is expected at 10 % lower
  growth velocities than in the field with an adjusting freeboard. The main effect of
  expansion is an increase in the plate spacing of ≈ 4 to 7 %, as long convection is
  not present.

- At early stages of rapid ice growth crystals are inclined in the vertical. Hence,
  a_0 obtained in horizontal thin sections close to the cooling interface, may be an
  overestimate compared to the true spacing. The bias can reach a factor of two near
  the freezing interface and still be near 10 % at a 5 cm distance.

- Thermal convection, or advection of heat from container walls, steepens the interfacial temperature gradient, and thus the solute gradient. This is expected to
  decrease the plate spacing. The quantification of this effect is difficult.

- An interaction between a first weak convective mode at $Ra_c \approx 40$ with wall heat
  fluxes is a possible variant of thermal convective influence.

All noted effects are indicative in the laboratory observations, when compared to the model predictions. The first two mechanism increase the plate spacing at large growth velocities. The second two have the potential to decrease the plate spacing at velocities where haline convection sets in. Most of them are not expected to occur during natural freezing of seawater, at least not in this pronounced way.

In figure 10.12 the field observations of sea ice plate spacings compiled in section 6.1 are compared to the model predictions in the natural sea ice growth regime. The laboratory data from Lofgren and Weeks for the corresponding salinity range are also plotted. Shown are the prediction without convective modification as a dashed curve, the model including convection with the most reliable $c_{Nu} \approx 0.15$ as a solid curve, and the convection model for $c_{Nu} \approx 0.30$ as a dotted curve. The large difference between all field observations of a_0 and the laboratory data has been mentioned earlier. It clearly supports the above contention of a considerable influence of thermal convection in the study by Lofgren and Weeks. It may be noted that the observations documented by Weeks and Hamilton (1962), shown as plus signs in figure 10.12, also correspond to laboratory experiments, yet under a different mode of freezing in a cold room, when thermal convection is not expected.

A final note is in order here. Although figure 10.12 clearly favours a rigid boundary convection model (solid curve) over the free boundary (dotted) curve, it is recalled that the interfacial temperature gradient has been assumed to be given by the approach from appendix B.6. The latter is an approximate model for the interfacial heat flux under steady growth conditions (figure B.6). If this gradient is lowered by a factor of two, then the observations would be matched better by the free boundary model. A more exact solution of the problem must therefore focus on the proper prediction of the temperature field and an effective thermal conductivity near the interface. The dependence of a_0 on the growth velocity would, in the natural growth regime, not change much. However, the convective transition at high growth velocities could be better understood, see figure 10.3. This is, of course desirable, as it would allow a better understanding of deviations from the ideal model due to natural variability, as summarised in the next paragraph. Detailed observations of the interfacial temperature regime and the convective transition appear to be reasonable means to proceed from here.

Mechanisms of natural variability

Unfortunately there are little observations near the proposed transition velocity of $\approx 1.6 \times 10^{-4}$ cm/s. The available observations, however, agree reasonably with the predictions over the whole regime of natural sea ice growth velocities. The agreement being confident, it is of interest to get an idea of the mechanisms that might create a variability in a_0 during natural growth:

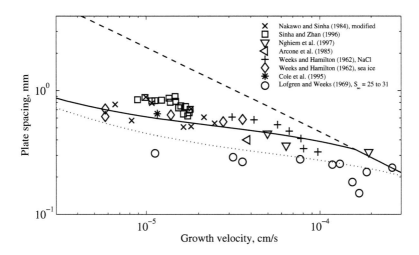

Fig. 10.12: Compilation of plate spacings observed in natural sea ice from chapter 6. The circles are the laboratory observations from Lofgren and Weeks (1969a) at comparable NaCl concentrations. The present model predictions are shown without convection (dashed line), the convective standard prediction with $c_{nu} = 0.15$ (solid curve) and for the extreme case $c_{nu} = 0.30$ (dotted line).

- *Thermal convection.* Thermal convection may be relevant in areas of large oceanic heat fluxes and a thermal stratification. Under natural conditions this thermal convection will normally be forced by the under-ice current and not necessarily have an effect comparable to that due to the largest scale circulation in a container. The effect on the plate spacing will depend on the thermal field near the interface and thus on the conductive and convective fractions of the heat flux. It will also depend on the interaction of the flow with the solutal flux boundary layer and the flow within the ice in relation to macroscopic brine channels. The present simple model can no longer be applied. The effective growth velocity reduction due to the oceanic heat flux, estimated in section 5.5 and appendix D for sea ice growth time-series, is typically of 0.3 to 0.5×10^{-5} cm/s. This indicates, at which freezing rate such an influence may be noticed. Although one intuitively expects a reduction of a_0 as in the experiments from Lofgren and Weeks, there is a need for advanced modeling and, in the first instance, systematic observations.

- *Crystal inclination.* Most observations in figure 10.12 refer to levels relatively far from upper surface, while for 10 cm thick ice the crystal inclinement is normally less than ten degrees (Weeks and Assur, 1964; Weeks and Ackley, 1986). Hence, this source of overestimate in a_0 should be neglegible in most samples of not too thin natural ice.

- *Coarsening.* It seems plausible that some of the brine layers become less easily detectable, if the ice is analysed at very low temperatures. It is unknown, if some plates merge during the cold winter period. Inclusions, however, can migrate during thermal cycling which will imply an effective increase in the plate spacing. The observations from Sinha and Zhan (1996) in figure 10.12 refer to the thickest and oldest ice and were obtained after the onset of warming. That they systematically exceed the predictions is qualitatively consistent with the noted coarsening mechanisms. Two bottom values from Sinha and Zhan (1996) at lowest growth velocities (not shown), and an observation from Nakawo and Sinha (1984), shown in figure 6.6, are very close to the predicted curve with $c_{Nu} \approx 0.15$ and ≈ 20 % lower than observations farther away from the interface.

- *Unsteady growth.* Unsteady growth will, if fluctuations are not larger than a factor of 2 or 3, result in an almost unchanged average equilibrium a_0, because the V-dependence is rather weak. Increasing velocities may, however, imply the regrowth of a skeletal layer. In such a situation the boundary layer scale might be less than D_s/V, reducing the effect of convection. This effect on a_0 may be more significant than the growth velocity variation, but is only expected under extreme conditions.

- *History effects.* The adjustment to a new spacing upon a sharp growth velocity drop may take some centimeters of growth (see the discussion for other alloys in section 8.3.5). The observations from Milosevic-Kvajic et al. (1973) shown in figure 9.7 may eventually be interpreted in terms of delayed adjustment on a scale 50 times the plate spacing. For sea ice this distance corresponds approximately to one skeletal layer. Some observations had been discussed in connection with figure 6.6 and the data from Nakawo and Sinha (1984). Also this effect is only expected under strong changes in the growth conditions.

- *Brine channel convective flow.* Desalination of young sea ice occurs due to periodic downward flow through large brine channels, which is compensated by upward flow through the plate lamellae. Also this leads to a relative reduction of the scale D_s/V by the upward motion. Observations by Wakatsuchi and Ono (1983) indicate that this volume flow increases with decreasing growth velocity, reaching $\approx 0.2\ V$ at $V \approx 2 \times 10^{-5}$ cm/s. This should imply a 20 % smaller scale than D_s/V. The effect on the plate spacing may eventually be found by applying a correspondingly smaller c_{Nu}. It would increase a_0 by ≈ 15 % at a growth velocity of $V \approx 2 \times 10^{-5}$ cm/s.

It appears that, among the noted mechanisms that might influence the interface thermodynamics and plate spacing, the convective motion within the ice is most relevant. It is related to the internal ice circulation and hence to the permeability and stability of larger brine channel systems, a process which is not well understood at present. The brine circulation estimates from Wakatsuchi and Ono (1983) indicate that this mechanism may change the plate spacing by more then 10 % at growth velocities below 3×10^{-5} cm/s. Wakatsuchi and Ono's estimates are based on an indirect method and mostly refer to laboratory observations. More observations are required to quantify this effect in dependence on growth conditions and evolution of brine channels.

A second major systematic deviation source might be the observation procedure, linked to coarsening of brine layers or they detection limit. These aspects may be addressed by more sophisticated analysis of observations. The expected behaviour to retrieve larger plate spacings from thicker and older ice is not inconsistent with observations. In view of the noted mechanisms and caveats, the model predictions are fully consistent with the available sea ice field observations of plate spacing in the natural growth regime.

Low growth velocities

A rigorous test of the present theory also requires observations at very low growth velocities. Although sparse, some data are available and shown in figure 10.13. The latter contains all field observations from figure 10.12 as crosses, as well as the laboratory data from Lofgren and Weeks. In addition it contains two observations. The squares are observations from the bottom of a 10 meter thick ice flow (Cherepanov, 1964). Due to the large timescale $H_i^2/\kappa_s \approx 3$ years for ice of 10 m thickness, the latter will likely grow with an almost constant growth velocity. Its winter upper limit, neglecting oceanic heat fluxes, has been estimated here to fall between 0.9 and 1.3×10^{-6} cm/s. It was assumed that the average annual ice surface temperature is between -15 and -19 °C (Doronin and Kheisin, 1975). The to date largest a_0 has been documented for the bottom of an Antarctic ice shelf, with an estimated growth velocity of 2 cm/year (Zotikov et al., 1980). An upper bound for the growth velocity has been taken from a recent numerical model study (Assmann et al., 2003).

The predicted curves in the figure are again the prediction without convection, dashed curve, and the convection model as the solid curve. The agreement of the predictions with this limited set of observations is very promising. Two aspects of the validity of the Nusselt number at low growth velocities are also pointed out by two additional curves in figure 10.12. The first addressees the sharp Nusselt number approach $Nu = V/V_{cr}$, introduced by equation 10.16, which is shown as a dash-dotted line. Its basic behaviour is not fundamentally different from the standard predictions, yet its agreement

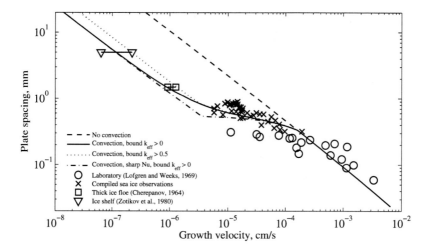

Fig. 10.13: Compilation of plate spacings observed in natural sea ice, showing all field observations from figure 10.12 as crosses and laboratory observations from Lofgren and Weeks (1969a) as circles, and two data points at very low growth velocities. The present model predictions are shown without convection (dashed line) and with convection for $c_{nu} = 0.15$ and no limit in k_{eff} (solid curve). For comparison a different Nusselt number parametrisation (dash-dotted curve) and the limit $k_{eff} < 0.5$ (dotted line) are shown.

with observations is not that good. Only at very low growth velocities it approaches the modified Nusselt number results. The second is a prediction where simply k_{eff} in equation 10.16 for the Nusselt number has been limited to a value of 0.5. The tentative argument for this condition is that the turbulent interface flux might be limited by a local constraint within the redistribution regime: When more solute is transported in plumes than distributed to its source layer, then the convection might be limited. This k_{eff} relates not to the final k_{eff} far from the interface and after delayed desalination, yet to the interface redistribution. This simulation is shown as a dotted curve. The limited observations do not allow a decision between them. The reality may fall between these predictions. In the shelf ice from (Zotikov et al., 1980) it was found that $k_{eff} \approx 0.1$ far from the interface, which is at least a lower bound. Due to the uncertain growth velocity it has, however, no potential to validate the interface restrictions.

However, for the range of realistic restrictions the predictions do not change severely. Figure 10.12 thus further promotes the model predictions of the sea ice plate spacing, extending the agreement over five orders of magnitude in the freezing rate. This result offers new perspectives in modeling and interpretation of microstructural observations of natural sea ice and marine ice shelves, in terms of their growth history and physical environment.

11. Convection in a porous medium

In the previous chapters it was shown that a simple modification of morphological stability theory can successfully predict the principle plate or brine layer spacing of sea ice. The predictability of the plate spacing is now taken advantage of in applying it to the problem of *internal* convective stability of the sea ice matrix. Internal convection, as already qualitatively identified in chapters 4 and 5 , and in the discussion of the sea ice skeletal layer, can now be discussed in a more quantitative manner.

The basic approach to be problem is that the brine salinity, increasing, due to decreasing temperature, upwards from the interface, is limited by a critical value. In the following this value is termed ΔS_{pc} to distinguish it in a first instance from ΔS_{sk}, the earlier discussed salinity increase at the top of the skeletal layer. If this salinity increase ΔS_{pc}, with respect to the interface, is exceeded, the sea ice brine becomes hydrodynamically unstable and can no longer be kept motionless between the platelets. Convective exchange with the seawater below brings less saline water into the skeletal layer, where it freezes and reiterates the process. To find ΔS_{pc} and H_{pc} linear convective stability theory for a porous medium will be applied here.

11.1   Horton-Rogers-Lapwood problem

Due to the pioneering work of these researchers Nield and Bejan (1999) have termed the problem of onset of convection in a porous medium the Horton-Rogers-Lapwood-problem. The nondimensional number which determines the onset of thermal convection in a saturated porous medium described by Horton and Rogers (1945) and Lapwood (1948). It is the porous medium Rayleigh number Ra_p

$$Ra_{p,T} = \frac{\alpha \Delta T_p g \Pi H_p}{\nu \kappa_p} \tag{11.1}$$

In this equation $\alpha \Delta T_p$ is the vertical density difference due to thermal expansion, g the acceleration of gravity, ν the kinematic viscosity, H_p and Π the height and vertical permeability of the medium, and $\kappa_p = K_p/(c_l \rho_l)$ an effective diffusivity, corresponding to the ratio of effective thermal conductivity K_p of the medium and the volumetric specific heat $c_l \rho_l$ of the fluid.

When the density is controlled by solutal buoyancy via the gradient $\beta \Delta S_p / H_p$, the corresponding equation may be written as

$$Ra_{p,S} = \frac{\beta \Delta S_p g \Pi H_p}{\nu D_s} \tag{11.2}$$

with solute diffusion coefficient D_s, imposed salinity difference ΔS_p and the coefficient of solutal contraction for the brine, $\beta = \rho_b^{-1} \partial \rho_b / \partial S_b$. The only difference to 11.1 for the problem of heat conduction is, that solutal diffusion through the solid has been assumed negligible and therefore no effective diffusivity appears.

Horton and Rogers (1945) demonstrated theoretically that, in analogy to Rayleigh's convection problem in a viscous medium (Rayleigh, 1916), there is a critical Ra_{pc} at which convection starts, and that its upper bound is

$$Ra_{p,c,max} = 4\pi^2 \approx 39.48 \tag{11.3}$$

In correspondence to the classical Rayleigh number problem this upper bound is given for two impermeable boundaries at constant temperature. Lapwood (1948) obtained an analysis for different boundary conditions and Nield (1968) presented Ra_{pc} for all possible combinations. Historical overviews, summaries and discussions on many aspects of porous medium convection, investigated since then in systems of small to geophysical scales, are found in Nield and Bejan (1999). The theory given and meanwhile confirmed by experiments, in practice the proper prediction of the permeability Π in equation 11.1 is one of the most challenging problems.

11.2 Darcy's law

For the one-dimensional stability problem the vertical permeability Π is frequently assumed to be given by Darcy's equation (Nield and Bejan, 1999)

$$\overline{W} = v_b W = -\frac{\Pi_v}{\mu} \frac{\partial p}{\partial z} \tag{11.4}$$

Here it is referred to the case of sea ice, where the brine volume fraction v_b corresponds to the porosity of the porous medium. Darcy's equation implies that the mass flow velocity $\overline{W}$ is proportional to the pressure gradient $\partial p / \partial z$, the permeability Π with dimension $length^2$, and the dynamic viscosity μ. Considering sea ice, where the density is governed by the brine salinity, the Darcy velocity $\overline{W}$ is related to the pressure gradient $\partial p / \partial z = g \rho_b \beta \Delta S_p$ via

$$v_b W = -\frac{\Pi_v}{\nu} g \beta \Delta S_p, \tag{11.5}$$

where $\nu = \mu / \rho_b$ is the kinematic viscosity of the fluid.

The applicability of Darcy's law to describe flow in a porous medium requires generally the Reynolds number $Re_p = Ua_\Pi/\nu$, based on the pore scale a_Π, to be of order unity or less (Nield and Bejan, 1999). A crossover to non-Darcy flow has been demonstrated to occur near $Re_p \approx 1.4$ to 1.5 for glass-water mixtures (Palm et al., 1972). $Re_p \approx 1$ to 10 is the range into which most observations for different materials fall (Dullien, 1979; Nield and Bejan, 1999). Niedrauer and Martin (1979) have observed typical velocities from 0.006 to 0.25 mm/s in the skeletal layer of sea ice and noted that for their observed maximum channel diameters of 1.5 mm and $\nu \approx 2$ mm$^2 s^{-1}$ the condition of Darcy-flow is reasonably fulfilled. However, the experiments by Niedrauer and Martin (1979) were conducted in a thin constricted growth cell. Velocities of brine streamers measured by Wakatsuchi (1977) below a growing ice skeletal layer were an order of magnitude larger and would have yielded a Reynolds number of ≈ 20.

When the deviation from Darcy flow is not two large, the onset of convection may still be described in the above manner, with a slightly increased critical Ra_{pc}. A number that is frequently used to evaluate the applicability of Darcy's law is the ratio H_p^2/Π or inverse Darcy number Da^{-1}. For the Hele-Shaw cell, the two-dimensional analog of a porous medium (Elder, 1967; Hartline and Lister, 1977), this effect has been quantified. Degan and Vasseur (2003) considered instead the width ratio L_p^2/Π and showed that a porous medium typically requires a value larger than 10^4, to be described closely by linear convective stability and Darcy's law. At Da^{-1} of 10^3 critical Rayleigh numbers are ≈ 10 % larger. An analysis by Frick and Clever (1980) and Schoepf (1992) indicates similar results, if the height H_p is used as the length scale. For brine channels and layers of diameter d_{cy} and width d_0, the inverse Darcy numbers are given by $32(H_p/d_{cy})^2$ and $12(H_p/d_0)^2$, respectively (section 3.3.2). Considering a sea ice skeletal layer of thickness 1 to 3 cm, the mobility ratio H_p^2/Π becomes ≈ 800 to 8000, if one identifies the pore scale with a brine channel of diameter $d_{cy} \approx 2$ mm. It becomes $\approx 2 \times 10^3$ to 2×10^4, if $d_0 \approx 0.25mm$ is identified with half the plate spacing of typically 0.5 mm (brine volume 0.5). The considerations indicate that Darcy's law is justified for the porous sea ice medium when considering the plate spacing, while one must be more careful once the flow in brine channels is involved.

The critical Rayleigh Number Ra_{pc} for the porous medium depends, analogous to the Rayleigh-Bénard problem for viscous fluids (see chapter 10), strongly on the boundary conditions which often are not ideal (Nield, 1968; Nield and Bejan, 1999). The boundary conditions most relevant for sea ice will be discussed below in section 11.5.2. Before this is done, a fundamental difference between sea ice and other porous media needs to be outlined.

11.3 Porous media with phase changes

Linear convective stability theory of porous media can, for example be applied to geophysical systems as snow, rocks and sands. When phase changes within the medium occur, like for magma-chambers, the earth core and sea ice, the problem is modified (Nield and Bejan, 1999). The basic difference is that molecular thermal properties in a *reactive* porous medium are neither homogeneous nor constant with time. Theoretical modifications have been initiated in the field of metallurgy (Chalmers, 1964; Flemings, 1974). In the metallurgical literature the critical 2-phase region, where reacting liquid and solid phases coexist, is often called a *mushy* zone. This phrase has recently also been used to describe the solidification of sea ice (Wettlaufer et al., 1997; Worster and Wettlaufer, 1997). A fundamental modification applicable to many situations is that the critical Rayleigh Number now is often evaluated from

$$Ra_p = \frac{\beta \Delta S_p g \Pi H_p}{\nu \kappa_b}. \tag{11.6}$$

Equation 11.6 differs from equation 11.2 in one respect: Instead of the solutal diffusivity D_s, it is now the thermal diffusivity κ_b of the liquid which stabilises the buoyancy gradient. It is still the solutal buoyancy which dominates the density field. The justification for this approach is the large ratio of thermal to solutal diffusivity in metals and aqueous solutions. If κ_b/D_s is large, a displaced particle will equilibrate thermally to its new level. With respect to the melting point it becomes over- or undersaturated, which leads to remelting of liquid (brine) or freezing of solid (ice). The principal effect of this equilibration process is that solutal buoyancy can be treated as being effectively dissipated by thermal diffusion. The applicability of equation 11.6 depends on the assumption that the liquid in the porous medium is everywhere in intimate contact with the internal surface of the solid, and that no sub-pore gradients in the concentration field occur. Numerical solutions of the linear stability problem for such a reactive porous continuum have been performed for slightly different boundary conditions (Worster, 1992; Chen et al., 1994; Emms and Fowler, 1994; Lu and Chen, 1997). The outcome of these studies is that, in a reactive, vertically inhomogeneous porous medium, the critical Ra_{pc} depends strongly on the distribution of the porosity, and the functional form of Π. Due to these factors a unique critical Ra_{pc} cannot be defined in the simple terms of equation 11.2. However, in the near eutectic approximation, when the brine volume fraction and permeability are almost uniform through the medium, the classical solutions are retained (Amberg and Homsy, 1993; Chen et al., 1994; Chung and Chen, 2000).

The absence of sub-pore gradients is unlikely to be true for sea ice, but it is a necessary assumption to proceed with a macroscopic model of the system. As will be seen in the following chapters, sea ice differs from a *mushy* zone due to its vertically oriented parallel plates, whose lateral freezing may lead to the formation of new micro scales.

Such sub-pore processes and the related thin boundary layers of salt are expected to fundamentally influence the permeability Π and the ice's convective behaviour, when they lead to bridging between the plates and the formations of other microscales than the plate spacing a_0. It needs therefore to be stressed, that a macroscopic porous medium approach for sea ice is based on the assumptions of (i) small sub-pore solutal gradients with respect to overall vertical gradients, (ii) large enough brine volume fraction, implying that lateral freezing irregularities do not change the microstructure fundamentally. These conditions are mostly fulfilled in the bottom *skeletal layer* and allow a treatment of the system's overall stability by a continuum model of transport properties.

11.3.1 Channel formation by convection

The theories considered so far only determine the onset of convection. Once this happens, the structure and properties of the porous medium are modified by remelting and freezing, leading to redistribution of the solute and changes in the liquid fraction. The mode of redistribution of solute, which creates channels and solute inclusions on a much larger scale than microstructural entrapment between the plates, is often termed *macrosegregation* (Chalmers, 1964; Flemings, 1974). It exists in sea ice in form of brine channels and, as noted earlier, determines the principal permeability of bulk ice with small brine volumes. In the solidified casting product these channels appear as regions of high local solute concentration, similar to brine channels in sea ice. Reviews of experiments on macrosegregation have been published in many different disciplines (Hellawell et al., 1993; Beckermann and Wang, 1995; Prescott and Incropera, 1996; Worster, 1997; Beckermann et al., 2000; Chen, 2001). All of them report problems in the predictability of convection and the need for further analysis and experiments.

Fowler (1985) was the first who developed a simple mathematical model for the formation of chimneys or channels, discussing their formation in terms of a second critical Rayleigh number, here termed Ra_{pcc}, which can be larger or smaller than Ra_{pc}, in dependence on the character of formation of channels. Similar to convection in viscous fluids (see the previous chapter), the prediction of channels formation requires a nonlinear analysis. An analogous problem appears in nonreactive porous media, when the Darcy-flow breaks down at large Ra_p or pore sizes (Palm et al., 1972). In reactive media the situation becomes much more complicated than the nonreactive stability and a first analysis was performed by Amberg and Homsy (1993). The nonlinear theory of channel formation is still an evolving disputed field of research (Emms, 1998; Loper and Roberts, 2001; Riahi, 2002; Chung and Worster, 2002). Also for sea ice a principle theory of brine channel formation is lacking, last but not least due to the lack in observations. For sea ice it can in addition be expected that the permeability anisotropy affects the problem, as it does for nonreactive porous media (Kvernvold and Tyvand, 1979). Anisotropy or inhomogeneity in transport properties or boundary conditions may also effect the process of channel formation (Nield and Bejan, 1999). Some aspects are noted in the next paragraph on the basis of recent studies with the binary $NH_4Cl\text{-}H_2O$.

The binary alloy $NH_4Cl\text{-}H_2O$

Aqueous ammoniumchloride ($NH_4Cl\text{-}H_2O$) may be, due to its popularity, used exemplarically to illustrate some of the faced problems. This transparent aqueous solution has received much interest as a metal analog in experimental and theoretical studies (Hellawell et al., 1993; Chen, 2001). $NH_4Cl\text{-}H_2O$ is a hyper-eutectic binary that releases light fluid upon solidification. During cooling from the bottom it freezes uwpards as a dendritic array which is compositionally unstable, while the overlaying liquid is thermally stably stratified. A detailed experimental study of the onset of convection in this solidifying porous medium was performed by Tait and Jaupard (1992a), who estimated a critical Rayleigh Number $Ra_{pc} \approx 25$. This was noted by the latter authors to be in reasonable agreement with the expectable value of 27.1 for the assumed boundary conditions (Nield and Bejan, 1999). The work from Tait and Jaupard has served as a database to validate theoretical predictions from different approaches (Worster, 1992; Emms and Fowler, 1994; Hwang and Choi, 2000), apparently with some success.

In some contrast, Chen and Chen (1991) estimated a critical Rayleigh Number of 2 to 2.5 from their experiments (if converted to equation 11.6), yet using a different permeability relation. Had they used the same approach as Tait and Jaupard (1992a), equation 11.12 to be discussed below, they would have found $10 < Ra_{pc} \approx< 13$, still a factor of 2 smaller. Tait and Jaupard (1992b) reported a range of Ra_{pc} between 70 and 90 for the onset of instabilities, for experiments where the boundary temperature was lowered during the freezing process. Different causes for the discrepancies have been suggested (Worster and Kerr, 1994; Chen et al., 1994) yet a clear experimental proof is lacking so far. Both experimental (Huppert and Hallworth, 1993) and numerical studies (Goyeau et al., 1999) indicate the strong dependence of the internal structure and permeability on boundary conditions, solid fraction and eventually time-dependent processes emphasised by Hwang and Choi (2000). The prediction of permeability from the liquid fraction, and of the formation and evolution of channels is a difficult task, when highly dendritic arrays like $NH_4Cl\text{-}H_2O$ are considered.

Finally, considering the question, if channels are initiated at Ra_{pcc} larger or smaller than the classical Ra_{pc}, the study by Worster and Kerr (1994) is of interest. They evaluated the appearance of chimneys in the work from Huppert and Hallworth (1993) by use of a slightly modified theoretical model from Worster (1992). It was estimated, that chimneys were typically observed at Ra_p values in the range 100 to 300, that is an order of magnitude larger than the same theory predicts for the onset of convection.

However, also this result is based on an uncertain parametrisation of the bulk permeability and does not consider the particular microstructure or the active fraction of the porous medium.

Sea ice brine channels

The existence of large vertical pores in sea ice is long known, e. g., (Hamberg, 1895), yet their quantitative description begun only some decades ago. Bennington (1967) and Lake and Lewis (1970) described brine channels and their networks in vertical sections of thick natural sea ice, while Cherepanov (1973) classified different sea ice types in terms of their dominant pore sizes. Eide and Martin (1975) and Niedrauer and Martin (1979) could visualise the flow related to channel formation in a thin laboratory growth cell. They obtained scales smaller than under natural conditions, presumably due to the laboratory confinement. Quantitative investigations on brine channel widths and distributions within young sea ice under natural growth conditions were published by Japanese researchers (Saito and Ono, 1977; Wakatsuchi and Saito, 1985), who also documented the related convective plumes in the water below (Wakatsuchi, 1977; Wakatsuchi and Ono, 1983). A review and discussion of several experiments was given by Wakatsuchi (1983). These first generation channels in the skeletal layer or in thin new ice show much smaller widths and spacings then the chimneys described by Bennington (1967) and Lake and Lewis (1970). In spite of the importance of brine channel scales and distributions, further observations are limited to a few studies (Cottier et al., 1999; Freitag, 1999; Tison and Verbeke, 2001), and a systematic framework of observations with varying growth conditions does not exist.

The comparability of experiments with NH_4Cl-H_2O , where more studies under controlled growth conditions have been performed, depends on the question, if NH_4Cl-H_2O and sea ice behave similarly in terms of their internal structure. The complexity of the results noted above indicates that the highly dendritic NH_4Cl-H_2O behaves differently. For the latter the dendritic substructure and the solid-fraction permeability relations appear to be the main unknowns in the problem. These make it difficult to exactly predict the convective behaviour of the NH_4Cl-H_2O binary system in experiments. Sea ice is, at least near the bottom, due to the general lack in sidebranching of secondary and tertiary arms, a simpler system, in particular as the substructure a_0 has now turned out to be predictable. However, so far it has not been analysed for sea ice, if brine channels form at Ra_p different from Ra_{pc} for convection. In thin ice the channels are found to exist at the surface and they frequently persist at deeper levels while the ice is growing (Wakatsuchi and Saito, 1985; Wettlaufer et al., 1997). Nevertheless does an *a posteriori* analysis of sea ice sections not give a simple answer on the critical Ra_{pcc} for channel formation. The extension of channels to the top surface may indicate that, in terms of Fowler's (1985) description, channel formation in sea ice is *subcritical*. That is, Ra_{pc} is the larger number and channel formation begins earlier, eventually triggered by the convection in the liquid, as proposed by Chen et al. (1994). It appears unlikely that channels, triggered by the onset of convection at Ra_{pc}, afterwards extend to the surface, because upward flow should decrease the solid fraction. However, channel dynamics can be complicated, as they may migrate horizontally (Niedrauer and Martin, 1979). As channels are likely local features operating at marginal stability, they may respond locally to expansion flow and fluid instabilities in the liquid (Solomon et al., 1999).

Clearly, a simple prediction of channel width and spacing in sea ice seems, based on the momentary database and the lack in a concise theoretical analysis, not feasible. However, as discussed below, equation 11.6 may be applied to evaluate the global convective threshold, as long brine can circulate freely between the plates, because in this *skeletal layer* regime the flow is not restricted to the larger channel features, let them exist or not. If they exist, the applicability of equation 11.6, with the plate spacing as the principal microstructure for Π, in principal requires that the convective onset is independent of their presence and strictly governed by the smallest microstructure. As downward channel flow always requires upward and lateral feeding flow through the plate laminate, such an assumption is not unreasonable. The remaining part of this chapter focuses on the description of Ra_{pc} and the question, at which level near the bottom of sea ice the 'normal' convection sets in between the plates. While it principally ignores brine channels, it should be kept in mind, that the channels may require a modification.

11.4 Permeability

As the discussion so far has shown, a proper evaluation of 11.6 implies the challenging question to predict of the permeability Π properly, and of cause the latter is also a central property for the thermo- and fluid dynamics of bulk sea ice. During the past decades permeability models have evolved from a mixed geometrical and phenomenological fundament and basic understandings (Carman, 1956) to more exact theoretical predictions involving pore structure, connectivity and hydrodynamics (Dullien, 1979; Happel and Brenner, 1986; Nield and Bejan, 1999). This theoretical basis is important to evaluate the parametric dependence of Π on microstructure and porosity brine volume v_b. To understand the approach made below for sea ice it is important to realise the approaches made so far for other porous media.

11.4.1 General permeability models

Most permeability laws are conceptually based on the law for viscous flow through pipes, empirically suggested by Hagen and Poiseuille and theoretically solved by Stokes, e.g.,

Lees (1914). In connection with Darcy's law 11.4 the simplest permeability for laminar flow through a bundle of parallel circular capillaries is given by

$$\Pi = \frac{D_c^2 v_b}{32},$$
(11.7)

where v_b is the volumetric cross section of the capillaries and D_c their diameter.

Isotropic quasi-granular media

The first concepts to modify equation 11.7 for natural porous media were suggested by Blake, Kozeny and Carman, mainly based on geometrical grounds, and for granular porous media like sands. They implied a modification of the Stokes drag in terms of an effective hydraulic radius and surface area of the particles. These approaches led to the form

$$\Pi = \frac{D_p^2}{36b_p} \frac{v_b^3}{(1 - v_b)^2}$$
(11.8)

which is now known as the the Kozeny-Carman equation (Carman, 1956). D_p is the average particle diameter and b_p a constant related to flow path and surface area. It is empirically found to be $b_p \approx 5$ for spheres and often some 10 to 20 % smaller for flat-sided particles, in agreement with simple geometrical considerations (Carman, 1956; Dullien, 1979). Due to the limited packing density of equi-shaped particles , equation 11.8 had been empirically based on porosities of $v_b < 0.5$. For fibre ensembles much larger porosities are possible. From geometric principles and using D_p as the fibre diameter

$$\Pi = \frac{D_p^2}{16b_p} \frac{v_b^3}{(1 - v_b)^2},$$
(11.9)

is the Kozeny-Carman equation for random fibres with a slightly larger $b_p \approx 6$ found in experiments for the range $0.5 < v_b < 0.8$. On the one hand this makes the permeability of random fibres, at the same brine volume, a factor two larger than for particle ensembles. On the other hand, was relation 11.9 found to break down at porosities above ≈ 0.8, where experiments with fibres show a strong increase of b_p (Carman, 1956).

The behaviour at large porosities due to modified Stokes-drag was first outlined by Brinkman (Carman, 1956), who derived the equation

$$\Pi = \frac{D_p^2}{18} \left(\frac{3}{4} + \frac{1}{1 - v_b} - \frac{3}{4} \left(\frac{8}{1 - v_b} - 3 \right)^{1/2} \right)$$
(11.10)

for flow around spherical particles in a porous medium. As shown by Carman (1956), equation 11.10 gives, when compared to the Kozeny-Carman equation 11.8, similar $4.3 < b_p < 5.5$ for the porosity range $0.5 < v_b < 0.8$, yet it demonstrates theoretically that at higher brine volumes the parameter b_p, and thus the relationship between Π and v_b, changes considerably. Further empirical studies have shown that a permeability law based on the Kozeny-Carman approach with constant b_p will be correct within 20 % up to a porosity of 0.9, but then changes very sharply (Dullien, 1979). The disadvantage of equation 11.10 is that it is not applicable below $v_b \approx 0.5$.

While the Kozeny-Carman equation 11.8 is conceptionally successful, a more detailed view on its performance indicates that b_p increases systematically when v_b is increased from 0.4. Many authors have, for particle size distributions, therefore suggested empirical permeability laws of the form

$$\Pi \propto D_p^2 v_b^{e_p} \tag{11.11}$$

An interesting result, emphasised by Dullien (1979), is that in the latter equation $e_p \approx 5.5$ provides a better empirical fit to observations over a wider porosity range $0.35 < v_b < 0.7$ than the Kozeny-Carman relation.

In general, all the simplified relations have a reasonable empirical basis and can be applied in the noted porosity regimes, yet bulk permeability estimates may easily be subject to an uncertainty of ± 50 %. More recent work on modifications of Brinkman's equation has been discussed, on the basis of numerical simulations, by Higdon and Ford (1996). Koponen et al. (1997) also illustrated the behaviour of the Kozeny-Carman scaling by comparison with numerical studies. The optimal parametric form between Π and v_b is also influenced by size distributions and packing (Dullien, 1979; Happel and Brenner, 1986) and therefore needs, in most natural applications, to be verified empirically for a particular material.

Another aspect not discussed so far is the formation of a disconnected porosity fraction. For isotropic media in three dimensions it was found to be relevant at porosities below ≈ 0.3 (Koponen et al., 1997), which points to the need of a modification of the permeability relationship, a possible path being percolation theory (Stauffer, 1991; Baker et al., 2002). The problem in two dimensions will be discussed to some degree in section 12.3.2, in connection with the bridging of sea ice brine layers. As long as only the rigorous convection in the unbridged high-porosity skeletal layer is of interest, this effect is not relevant.

Considering the convective stability of porous media, the principle difference between the Kozeny-Carman approach and the Brinkman relation resembles the importance of hydrodynamic boundary conditions at high porosities modifications (Nield and Bejan, 1999). Although the difference is moderate, the inclusion of the Brinkman equation may decrease the critical Rayleigh number for convective onset by typically 30 % (Emms and Fowler, 1994; Lu and Chen, 1997).

Anisotropic media: flow along cells or dendrites

A situation comparable to the capillary bundle permeability 11.7 is found when the flow parallel to fibres is considered. Experiments and early simplified analytical approaches indicate that the parallel fibre permeability equals the Stokes-Hagen-Poiseuille result near $v_b = 0.8$, increasing towards a by a factor of two larger Π at low v_b, while at larger v_b, when the fibres take a more open arrangement, the experimental fibre permeability becomes approximately three times smaller (Carman, 1956). Of course, above the hexagonal circle packing $v_b = pi/6\sqrt{3} \approx 0.907$ the computation of a capillary bundle limit makes little sense. A frequently used approximation of the permeability of flow along parallel circular solid cylinders Happel and Brenner (1986) is

$$\Pi = \frac{D_c^2}{32}\left(-2\,ln\,(1 - v_b) + 4\,(1 - v_b) - 3 - (1 - v_b)^2\right). \tag{11.12}$$

It takes a similar form as the Brinkman equation for spheres. In the above mentioned freezing experiments of $NH_4Cl\text{-}H_2O$ at low solid fractions (Tait and Jaupard, 1992a), the application of equation 11.12 gave a critical $Ra_{pc} \approx 25$, in reasonable agreement with later simulations (Worster, 1992; Emms and Fowler, 1994; Hwang and Choi, 2000). However, as noted above, the ranged spanned by several experiments with this binary (Chen and Chen, 1991; Tait and Jaupard, 1992b; Worster and Kerr, 1994) appears to be $10 < Ra_{p,c} < 80$. This may indicate, that the neglect of a more detailed side branch morphology in equation 11.12 is a serious problem and makes its general application to dendritic media unreliable. Direct evidence can be found in recent numerical simulations (Goyeau et al., 1999).

11.4.2 Permeability of the sea ice skeletal layer

From the preceding summary the main problems in the predictability of the permeability of porous media may be summarised as

1. A deviation from Darcy's law due to large pore scale sizes

2. Complex size distributions, packing and surface area of particles or pore diameters

3. Disconnected brine volume fractions, not participating in the throughflow

4. The transient character of the pore structure (reactive porous medium), or a heterogeneous packing or size distribution (non-reactive)

5. Sidebranching of dendrites

Bulk sea ice

The first four mechanisms make the general longterm prediction of the permeability of sea ice a difficult task. This is evident in the large scatter in permeability measurements (Ono and Kasai, 1985; Freitag, 1999). Attempts have been made to correlate these measurements via the parametric form of the Kozeny-Carman-equation (Maksym and Jeffries, 2000), by a power law of the form 11.11 or by a percolation analogy (Golden et al., 1998). However, none of these approaches accounts consistently for the above mechanisms. Physical models of the formation or evolution of larger brine channels (section 11.3.1) are still lacking and not included in these approaches. As the brine channels set the limits on the permeability Π due to a variant of equation 11.7, their understanding is essential to describe bulk sea ice is fundamental. Their evolution and disconnection, implying how D_c evolves with v_b in equation 11.7 is not well understood or documented (Freitag, 1999). Due to the limited data obtained so far the suggested correlation approaches to Π should be viewed as strictly empirical and of limited generality. The most reliable form so far seems to be equation 11.11, with $e_p \approx 1.6$, but it requires the determination of the disconnected pore volume and still leaves an unexplained variability of two orders of magnitude in Π (Freitag, 1999).

The skeletal layer as a Hele-Shaw cell

The regime of interest here, in the scope of early desalination, is the *skeletal layer* near the ice-water interface and not the bulk ice. It is normally defined as the lamellar regime where brine can circulate freely between the plates, and was described in some detail in chapter 3. Here it suffices to say that its permeability is not expected to be affected by the first four above aspects. The first morphological transition in the pore structure is the bridging of the plates and takes, by definition, place above the skeletal layer. Due to the large anisotropy in the growth rates normal and parallel to the c-axis, neither sidebranching appears to be a problem, at least not under most natural growth conditions. The permeability near the bottom of sea ice may then reasonably be approximated by an array of slots or Hele-Shaw cells. For this setting the viscous drag is analytically given (Lamb, 1932) as

$$\Pi_d = \frac{v_b d^2}{12},\qquad(11.13)$$

where d is the width of the brine layers. The form is similar to equation 11.7. However, due to the relation $v_b = d/a_0$ between brine volume, brine layer width, and plate spacing one now obtains

$$\Pi_d = \frac{v_b^3 a_0^2}{12}.\qquad(11.14)$$

This is a concise expression for the permeability near the ice-water interface, which does not require any empirical adjustment. This result cannot be stressed enough: the bottom regime of sea ice is essentially a medium for which the permeability is analytically predictable, if the plate spacing a_0 is known. It is an ensemble of Hele-Shaw cells, which each for itself resembles a two-dimensional porous medium. For other porous media with phase transitions, sometimes being classified as 'mushy layers', the preference of certain permeability models has always to be viewed with caution. Applications of the Hele-Shaw permeability 11.14 to fractured rocks have been discussed by Phillips (1991). The only principle limitation is that, for the Hele-shaw viscous approximation to be valid, the length and height H_p of the brine layers must be much larger than their width d (Hartline and Lister, 1977). The approximate criterion $H_p/d > 20$ given by Frick and Clever (1980) is, taking $0.3 < a_0 < 1$ mm and a brine volume of order one for natural sea ice, fulfilled at distances more than 5 to 20 mm from the interface.

A principle test of the relation has been made by comparison with young ice permeability measurements obtained by Freitag (1999) during the INTERICE project. His permeability shows considerable scatter, eventually indicating the presence of two principle types of ice, due to sampling before and after a warming event. The group with the larger permeability is here suggested to be related to the skeletal regime of unbridged brine layers, in particular when evaluated at brine volumes between 0.15 and 0.2. These brine volumes refer, recalling the discussion from chapter 5, likely to the top of the skeletal layer but still to locations within. If compared to equation 11.14, the latter high permeability observations from Freitag (1999) thus indicate a range $0.30 < a_0 < 0.60$ mm for the plate spacing. The average relation given by Freitag may be written as

$$\Pi_d = (2.0 \times 10^{-8}) v_b^{3.1}. \tag{11.15}$$

Hence the empirical relationship between Π_d and the brine volume v_v is close to the theoretical one. Furthermore it gives, by comparison to equation 11.14, a range $0.45 < a_0 < 0.49$ mm, if evaluated by assuming the range $0.2 < v_b < 1$. This is essentially the plate spacing expected for the growth velocities of 2 to 5 cm d^{-1} documented for the ice samples (Freitag, 1999; Eicken et al., 1998). It is worth a note, that the permeability experiments were obtained from samples taken at different stages, which may explain some of the scatter in Π_d. The thin ice cover in the tank first grew to 20 cm thickness and then melted back to 12 cm. Any further discussion would require a detailed analysis of the microstructure of the samples. However, the limited data indicate that equation 11.14 describes the permeability near the bottom of sea ice in a reasonable manner.

11.5 The critical Rayleigh number of the sea ice skeletal layer

11.5.1 Evaluation criteria of Ra_p

To evaluate the Rayleigh Number of the skeletal layer, Worster (1992) suggested to compute averaged properties of v_b and a_0 for the sea ice skeletal layer. A different approach has been proposed by Beckermann et al. (2000). The latter authors computed *local* Rayleigh numbers by averaging the brine volume from the interface downwards (solidification was upwards) into the mushy layer, and inserted it into an empirically evaluated Kozeny-Carman relationship (equation 11.8). The concept yielded, in comparison with experimental data on freckle formation, a predictive criterion in terms of Ra_p. On the other hand, Yang et al. (2001) compared different datasets of freckle observations in Pb alloys. They found the best consistency when κ_l/V was used as the characteristic length scale to compute the Ra. Ramirez and Beckermann (2003) showed, that the most successful method depended on the alloy in question. Absolute values of critical Rayleigh Numbers were different and apparently dependent on the assumed permeability relationship.

The comparison indicates the problem to use a consistent scale in evaluating Ra_p. Worster (1992) proposed the critical length scale κ_l/V. As mentioned above, this gives no simple critical Ra_p in terms of the classical convective stability theory, if the solid fraction is non-uniform, which is the case for sea ice. For example, for growth velocities of 10^{-4} to 10^{-5} cm/s, one obtains length scales from ≈ 13 to 130 cm for sea ice. This exceeds the skeletal layer extent, where the main drop of the solid fraction takes place, by far.

In the following, the concept of a local Ra_p is used. However, in contrast to the mentioned approaches (Beckermann et al., 2000; Ramirez and Beckermann, 2003), the local Ra_p is computed from the local brine fractions, not from its average value integrated from the interface. The use of the local solid fraction in the local Rayleigh Number appears intuitively more consistent than the use of average properties. This is illustrated by figures 11.1 and 11.2. The first one shows the equilibrium phase distribution of an NaCl solution at $S_\infty = 35$ ‰ NaCl, frozen from a cold boundary at -20 °C. As discussed in the appendix B.6, the temperature distribution is expected to be slightly nonlinear and the simplified solutions have been applied here. The second figure 11.1 shows the local Rayleigh number. It is computed from equations 11.6 and 11.14, by using ΔH from the bottom as length scale, and the form 11.14 for the permeability with local brine volume fractions. The importance to properly account for the temperature dependence of the viscosity (appendix A) is pointed out here.

Figure 11.2 shows that the local Rayleigh number has a pronounced maximum at intermediate depths. This situation is general for saline NaCl solutions, if not freezing

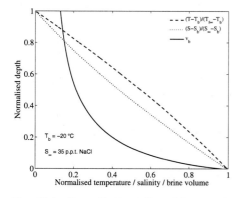

Fig. 11.1: Normalised profiles of brine volume, salinity and temperature of ice growing from an S_∞=35 ‰ NaCl aqueous solution from a cold boundary at −20 °C, the solution being at the melting point $T_f(S_\infty)$.

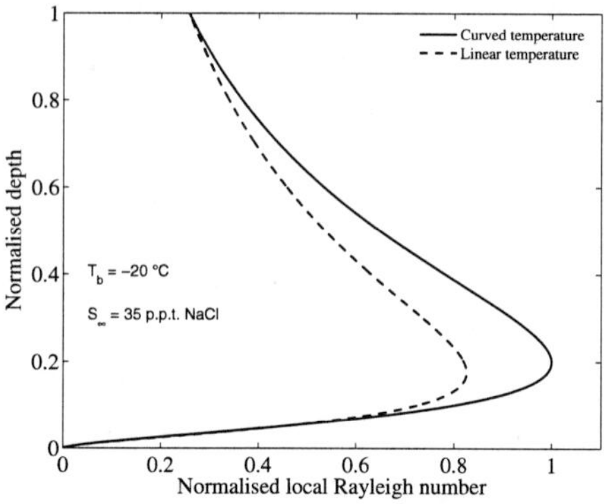

Fig. 11.2: Local Rayleigh number Ra_p, based on the distance ΔH from the bottom, for the slightly nonlinear profiles in figure 11.1. For comparison Ra_p for a linear temperature profile is shown. All values are normalised by the maximum from the nonlinear profile.

is obtained at very high temperatures and close to the eutectic composition. In the computation of Ra_p the bulk ice salinity and plate spacing have been assumed constant. As, in reality, for conditions of a decreasing growth velocity, both are decreasing upwards, the maximum can be expected to be even sharper. The arguments for the use of the local Rayleigh number, based on the local brine volume, as the basic criterion for the onset of convection, are (i) the general decrease in the brine volume towards the top and (ii) the quasi-linear increase in the Rayleigh number. This situation is thought to resemble the classical problem. However, to make a quantitative identification with the classical problem possible, one must consider the type of boundary conditions that prevails at the maximum within the porous medium. This will be discussed next.

11.5.2 Plausible boundary conditions for growing sea ice

A complete set of critical Rayleigh Numbers at the onset of convection in an isotropic and homogeneous porous medium was given first by Nield (1968) and, with a few corrections, by Nield and Bejan (1999).

The combinations which are thought to be relevant for sea ice are reproduced from Nield and Bejan (1999) in table 11.1. In the following they are discussed in terms of the lower boundary, the ice-water interface, and the upper boundary, from where the convection sets in:

Critical Ra_{pc} in a porous medium

Upper P	Lower P	Upper flux	Lower flux	Ra_{pc}	ω_{pc}	comment
IMP	FREE	COND	COND	27.1	2.33	whole layer convects
IMP	FREE	INSU	COND	17.65	1.75	whole layer convects
FREE	FREE	COND	COND	12	0	lower part convects
FREE	FREE	INSU	COND	3	0	lower part convects
FREE	FREE	INSU	INSU	0	0	lower part convects

Tab. 11.1: Critical Ra_{pc} and nondimensional wave number ω_{pc} in an isotropic porous medium after Nield (1968). Only combinations of boundary conditions relevant for growing sea ice are shown. Dynamic conditions: IMP = impermeable, FREE = stress free; salt flux conditions: COND = constant salinity, INS = constant salt flux.

- *Ice-water interface.* The permeability boundary condition at the bottom of ice is least debatable. Here the ice-water interface should be close to a FREE boundary. The temperature will be at the freezing point, or more accurate, at the marginal stability limit from stability theory, which is slightly lower for seawater. This boundary condition, a constant temperature and salinity, is thus CONDucting.

- *Convection level.* As pointed out it is, under normal freezing situations, expected that the convective stability sets in from the lower part of the skeletal layer. First, it is therefore suggested, that this convection level must be also treated as a FREE boundary, if the instability is described by the maximum of Ra_p. However, in the following also the rigid approximation is considered, which is relevant when the instability indeed comprises the whole layer. Second, the convection from within the skeletal layer further implies, that the concentration boundary condition is no longer conducting, even if the ice surface temperature is hold constant in the laboratory. The effective boundary condition at this level can be expected to take an intermediate value between the insulating and conducting extremes. Hence, treating the upper level of convection as free, due to table 11.1 the critical Ra_p is expected in the range $3 < Ra_{pc} < 12$; treating it as rigid one would expect $17.65 < Ra_{pc} < 27.10$.

- *$Bi \approx 1$ approximation.* The mixed thermal boundary condition at the upper convection level can be determined as a function of the Biot number Bi (Sparrow et al., 1964). For a convecting system the latter may be defined as the ratio of the external to internal heat flux derivatives dQ/dT near the interface (Sparrow et al., 1964; Nield, 1967). For a level within the ice, if there is no jump transition in brine volumes, a value of $Bi \approx 1$ the seems to be consequent. The corresponding critical $Ra_p \approx 20.5$ has been simulated by Wilkes (1995) for $Bi = 1$ for the case of a rigid upper boundary. For a free upper boundary and the present set of thermal conditions no results are available. However, for the case with two rigid boundaries simulations have been performed for porous media (Wilkes, 1995; Kubitchek and Weidman, 2003) and viscous fluids (Sparrow et al., 1964). Comparison indicates a general result for both media and the case of two rigid pressure boundary conditions: If the insulating condition is relaxed from $Bi = 0$ to $Bi = 1$, the critical Rayleigh numbers increases by ≈ 0.25 times the difference between the insulating and conducting Rayleigh numbers. Assuming that this condition also holds when both boundaries free, one may obtain $Ra_{pc} \approx 3 + 0.25 * (12 - 3) \approx 5.3$ as the critical Ra_p for two free boundaries. This value should be verified by simulations. It is further noted that the wave numbers at instability, with this $Bi \approx 1$ condition at the upper boundary, are expected to be close to one (Wilkes, 1995).

Having given some arguments for the choice of free boundary conditions for a partially convecting sea ice skeletal layer, a note on the high porosity lower boundary is in order. As mentioned above, Darcy's law 11.5 breaks down at high porosities. To account this at the boundary of the porous medium and the underlying fluid is problematic (Nield and Bejan, 1999). A study of different boundary conditions for a mushy layer has been performed by Lu and Chen (1997), and shows a decrease of the critical Rayleigh number when the boundary conditions at the mush-liquid interface are modified. The present heuristic approach to handle the system in a simplified manner cannot account for mush-liquid interactions. One may, however, reduce this aspect to the question, under what conditions the lower boundary can be assumed as conducting, defined by marginal equilibrium. If this condition is relaxed to an insulating boundary, the medium convects at zero Rayleigh number (table 11.1). For two mixed boundary conditions, with each $Bi \approx 1$, again only computations for the rigid case exist and give $Ra_{pc} \approx 9.6$ (Wilkes, 1995). With the above suggested projection to free boundaries the expected critical Rayleigh number would take a value slightly above one.

To conclude, a plausible critical Rayleigh number for convective instability, when convection is restricted to a fraction of the sea ice skeletal layer, is ≈ 5.3. If almost the whole layer convects, or if arguments for a rigid upper boundary condition can be given, then $Ra_p \approx 20.5$ can be expected. As will be discussed next, there is a second modification that likely applies to sea ice.

11.5.3 Anisotropy

Another factor that may influences the critical Ra_{pc} is the anisotropy of the diffusivity and the permeability. Considering the specific case of sea ice, where crystals are vertically oriented, a reasonable ad hoc estimate for the anisotropy of vertical to horizontal anisotropy is $\Pi_v / \Pi_h \approx 2$, where Pi_h is the horizontal permeability (Lister, 1974; Phillips,

1991). It results from the assumption of random orientation of the involved crystals. Kvernvold and Tyvand (1979) pointed out that the theoretical solution also involves the consideration of an anisotropic diffusivity κ_{lv}/κ_{lh}. They showed that, for the case of impermeable and conducting boundaries, the critical Rayleigh number $Ra_{pc} = 4\pi^2$ based on the vertical permeability is changed to Ra'_{pc} due to

$$Ra'_{pc} = Ra_{pc} \left(\frac{1}{2} min \left(\left(\frac{\Pi_v}{\Pi_{hx}} \frac{\kappa_{lhx}}{\kappa_{lv}} \right)^{1/2}, \left(\frac{\Pi_v}{\Pi_{hy}} \frac{\kappa_{lhy}}{\kappa_{lv}} \right)^{1/2} \right) + \frac{1}{2} \right)^2, \qquad (11.16)$$

where the indices $_{hx}$ and $_{hy}$ refer to the horizontal directions. If there is no prefered direction, then $\Pi_v/\Pi_{hx} = \Pi_v/\Pi_{hy} \approx 2$ for the present case. The equation implies further that the critical Ra_p is unaffected, if the same anisotropy pertains to the diffusivity. For a porous medium with phase changes the latter is, however, related to the assumed equilibration of the brine salinity by melting or freezing. Therefore κ_{lv}/κ_{lh} should be, in the context of the effective thermal dissipation approach, taken as unity. Hence, with only a permeability anisotropy, it follows from equation 11.16 that

$$Ra'_{pc} = Ra_{pc} \left(\frac{1}{2} \left(\frac{\Pi_v}{\Pi_{h,max}} \right)^{1/2} + \frac{1}{2} \right)^2, \qquad (11.17)$$

For sea ice, $\Pi_v/\Pi_{h,max} \approx 2$, this implies that the critical Rayleigh number, based on equation 11.14 for the vertical permeability Π_v, should be increased by a factor 1.457.

The suggested permeability anisotropy of 2 is an ad-hoc assumption. Observations of sea ice permeability anisotropy are sparse, and those for the skeletal layer even sparser. A few measurements from Freitag (1999) are available. The latter author reported one sample of young ice for with $\Pi_v/\Pi_{hx} \approx 2.1$ and $\Pi_v/\Pi_{hy} \approx 13$. This would, from equation 11.17, give an increase in the critical Ra_p by a slightly larger factor of 1.5. It is unlikely that the few other observations reported by Freitag, with up to two orders of magnitude larger anisotropies, resemble the structure of the skeletal layer. The noted observation is in agreement with the present conjecture.

Wave length and grain size

The simplistic random distribution approach made here can be expected to be valid, if a large number of randomly oriented grains is involved. In practice, this means that the critical wavelength at the onset of convection must be much larger then the long horizontal extension of the brine layers and thus, than the length of the crystals. Some estimates can be made. The first generation brine channels in thin ice are typically spaced by 1 to 5 cm (Wakatsuchi, 1977; Niedrauer and Martin, 1979; Wakatsuchi and Saito, 1985; Wettlaufer et al., 1997). If brine channels, associated with downward flow, are at least initially linked to the convection, then their distance should reflect the wave length of the convective flow in the skeletal layer. One may further consider that the onset of convection is, according to table 11.1, expected at certain wavelengths. For the assumed upper level Biot number of unity, instabilities should have a wave number near 1 (Wilkes, 1995), a value found in other studies when the boundary conditions are relaxed to yield $Ra_{pc} \approx 5$ (Emms and Fowler, 1994; Chen et al., 1994). The permeability anisotropy increases this wavenumber by the factor $(\Pi_v/\Pi_{hx})^{1/4}$, if Π_{hx} is the horizontally preferred direction (Kvernvold and Tyvand, 1979; Phillips, 1991). An anisotropy of 2 then implies a wave length ≈ 5 times the convective layer depth. The latter is, taking the results of the following sections in advance, only a fraction of the skeletal layer and typically 0.5 to 1 cm. These considerations indicate a convective wave length of 2.5 to 5 centimeters, in apparent agreement with brine channel spacings.

Hence, the wave length of channel distance appears to be several times larger than the typical long dimension of crystals, which is 0.3-1 cm for young ice of thickness less than 20 cm (see section 5.6). This indicates that the convective stability involves several crystals. It does, however, not involve many crystals. In this connection it is important to note, that crystals in sea ice are not randomly oriented, but are often locally aligned, with several crystals of similar orientation in packets. Moreover, very strong alignment is observed when a global fluid flow below the ice exists (Langhorne, 1986; Stander and Michel, 1989). Under such conditions the critical Rayleigh number is much less effected by the anisotropy of permeability. Ra_p might then just increase by a factor proportional to the inverse cosine of the inclination of the crystals within the packets, e. g., (Phillips, 1991).

11.5.4 Expected Ra_{pc}

From the above considerations it is expected that the critical Rayleigh number for the onset of convection in a sea ice skeletal layer can take the values

- $R_{pc} \approx 5.3$ for strongly aligned crystals.

- $R_{pc} \approx 5.3 \times 1.457 \approx 7.7$ for randomly oriented crystals.

Corresponding upper limits $20.6 < R_{pc} < 30$ may be approached when the convecting boundary layer operates close to a rigid upper boundary. These do not appear likely, when convection sets in from a level within the skeletal layer. It is recalled that these numbers are based on the assumption of free-slip boundary conditions of a partially convecting Hele-Shaw cell, which are thought to represent the sea ice skeletal layer quite reasonably. The lower flux boundary condition may, however, be imagined to relax towards a more insulating condition, and $R_{pc} \approx 1$ can be expected, if the same mixed boundary conditions apply at the top and bottom.

11.6 Comparison with experiments from Wettlaufer et al. (1997)

Wettlaufer et al. (1997) performed freezing experiments of NaCl solutions at different salinity in the laboratory. These are now compared to the proposed theoretical critical Rayleigh numbers. The ice in the experiments from Wettlaufer et al. (1997) was grown downwards into saline solutions of different initial salinities C_∞. Cold boundary temperatures of -10, -15 or -20 °C were constant during an experiment and controlled by a coolant system. Ice growth was observed visually through the side of the tank. At the same time the salinity of the solution below the ice was observed continuously. The time, when this salinity started to increase was associated with the onset of convection. The thickness, at which this convective transition was observed, is shown in figure 11.3 after Wettlaufer et al. (1997). One observation at 35 ‰ and -20 °C has been taken as slightly larger than reported by (Wettlaufer et al., 1997), which is more consistent with their figure 14.

For the interpretation of the experiments it is important to note, that in most experiments freezing was initiated, when the bulk solution was still above its equilibrium freezing temperature $T_f(C_\infty)$. The superheat θ_∞, defined in its conventionally used form as

$$\theta_\infty = \frac{T_f(C_\infty) - T_\infty}{T_b - T_f(C_\infty)}, \tag{11.18}$$

is shown in figure 11.4, where T_b is the temperature at the cold boundary and T_∞ the temperature of the melt when the freezing was started. The nondimensional superheat was considerable in the runs with concentrations from 70 to 140 ‰. Several aspects are noteworthy here. The superheat θ_∞ shown in figure 11.4 is normalised with the temperature difference to the solid boundary, while $\theta_{\Delta H}$, to be considered below, is based on the temperature difference to the level of the maximum Rayleigh number. The thermal expansion behaviour of saline solution implies that the thermal stratification is stable for the runs at lowest concentrations, while it is unstable for all other runs. In the appendix B.5 it was found that a simple model for freezing of saline solutions could predict the growth rates of the low superheat runs properly, while in all high salinity runs the growth rate was considerably overpredicted (figure B.5). This overprediction is associated with thermal convection, described by Wettlaufer et al. (1997) in some detail.

11.6.1 Local Rayleigh number

The procedure made here is to evaluate the maximum local Rayleigh number of each profile at the onset of convection, as described in section 11.5.1 and illustrated in the above figures 11.1 and 11.2. As discussed above, a slightly nonlinear temperature profile

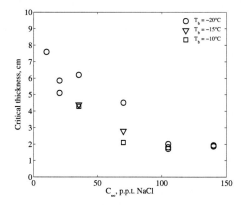
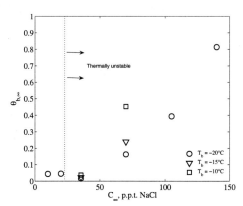

Fig. 11.3: Critical ice thickness observed by Wettlaufer et al. (1997) at the onset of rigorous convection for different boundary temperatures T_b and bulk solution salinities C_∞.

Fig. 11.4: Initial superheat θ_∞ in the experiments from Wettlaufer et al. (1997) for different boundary temperatures T_b and bulk solution salinities C_∞.

is computed from T_b and T_f via the method proposed in the appendix B.6. The nonlinearity is consistent with temperature time series at different thermistor levels shown by Wettlaufer et al. (1997).

To compute dimensional Rayleigh numbers via the permeability relation 11.14 the plate spacing a_0 must be known. The latter is predicted from the marginal morphological stability validated in section 10. This requires the growth velocity V. It is computed for a given level z measured from the top of the ice via $V = 0.5\eta_*^2/z$, where $\eta_* = H(t)/t^{1/2}$ has been obtained from the ice thickness time series in figures 13 and 14 from Wettlaufer et al. (1997) before the onset of convection. Growth rates for the experiments with $T_b = -10$ and -15 °C at 35 ‰ were not given in the original work, yet those at -15 °C could be obtained from Wettlaufer et al. (2000). These values are shown in figure 11.5, where they have been normalised by $(2K_i\Delta T_b/L_v)^{1/2}$, which is the growth rate of pure ice for a linear temperature gradient $\Delta T_b/\Delta z$. The fact that growth rates at larger salinities are much less is related to superheat in these experiments and will be discussed below. The comparison with the growth rate predictions from several simple models in the appendix (figure B.5) gives some confidence to these numbers for the experiments with small superheat. The accuracy at which η could be extracted from the figures is approximately $\pm$ 5 %. The error in the growth velocity is probably less than 10 % and very likely less than 20 %.

Having determined $V(z)$ as a funtion of the level in the ice, the plate spacing $a_0(z)$ is obtained as a function of depth. With these basic informations, profiles of the local

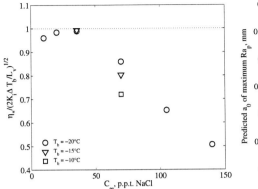

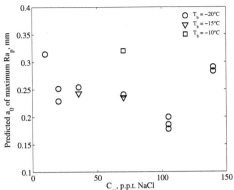

Fig. 11.5: Growth rate coefficient η_* in the experiments from Wettlaufer et al. (1997), normalised by $(2K_i\Delta T_b/L_v)^{1/2}$. The corresponding V is used to predict the plate spacing.

Fig. 11.6: Predicted plate spacing a_0 at the level of maximum Ra_p in the experiments from Wettlaufer et al. (1997).

permeability and the local Rayleigh number, with ΔH measured from the bottom as the length scale, can be calculated for each of runs at the onset of convection. These all look similar to figure 11.2, but are steeper towards the top, because the plate spacing is nonuniform and increasing downwards. The plate spacing at the level of the maximum in the local Rayleigh number is shown in figure 11.6.

11.6.2 Maximum Ra_p

The maximum local Rayleigh numbers at the onset of convection are shown in dependence on solution salinity and boundary temperature in figure 11.7. The three observations for the thermally stable low salinities runs, denoted with crosses, fall considerably below all other observations. This will be discussed below. The first focus here is on the salinity runs with 35 ‰ and larger. The proposed theoretical range $5.3 < Ra_{pc} < 7.7$ is indicated. Although the observations indicate considerable deviations from the proposed range, the agreement may be termed reasonable. From the runs at 35 and 70 ‰ no clear dependence of Ra_{pc} on the boundary temperature T_b can be identified. This indicates that the variation at these runs may be interpreted as observational scatter. It is noted, that for the runs at 105 and 140 ‰ only a single growth rate η was determined from the figures from Wettlaufer et al. (1997), which might have underestimated the actual experimental scatter at these salinities. The data are now discussed in some more detail in terms of systematic causes of the variability.

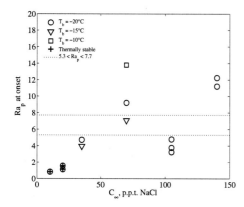
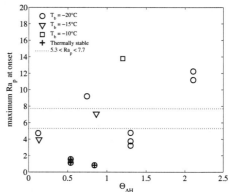

Fig. 11.7: Maximum Rayleigh number based on the distance ΔH from the bottom, in dependence on solution salinity and boundary temperature T_b in the experiments from Wettlaufer et al. (1997).

Fig. 11.8: Maximum Rayleigh number based on the distance ΔH from the bottom, in dependence on $\theta_{\Delta H}$, which is the melt superheat normalised by the temperature difference between the interface and the level of maximum Ra_p.

Dependence on C_∞

There is no clear salinity dependence visible in figure 11.7. However, in figure 11.6 it is seen, that for the runs at 105 ‰ a considerably smaller plate spacing than for all other runs has been predicted. One may therefore consider that the prediction at this salinity is, for some reasons, an underestimate. Raising a_0 by $\approx$ 30 to 50 % towards the levels of the other runs would yield higher Ra_{pc} within the range of the rest of the data. However, due to the systematic difference for all three runs at 105 ‰ at the same forcing conditions, it appears unlikely that uncontroled initial conditions, have caused the disagreement. Looking at figure 9.9 from chapter 9, there is neither an indication that something is wrong with the model prediction of a_0 at this concentration.

Superheat $\theta_{\Delta H}$

The superheat $\theta_{\Delta H}$ considered here differs from equation 11.18, in that T_b is replaced by the temperature at the level of the maximum Rayleigh number. The resulting $\theta_{\Delta H}$ is considerably larger than θ_∞. The dependence of the critical Ra_{pc} on $\theta_{\Delta H}$ with respect to its convecting level is shown in figure 11.8. No clear dependence of the critical Rayleigh number on the superheat is relevant. A decrease in Ra_{pc} with increasing $\theta_{\Delta H}$ would be expected, if the average solid fraction of the porous medium had been used in connection with its overall height (Worster, 1992; Emms and Fowler, 1994; Hwang and Choi, 2000).

When the unaveraged system properties and the length scale κ_l/V are used, the opposite is the case (Worster, 1992; Chen et al., 1994). The latter authors' model calculations indicate an increase in Ra_{pc} by a factor of ≈ 2 for an increase in θ_∞ from small values to 1.

In the present evaluation, because of the local Rayleigh number computation from ΔH, other quantitative dependencies can be expected. Two main mechanisms associated with the superheat need to be considered in the correct evaluation of the data. First, superheat changes the basic temperature profile in the ice, because of the thermal convection in the liquid. This makes the temperature gradient steeper and more linear. However, as seen in figure 11.2 this may only change Ra_{pc} by ≈ 20 % at 35 ‰ and -20 °C, and the difference becomes smaller at larger concentrations (not shown here). However, under unstable thermal stratification superheat will lead to thermal convection. The main change due to thermal convection is thought to be an effect that has been conjectured earlier, in connection with laboratory observations of the plate spacing (section 10.4). The steepening of the temperature gradient at the interface leads to a decrease in the plate spacing at the same growth velocity. This implies that plate spacings are overpredicted by the present model for a_0, which does not account for thermal convection. For a constant influence of thermal convection this effect will be more pronounced at low growth velocities, see figure 10.9 from the previous chapter.

To investigate the aspects of thermal convection further, Ra_{pc} is shown in figure 11.9 versus the growth velocity the ice had at the predicted critical level of maximum local Rayleigh number. Disregarding again the thermally stratified runs at low salinities, the data appear now more or less separated in two groups. The low velocity group represents the values of largest Ra_{pc}. Comparison to previous figures shows that this group consists of the two runs with largest superheat, at a salinity of 140 ‰, and the run with the highest boundary temperature $T_b = -10$ °C at 70 ‰. For these runs figure 11.5 (and B.5 in the appendix) indicated considerable effects of thermal convection on the growth rates, reducing the latter to 50-70 % of the expected values without convection. While the effect is in principal also expected at 105 ‰, it will be less pronounced, because the growth velocities were a factor of two larger at the critical level. This may explain qualitatively the larger Ra_{pc} at the three points of interest. Hence, there is only an apparent velocity dependence of the critical Rayleigh number. It appears that also these observations, similar to the plate spacing observations from Lofgren and Weeks (1969a), for example figure 10.12 in the preceding chapter, are strongly influenced by thermal convection. Only the evaluation of the runs at 35 ‰, which were little effected by thermal convection, can be expected to represent unambiguously the onset of convection.

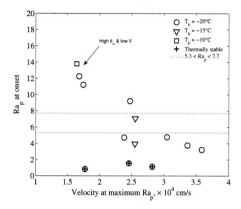

Fig. 11.9: Maximum Rayleigh number based
on the distance ΔH from the bottom, in dependence on the growth velocity at the level
of the maximum Ra_p.

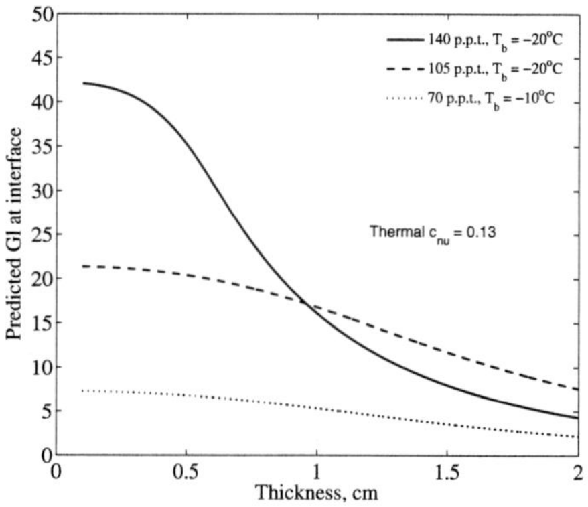

Fig. 11.10: Liquid temperature gradient at
the interface, simulated via thermal turbulent convection with $c_{nu} = 0.13$ for several
freezing runs from Wettlaufer et al. (1997).

Thermal convection: Revised plate spacings and permeabilities

Thermal thermal convection could not be corrected in the experiments from Lofgren and
Weeks (1969a) discussed in the previous chapters, because the initial conditions were
not known. Wettlaufer et al. (1997) have documented the superheat and also discussed
the retarding effect of thermal convection on the growth rate. Here, simulations of the
thermal convection have been performed based on the initial superheat in each of the
runs. The simulation has to be performed time-dependent, because the limited heat in
the reservoir is removed by the process. In this way the Nusselt number and interfacial
temperature gradient were calculated at each depth increment. The parameterisations are
essentially the same as described in the previous chapter for haline melt convection into a
seminfinite layer. The constant c_{nu} is expected to be slightly smaller for a Prandtl number
of order 10 and should take a value of 0.13 to 0.15 for a rigid interface (section 10.2.3).
Wettlaufer et al. (1997) concluded from simulations, that $c_{nu} \approx 0.30$ was necessary to
explain the time-temperature curves in the liquid, and suggested a modified empirical
growth relation. Some aspects have been discussed in appendix B.5. In the present
calculations $c_{nu} \approx 0.13$ was used. It is, however, noted that the use of $c_{nu} \approx 0.30$ leads
only to small changes in the following results. The reason is that smaller c_{nu}, while
predicting only half the interface gradient, removes the reservoir heat more slowly and
keeps the heat flux for a longer time. The effects almost cancel each other at the critical
thicknesses. Figure 11.10 shows the evolution of the simulated interfacial temperature
gradient in dependence of thickness. At the lowest salinities small fluxes may be present,

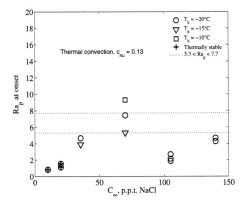

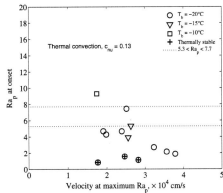

Fig. 11.11: Ra_{pc} versus C_∞ as in figure 11.7, but with corrected plate spacings, computed with the temperature gradient Gl in the liquid.

Fig. 11.12: Ra_{pc} versus the growth velocity at the critical level, as in figure 11.9, but with corrected plate spacings, computed with the temperature gradient Gl in the liquid.

which was neglected due to the stable stratification.

The liquid temperature gradient Gl implies, according to the theory in chapter 8, slightly smaller plate spacings. This changes the permeability. The modified evaluated critical Rayleigh numbers obtained in this way are shown in figures 11.11 and 11.12. The data appearance has changed. The observations collapse much better and the scatter is considerably reduced. The agreement of the runs at 35, 70 and 140 ‰ with the heuristically proposed Rayleigh number range also improves. The results for the low salinity runs remain unchanged due to the lack in thermal convection. However, the three runs at 105 ‰ appear now at a similar lower Ra_{pc} level, compared to the rest of the data.

In the following all further relations are shown for the predictions of a_0 corrected due to thermal convection.

Stable temperature stratification

Having found a reasonable collapse of the experimental data at salinities with an unstable temperature stratification, it is now turned to the low salinity runs, denoted with crosses in all preceding figures. The respective critical Rayleigh numbers in figure 11.11 are 0.8, 1.1 and 1.6, being considerably lower than all other observations, except the runs at 105 ‰. These runs do not indicate considerable superheat (figure 11.8), nor does the above thermal convection argument apply in their case, as the growth rates are almost unreduced (figure 11.5). Nevertheless, there are other mechanisms that may be considered:

1. The present calculations have ignored the redistribution of solute due to expansion. At lowest salinities, where the solid fractions are highest, this might redistribute solute to the bottom fractions of the ice and increase the liquid fractions.

2. Chen et al. (1994) showed in a linear stability analysis of NH_4Cl-H_2O that, for intermediate superheats and a stable thermal stratification a double-diffusive oscillatory convection mode appears. This mode triggers the onset of convection in the boundary layer below the ice.

Considering the first argument, some estimates in the lines of figure C.1, appendix C, have been made. The concept of reduced redistribution of salt upon expansion, where fluid columns in the ice are thought to be displaced downwards before remelting of ice, would at most give an increase in the brine volume by 20 %. It cannot explain the by a factor of 3 to 6 smaller Ra_{pc}. Solid fractions have been observed for a mushy layer of alcoholic solutions with expansion $\rho_i/\rho_b \approx 0.74$, large compared to seawater (Chiareli and Worster, 1992). These indicate that the effect of expansion on the evolution of the solid fraction near the mush-liquid interface is small. The aspect deserves further studies in connection with other boundary conditions, yet appears to be of minor relevance.

The results from Chen et al. (1994) refer to aqueous NH_4Cl-H_2O solutions. From their graphs it seems that the double-diffusive mode requires the parameter $m\alpha/\beta$ to be positive and of order one. The latter is based on liquidus slope m and thermal and haline expansion coefficients α and β. Double-diffusion may allow the possibility for fingers to rise into the ice, contrasting the strict solutal convection in the melt. For NaCl the corresponding stable thermal and unstable haline stratification is present at low salinities, but $m\alpha/\beta$ is less than 10^{-2}. While the effectivity of salt fluxes is strongly enhanced by the double-diffusive mechanism (Turner, 1973), the haline stratification is also very strong in the saline ice, m being small. One may think of a change in the interfacial flux condition from conducting to intermediate. This would, as estimated above, be sufficient to change the Rayleigh number Ra_{pc} to small values of order one. Also this problem requires further investigation, but a clear argument for a triggering of convection by such a mode is, due to the numerically small value of $m\alpha/\beta$, not given.

Grain size

The next question to be considered is the mode of convection in terms of grain size. To do so, the critical Rayleigh number has been plotted in figure 11.13 against the ratio $\Delta H/a_0$. It is first noted, that this ratio is, if divided by the local brine volume, for all experiments larger than 30. Hence the limitations due to too small cell heights mentioned above (Frick and Clever, 1980; Schoepf, 1992) do not apply here. The graph may then be further interpreted as follows.

Some grain size observations have been discussed in section 5.6.1. In a few studies, where the plate spacing and grain size have been determined simultaneously (Weeks and Hamilton, 1962; Tabata and Ono, 1962; Cole et al., 1995), the ratio L_{gr}/a_0 is typically in the range 15 to 25, L_{gr} being the grain length normal to the horizontal c-axis direction. The estimate accounts for a factor $2^{1/2}$ to convert from average grain size to L_{gr}. While the noted observations refer to field sea ice, grain sizes of 5 to 10 mm have also been obtained in laboratory studies for ice of thickness of a few centimeters (Wakatsuchi and Saito, 1985). A single photograph from Wettlaufer et al. (1997) indicates comparable sizes. As discussed above, the wave length λ_{pc} of the convective cells can be expected to be ≈ 5 times the convective layer depth. The ratio $\Delta H/a_0 \approx 3$ to 5 may then typically be associated with convection within a single grain. Figure 11.13 thus indicates, that in most experiments convection is expected to have set in across several grains. However, in this respect the observations at low salinities appear as the group closest to the single grain size.

One may also consider that the grain size evolves independent of a_0. An approximate formula for L_{gra} in dependence of depth, which represents most thin ice observations mentioned in section 5.6.1, is $L_{gra} \approx 0.5 + z/15$ cm, with z the level from the top of the ice in centimeters. An approximate check could be made against a single micrograph from Wettlaufer et al. (1997) showing a horizontal cross section at 11.5 cm depth of a freezing run at 35 ‰: The visible grain lengths are between 1 and 2 cm and consistent with the predicted $L_{gra} \approx 1.3cm$ from the scaling. Assuming also here the instability wavelength to be $\lambda_{pc} \approx 5\Delta H$, one may interpret $N_g = 5\Delta H/L_{gra}$ as the grain number of a convection cell at the level of maximum Ra_{pc}. It is shown in figure 11.14 and indicates some relation between the onset of convection and the number of grains comprised by the expected convection cells. Further, if one imagines, that a convective system with downflow in brine channels will consist of double-rolls or hexagonal cells, then the fluid will only have to cross half the cell in the horizontal. Hence, the easy convection, without crossing between grains, is represented by $N_g \approx 2$ in figure 11.14.

The appearance is similar as for figure 11.13. The observations eventually indicate, as one expects, an increase in Ra_{pc} with grain number. The low salinity runs are closest to the easy flow within one grain of parallel plates. Most experiments appear to represent convective motions across several grains. In spite of the approximate agreement in grain size from the experiment at 35 ‰ with the present estimate, one may ask if the convective laboratory conditions could be different. It is well known that fluid flow has the tendency to orient grains parallel to each other (Langhorne, 1986; Stander and Michel, 1989). In the case of turbulent thermal convection, at high salinities, very likely no organised orientation may be expected. This might have been different during the experiments with

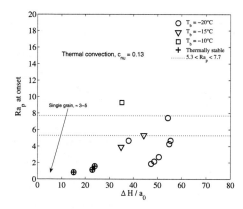

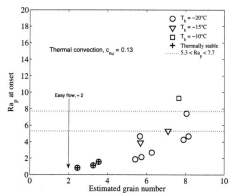

Fig. 11.13: Ra_{pc} versus $\Delta H/a_0$, which may be expected as proportional to the number of rains involved in convective cells.

Fig. 11.14: Ra_{pc} versus grain number along a convection cell, estimated from the equation $N_g = 5\Delta H/(0.5 + (H_{cr} - \Delta H)/15)$.

stable thermal stratification. If in these experiments a global thermally driven circulation might have been present due to sidewall effects is unknown. However, as the random influence of thermal convection can be expected to have been absent from the beginning, grain size might have been evolved differently in the low salinity experiments. This could come in addition to the smaller grain number and essentially imply a convection event interior to a single grain (or between two grains). It is further worth a note in this context, that Stander and Michel (1989) observed a maximum crystal alignment response during the freezing of saline ice at salinities near 10 ‰. This indicates that alignment and grain size selection might be salinity-dependent. Theory and observations are lacking here. The discussion points to the importance to obtain texture analysis of grain size and plate spacing.

Taking the dataset as a whole and interpreting it in terms of an increase in stability with grain number in a convection cell, the single grain Ra_{pc} being of order oner, the increase suggested by figure 11.13 towards across-grain convection is a factor of 5 to 10. This is much larger than the factor 1.457 that follows from equation 11.17 under the assumption of a horizontal of a vertical anisotropy of 2 in the optimum horizontal direction. It would imply an anisotropy of 10 to 30 comparing vertical and maximum horizontal permeability. This appears to be rather large for the high-porosity skeletal regime with freely circulating brine, also in comparison with the few observations available (Freitag, 1999).

Hence, although the hypothesis of a grain size influence is not inconsistent with the data appearance, the dependence is stronger than expected on first grounds. It should be pointed out, that the present estimate of the grain number within a convection cell is

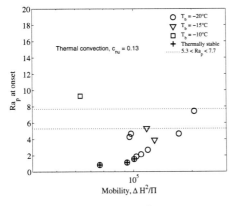
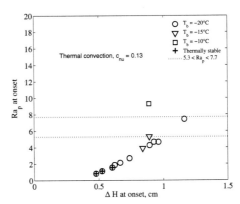

Fig. 11.15: Ra_{pc} versus $(\Delta H^2/\Pi)$ at the level of maximum Rayleigh number. the may be expected as proportional to the number of rains involved in convective cells.

Fig. 11.16: Ra_{pc} versus thickness ΔH of proposed convective layer (distance of $Ra_{pc} = max(Ra_p)$ from the interface).

tentative and actual measurements do not exist. It depends, moreover, on the assumed wave length $5\Delta H$. It seems therefore useful to investigate the variability of ΔH in some more detail.

Mobility

A conventional parameter to classify motion in a porous medium is the mobility ratio or inverse Darcy number $\Delta H^2/\Pi$. It is a measure of the validity of the assumption of Darcy-like flow in a porous medium. In figure 11.15 the critical Rayleigh number is shown versus this parameter. It is seen that the mobility ratio is quite large with values around 10^{-5}. A Rayleigh number dependence that might be expected on the basis of this parameter would be an increase in Ra_{pc} towards lower $\Delta H^2/\Pi$. The observations do not support this. This is not surprising, as for a value of 10^4 only minor deviations are expected when considering a porous medium (Schoepf, 1992; Degan and Vasseur, 2003) [1]. The effect might, however, be enhanced by anisotropy (Degan and Vasseur, 2003). From these considerations it seems unprobable that the single observation of largest Ra_{pc}, at lowest $\Delta H^2/\Pi$, results from a non-Darcy flow effect.

There is a different mobility effect that one might expect. It is related to the interaction of the porous medium with the fluid layer below and considered as follows. The most trustworthy evaluations of Ra_{pc} appear to be the freezing runs at 35 ‰, when almost

---

[1] The results from Worster (1992), who shows a dependence of the critical Ra_{pc} at a mobility $H*^2/\Pi$ as large as 10^5, are not comparable, because his length scale $H*$ corresponds to κ_l/V and not to the actual convective layer depth.

no thermal convection should have been present. These gave values of Ra_{pc} between 4 and 5, which is slightly lower than the value of 7.7 conjecture for cross-grain instabilities. However, the conjecture has assumed a porous medium that does not interact with fluid flow in the melt below, letting only the boundary being permeable. Lu and Chen (1997) compared results obtained with the conventional Darcy-flow and non-slip boundary condition in the melt below the mushy layer to alternative approaches. Depending on parameter ranges, they found a reduction of the critical Rayleigh number of 10 to 40 %, when either a Brinkman-extended Darcy flow law or a Beavers-Joseph boundary condition were used at the interface. A reduction of the classical Rayleigh number by ≈ 30 % has also been predicted by Emms and Fowler (1994), accounting for a non-stagnant finger-like convection in the liquid, with a further 10 % reduction, when nonlinearity is introduced into the problem (Emms, 1998).

To consider the latter effect the critical Rayleigh number has been plotted against the thickness of the convecting layer, that is the distance from the interface on which Ra_{pc} is based, in figure 11.16. First, it is seen that ΔH is between 0.5 and 1.2 cm for all experiments. The appearance of the data in this plot are interesting, as they indicate an increase in Ra_{pc} with ΔH. Also, this is the first data appearance, where all data points collapse to a single curve. A possible interpretation is that, the thinner the convecting skeletal layer, the larger the influence of the unstable boundary below the ice. A comparable problem has been studied for the case of a porous medium with an overlying fluid layer (Chen and Chen, 1988; Chen et al., 1991) for different ratios of these layers. For a fluid to porous medium thickness ratio of 0.1 the critical Rayleigh number was calculated to decreased by a factor of two with respect to the case of an infinitely deep porous medium. At thicker fluid layers the critical Ra_{pc} for the porous medium dropped very rapidly to extremely small values. A comparison can be made with the convective solutal boundary layer discussed in the previous chapter. From the proposed critical velocities of solutal convection (section 10.3.3) the convective layer scale, if based on D_s/V_{cr}, is in the range 0.3 to 0.4 mm. Hence, while ΔH changes from 1 to 0.5 cm, the ratio of these layers approaches a value of the order of 0.1.

It must be pointed out, that the simulations (Chen and Chen, 1988; Chen et al., 1991) refer to an *overlying* fluid layer. A comparison to a vigorously convecting turbulent boundary layer of the same thickness is tentative and may not be justified. However, so far there appears to be no binary solidification model, that has addressed these question in detail. In the 'mushy layer' model formulation by Worster (1986, 1991, 1992) the relative influence of the solutal boundary layer is related to the diffusivity ratio D_s/κ_l. Also here the critical Ra_{pc} decreases, when D_s/κ_l increases. However, as will be addressed in the conclusion, the boundary condition is different from which is proposed here. Moreover, does this model not have a turbulent boundary layer. The problem requires more work, eventually in the lines of Emms and Fowler (1994) and Lu and Chen (1997), by modifying the boundary conditions and focusing on the interface in more detail as these models so far have done. Nevertheless, figure 11.16 indicates a much better consistency of the apparently uncorrelated data in earlier figures, and its qualitative explanation is reasonable. If the ad-hoc approach of depth ratios is applicable, then the freezing runs at 10 and 20 ‰ are likely to have convected at a 2 times smaller Ra_{pc} then conjectured. If this effect is combined with the described grain size effect, it seems almost capable to explain the lower values of Ra_{pc} at the low salinity runs and gives also a plausible explanation for the deviations at the concentration 105 ‰. A further discussion requires more observations and specific theoretical work.

11.6.3 Conclusion on experiments

The discussion so far has indicated the consistency of the maximum local Rayleigh number approach to predict the onset of connection from growing saline ice. In most runs the onset of convection is related to a condition close to the heuristically proposed range $5.3 < Ra_{pc} < 7.7$. There was a discrepancy between observations and predictions of the critical Ra_{pc} at the low salinity runs (10 and 20 ‰) with thermally stable stratification, and the runs at one of the high salinities of 105 ‰, which convected at smaller Rayleigh numbers. Considering the variation at each salinity this difference is significant. The possibility of failure in the prediction of the plate spacings cannot be excluded. The novel theory has, however, been properly verified in the previous chapters, and no systematic differences of the predictability at different salinities had been found. In view of figure 10.10, even with some unaccounted thermal convection present, one should not expect prediction failures that lead to Ra_{pc} differing by a factor of 4. For the low salinity runs eventually a different mode of convection due to the thermal stratification might be expected. Moreover are these runs closest to the limit of convection internal to a single crystal. Assuming the range $5.3 < Ra_{pc} < 7.7$ also this would be insufficient to explain the very small Ra_{pc}. Several suggested ideas about a possible grain size dependence of Ra_{pc} can only be verified by texture analysis of samples, in connection with similar experiments, but they are qualitatively consistent with the data appearance.

None of the suggested explanations apply to the runs at 105 ‰, which indicate Rayleigh numbers a factor of two below most of the other thermally unstable runs. These runs appear also slightly anomalous when considering the ice thickness at the onset of convection, retrieved by Wettlaufer et al. (1997), here shown in figure 11.3. They are distinguished from all other runs, in that convection was initiated at considerably larger growth velocities, see figure 11.12. Why essentially at this concentration an anomalous convection behaviour could be expected, is neither apparent from molecular properties nor from the thermal convection forcing.

The most likely explanation for the discrepancies appears to be the one discussed on the basis of figure 11.16. Interaction with the fluid layer below, not included in the present heuristic estimates, leads to a decrease in the critical Rayleigh number. Theoretical analyses of this effect (Chen and Chen, 1988; Chen et al., 1991) indicate a strong decrease in Ra_{pc} when the fluid layer in the fluid becomes one tenth of the convecting porous medium. Identifying the fluid layer with the turbulent boundary layer due to solutal convection, one finds, in the experiments, similar layer ratios in those runs for which the smallest Ra_{pc} was estimated. The results strongly emphasise the role of the coupling between convection in the liquid and the porous medium. Future studies should focus on finding a simplified theoretical solution to thgis problem. The approach from Emms and Fowler (1994) appears relevant in this context.

It is finally noted that the results have been obtained by assuming a parabolic growth law for the ice at early growth stages. This is reasonable for the experiments without convection. However, for the high salinity experiments with vigorous thermal convection, the interfacial temperature gradient was rapidly changing initially (figure 11.10). This changes the growth law to a more complex one, and the evaluated plate spacings and maximum Ra_{pc} may also change. To obtain an idea of the potential uncertainty of the present approach, the growth rate has been simulated in a fully self-sustained manner. The formulation B.35, as summarised in appendix B.6.2, has been applied to predict the growth rates. It was found that a good agreement with the observations from figure 11.5 could be obtained with an average conductivity from serial and geometric conduction (equations B.24 and B.25). Such a model is plausible due to the expected initial random growth of crystals. In this way the thickness after 100 to 200 hours was predicted within 8 % for any run, when thermal convection was included with $c_{nu} = 0.13$. The simulation can be expected to account better for the changing contribution of thermal convection. On the other hand, the model depends on the assumption about the thermal conductivity at the top and is an approximate solution.

The completely self-sustained simulation results are shown in figures 11.17 and 11.18. The first figure shows that the data appearance does not change considerably. However, the estimates of Ra_{pc} at the concentrations with thermal convection present have all increased. This can be attributed to the effect of a decaying influence of thermal convection, which changes the local Rayleigh number maximum closer to the interface, where a_0 is larger. As seen in figure 11.10, Gl changes rapidly near 1 cm from the interface, where the maximum Ra_{pc} was situated in these freezing runs. The values at 105 ‰ are still lowest. However, the comparison indicates that unaccounted peculiarities of the growth process, or the thermal convection parametrisation, have some potential to change the estimates. Comparison of figures 11.11 and 11.17 indicates the uncertainty of the approach in terms of the predictability of growth rates. Figure 11.18 shows not that nice collapse of the data as figure 11.16. However, the overall qualitative interpretation is the same, with an indication of a decrease in Ra_{pc}, when the convecting layer depth is small.

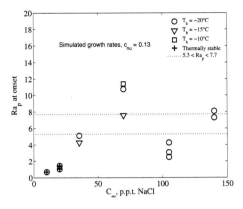

Fig. 11.17: Ra_{pc} versus C_∞ as in figure 11.11, but obtained via simulated growth rates including the variable effect of thermal convection.

Fig. 11.18: Ra_{pc} versus thickness ΔH of proposed convective layer, as shown in figure 11.16, but obtained via simulated growth rates.

11.7 Summary and discussion

11.7.1 Interpretation of porous medium stability for sea ice

A review of a number of studies of convective stability theory for a porous medium with and without phase changes has been given. It was concluded that many of the discrepancies encountered so far with the predictability of strongly dendritic arrays, are most likely attributable to uncertainties in the prediction of the vertical permeability. Due to the strictly plate-like structure of sea ice and saline ice at not too large supercoolings, these problems are not expected for the sea ice skeletal layer. The vertical permeability Π_v of the latter corresponds in principal to an array of Hele-Shaw cells and is given by a simple analytical expression in dependence on the plate spacing a_0 and the brine volume v_b. The present theory of morphological stability to predict the plate spacing allows thus for a proper prediction of the most important variables of the problem.

Next, it has been shown that the typical freezing conditions of the sea ice skeletal layer likely imply the onset of convection from within the layer. In contrast to the more complete formulation of the convective stability by mushy layer theory discussed above, a convection criterion of the local maximum Rayleigh number has been considered. Extending earlier work it has been argued that, for a solid fraction decreasing in the direction of increasing local Ra_p, such a condition should be evaluated with the local solid fraction.

Based on this approach the system of the solidifying sea ice skeletal layer has been discussed in terms of the proper boundary conditions that can be expected to apply in the classical theory without phase changes. This analysis, taking anisotropy and most relevant boundary conditions into account, have led to the conjecture that convective instability should be expected at a local maximum Rayleigh number of $5.3 < Ra_{pc} < 7.7$. These low values are proposed to apply, because the upper and lower boundary conditions in the modified problem become effectively free, when considering the local Ra_p of a partially convecting porous medium. These values are in contrast to a value of 27.10 proposed in many earlier studies on the basis of a rigid conducting upper boundary condition.

11.7.2   Experiments

The analysis of a set of experiments from Wettlaufer et al. (1997), who observed the onset of strong convective salt fluxes from freezing NaCl brine skeletal layers, has resulted in a reasonable confirmation of the simplified approach. The initial scatter in the evaluated critical Rayleigh number was considerably reduced, when thermal convection was accounted in the prediction of the plate spacings. This was possible, because the early growth rate and the initial conditions were properly documented for each experiment. Investigation of different parametric relations indicated that the observations might be divided into three groups:

1. The main group, comprising most freezing runs at high salinities with unstable thermal stratification. Here the estimates of Ra_{pc} fell between 4 and 10. No clear trend of Ra_{pc} versus plate spacing, suspected grain size variation, superheat or salinity was evident.

2. An exception to group 1., comprising all three experiments at an intermediate concentration 105 ‰ and significantly smaller $1.9 < Ra_p < 2.7$. During these runs convection appears to have occurred at levels of considerably smaller plate spacing (growth velocity) than in all other runs. No other distinct characterisation of this run, in terms of boundary conditions and critical variables, was found.

3. Low salinity (10 and 20 ‰) solutions, which froze at small superheat but with stable thermal stratification. This ice convected at the lowest Rayleigh numbers between 0.8 and 1.6. It was found to be separated from all other experiments in terms of the ratio of convecting layer thickness ΔH and plate spacing a_0.

For group 3. several possible mechanisms for the smaller Ra_{pc} were identified: (i) the possibility of enhanced double-diffusive fingering convection due to interaction of thermal and solutal diffusion, eventually modifying the interfacial flux condition, and (ii) the tendency of these observations to be much closer to the expected convection within a single grain. Observations of concentration-dependent texture evolution, as performed by Stander and Michel (1989), may strengthen this difference. If the latter interpretation is relevant, the critical Ra_{pc} for a single sea ice crystal would likely be of order one. This is smaller than the above suggested value of 5.3 and would, in terms of the present heuristic approach, imply that the thermal boundary condition at the freezing interface is not conducting. As a grain number dependence of the rest of the data is not evident, the possibility of a different convection mode appears more likely. Although the stable thermal stratification is very weak, it might enhance the interaction with the solutal boundary layer underneath the ice.

A correlation that appears to explain the overall data dependence is the decrease in Ra_{pc} with derived convecting layer thickness in figure 11.16. This figure indicates a plausible explanation of the anomalies in the runs at 105 ‰ and 10 to 20 ‰ in terms of a relative increase in the solutal boundary layer thickness compared to the convecting layer thickness. Considering these ideas, the observations appear in reasonable agreement with the heuristically suggested values. Taking the most trustworthy runs at 35 ‰, with $\Delta H \approx 1$ cm, for which theory predicts a critical $D_s/V_{cr} \approx 0.033$ cm, one obtains a ratio of boundary layer to convecting skeletal layer thickness of ≈ 0.03. A reduction from the tentative value of $Ra_{pc} \approx 7.7$ to the observations between 4 and 5 is consistent with simulations from Chen and Chen (1988). It must, however, be mentioned, that these refer to an overlying fluid layer. The aspect deserves further investigations.

Self-sustained simulations of the growth rate, and a repeated evaluation of Ra_{pc}, indicate the uncertainty in the method. When thermal convection is present, the reconstruction of a plate spacing profile and the maximum Ra_{pc} is more uncertain. However, the principle results are not changed much, giving credence to the proposed methodology. In conclusion, it seems possible to characterise and predict the instability of the sea ice matrix during growth by a relatively simple criterion of a critical Ra_{pc} between 5 and 8. The latter value can be lowered due to the turbulent haline convection below the porous medium. The following consideration of melt convection below the skeletal layer appears relevant in this context.

11.7.3 Porous medium versus melt convection

It has earlier been suggested, that salt-finger convection is responsible for triggering convection in a mushy layer (Hellawell et al., 1993; Chen et al., 1994). Chen et al. (1994) proposed that the solutal convection must become vigorous when this shall happen. Experiments, however, did not reveal the formation of new channels upon shear flow or normal flow imposed in the liquid (Solomon et al., 1999). Suction applied at the bottom of the porous medium was sometimes suited to create channels (Hellawell, 1987) and sometimes not (Chen, 1995). All these approaches to create artificial flow may not necessarily resemble the turbulent transitions in a boundary layer and their global effect on the boundary condition of the porous medium. This boundary condition in terms of salt flux and redistribution may be thought as the key to a simplified stability criterion. Even for a frequently used system as NH_4Cl-H_2O observations do not reveal a clear picture about the relative roles of (i) internal channel formation and (ii) interfacial triggering by solutal melt convection. Theoretical models (Amberg and Homsy, 1993; Worster, 1997; Chung and Chen, 2000; Emms, 1998) are lacking a systematic database for their piecewise validation. Experiments on the possible interaction during freezing in saline solutions are also lacking. The discussion of the observations from Wettlaufer et al. (1997) is valuable in this context, as the control parameters have been modeled and described properly. Another aspect of these observations will now be considered with respect to the onset of convection.

Most theoretical approaches discuss the existence of the two modes of haline convection obtained from the frozen-time analysis, e.g., Worster (1992), and do not describe the evolution of a turbulent state in the liquid boundary layer. The only studies to date that have addressed this problem are by Emms (Emms and Fowler, 1994; Emms, 1998). In the present work a reasonable parametrisation of the latter process has been validated in the previous chapter, giving the critical growth velocity for the onset of turbulent solutal convection in the melt. These curves, computed from equation 10.14 and already shown in figure 10.2, are redrawn here in figure 11.19. Also shown are the growth velocities at the time of onset of convection in the experiments from Wettlaufer et al. (1997) in dependence on the solution salinity C_∞. The symbols refer to the different boundary temperatures from the figures above.

The V_{cr} of boundary layer convection is drawn for three different critical Rayleigh numbers and the corresponding Nusselt number, as they have been discussed in chapter 10.14. It is seen that most observations, also the low salinity runs with considerably smaller Ra_{pc}, lie close to the curve $c_{nu} = 0.19$, which corresponds to a critical Rayleigh number of 146. It is recalled that the latter is the typical value of occurrence of velocity perturbations at the interface (section 10.2.2). Three observations are exceptional and fall on the dotted line corresponding to $c_{nu} = 0.30$ or $Ra_c \approx 37$. Also this value has been discussed in the previous chapter 10, representing eventually the value of the very first onset of any fluid motion, and being indicated as a change in the plate spacing from pure diffusion (figure 10.3). It may, however, also be interpreted as the salt flux condition for a free boundary, which has remained an open question in the previous chapter. These three observations at high c_{nu} correspond to the freezing runs at 105 ‰, which remained to some degree anomalous in the interpretation of the porous medium Rayleigh numbers from the preceding section.

One may speculate that these two groups represent two groups of observations, both associated with transition at the interface of the porous medium. This would be an explanation for the much lower Ra_{pc} for the freezing runs at 105 ‰ from discussed above.

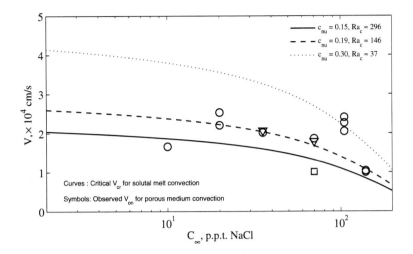

Fig. 11.19: Growth velocities at the time of onset of convection in the experiments from Wettlaufer et al. (1997), overlayed on the critical velocities for the onset of convection in the liquid (equation 10.14) for different critical Rayleigh numbers. The symbols refer to the different boundary temperature from the figures above.

If there are distinct transitions at the given Rayleigh numbers, as they have been observed in other systems, the corresponding reorganisation of the flow might be associated with a forcing that might trigger convection at low Ra_{pc}. The figure would then also explain the much lower Rayleigh numbers $0.8 < Ra_{pc} < 1.6$ at the low salinity freezing runs at 10 and 20 ‰, which have remained unexplained in the discussion so far. These ideas should be understood as complementary to the hypothesis from figure 11.16, that a relatively thick fluid layer may lower the critical Rayleigh number in interaction with the porous medium. Figure 11.19 may be interpreted in the way that at distinct convective transition the hydrodynamics are particularly suited to resemble the interaction with the convecting layer.

The possible relevance of figure 11.19 needs to be investigated in future studies. It indicates, that the instability of the sea ice skeletal layer can be triggered at distinct transitions of the haline finger convection below, if the Rayleigh number is relatively small, in the range 1 to 3. It must be noted that the prediction of the critical growth velocity of convection in the melt via equation 10.14 is based on the assumption of a marginal stability condition at the interface. Via this interfacial condition the skeletal layer and the solutal boundary layer are coupled. It is not yet clear, how this coupled system is best described or parametrised in the interfacial regime, where supercooling and melt convection interact with the cellular morphology. The evaluation of the system in terms of the present local Rayleigh number approach is very simple and promising. Even if the critical Rayleigh number will not be constant with changing growth conditions, it should be possible to parametrise this change effectively. In this respect theoretical work and, of course, more observations are needed. An important aspect of future discussions is certainly the understanding of the behaviour of brine channels in the problem.

11.7.4 Models of binary solidification

Models of binary solidification implicitly account for the effect of nonuniform solid fraction in that they predict an increase in the critical Rayleigh number, if based on the whole thickness (Worster, 1992), or only treat a fraction of the thickness in the evaluation of Ra_{pc} (Amberg and Homsy, 1993; Roberts et al., 2003). An interesting future task is to consider, how well the heuristic maximum Rayleigh number criterion corresponds to the onset of convection when simulated with such models under different boundary conditions, and in different parameter regimes. This criterion is very simple and might be used effectively. The limited simulations from Beckermann et al. (2000) indicate, for an evaluation via the average permeability, a reasonable consistency.

The above evolved questions of interaction between the convection in the melt, in particular with a convecting plume boundary layer, has been addressed very little in model formulations, with the exception of the work by Emms and Fowler (1994). However, all formulations have one flaw in common: They need to specify a marginal stability condition at the macroscopic mush-liquid interface and must assume the lack in sub-pore gradients. To do so, the marginal stability criterion is generally set to the constitutional supercooling condition 7.19 on the liquid side (Hills et al., 1982; Fowler, 1985; Worster, 1986). However, the analysis of observations and interface modeling in chapters 7 through 9 strongly supports that this is not reasonable. Instead the more stringent Mullins-Sekerka condition of constitutional supercooling, equation 7.20, should be applied. The latter is the key to the successful plate spacing prediction in the present work, and it results in a reliable parametrisation of turbulent convection that was also approximately validated (chapter 10). If this interface condition would be applied, many results of the models can be expected to change. In general it is concluded, that the parametrisation of interface processes in binary solidification models is a very important future task for a better understanding of coupled convective stability as indicated in, for example, figure 11.19.

11.7.5 Outlook

In conclusion, there is strong evidence that the present approach of a local critical Rayleigh number, with a numerical value of $Ra_{pc} \approx 5$ to 8, given by a heuristic set of boundary conditions, is very useful to make progress in the evaluation of the early growth of ice from seawater and saline solutions. For binary alloys in general the permeability law is less clear, but the approach might improve the derivation of thresholds and scalings in general. In its essence, it is the modification of the maximum local Rayleigh number criterion for the onset of convection from Beckermann et al. (2000) in terms of local properties. The proposed calculation of Ra_p from local properties is suggested to be physically more consistent than bulk permeability approaches. The onset of convection from only a part of a 'mushy layer', the near-interface high porosity fraction, has been documented earlier on the basis of experiments with NH_4Cl-H_2O (Chen and Chen, 1991). The present chapter has demonstrated, that freezing saline solutions can be treated by rather realistic approximations, resulting in a formulation which allows sensitivity studies for a better understanding of the convective instability problem.

The analysis of freezing NaCl solutions provides relevant quantitative insight into the problem because

- molecular transport properties and their liquidus dependence are well known,

- a novel morphological stability approach to the plate spacing allows a rather accurate prediction of the microstructure,

- the permeability law of a Hele-Shaw cell is simple and corresponds close to the reality of saline ice.

It seems likely that the critical Rayleigh numbers derived above correspond much closer to the intrinsic ones than the large spectrum that has been proposed for the highly dendritic NH_4Cl-H_2O (Chen and Chen, 1991; Tait and Jaupard, 1992a,b), where permeability laws are much less self-evident. However, also the latter studies may be taken as an indication, that the critical Ra_{pc} of the porous medium may strongly depend on the coupling to the boundary layer convection. For further evaluation of this aspect, properly controlled freezing experiments with saline solutions, like those performed by Wettlaufer et al. (1997), need to be combined with texture analysis of the ice, and controlled variation of hydrodynamic conditions. In this way the particular role of several processes emphasised above may be determined. Due to their predictable microstructure, saline solutions are potentially important to proceed with the understanding of convective instabilities and channel formation in solidifying two-phase mixtures.

In the following final chapter some further support of the present results is given in section 12.2.

12. Applications

Having formulated and validated a framework of morphological and convective instabilities during the growth of saline ice, this final chapter presents some applications to
problems of natural ice growth. The first section, recalling the discussion from chapter 4,
presents the solution of a relevant problem during freezing of natural waters in general:
The planar-cellular transition. It is predicted, at which salinity the transition will take
place. In the second section the main results from chapters 10 and 11 are combined
to describe the skeletal layer of sea ice, discuss its thickness and modes of desalination.
Finally a solution for the problem of stable salt entrapment in sea ice is sketched.

12.1   Planar cellular transition

The cellular freezing of ice from saline solutions and predictability of microstructure has
been a central part of the present work. In the early chapter 4 also planar freezing has
been considered. In nature this is the freezing regime of lake ice. Under pure diffusive
conditions also lake ice would be expected to freeze cellular (chapter 7), but this state
is never reached in downward freezing, because natural convection effectively transports
the rejected solute away from the interface. This has been pointed out earlier by many
authors on the basis of experiments (Shumskii, 1955; Knight, 1962b; Wintermantel and
Kast, 1973; Weeks and Ackley, 1986), yet a quantitative discussion has not been given
yet. The present work allows a quantitative description of the planar-cellular transition
by coupling free convection scalings from chapter 10 with morphological stability theory.

In laboratory observations the transition is normally identified as a change to a milky
appearance of the ice, or by a sharp increase in k_{eff} due to interstitial solute incorporation
(Quincke, 1905b; Weeks and Lofgren, 1967; Kvajic and Brajovic, 1971; Wintermantel and
Kast, 1973). Systematic observations of the water salinity, at which the transition occurs
in natural brackish water, are lacking, but some bounds have been reported. It is likely
that the transition occurs at concentrations below 3.6 to 4.1 ‰, for which the cellular
plate substructure was regularly present in Baltic sea ice (Gow et al., 1992). According to
Weeks and Wettlaufer (1996) a transitional salinity of 2 ‰ has been reported by Russian
scientists. Observations from these studies have not become available. In laboratory
experiments almost any transitional salinity of up to 50 ‰ has been deduced, if only the
stirring was large enough (Weeks and Lofgren, 1967; Kirgintsev and Shavinskii, 1969b;

Wintermantel and Kast, 1973; Kvajic and Brajovic, 1971; Ozum and Kirwan, 1976; Gross et al., 1977; Cragin, 1995; Matsuoka et al., 2001). It is therefore expected that also the transition in natural waters depends on the growth conditions and the degree of natural stirring and convection.

A similar transition has been observed for the crystal orientation. The generally horizontal c-axis in sea ice (Hamberg, 1895), contrasts the occurrence of both orientations in lake ice (Shumskii, 1955; Knight, 1962b; Michel and Ramseier, 1971). Some limited field and laboratory studies of brackish water ice (Palosuo, 1961; Kawamura et al., 2002) indicate, that the c-axis orientation changes from primarily vertical to horizontal, most likely in a salinity range between 0.5 and 1.3 ‰. For freezing experiments with NaCl solutions Palosuo and Sippola (1963) reported on the shift from a strict vertical to increasing horizontal c-axis orientation. In small laboratory containers the latter was observed at water salinities between 0.1 and 1 ‰. Transition salinities were the lower the larger the cooling rate. The limited observations indicate that orientation changes parallel the onset of cellular growth. An understanding of the planar-cellular transition may therefore provide ideas about the orientation selection in fresh lake ice.

12.1.1 Interface concentration and constitutional supercooling

The following derivation assumes, for simplicity, that the temperature gradient in the water is negligible. From chapter 7 the condition 7.20 for morphological instability may be written as

$$C_\infty \frac{S_\Gamma}{k} > 1.32 \ ^\circ/_{\circ\circ}, \tag{12.1}$$

where the concentration on the right hand side follows from inserting molecular properties into equation 7.20. These were evaluated as $L_v = 305.6$ kJ/kg, $K_i = 2.14$ W/(mK), $K_b = 0.560$ W/(mK), $m = -0.0602$ K/‰ and $D = 7.03 \times 10^{-6}$ cm^2/s at the corresponding liquidus temperature and concentration (appendix A). As shown in figure 7.5 the stability function S_Γ from Sekerka (1965) is close to 1 for natural freezing velocities and the expected planar k. Furthermore, for the transient case equation 7.31 replaces the equilibrium freezing condition 7.20 for morphological instability which is accounted by setting $C_{int}(t) = C_\infty/k$. Hence

$$C_{int}(t) > 1.32 \ ^\circ/_{\circ\circ} \tag{12.2}$$

becomes the simplified condition for the planar-cellular transition in natural waters. At a concentration larger than 1.32 ‰ NaCl the interface will always become unstable. According to 7.20 this value may be increased by a temperature gradient in the water, in particular at low growth velocities. In the absence of a temperature gradient it does not depend on V. The question to be answered is, up to which solute concentration natural convection can suppress the build-up of a sufficiently large interface concentrations.

12.1.2 Diffusive solute redistribution

To predict the evolution of $C_{int}(t)$ with time a model for the diffusive boundary layer evolution is needed. In case of a constant solute flux the latter evolves due to (Carslaw and Jaeger, 1959)

$$C_{int}(t) = C_\infty \left(1 + \frac{2\rho_i}{\sqrt{\pi}\rho_b} V (1-k) \left(\frac{t}{D}\right)^{1/2}\right),\tag{12.3}$$

where ρ_i and ρ_b are the densities of pure ice and the solution. This form is inaccurate for natural growth conditions at low salinities, as due to increasing C_{int}/C_∞ the salt flux at the interface is not constant. A frequently used simple approximate solution to the problem at large times (Tiller et al., 1953) is

$$C_{int}(t) = C_\infty \left(1 + \frac{1-k}{k} \left(1 - exp\left(-k\tau\right)\right)\right)\tag{12.4}$$

with the normalised timescale τ

$$\tau = \left(\frac{\rho_i V}{\rho_b}\right)^2 \frac{t}{D}.\tag{12.5}$$

The exact solution has been given by Smith et al. (1955) as

$$C_{int}(t) = \frac{C_\infty}{2k} \left(1 + erf\left(\frac{\tau}{4}\right)^{1/2} + p\, exp\left((k-1)\, k\tau\right) erfc\left(p\left(\frac{\tau}{4}\right)^{1/2}\right)\right)\tag{12.6}$$

where

$$p = (2k - 1).\tag{12.7}$$

The evolution in the relative interfacial concentration C_{int}/C_∞, obtained with these three diffusion models, is shown in figure 12.1 for short time scales, and in figure 12.2 for long times, on a linear and logarithmic scale, respectively. It is seen that the constant salt flux condition is only reasonable for short times, when the approximation (12.4) from Tiller et al. (1953) cannot be used. At larger time the opposite is the case.

Figure 12.2 shows that lake water, with $C_\infty \approx 0.1$ ‰, requires, even during rapid growth, a time of $\approx 10^4$ seconds to reach the critical concentration $C_{int} \approx 1.3$ ‰. One may also deduce from equation 12.1 that a rather dilute solution with $C_\infty \approx 0.0013$ ‰ might become unstable if diffusive equilibrium is reached. However, this according to figure 12.2 this implies, even during rapid freezing, ≈ 20 days and 2 meters of ice growth. In a closed container this will, of course, be different, as the released solute concentrates the solution and changes C_∞.

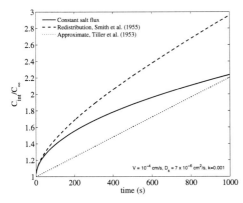

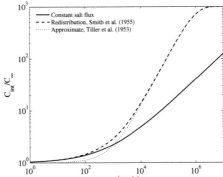

Fig. 12.1: Relative increase in the interfacial concentrations with time due to different diffusion models for the growth conditions indicated in the figures and short times. The dashed line is the exact solution.

Fig. 12.2: Relative increase in the interfacial concentrations with time due to different diffusion models for the growth conditions indicated in the figures and large times. The dashed line is the exact solution.

12.1.3 Critical time of convection

The next step is to introduce the limiting effect of convection into the problem. A solution for the critical time of convective onset is, as described in chapter 10, the equation

$$t_{cq} = 21 \left(\frac{\nu \rho_b}{g \beta V (1 - k) C_{int} \rho_i} \right)^{1/2} , \tag{12.8}$$

slightly modified after the nonlinear analysis by Foster (1972), who gave a prefactor of 14 for a free boundary condition at the freezing interface [1]. In the underlying heuristic model of intermittent convection the interface concentration increases by pure diffusion until t_{cq} is reached. Then a plume of dense fluid breaks off and the process starts again.

Equation 12.8 is valid for a constant salt flux, for which it can be combined with 12.3 to predict the interface condition at the onset of convection, in dependence on the growth velocity. The latter has been given, with a higher prefactor of 27, as equation 10.10. As equation 12.8 is valid for a constant salt flux, the equations have been solved iteratively and by assuming an effective interface concentration $C'_{int} = (C_{int} + C_\infty)/2$ in equation 12.8. The resulting critical time and interface concentration are shown for different far-field concentrations in figures 12.3 and 12.4. As mentioned earlier, the time of ≈ 60 seconds, observed by (Foster, 1969) for the onset of convection in experiments with $C_\infty \approx 25$ ‰, is in agreement with the predictions of equation 12.8, if the total salt

---

[1] It is noted that equation 10.7, first suggested by Foster (1968b) on the basis of a linear analysis, implied a larger prefactor of 27

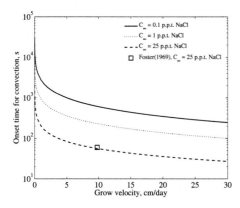
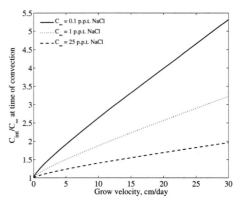

Fig. 12.3: Critical time at the onset of convection obtained from equation 12.8 for different solution salinities.

Fig. 12.4: Critical interfacial concentration at the onset of convection corresponding to equation 12.3

flux is used. Using the total salt flux implies an early stage of planar growth, which is not unreasonable because a cellular field still has to evolve. This observational point is shown in figure 12.3. The agreement with the prediction is very good.

12.1.4   Critical concentration

Equations 12.6 and 12.8 are readily applied to compute, at a given growth velocity, the critical far-field concentration C_∞ at which the interface concentration C_{int} exceeds the cellular instability value of 1.32 ‰. This prediction is shown in figure 12.5 for the prefactors of 21 and 14 in equation 12.8, likely resembling rigid and free boundary conditions. It can be compared to a number of laboratory studies where the transition has been documented during downward freezing (Weeks and Lofgren, 1967; Kvajic and Brajovic, 1971; Killawee et al., 1998). The observations from (Weeks and Lofgren, 1967) show the least agreement. As mentioned earlier, the freezing procedure applied by these authors often involved rigorous thermal convection. While in the experiments shown in figure 12.5 the temperature stratification was stable with respect to the freezing interface, it cannot be excluded that, during freezing in a cold room, cooling from below and the sides has driven thermal convection. The observations from (Killawee et al., 1998) are likely not affected by strong thermal convection. In these experiments the growth velocity was constant, and the changing concentration was monitored. The slightly lower transition velocity in the experiments from Kvajic and Brajovic (1971) is not unexpected, because the diffusivity of NaOH is approximately 60 % larger than for an NaCl solution, used in the predictions.

The reasonable agreement between predicted and observed transition salinities indi-

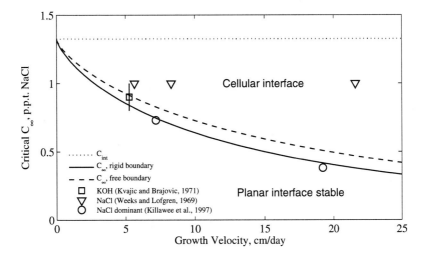

Fig. 12.5: Planar-cellular transition during downward freezing of saline solutions in several laboratory studies(Weeks and Lofgren, 1967; Kvajic and
Brajovic, 1971; Killawee et al., 1998) compared to the present predictions
of critical interface (dotted line) and bulk far-field concentrations (dashed
and solid curves). The circles are the most reliable observations in terms of
control of the growth conditions.

cates that the effect of convection can be modeled with the assumed simple boundary
layer approach, on the basis of intermittency. There is another condition that might
be relevant under certain freezing conditions. Merchant and Davis (1989) discussed the
question, if the transition to cellular growth is subcritical or supercritical, corresponding
to deep grooves or shallow cells. They found several regimes, of which one is always supercritical, which means that small cells do not evolve into grooves but fade away. This
regime is dependent on the thermal conductivity ratio of ice and the saline solution and
on the distribution coefficient k. From their figure 7, one obtains that the transition in
the ice-water system requires $k > 0.12$ to become subcritical. Associating k with the effective interfacial k in the present macroscopic approach of marginal stability, this implies
$C_\infty > 0.2$ ‰, and thus should not have be a limitation in the experiments. However,
it is interesting that Weeks and Lofgren (1967) indeed documented the transition to a
non-cellular ice in all runs near an effective $k_{eff} \approx 0.12$, rather independent on growth
velocity. This aspect deserves further studies.

12.1.5 Discussion

Natural ice growth velocities are normally limited by ≈ 10 cm/day. This implies, according to figure 12.5, a planar-cellular transition in the concentration regime $0.7 < C_\infty < 1.3$ ‰. The above mentioned salinity of the transition from vertical to horizontal c-axis orientation in brackish water ice is remarkably similar (Palosuo, 1961; Kawamura et al., 2002), indicating an intimate connection between cellular growth and preferred crystal orientation. Such a connection can likely be understood in terms of the large anisotropy in the free growth rate of ice crystals at low supercooling (Hillig, 1958; Michaels et al., 1966). It is expected that, when dynamical cellular growth exists, this growth advantage parallel to the basal plane involves particularly rapid geometric selection. Simultaneous observations of the crystal orientation and planar-cellular transitions are not available. These might provide a clue for the control of crystal orientation in lake ice sheets which is yet not fully understood (Shumskii, 1955; Knight, 1962b; Michel and Ramseier, 1971).

The ingredients of the proposed prediction of the onset of cellular growth during freezing of natural waters are three equations: (i) constitutional supercooling from morphological stability theory, (ii) an exact diffusive solute redistribution, (iii) an intermittent turbulent solutal convection model. The main results of the approach are:

- The low growth velocity threshold for cellular instabilities from morphological stability theory is ≈ 1.3 ‰ NaCl.

- The dependence of the cellular transition far-field salinity C_∞ on the growth velocity was predicted in agreement with observations.

- A simple concept of intermittent diffusive boundary layer growth appears to be a reasonable approach to predict maximum interface concentrations.

- The change in crystal orientation from primarily vertical to horizontal and the planar-cellular transition appear to be closely related.

The quantitative results apply to solutions with NaCl being the dominating salt. They are therefore relevant for the freezing of most natural brackish waters.

12.2 Sea ice skeletal layer and desalination

The previous chapter has indicated, that the onset of internal convection from growing sea ice may be described by a critical local Rayleigh number. The range $5.3 < Ra_{pc} < 7.7$, proposed from a simplified analyses of the boundary conditions, agreed reasonably with observations. This condition refers to the onset of convection. The important question, how the system evolves after the onset of convection, requires a nonlinear analysis and the description of the formation of brine channels. The difficulties of a theoretical analysis have already been pointed out. Here some alternative observations are explored. It will be considered, on the basis of nondestructive salinity profiles, if the system evolves by keeping a kind of marginal equilibrium Rayleigh number.

12.2.1 Maximum local Rayleigh number

The only available set of observations that can be used in this respect are the non-destructive ice salinity observations from Cox and Weeks (1975) which were discussed in chapter 5. The observations consist of 2 timeseries of vertical ice salinity and temperature profiles. As also growth velocities are well documented, the procedure from the preceding chapter can be repeated: (i) compute the plate spacing a_0, (ii) calculate the vertical brine volume profile, (iii) calculate the Hele-Shaw cell permeability and the local Rayleigh number.

The resulting Ra_p is shown in figures 12.6 and 12.7 for two fixed levels, 1.3 and 3 cm, against the growth velocity. Each point reflects the local Rayleigh number of one salinity profile. The average Rayleigh numbers indicated in the figure have been computed by excluding the values during low growth velocities, when thermal convection was present, growth was very slow, and temperature gradients very uncertain (see section 5). Figure 12.6 shows the results for the level of 1.3 cm, where the average of the local Ra_p for all runs has a maximum of 5.6 ± 2.9. The maximum Rayleigh number for each profile has also been computed, followed by averaging of the maxima. This gives a very similar average of 6.0 ± 2.8 and an average distance $\Delta H_p 1.3 \pm 0.4$ cm from the interface. These numbers correspond to the proposed critical Ra_{pc} for the onset of convection and agree with the results from the preceding chapter.

As discussed in chapter 5, one should consider several uncertainty sources of the dataset from Cox and Weeks (1975) in the interpretation: (i) the salinity observations were deconvolved from coarse resolution sampling and presented at 1 cm vertical resolution, (ii) the temperature gradient was also coarsely resolved, (iii) the interface position could was measured with precision ± 0.5 cm, (iv) at later stages thermal convection halted the growth. According to the analysis in chapter 5 the interfacial gradient is reasonable to within 10 to 20 %, as long one focuses on growth velocities above 2 cm/day, below which thermal convection effects become relevant. To minimise the effect of errors in the interface position and its rounding off, ΔS_b was computed with respect to $S_b(T)$ at the estimated interface and not with respect to S_∞ in the liquid below. The variability in Ra_p may still be attributed to the noted factors. The average from 15 profiles appears, from the overall data appearance, reasonable. The absolute values of the level 1.3 cm and the maximum Rayleigh may, however, be affected by a temperature distribution close to the interface that differs from the slihtly nonlinear gradient estimates from Cox and Weeks.

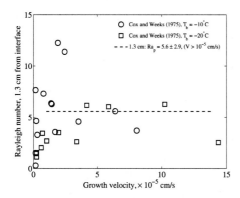
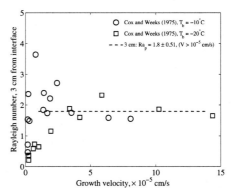

Fig. 12.6: Local Ra_{pc} in the nondestructive salinity profiles from Cox and Weeks (1975) at a distance 1.3 cm from the interface, where Ra_{pc} was, on average a maximum. Due to thermal convection only growth velocities above 10^{-5} cm/s are averaged for an indication of the best estimate, the dashed line.

Fig. 12.7: Local Ra_p as in figure 12.6, yet at a distance of 3 cm from the interface, corresponding to the typical upper level of the skeletal layer.

This possible bias has been estimated to be likely less than 10 %, in terms of Ra_{pc} and ΔH_p.

In figure 12.7 the corresponding Ra_p at 3 cm from the interface has been plotted. This is the typical upper level of the skeletal layer. Here the local Ra_p is a factor of 3 smaller than the maximum. Interestingly, this value is rather stable. This gives further credence to the reliability of the maximum Ra_p. It appears that the skeletal layer operates at the same maximum Rayleigh number which limits the onset of convection.

Finally it is recognised that, at low growth velocities, the maximum Rayleigh number drops to low values. This can be related to the effect of thermal convection described in chapter 5. The latter steepens the temperature gradient at the interface, while the growth velocity stays small. Under such conditions the skeletal layer becomes stabilised. Desalination, however, continues on the basis of the liquidus temperature gradient.

12.2.2 Convecting layer thickness

Assuming a fixed critical Ra_{pc} it is straightforward to solve equation

$$Ra_{pc} = \frac{\beta \Delta S_b g \Pi \Delta H_p}{\nu \kappa_b}. \tag{12.9}$$

in dependence on the growth velocity for ΔH_p and ΔS_b. This is possible because $V \sim \Delta T / \Delta H_p \sim \Delta S_b / \Delta H$. However, one must assume, that v_b, which enters Π, is only a function of temperature. This was justified when analysing the experiments from Wettlaufer et al. (1997) in the previous chapter, because the salt had remained essentially trapped in the ice, that is $k_{eff} \approx 1$. Calculations are first performed in this way and then extended to a more realistic skeletal layer. It is now recalled that the turbulent solutal convection model in the liquid implies a decreasing k_{eff} at lower growth velocities (figure 10.5). Therefore the brine volume in equation 12.9 must be scaled by this predicted k_{eff}. In this way the coupled system contains the essential physics on the basis of the two convection processes and morphological stability theory, which gives a_0 and Π. No additional parameters than the two critical Rayleigh numbers need to be introduced. To keep things simple, it is in the following assumed that a_0 is constant within the convecting skeletal layer, which corresponds to the assumption of a quasi-constant growth velocity for which a_0 is predicted.

The predicted critical convecting layer thickness is shown in figure 12.8 for two critical Rayleigh numbers. A slight nonlinearity has been implemented for the temperature gradient. To do so, it was assumed that a similarity to the temperature distribution B.6 holds in the skeletal layer, if only the brine volume change is similar. The same calculations have also been done for a linear temperature gradient, shown as a dotted curve. The curves cross at some velocity due to two effects: The steeper linear temperature gradient implies a larger ΔS_b, yet also a lower brine volume. The former effect becomes dominant at low growth velocities, when brine volumes are larger. The nonlinear basic temperature profile represents likely the more realistic one. The main increase of ΔH_p towards low velocities is caused by the decrease of k_{eff} due to turbulent convection. In reality one may eventually expect a limitation, of turbulent convection, when k_{eff} near the interface drops below a certain value. For example, for the alloy NH_4Cl-H_2O , it has been observed that the formation of channels and rigorous porous medium convection halts the turbulent plume convection in the melt (Chen et al., 1994). However, this must not be the case for seawater. The predictions made here are notably based on the assumption that k_{eff} decreases with decreasing growth velocity, due to turbulent convection as described in chapter 10. While this assumption yielded agreement with observed plate spacings over a wide range of growth velocities (figure 10.12), it is recalled that the detailed system behaviour near the interface needs to be investigated.

The obvious outcome of the predictions is a convecting layer thickness which varies little over a wide regime of growth velocities. The high velocity stability limits have been indicated as circles. At larger velocities the skeletal layer will not convect internally. The convecting layer thickness decreases from its high velocity limit, reaches a wide plateau, and then increases again at low growth velocities. These results may be interpreted with respect to two natural situations:

- The convecting layer thickness resembles a quasi-constant skeletal layer at the bot-

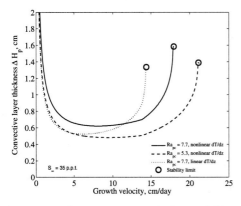
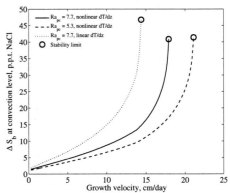

Fig. 12.8: Convecting layer thickness ΔH_p for different critical porous medium Rayleigh numbers and $C_\infty = 35$ ‰ in dependence on the growth velocity of the ice.

Fig. 12.9: Brine salinity difference ΔS_b between the interface and the top of the convecting layer of thickness in figure 12.8.

tom of growing ice.

- The onset of internal convection during early growth of thin ice takes almost always place, when the ice is less than a centimeter thick.

The corresponding brine salinity difference ΔS_b, between the interface and the top of the convecting layer, is shown in 12.9. This will be discussed later. First, some observations of the thickness are discussed.

Skeletal layer thickness

The data from Cox and Weeks (1975) can be compared to the predictions by associating the convecting layer thickness with the depth of the maximum Rayleigh number identified above. Doing so one difference must be recognised. The predictions assume that, in a thin layer close to the interface, the sea ice has essentially the bulk salinity of the seawater. However, at the Rayleigh number maxima within the ice, corresponding to figure 12.6, k_{eff} is found already to be reduced to ≈ 0.6 to 0.7. This appears to be a consequence of the evolution of the system after onset of convection. To allow a comparison with the ΔH_p values estimated from the Cox and Weeks study, calculations have also been performed with k_{eff} reduced by a factor 0.7. The role of the increasing salinity in the experiments is also considered, by performing the calculations for both 35 and 50 ‰ NaCl. The results are shown in figures 12.10 and figure 12.11.

The higher salinity implies a slightly smaller convecting layer and salinity difference, because brine volumina near the interface are larger. A much larger effect comes from the smaller k_{eff}, which reflects an active stage of convection. Decreasing k_{eff} in turn increases the convecting layer thickness. A simple scaling analysis of equation 12.9 may illustrate this. For a constant Ra_{pc} one has $\Delta S_b \Delta H_p \sim \Pi^{-1} a_0^{-2} v_b^{-3}$. Approximating a linear liquidus gradient with $\Delta S_b \sim V \Delta H_p$ and a dependence $a_0 \sim V^b$, with $b \approx -0.3$ over much of the natural growth regime (figure 10.12), one gets

$$\Delta H_p \sim v_b^{-3/2} V^{-0.2}. \tag{12.10}$$

In this form it is seen that the response of ΔH_p on V alone is weak. If now k_{eff} decreases due to the onset of salt fluxes, v_b is effectively decreased by this factor, and the convection level moves upwards to larger ΔH_p. The dependence is stronger than linear. As mentioned above, the observations from Cox and Weeks (1975) are coarsely resolved. Considering the scatter in figure 12.6 for the Rayleigh number, the agreement of predicted and observed ΔH_p in figure 12.10 is, for a modified k_{eff}, reasonable. The corresponding brine salinity difference ΔS_b, in figure 12.11, is also consistent with the theoretical predictions. One point to be discussed below emerges already here: ΔS_b and ΔH_p are larger for lower solution salinities, because brine volumes are smaller. This supports the conclusion drawn in chapter 5: The observations from Cox and Weeks (1975) do not resemble sea ice growth conditions at low growth velocities, when S_∞ was large. An additional already discussed flaw at low growth velocities is the rigorous thermal convection in these experiments. It is plausible that the difference between observations and predictions at lower velocities is due to the latter effect, implying smaller plate spacings than predicted.

A few other observations in support of a thin layer of rigorous convection are worth mentioning. The skeletal layer thickness estimates for the thinnest ice, summarised in section 3.1.2, were in the range 0.5 to 1 cm. These should be closest to the predictions. As another indirect determination, laboratory experiments by Wolfe and Hoult (1974) are notable. Oil, released below sea ice of 12 to 16 cm thickness, was later mostly found in the lower 0.5 cm of the skeletal layer. Another supporting observation has been made by Haas (1999). He noted that, during the freezing-in of a thermistor into sea ice grown in a large tank, the temperature fluctuations ceased some 4 mm from the interface. Haas interpreted this as a considerably thinner skeletal layer than normally assumed. Accounting, on the basis of appendix B, for a weaker bottom temperature gradient than assumed by Haas, the transition in his experiments appears to have occurred between 0.5 and 1 centimeter from the interface. Finally, it is notable as a general result, that the weak growth velocity dependence of the convecting layer thickness parallels the quasi-constant skeletal layer thickness of sea ice (chapter 3).

Onset of convection during early growth

Figure 12.8 implies that also the onset of convection, during the very early growth of a thin ice skim, is expected at a thickness of 0.5 to 1 centimeter, irrespective of growth rate,

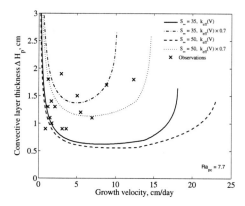
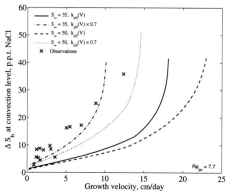

Fig. 12.10: Predicted convecting layer thickness ΔH_p in comparison with observed maximum Rayleigh number levels in the nondestructive profiles from Cox and Weeks (1975).

Fig. 12.11: Brine salinity difference ΔS_b between the interface and observed maximum Rayleigh number levels in figure 12.10 along with the predictions.

as long the latter is constant. For a decreasing growth rate the onset will be delayed, because the plate spacings at the upper boundary are smaller. For example, if the growth follows the law $H_i = \eta(\kappa_i t)^{1/2}$, then the apparent average V of ice grown to thickness H_i refers to the level $H_i/2$. However, under typical Arctic conditions the early growth rate will often be close to constant, with a rapidly changing surface temperature (Devik, 1931; Maykut, 1986). One may thus not expect a large deviation from the predictions. In fact, essentially all observations of early ice growth of some 0.5 to 1.5 centimeters, some of which were discussed in section 4, show that considerable desalination has taken place (Johnson, 1943; Tsurikov, 1965; Wakatsuchi, 1974a; Melnikov, 1995)[2].

According to figure 12.8, a larger convective layer thickness is expected only at very high growth rates. These were realised in experiments from Wettlaufer et al. (1997), discussed in the previous chapter, due to freezing on to a cold plate at quasi-constant temperature. The latter authors interpreted their results in terms of bulk properties and a stability condition, from which they later deduced the onset of convection in a natural lead at thicknesses above 3 cm (Wettlaufer et al., 2000). However, the derived stability function does not account for (i) the strong temperature-dependence of the kinematic viscosity, (ii) the growth velocity dependence of the plate spacing, (iii) the restriction of convective motion to a fraction of the ice thickness in this case and (iv) the effect of thermal convection, which were all found to be relevant in the previous chapter. The

---

[2] That this is not an effect of possible brine loss during sampling alone, is evident from simultaneous water salinity observations by Wakatsuchi (1974a, 1977)

empirical approach from Wettlaufer et al. (1997, 2000) therefore may at most be expected to give approximate results.

One finally has to consider that initial growth in nature often leads to vertically inclined crystals, which prevail down to 0.5 to 1 cm thickness Weeks and Assur (1964); Weeks and Ackley (1986). If this is accounted in a stability criterion, convection would also be delayed to larger thickness. Therefore it is not unexpected to observe the onset of convection from naturally growing ice at a slightly larger thickness than predicted by 12.8.

12.2.3 Desalination effects

Figures 12.10 and 12.11 have indicated that the system, after the onset of convection, changes towards an actively convecting stage. Both the observed convecting layer thickness and the brine salinity difference ΔS_b appear to be larger than predicted by the basic theory. This is not unexpected, as the convection leads to exchange of brine with the seawater below. Subsequent freezing implies a decrease in k_{eff} and stabilisation. To obtain a better quantitative understanding, in figure 12.12 the predictions are made for several fixed values of k_{eff}. The influence of the boundary layer convection on k_{eff}, which produces the sharp rise in ΔH_p at low growth velocities (figure 12.8), has been excluded here to emphasise the internal changes.

In figure 12.12 it is seen, how a decreasing k_{eff} may raise the critical convection level. The following balance may be thought to determine the actual level of maximum Ra_p. First, if the convective desalination is two rapid, then the growth rate cannot follow up and desalination stops. Second, if the growth rate is too rapid, then desalination near the interface is not large enough and the convection level stays closer to the latter. One may think of a similar intermittent process as for compositional convection in the liquid. If one identifies some equilibrium level of these processes with the maximum Ra_p, then in the experiments the latter was apparently operating near $k_{eff} \approx 0.6$ to 0.7 and with a convective layer of thickness ≈ 1.3 cm. This is consistent with the curves in figure 12.12. It is also seen that for $k_{eff} < 0.4$, corresponding to observed values above the skeletal level in the experiments, the convecting layer thickness increases more rapidly. Already at $k_{eff} \approx 0.5$ the predicted convection level is ≈ 3 cm. Hence, the skeletal layer cannot convect from above. Accordingly, in the experiments from Cox and Weeks, the local Ra_p at the 3 cm level was below the convection threshold (figure 12.7).

It is further interesting, that the corresponding critical ΔS_b, shown in figure 12.12, is always limited to ≈ 40 to ≈ 45 ‰, irrespective of the assumed value of k_{eff} and the growth conditions. This is an important constraint of the system. It is worth a note, that Melnikov (1995) found brine salinity fluctuations in the range $25 < \Delta S_b < 50$ ‰ from sampling of the bottom 6 to 8 cm portion of young ice. However, this is not strictly the

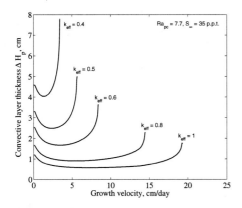

Fig. 12.12: Change in convecting layer thickness ΔH_p with a decrease in k_{eff} in the ice.

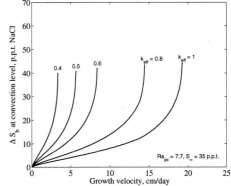

Fig. 12.13: Brine salinity difference ΔS_b corresponding to a change in k_{eff} which is constant through the skeletal layer.

skeletal layer

That the maximum Ra_p is apparently operating near $k_{eff} \approx 0.7$ is thus, from the graphs in figure 12.12, in principal plausible. To predict this thickness one needs to model the process in detail, but some simple considerations are possible here. In reality, k_{eff} does not drop homogeneously within the skeletal layer to the values assumed in figure 12.12, yet it decreases with distance from the interface, reaching values between 0.1 and 0.5 at the top of the skeletal layer. As the increase in ΔH_p in figure 12.12 is nonlinear, while k_{eff} in fact changes almost linearly (see chapter 5), there must be a limit. Regarding the convection as an intermittent exchange of brine, one may expect $k'_{eff} \sim S_\infty/S_b$. Here the assumption

$$k'_{eff}(V) = k_{eff}(V)\frac{S_\infty}{S_\infty + \Delta S_b} \tag{12.11}$$

is made, which implies that a convective event ultimately exchanges the brine of salinity $S_b = (S_\infty + \Delta S_b)$ against seawater of salinity S_∞. Of course, this approximation can only expected to be valid in the convecting regime.

Figure 12.14 shows, how the predicted convecting layer thickness changes on the basis of this simple assumption. The focus is now on the solid and dashed curves for $Ra_{pc} = 7.7$. The dashed curve represents the k_{eff} solution, modified by equation 12.11, and yields larger ΔH_p. The stabilisation of the convecting layer towards thicker values thus gives, as expected from the discussion above, values closer to the observations. In this system however, there is no longer a free parameter (if Ra_{pc} is given). Notably, the larger thickness branch is limited to lower growth velocities. One may interpret the curves in the following way: Along the branch for the onset of convection, with unmodified k_{eff}, there is always the potential for convective instability closest to the interface. When convection becomes rigorous enough and exchanges the fluid, the instability level will eventually be moved up some millimeters. For the assumed functionality of $k_{eff}(V)$ this branch does not start above $V \approx 7$ cm/day. As it is unlikely, that the dependence of k_{eff} is stronger than in case of a complete solute exchange, this branch may represent an upper bound on ΔH_p. Finally, the discussion indicates, that the ice may may become unstable at two levels at the same time: If ice growth proceeds after a convection event without considerable brine loss, the critical local Rayleigh number may be reached simultaneously at both levels. Such a situation would eventually resemble a more rigorously convecting layer, with nonlinear effects propagating further upwards. In the following, this idea is illustrated in connection with some limited observations on brine channels.

12.2.4 Brine channels and bridging

The simple relations for the convecting layer thickness and their observational validation may encourage to focus on parameterizations of the effective salt flux, also here in terms of relations suggested for classical porous media (Palm et al., 1972; Nield and Bejan, 1999). Two phenomena complicate the problem of convection near the interface. First, as described in chapter 3, it is known that at the top of the skeletal layer bridging between the plates takes place. This implies a decrease in the permeability. Even if such bridging does not take place in the convecting layer, what is not definitely known, it will have remote effects. Such remote effects will be relevant for the second and major aspect of complexity: The formation of brine channels.

The existence of brine channels in ice of any thickness is well known (appendix C.3), yet their dynamics is, as mentioned in the previous chapter, not well understood. Among the few observations of brine flow from brine channels, the results from Wakatsuchi and Ono (1983) may be compared to the present predictions. The latter authors conducted an experiment, where brine excluded from growing sea ice was sampled by some conical device. The maximum ΔS_b measured in a deflatable balloon at the bottom of this brine sampler are shown as plus signs in figure 12.15. Also shown are the two predicted limiting branches corresponding to figure 12.14 for ΔH_p. Wakatsuchi and Ono (1983) further provided estimates of corrections of these brine salinities, accounting tentatively for some overflow from the samplers. However, at growth velocities below ≈ 7 cm/day these values are likely incorrect. This is indicated by a comparison of the surface temperatures (and corresponding equilibrium S_b) that the ice needs to have for the reported growth velocities and thicknesses, which were also provided. However, as discussed below, the overflow assumption might be justified at high growth velocities. The observed brine salinities are *de facto* lower bounds on the outflow salinity of brine streamers. The observations from Wakatsuchi and Ono (1983) are unique in this respect, because they are nondestructive.

It is seen that, at high growth velocities, the observed brine salinities are close to the predictions, while at low growth velocities, they are not. In the latter regime the brine apparently stems from levels much farther away from the interface. An important complementing information to interpret this result comes from an earlier study by Wakatsuchi (1977), who studied brine streamers at two different growth velocities:

- At 12.2 cm/day, Wakatsuchi observed that convecting brine excluded from the sea ice remained in an a few centimeters thick high salinity layer below the ice. From this layer a large number of plumes emerged, a process which which mixed the water below homogeneously.

- For a growth velocity of 7.7 cm/day the plumes were thicker and fell directly into the main water body. These plumes always escaped from the outlets of brine channels at the interface and created, in the long run, a stratified water body.

If these circulation types are considered in connection with the brine salinity estimates from Wakatsuchi and Ono (1983), it appears possible, that at large growth velocities their overflow correction is justified. The estimated brine salinities (crosses in figure 12.15) agree indeed reasonably with the present predictions for $Ra_{pc} = 7.7$. At low growth velocities, however, the falling streamers should have entered the lower balloon of the sampler almost unmixed. Here the actually observed ΔS_b is likely the relevant measure. As mentioned above, the values of ΔS_b corrected by Wakatsuchi and Ono (1983) at low growth velocities are physically not consistent with the growth rate and thickness of this ice.

The evidently increasing contribution of larger brine channels to the flux from the ice, while the growth velocity decreases, implies a fundamental change in the desalination mechanism. A possible interpretation can now be offered on the basis of the two noted instability curves in figures 12.15 and 12.14, and eventually the dotted curve for $Ra_{pc} = 5.3$. The modified k_{eff}-branch, suggested from the intermittent brine exchange model, just becomes possible at growth velocities below ≈ 7 cm/day. When it exists, two conditions may be relevant to trigger the existence of brine channels. First, it implies a larger thickness, and thus should be linked to an intergranular mode of convection. As grain boundaries are preferential sites for the formation of brine channels (Wakatsuchi and Kawamura, 1987), this mode will trigger their formation. The second aspect is the mentioned possibility of more rigorous convection events in case of the two stability curves, which might trigger each other.

Wakatsuchi and Ono (1983) presented their data in a way that suggested a gradual transition between the regimes. However, comparing the observations with the stability curves, it is seen that such an assumption is only based on a single observation in the growth velocity range 6 to 10 cm d^{-1}. The observations may be also interpreted as a

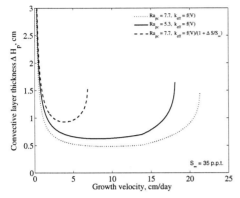

Fig. 12.14: Convecting layer thickness ΔH_p for the onset of convection (solid curve), compared to a modified k_{eff}, being inversely proportional to the brine salinity S_b (dashed curve). The onset curve for a lower $Ra_{pc} = 5.3$ is also shown.

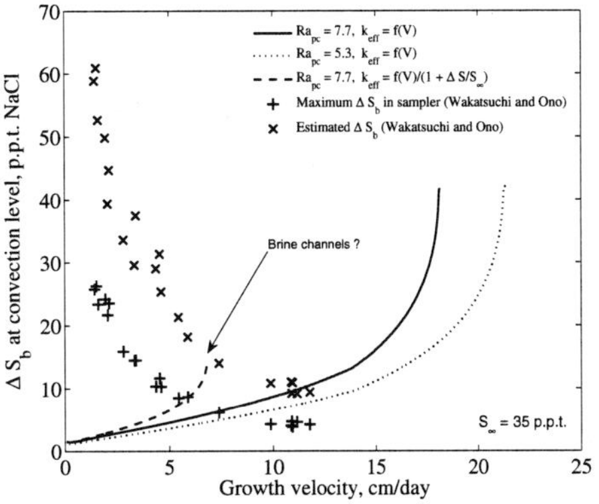

Fig. 12.15: Brine salinity difference ΔS_b corresponding to the standard and modified k_{eff} cases in figure 12.14. The plus signs are the brine salinities observed by Wakatsuchi and Ono (1983) in samplers below sea ice. The crosses were estimated by these authors using a mixing correction.

sharp increase in ΔS_b at a transition velocity which appears to be close to the onset of the second stability branch. It seems therefore that there is experimental support for the transition to a desalination mode that increasingly invokes the formation and activity of larger brine channels.

Two questions of particular importance may be noted here. The first one is, if the lower skeletal layer mode of convection persists in the presence of brine channels. The increase in brine salinity with decreasing growth velocity is most likely interpreted as a gradual transition between these modes. As the samples all had a similar thickness of ≈ 5 cm, this appears not to be a consequence of a maturing process. A shut-down of the lower skeletal convection might be related to the upward movement of seawater that needs to compensate the downward movement of brine through the channels, as it is expected from theoretical considerations (Worster, 1991; Amberg and Homsy, 1993), and has also been observed (Niedrauer and Martin, 1979)[3]. However, if this upward flow actually shuts down the convection is unclear. It appears also plausible, that this mode persists and triggers the convection in larger channels from below. The local Rayleigh number in the profiles from Cox and Weeks, shown in figure 12.6, remains near the

---

[3] This compensating flow has been described by Niedrauer and Martin (1979) in a thin growth cell. However, two-dimensionality, confinement and wall heat fluxes make a detailed comparison to real sea ice problematic.

marginal stability limit at velocities down to 1 cm/day, which indicates the persistence of the lower skeletal layer mode. The drop at very low velocities may, as mentioned, be related to thermal convection and the very slow growth.

The second question is concerned with the critical Rayleigh number. As discussed in the previous chapter, a value of 5.3 might be expected for an *intragranular* convective instability, while 7.7 is a plausible value for *intergranular* convection with cells over several grains. The latter number is very likely the appropriate one for brine channels. The former may be expected to prevail at high velocities, when grain sizes are small (Wakatsuchi and Saito, 1985). A transition between these modes in the lower convecting layer would also lead to a rise in brine salinities and convecting layer thickness. The observations in figure 12.15 are too sparse for such an evaluation.

It is finally noted that two other studies have been performed, where information on brine channels salinities were obtained. The mentioned optical observations of streamers made by Wakatsuchi (1977), at velocities of 7.7 and 12.2 cm/day, have been related to salinity differences based on salt budget calculations. They yielded ΔS_b of ≈ 2 and 4.3 ‰. The lower value was observed at the higher growth velocity. The values are very approximate. The budget estimates should on the one hand be reduced due to brine loss during sampling, but on the other hand be increased, because the brine release likely did not start before the finally 1.5 cm thick ice had reached a thickness of 0.5 cm. With the latter correction one obtains approximate agreement with the later study from Wakatsuchi and Ono (1983). A second observation was documented by Farhadieh and Tankin (1975) during freezing of seawater in a thin growth cell. From interferometry these authors derived the center salinity $\Delta S_b \approx 6$ ‰ for the first plumes released from the ice. The ice growth rate was only documented in nondimensional units, yet appears to have been in the range 10 to 15 cm/day, at the onset of pluming. Growth was, however, accelerating after a remelting phase, which makes predictions of plate spacings less reliable. Also due to the high thermal gradient, thermal convection and superheat in the cell the results are not directly comparable to the calculations made here. Similar experiments with systematically controlled growth conditions could be useful to obtain valuable information on the problem.

Disequilibrium

All calculations done so far have been based on the assumption that the brine equilibrates by freezing and therefore is always at its local liquidus temperature. Under stagnant conditions this is, due to $\kappa_l/D_s \approx 200$ at the freezing point of seawater, a reasonable condition, if sub-pore gradients are neglected. The latter condition becomes eventually problematic, when considering the lateral freezing of plates at high brine volume fractions. This aspect deserves further studies. More severe is likely, that the assumption of thermodynamic equilibrium becomes less reasonable once quasi-steady convection is established. This is backed by some observations. Ono (1967) performed an experiment that allowed him to observe the temperature difference between ice and artificial brine pockets of 2 mm diameter. Heating the ice, he observed that the brine took almost an hour to equilibrate. A similar disequilibrium observation was made by Melnikov (1995). He reported on more than 50 % undersaturated (supercooled) brine samples obtained from young first year sea ice. The latter observations probably also refer to larger brine channels. The periodicity of flow events has been reported to be of the order of 1 hour (Eide and Martin, 1975; Melnikov, 1995). This indicates that disequilibrium might be a regular situation. Due to the coupled nature of the circulation between brine layers and brine channels, a complete lack of disequilibrium for the brine layer network of smaller pores seems unprobable. The effect of upward flow and supercooling may, for example, create an increase in the intrinsic local Rayleigh number in the dataset from Cox and Weeks, because the actual brine volumes are larger. It might then also raise the estimated critical skeletal layer depth of 1.3 cm. This question is clearly important and should be addressed in future studies.

12.2.5 Conclusion

It has been analysed, if the heuristic convective stability approach for the sea ice skeletal layer, discussed in detail in chapter 11, can be extended to describe the behaviour of an internally convecting sea ice skeletal layer, after the onset of convection. A comparison with nondestructive salinity profiles from Cox and Weeks (1975) indicates, that this is the case. The assumption of marginal stability and intermittent convection explains the quasi-constant thickness of the sea ice skeletal layer to reasonable degree, analogously to the intermittent solutal convection in the liquid boundary layer. The consistency of the predictions, based on a coupling of morphological stability, melt convection and porous medium convection, is further supported by a number of independent observations of ice thicknesses and salinities.

The critical thickness for the onset of convection during the early growth of sea ice under typical natural growth conditions, is readily formulated on the basis of the approach from chapter 11. Convective onset is predicted to take place at a thickness of less than 1 cm, and to depend little on growth velocity. Also this is consistent with observations. The theory also allows a reasonable prediction of the onset of convection from leads. Earlier empirical formulations (Wettlaufer et al., 2000) are less reliable, as they only refer to a plate spacing and permeability during rapid ice growth in the laboratory.

Theoretical predictions were, for large growth velocities, shown to be consistent with the salinity of brine streamers observed by Wakatsuchi and Ono (1983). On the other hand, at lower growth velocities, the excluded brine apparently derived from levels much deeper than the skeletal layer. This indicates the increasing importance of brine channels and a different mode of desalination. A simplified approach has been used to describe the expected change in the skeletal layer thickness, after the onset of convection, in terms of intermittent brine exchange. It implies a change in the level of maximum local Rayleigh number and thus the thickness of the actively convecting layer. The resulting instability curve has its origin close to the growth velocities, where brine channel dominated convection has been observed. Qualitative arguments have been presented, why such a transition might occur. The principle mechanisms may be summarised as follows

1. The onset of convection from the skeletal layer takes place at $\Delta H_p \approx 0.5$ to 0.7 cm.

2. Convective brine exchange with the seawater decreases k_{eff} and thus raises ΔH_p.

3. The convection becomes intermittent and operates at marginal stability close to the critical Rayleigh number. While at high velocities the convection level remains close to the interface, a second mode can reach some millimeters further up into the ice when the growth velocity becomes smaller.

4. The convecting layer triggers the formation of brine channels at ΔH_p. The higher this level, the more rigorous the convection and possibly the formation of brine channels.

5. Convecting brine channels drag fluid through the brine layers upwards. Lateral freezing of the upward flowing brine decreases the solid fraction and permeability. The interface must advance to increase the brine salinity again. The process continues to be intermittent.

While the approach presented here is simplified, it appears to be a reasonable working hypothesis of more advanced theoretical models. Some questions may be addressed by proper experiments on the basis of the stability curves. In order to proceed with the dynamics of brine channels, one needs to understand, if their intermittent flow is triggered from above or from near the interface. Expansion flow, reviewed in appendix C, may be relevant and its role needs to be determined. Another unknown factor is the onset of bridging and subsequent decrease in the permeability above the skeletal layer.

12.3   Prediction of the stable salinity of growing sea ice

The final section of this chapter will present a hypothesis, how the stable entrapment k_0 in sea ice can be predicted on the basis of the theories from the present work. To outline this approach the conceptual graph of desalination regimes, figure 3.1 from chapter 3, is

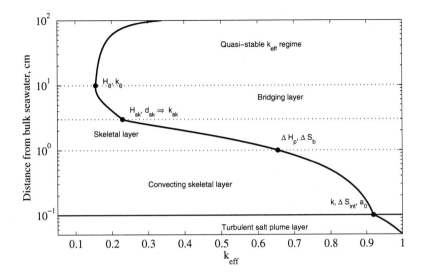

Fig. 12.16: Conceptual desalination sketch of sea ice near the ice water interface (solid horizontal line). The different regimes are explained in the text.

redrawn here as figure 12.16 with additions and on a logarithmic depth scale. It shows a typical normalised salinity profile of sea ice which has grown to 100 cm without severe warming events. The figure is no simulation, but the changes in k_{eff} near in the bottom layer regimes are typical for an ice growth velocity of ≈ 1 cm/day.

To describe the different desalination regimes in the perspective of the present work, the solute distribution equation 5.11 from chapter 5 is recalled:

$$k_{sk} = r_\rho r_{gra} \frac{d_{sk}}{a_0} \left(1 + \frac{\Delta S_{int}}{S_\infty} + \frac{\Delta S_{sk}}{S_\infty}\right). \tag{12.12}$$

It gives the the effective solute distribution coefficient at any $sk-$level. Here, as discussed in earlier sections, the intrinsic upper level of the skeletal layer is associated with k_{sk}, although desalination still proceeds above it. It is pointed out, that the depth of this level does not appear in equation 12.12, contrasting the first approach in chapter 5, where the latter was fixed at 3 cm. Instead the salinity increase ΔS_{sk} appears. As outlined in the following, the latter will be associated with the convective stability limit from the preceding section 12.2. It will further argued, why this is a physically justified approach, and finally leads to a reasonable prediction of the quasi-stable k_0. If this is accepted for the moment and the discussion from section 5.6.1 is recalled, then all variables and parameters on the right hand side of equation 12.12 turn out to be predicable, with one exception: d_{sk}, the brine layer width at the skeletal transition. At the moment it suffices to say, that the latter has been found to be rather constant, taking a value of $d_{sk} \approx 0.11$ mm. Before this is outlined below, the solutions for the remaining variables in equation 12.12 are sketched as follows.

Grain size and density parameters r_{gra} and r_ρ

The parameters r_ρ and r_{gra} are related to grain size and brine density and both slightly larger then one, see section 5.6.1. It turns out that, when all factors are included in the following calculations, r_ρ is almost independent of the growth velocity in the range below 20 cm/day: It is between 1.10 and 1.11. The grain size parameter accounts for finite crystal lengths. It is not predicted from the present theory, but can, on an empirical basis, be expected to vary between 1.02 for low growth velocities and 1.07 for vary rapid growth. It is henceforth set to 1.03.

Interface: k, ΔS_{int} and a_0

One main result of the present work is the prediction of k, ΔS_{int} and a_0 at the interface, due to the interaction of marginal morphological stability and turbulent plume convection in the solutal boundary layer, the lowermost regime in figure 3.1. From chapter 10 it is recalled, that k reflects the interfacial brine volume, while ΔS_{int} is small, only of the order of 1 ‰. These two variables are transient and do not play much a role alone, but they are coupled to the prediction of the plate spacing a_0. Its selection takes place at this level, fixing the principal skeleton of sea ice and entering fundamentally as a linear factor in equation 12.12.

Convecting skeletal layer: ΔS_{sk}

An important outcome of the preceding section 12.2 was the applicability of a marginal porous medium stability condition in the lower skeletal layer. This allows the prediction of the critical height ΔH_p and brine salinity increase ΔS_b that correspond to the maximum in the local Rayleigh number. These values are predicted at a distance of somewhat less than 1 cm from the interface for the convective onset (figure 12.8) and appear to raise to ≈ 1.3 cm under steady convective conditions (figure 12.10) observed. This is lower than the normally assumed skeletal layer thickness of 3 cm. However, the latter value is relatively uncertain. In the following it is assumed that

$$\Delta S_{sk} = \Delta S_b \tag{12.13}$$

limits the salinity increase in the skeletal layer, with ΔS_b being given in dependence on the growth velocity, by equation 12.9. Should the intrinsic skeletal layer, for some reasons like supersaturation or convective redistribution, have a larger thickness, one may argue that ΔS_b still represents a reasonable measure of the source salinity at the upper limit of the skeletal layer. This is due to the situation described above, that convection feeds laterally into downward convecting brine channels. This drives an upward flow through the brine layers thus carries less saline solute to the top of the skeletal layer. In dependence on the flow, the effective ΔS_{sk} advected to the level H_{sk} might be smaller than ΔS_p. If the convection is linked to the secondary convection regime discussed in the previous section, then ΔS_b might be larger, see figure 12.15. Here the basic value of ΔS_b, shown in figure 12.9 for $Ra_{pc} = 7.7$, is used. Uncertainties will be discussed below. Their magnitude, and thus the question of the detailed circulation in the skeletal layer, becomes only critical at very large growth velocities.

Onset of bridging: d_{sk}

The remaining unknown in equation 12.12 is the brine layer width at the top of the skeletal layer: d_{sk}. Its main characteristic discussed in chapter 3 is recalled: This is the brine layer width at which bridging between the plates starts. The bridging transforms them from layers into sheet-like inclusions. The skeletal layer is essentially defined by this transition, wherever it might take place. The, so far limited, observations have located it 2 to 4 cm from the interface (Stander, 1985).

Mechanisms that might explain the critical d_{sk} have bee considered in section 3.3 and it has been argued that minimisation of surface tension, as once proposed by Anderson and Weeks (1958), is no physical reliable concept in this case. The transition must be related to the principles of circulation outlined above, connected to lateral freezing on the basal planes of the plates. The upward flow through the brine layers implies constitutionally supercooled liquid brine and hence lateral freezing. This is the way by which sea ice looses its salt, and it reiterates, until the upward flow meets too large a resistance. The bridging of brine layers is thus another instability problem, where fluid flow, advection of supercooled liquid, local constitutional supercooling, and the formation of perturbations on the basal planes interact. The difficulties to make progress here may be realised by recalling some aspects from the first sections of chapter 8: (i) microscopic solvability theory and pattern formation are already difficult topics in the absence of solutal effects, which add in the present case; (ii) formulations need to account for the extreme anisotropy of the solid-liquid interfacial energy and free growth rate of ice; (iii) observations of the growth rate normal to the basal plane at low supercoolings are difficult to control and rare.

A simple solution to the morphological bridging problem is not at hand. However, it has been possible to give a quantitative estimate of d_{sk} on the basis of observations of the sea ice microstructure. The procedure is outlined in appendix F. Although microstructural data are seldom available within the skeletal layer, it turns out that the main features of inclusions, their width l_d, length l_g and aspect ϵ_h, are conserved also far away from the interface. It was then possible, by intercomparison of several statistical analyses (Perovich and Gow, 1996; Cole and Shapiro, 1998; Eicken et al., 2000), to obtain a basic result which is proposed here as follows. The skeletal transition can be characterised by a horizontal length l_{gsk}, an aspect ratio ϵ_{hsk} and a width l_{dsk} of inclusions which are given as

$$l_{gsk} \approx 0.44 \pm 0.04 \quad \text{mm}, \qquad l_{dsk} \approx 0.11 \pm 0.01 \quad \text{mm}, \qquad \epsilon_{hsk} = l_{gsk}/l_{dsk} \approx 4. \quad (12.14)$$

At the onset of bridging, to which the scalings 12.14 apply, l_{gsk} may be identified with the horizontal wavelength of dominant perturbations, while l_{dsk} essentially corresponds to the width d_{sk} of undeformed brine layers. These scales were found to be remarkably constant, when hysteresis effects of some laboratory protocols were taken into account. There is no indication, that they depend on the growth velocity of ice.

Two other justifications in support of the above length scales are mentioned. The first is an indirect approach to estimate d_{sk} from strength tests results, modified after Weeks and Assur (1964). This method is described in appendix F.4. It implies a confirmation of d_{sk} by a completely different indirect method resulting in the range $0.09 < d_{sk} < 0.13$ mm.

The second is a theoretical justification for the observed aspect ratio. Whatever happens in detail in the brine layers, it can be conjectured that they in principal are made up of two lateral boundary layers, where freezing takes place, and a fluid flow regime in the center. In this respect, they resemble the case of *viscous fingering* of miscible fluids in a Hele Shaw cell (Saffman and Taylor, 1958). The problem of wave length selection of such a configuration has been discussed by Paterson (1985). He showed in a simple analysis that, for the case of large initial perturbations, an instability wavelength of 4.0 times the cell width can be expected. This corresponds to $l_{gsk}/l_{dsk} \approx 4$ in the above notation. Paterson's work has been extended to more complex situations with throughflow, when the wavelength may differ from this value (Snyder and Tait, 1998; Fernandez et al., 2001; Graf et al., 2002). However, for the original case described by Paterson, a circular patch of viscous fluid, his own observations and later experiments (Maxworthy, 1989; Holloway and de Bruyn, 2005) reasonably confirm his first prediction. Such a radial geometry may be associated with the large scale streamlines in the skeletal layers of sea ice.

The tentative interpretation of observations and theory is now the following. The resonance wavelength for the formation of instabilities within the brine layers is approximately four times their width. If this wavelength is triggered by a freezing instability on the basal plane, then the system may be expected to respond by bridging of the brine layers. This may be used as a working hypothesis to discuss the bridging process of brine layers in future work.

12.3.1 k_{sk} at the bridging transition

The problem of the prediction of the effective solute distribution coefficient k_{sk} at the top of the skeletal layer is now closed. The further argumentation is also consequent. As bridging changes the permeability, any further desalination will be constrained by its degree. It will, however, still take place to some degree within the bridging layer. Hence, k_{sk} obtained from equation 12.12, should be an upper limit of the stable salinity.

The calculated k_{sk} is now compared to observations. The problems to observe k_{sk} at the top of the skeletal layer were discussed in chapter 5, where instead a reference dataset of k_0' at 6 cm from the interface has been compiled and critically discussed (section 5.5.6). It is recalled, that these observations refer to properly controlled growth conditions, but some may have been affected by moderate under-ice fluid flow. Details were given in figure 5.16 in chapter 5. In figure 12.17 observations are shown as symbols, except for the short dash-dotted line which resembles the field observations from Nakawo and Sinha (1981). The predictions are shown for two critical Rayleigh numbers Ra_{pc} of the convecting skeletal layer, and for the assumption $\Delta S_{sk} = \Delta S_{sk}$ for $Ra_{pc} = 7.7$, which is the standard case considered here.

It is seen that the predictions bound the observations from above. This is expected, because they refer to the skeletal transition, while the observations stem from a distance 6 cm from the interface, within the bridging layer. The influence of ΔS_{sk} on k_{sk} is only moderate at these growth velocities. Hence, the only parameter that is not theoretically founded in the prediction, is the empirically determined $d_{sk} = 0.11$ mm. The latter is, however, thought to be valid within 10 %. The overall data appearance is rather well reproduced. One would expect, that k_{sk} should be closest to k_0' at large growth velocities, because the time for desalination between H_{sk} and H_0' is shortest. This is the case. It is also reasonable that the observations from Nakawo and Sinha (1981) are relatively close to the predictions: The growth velocity was rather constant during growth, and no major warming events took place. In other experiments growth was decelerating with more time for desalination between H_{sk} and H_0'. The lowest k_0' value are the circles and squares from the laboratory study by Cox and Weeks (1975), also evaluated at 6 cm. Due to effect of thermal convection (figure 5.9) only observations above 3 cm/day are shown.

All simulations have been performed for seawater properties. The theory gives now the possibility to calculate the difference in k_{sk} with respect to NaCl solutions and for different salinities. The results are shown in figure 12.18, where all predictions have been normalised by k_{sk} for seawater at 35 ‰. It is recalled that, when only the liquidus slope was considered in figure 5.15, a slightly larger k_{sk} for NaCl at the same concentration appeared. This effect is almost completely neutralised by a slightly larger haline contraction of seawater. The predictions at 35 ‰ are almost identical. The figure also illustrates

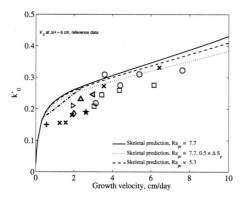

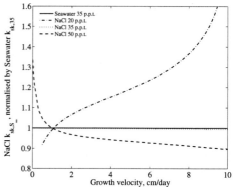

Fig. 12.17: Predicted k_{sk} at the top of the skeletal layer in comparison to the reference data compiled in section 5.5.6, see figure 5.16. Predictions are shown for two Rayleigh numbers, and for the case where only a salinity increase of $1/2\Delta S_p$ has been assumed

Fig. 12.18: Predicted k_{sk} at the top of the skeletal layer for different NaCl solutions salinities S_∞, normalised by the predicted value for seawater with 35 ‰. The solid horizontal line is the seawater reference = 1.

a complex concentration dependence. At 50 ‰, k_{sk} is smaller at large growth velocities. This explains, why the observations from Cox and Weeks (1975) fall slightly below the rest of the data in figure 12.17. At low growth velocities the situation is the opposite. While the observations have been omitted here due to unknown effects of thermal convection, a look at figure 5.15 indicates, that also this is plausible. Hence, the different appearance of the Cox and Weeks data with respect to real seawater, suspected from the comparison of observations in chapter 5, is supported by the theoretical predictions. For a lower salinity of 20 ‰, a considerably larger k_{sk} is predicted. Qualitatively this agrees with the discussion in section 5.5.6, but observations are lacking.

It can be concluded that a combination of theoretical models of microstructure and convection with a heuristic, but empirically founded bridging approach, reasonably predicts the structural brine entrapment in sea ice. The main variability derives from two factors: (i) the variability in the plate spacing with growth velocity and (ii) the limit in the salinity increase ΔS_p in an actively convecting skeletal layer. The latter effect is weak at low and moderate growth velocities. This means that a detailed knowledge of the convective circulation is not necessary to derive the salinity entrapment in sea ice growing at moderate velocities.

12.3.2 Stable k_0 of sea ice

It is, of course, desirable to obtain the delayed desalination above the skeletal layer. A simulation of this process is not expected to be straightforward, as in this regime the permeability changes by the bridging process, while at the same time brine channels evolve. However, a formulation for the lower bound of k_0' can be derived in a simple manner. The aspect that can be explored is again the Hele-Shaw cell character of the brine layers. The twodimensional situation encourages the application of a simple percolation approach to the problem.

Twodimensional percolation threshold

The basic model is recalled: Brine channels will drive an upward flow through the brine layers, as long the latter are permeable. As the brine layer throughflow is principally a two-dimensional process, with bridging instabilities considerable smaller than the grain size, it is a reasonable concept to describe the transition between H_{sk} and H_0 by continuum percolation in two dimensions. The limiting brine fraction at which a two-dimensional system attains its percolation limit is reasonably established in the range $0.66 < f_{p2} < 0.68$, when based on percolation of inter-penetrating squares or disks (Quintanilla, 2001; Baker et al., 2002; Yi and Sastry, 2004).

However, f_{p2} depends to some degree on the mode of transition or shape of inclusions. When aligned squares are considered a value of $f_{p2} \approx 0.63$, slightly lower than for overlapping spheres, is expected (Baker et al., 2002). On the other hand, f_{p2} increases when a size distribution of disks is present Quintanilla (2001), yet is limited to $v_{pc2} < 0.71$ if disk sizes are not varying by more than a factor of three. A more severe effect can be expected from anisotropy of the pattern. If continuum percolation is based on overlapping ellipses of, for example, an aspect ratio of 4, the transition changes to $f_{p2} \approx 0.50$ (Xia and Thorpe, 1988; Yi and Sastry, 2004). This, of course, can be understood in terms of easier flow through channelising structures. If the latter effect might be relevant for sea ice, depends thus on the vertical-horizontal anisotropy of inclusions within the brine layers. The latter may eventually be associated with the ratio of major inclusions lengths obtained from individual vertical and horizontal section analysis. For cold first-year ice this value appears to be close to 2, while for young ice, more representative of the skeletal transition a value between 1.1 and 1.5 seems more appropriate (Eicken et al., 2000). The only intrinsic size distribution reported so far within the plane of a brine layer indicates a maximum at an aspect of 1 to 2 (Cole and Shapiro, 1998), but was obtained at the end of the season. It may therefore reflect other redistribution processes driven by thermal cycling. Assuming, from these considerations, an anisotropy of 1.5 to 2 for the bridging instabilities, one obtains from (Xia and Thorpe, 1988) the range $0.63 < f_{p2} < 0.66$. As on the other hand, for a distribution of pattern sizes and a 10 % fractional area of 3 times larger inclusions, which approximates what is found for sea ice (Eicken et al., 2000), a 0.03 larger v_{pc2} may be expected (Quintanilla, 2001), it is suggested that

$$f_{p2} \approx 0.67 \pm 0.04 \qquad (12.15)$$

most likely represents the situation of present interest.

A lower stable k_0 bound

The lower bound for the stable k_0' is then simply obtained from

$$k_0 > f_{p2}k_{sk} \approx 0.67k_{sk} \qquad (12.16)$$

and follows from the assumption that the same fluid from the skeletal layer that is advected to the bridging transition level H_{sk} is also advected into the bridging layer between H_{sk} and H_0. Considering the intermittent flow driven by brine channels and the supercooling observed by Melnikov (1995) this is no unreasonable assumption. It implies a further decrease in the ice salinity by ≈ 30 %, and is in principle consistent with observations discussed in chapter 5. No detailed comparison can be made, because observations of the intrinsic skeletal value k_{sk} do not exist.

The observations from figure 12.17 are shown with the predictions for $Ra_{pc} = 7.7$ in figure 12.19 for an extended range in growth velocities. Due to the salinity dependence demonstrated in figure 12.18, data from Cox and Weeks are only shown for $S_\infty < 40$ ‰. Notably, the limit for brine loss from ice is seen at a growth velocity of ≈ 18 cm/day, which is seldom realised in nature. More important is, that now k_{sk} is interpretable as an upper bound of the stable ice salinity, while $k_0 > 0.67 \times k_{sk}$ is the lower bound. According to the bridging layer concept, based on the general appearance of salinity profiles, this level is ≈ 10 cm from the interface. As the observations are evaluated at 6 cm, they should fall mid into the bridging layer, between the bounds. This is indeed the case.

The agreement between theory and observations is very reasonable. The *percolation transition* level H_0 seems to be the limit for gravity-driven salt-fluxes from growing sea ice. The stable salinity of sea ice during the growth phase appears to be regulated by the structural entrapment between plates and a twodimensional percolation concept. It is, of course, possible that more salt can be excluded by warming events which may be expected to restructure the brine layers, or enhance brine channels convection directly. Most observations, that appear in figure 12.19 near the lower bound on k_0, are from field conditions with intermediate warming events. Such effects need to be discussed in more detail and demand for model studies. The present percolation approach is simple and a reasonable basis to further studies on pattern formation and permeability of the bridging layer.

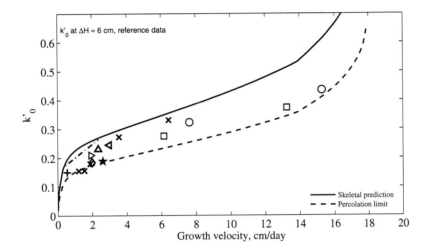

Fig. 12.19: Bounds on sea ice solute distribution. Predicted bridging transition k_{sk} (solid line) and the lower limit of the stable k_0 (dashed line). Proposed reference data (from figure 5.16) of k'_0 at ≈ 6 cm interface distance for growing sea ice. The short dash-dotted represent the study from Nakawo and Sinha (1981). Crosses: present work. INTERICE data are from Eicken (2003) (diamond) Cottier et al. (1999) (pentagram) and Tison et al. (2002) (triangles). Plus sign: Field data from Antarctic (Wakatsuchi, 1974b). The NaCl data from Cox and Weeks (1975), circles and squares, have been restricted to $V < 3$ cm/s and $S_\infty < 40$ ‰.

Appendix

A. Thermodynamic properties of ice and brine

Water and supercooled water

During natural lake and sea ice growth, mostly ice temperatures between 0 and −30 °C are accessed(Malmgren, 1927; Zubov, 1943; Doronin and Kheisin, 1975). This appendix summarises the present knowledge on the thermal properties of ice and NaCl brine in this temperature range and makes a simple extension to seawater brine. Some of the following dependencies have so far only been treated approximately in sea ice models.

Brine properties below the melting point of pure water depend to a large degree on the thermodynamic properties of supercooled water. The first extensive critical summary of the properties of water (Dorsey, 1940) referred mainly to temperatures above 0 °C. Since then thermodynamic models of increasing complexity were formulated to explain the properties of water (Franks, 1982). A general theory of the thermodynamic properties of pure water (Hill, 1990) has become the basis for international standards of heat capacity and density of water above the freezing point (IAPWS, 1996). An extension into the supercooled regime requires a more fundamental thermodynamic theory of water and its anomalies, with particular emphasis on the system's behavior near a critical temperature of ≈ −46 °C (Angell, 1983; Speedy, 1987; Sciortino et al., 1997; Kiselev and Ely, 2003). Supercooled water may exist down to the homogeneous nucleation temperature of ≈ −40 °C (Turnbull, 1950; Fletcher, 1970; Rasmussen and MacKenzie, 1972; Koop, 2004), and its thermodynamic properties have now been measured with high accuracy down to −35 °C (Speedy, 1987). Extrapolation below this temperature must involve thermodynamic modeling. Recent models of supercooled water have been discussed with respect to ice nucleation theory (Koop, 2004) and thermodynamics of electrolytes (Archer and Carter, 2000), indicating several uncertainties and open questions. More recent molecular dynamics simulations of water, e.g., Nada et al. (2004), are promising tools towards a fundamental understanding but still need to be developed.

Brine thermodynamics

A consistent thermodynamical theory for thermodynamical properties of water in the presence of solutes has been established by Pitzer et al. (1984). Clarke and Glew (1985) have provided an analysis of Pitzer's approach in the temperature regime near 0 °C and tabulated the corresponding equilibrium freezing temperature of NaCl brine. Archer

(1992) further included the pressure dependence up to 100 MPa. Model refinements have been proposed by Akinfiev et al. (2000) and Archer and Carter (2000), with emphasis on low temperatures and supercooled solutions, where observations are lacking. Thermodynamic model formulations have been proposed for aqueous NaCl in agreement with thermodynamic property data down to -20 °C (Archer, 1992; Archer and Carter, 2000; Akinfiev et al., 2000). Here, for the sake of simplicity, empirical formulas of the temperature and salinity dependence are derived from the available data. Some property dependencies can be linearised for the limited temperature range of interest, some can be neglected, and others are based on little experimental data.

Seawater

Also for seawater brine, the derivation of thermodynamic properties from Gibbs thermodynamic potentials (Millero and Leung, 1976; Feistel and Hagen, 1998) is more consistent then the set of parameterisations proposed in the following appendix. However, the required proper extensions to low temperatures are not available yet. Semi-empirical treatments of seawater as a multicomponent solution (Millero, 1974; Spencer et al., 1990; Tischenko, 1994) are promising but have not been validated in detail. Most seawater property studies have been restricted to the natural salinity range 0 to 40 ‰ and temperatures above 0 °C (Millero and Leung, 1976; Millero and Sohn, 1992). Below -10 °C, where the thermodynamics of seawater brine start to deviate increasingly from an NaCl solution, validation data are almost completely lacking. In the following, the differences of some seawater property data, compared to aqueous NaCl solutions, are only approximated in an ad-hoc sense. Seawater is treated as a three-component system of 24/33 parts NaCl, 5/33 parts $MgCl_2$ and 4/33 parts Na_2SO_4, and it is averaged over the different property dependencies or, more accurately, their anomalies with respect to pure water. In most cases such a test will be found to be within the error bounds of the available data, but validation at low temperature often has not been possible.

Pressure dependence

The present appendix gives thermal properties at 1 bar pressure, relevant for the freezing of floating ice at the ocean surface. When submarine ice shelves or thick polar ice sheets are considered, pressure effects on the thermodynamic properties may increase in relative importance. Thermodynamic property studies of aqueous NaCl (Pitzer et al., 1984; Archer, 1992) and of seawater (Millero and Leung, 1976; Fofonoff and Millard, 1983) cover the pressure and temperature ranges down to 100 bar and 0 °C, with an extension for seawater brine to -2 °C (Feistel and Hagen, 1998). Future predictions of the freezing point depression due to salt and pressure must consider the further development of the theory of supercooled water and homogeneous nucleation(Angell, 1983; Speedy, 1987; Koop, 2004).

A.1 Properties of ice

The most probable values of the thermal properties of pure ice have only slightly changed in standard books and references during the past century (Barnes, 1928; Dorsey, 1940; Shumskii, 1955; Pounder, 1964; Fletcher, 1970; Hobbs, 1974; Petrenko and Whitworth, 1999). In the following the dependencies in the range 0 to −30 °C are summarised.

A.1.1 Isobaric specific heat

References on the isobaric specific heat of ice are discussed in standard books (Dorsey, 1940; Hobbs, 1974; Petrenko and Whitworth, 1999). Suggested empirical relations differ slightly (Pounder, 1964; Speedy, 1987; Fukusako, 1990; Yen et al., 1991). In contrast to the specific heat of liquid water no international standard of the specific heat c_{pi} of ice at 0 deg exists. The temperature dependence

$$c_{pi} = 2.10\,(1 + 0.0034\,T) \tag{A.1}$$

represents the data from Giauque and Stout (1936) from 0 to −50 °C and most other references noted above within 0.5 %. c_{pi} has unit $kJK^{-1}kg^{-1}$ and T is given in °C.

A.1.2 Density

Hobbs (1974) summarised measurements of the ice density at 0 °C between 916.4 and 916.7 kg m^{-3}. Barnes (1928) had concluded, in his earlier review, that the most accurate measurements suggest a value between 916.7 and 916.8 kg m^{-3}. The weak temperature dependence down to -60 °C may be approximated as

$$\rho_i = 916.7\,(1 - 0.00014\,T) \tag{A.2}$$

with ρ_i in kg m^{-3} and T in °C. The linear normalised temperature dependence suggested by other authors based on literature sources is between -0.000153 (Fukusako, 1990) and -0.000117 K^{-1} and (Pounder, 1964), indicating an uncertainty of 0.03-0.05 % at -20 °C. Petrenko and Whitworth (1999) propose a theoretical value of 919.7 kg m^{-3} at a temperature of -20 °C, which is 0.05 % larger than 919.27 kg m^{-3} from equation (A.2). Down to -30 °C the relative change in ice density is almost negligible compared to the change in other thermal properties.

A.1.3 Thermal conductivity

Above a temperature of about 70 K, the thermal conductivity K_i of pure ice can be expected to be inversely proportional to the absolute temperature (Fletcher, 1970; Hobbs, 1974; Petrenko and Whitworth, 1999). For temperatures down to 230 K a linear form

$K_i(T) \sim (1 - cT)$ with T in °C is justified, and has been proposed by many authors (Shumskii, 1955; Pounder, 1964; Hobbs, 1974). Slack (1980) critically discussed recent data sources, comparing them to Powell's (1958) summary of older data, and showing that the temperature dependence near the freezing point is slightly stronger than an inverse absolute temperature relationship. He suggested a best estimate of $K_i \approx 2.14 \, Wm^{-1}K^{-1}$ at 0 °C. The values tabulated by Slack may be linearised down to -40 °C in the form

$$K_i = 2.14 \, (1 - 0.006 \, T), \tag{A.3}$$

with K_i in $Wm^{-1}K^{-1}$. Down to -40 °C a very similar temperature dependence of K_i is present in the high accuracy data from Andersson and Suga (1994).

Equation (A.3) should be interpreted as an average K_i for random c-axis orientation. Barnes (1928) and Shumskii (1955) both summarized studies on the anisotropy, K_i normal to the c-axis being $\approx 5\%$ larger than parallel to it. In his review Powell (1958) arrived at a similar conclusion based on more recent measurements from Landauer and Plumb cited therein. Petrenko and Whitworth (1999) hold that there is no properly established anisotropy.

A.1.4 Thermal diffusivity

The thermal diffusivity κ_i of ice can be computed via

$$\kappa_i = \frac{K_i}{c_{pi}\rho_i} \tag{A.4}$$

in units m^2s^{-1} from equations (A.1), (A.2) and (A.3). At 0 °C its value is $\kappa_i \approx 1.114 \times 10^{-6}$ m^2s^{-1}.

A.1.5 Latent heat of fusion

Isothermal L_f of pure water

Accurate measurements of the isothermal latent heat of fusion have been available since more than a century (Barnes, 1928; Dorsey, 1940) and $L_{f0} \approx 333.5 \, J \, kg^{-1}$ is the presently accepted value at 0 °C (Hobbs, 1974; Petrenko and Whitworth, 1999). Integrating the specific heat of supercooled water (equation A.29) relative to the specific heat of ice (equation A.1), gives the change in L_f with temperature (Speedy, 1987). Down to -30 °C, it can be expressed in the form

$$L_f = 333.5 \, (1 + 0.00598 \, T + 0.0000422 \, T^2) \tag{A.5}$$

in units kJ kg^{-1}, with T in °C. This equation gives L_f with an accuracy of better than 0.15%. The slight nonlinearity of the slope seen in Figure A.1 resembles mainly the nonlinear change in the specific heat of supercooled water. Hansen et al. (1997) devised an indirect determination of L_f, combining NMR and calorimetry. They obtained a 3.5 % larger value at -20 °C, implying a weaker temperature dependence, while compiled literature data shown by them agree closely with equation (A.5).

Isothermal $L_{f,NaCl}$ of aqueous NaCl solutions

Anderson (1966) pointed out that the latent heat of fusion of a NaCl solution or seawater must be different from that of pure supercooled water. The difference is related to the fact that the partial molal relative enthalpy of water in solution, L_1 in units of J/mol H_2O, takes positive values at high concentrations. L_1 is the difference between the enthalpy evolved per mole of pure water added to a solution without changing its solute concentration, and the molal heat content of pure water. As L_1 is a relative measure compared to the pure solution, this implies a larger isothermal latent heat of fusion $L_{f,NaCl}$

$$L_{f,NaCl} = L_f + 0.05551\, L_1 \tag{A.6}$$

given in kJ kg^{-1}. Anderson (1966) and Cox and Weeks (1975) calculated, for an NaCL solution at -20 °C and a salinity 220 ‰, a barely 1.5% larger $L_{f,NaCl}$ compared to supercooled pure water, yet they used enthalpy data obtained at 25 °C by Randall and Bisson (1920). Estimates of L_1 obtained by Rümelin (1907) at a lower temperature of 14 °C are almost a factor of 2 larger. Today it is well established that the partial molal enthalpy of water strongly depends on temperature. Clarke and Glew (1985) and Millero and Leung (1976) tabulated L_1 for NaCl and seawater down to 0 °C.

In the present study L_1 has been evaluated below 0 °C in two ways. First, the tabulated values of L_1 above 0 °C from Clarke and Glew (1985) were extrapolated by polynomial fits to the liquidus relation, up to the eutectic point near 235 ‰. Second, L_1 was obtained from the tabulated activity coefficients a_w of water, estimated by Thurmond and Brass (1987) from heat capacity measurements down to -40 °C. This was done from the classical thermodynamic relation

$$L_1 = R\,\theta^2 \frac{\partial(\ln a_w)}{\partial \theta}, \tag{A.7}$$

where θ is the absolute temperature, R the molar gas constant, and L_1 is given in J/mol H_2O. These values can be fitted by the polynom

$$0.05551\, L_1 = -0.111\, T + 0.2746\, T^2 + 0.0173\, T^3 + 4.82 \times 10^{-4}\, T^4 + 7.52 \times 10^{-6}\, T^5. \tag{A.8}$$

Equation (A.8) is shown in Figure A.2 in comparison to the graph obtained when extrapolating L_1 from Clarke and Glew (1985). The increasing deviation of the extrapolated values from the equation A.8 below -2 °C indicates the uncertainties when extrapolating thermodynamic properties into the anomalous regime of supercooled water. As the solution heat capacities from Thurmond and Brass (1987) and Archer and Carter (2000) agree, below 0 °C, within 2-3 %, only minor corrections to equation (A.8) can be expected. Combining the latter with equation A.5, the following approximation of the isothermal latent heat of fusion of NaCl solutions at the liquidus temperature is suggested:

$$L_{f,NaCl} = 333.5 \left(1 + 0.00521\, T_f + 0.000525\, T_f^2 + 0.0000202\, T_f^3\right). \qquad \text{(A.9)}$$

It shown in Figure A.1. $L_{f,NaCl}$ may differ by more than 10 % from L_f of pure supercooled water. In problems of phase transitions and interface stability this difference is important. Equation A.9 holds down to the eutectic temperature of -21.3 °C. An extension to lower temperatures based on heat capacity data from Archer and Carter (2000) would have to consider eutectic precipitation.

$L_{f,sw}$ of seawater

There appears to be a lack in indirect and direct experiments of $L_{f,sw}$ for cold, high salinity seawater brine. Values of L_1 have been given by Millero and Leung (1976) for seawater, based on observations of the specific heat of seawater for the salinity range 0 to 40 ‰ and temperatures above 5 °C. These are approximately 2/3 of the values of aqueous NaCl, given by Clarke and Glew (1985). The lower seawater L_1 is consistent with a different behaviour of other solutes, as Calcium and Magnesium have negative L_1 in the above sense (Rümelin, 1907; Tischenko, 1994). Considering natural seawater at its freezing point of $\approx$ -1.9 °C, one may first obtain, from equation A.9, a value of $L_{f,NaCl} \approx 330.8$ kJ kg^{-1}. This compares to 329.7 kJ kg^{-1} for pure water. Assuming a 2/3 smaller L_1 the expected value for seawater is $L_{f,sw} \approx 330.4$ kJ kg^{-1}. For cold seawater brine the ad-hoc correction may be less reliable, in particular the precipitation onset of $Na_2SO_410H_2O$ near -6.3 to -8°C, see section A.4.

Effective L_{eff} in a closed system

In geophysical problems, like the freezing of sea ice, the most important variable to be known is often the overall energy change of the ice-brine system during cooling and internal freezing. The local phase transitions within the ice, related to the local isothermal heat of fusion, are less important. The bulk internal energy of fusion also includes the internal energy changes due to the increase in solute concentration and therefore differs from the isothermal latent heat of fusion. Combining equations A.5 and A.1 for L_f and specific heat of pure ice with equations A.30 and A.2.3 for the specific heat and relative volume of brine at the liquidus (section A.2.2), one may obtain an effective latent heat L_{eff}. The latter needs to be defined with respect to a reference temperature,

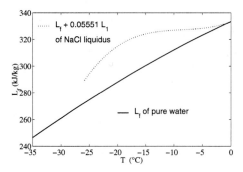

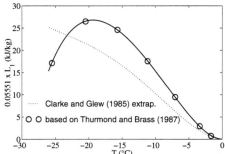

Fig. A.1: Isothermal latent heat of fusion of pure water (solid line) and of NaCl solutions at the liquidus temperature (dotted line)

Fig. A.2: Relative (with respect to pure water) partial molal enthalpy of water in NaCl solutions at the liquidus temperature. Solid line is based on activity coefficients from Thurmond and Brass (1987), dotted line extrapolated from tabulated values above 0 °C (Clarke and Glew, 1985).

for example, the seawater freezing point being relevant in many applications. Simplified parameterisations of L_{eff} are desirable in sea ice model studies.

Due to the relatively small specific heats of ice and water the reduction of L_{eff} for sea ice takes place in almost linear proportion to the brine fraction (Buchanan, 1887; Pettersson, 1883). A simple bulk formula, which accounts for this dependence has been first suggested by Malmgren (1927). It takes the simple form

$$L_{eff} = L_{ref}\left(1 - b_1 T_i + b_2 \frac{S_i}{T_i}\right), \tag{A.10}$$

where b_1 and b_2 are positive constants, and S_i and T_i are salinity in ‰ and Celsius temperature of the ice-brine mixture. Malmgren's approach has been adopted by other authors who proposed somewhat different constants b_1 and b_2 (Schwerdtfeger, 1963; Ono, 1968; Maykut and Untersteiner, 1971). In these studies the thermal properties of brine and ice, in particular the seawater liquidus and relative brine volume, were only roughly approximated. For practical purposes the form (A.10) has been analysed based on the thermal property relations from the present appendix and assuming $L_{ref} \approx 330$ kJkg^{-1} for seawater. For natural ranges $-3 < T_i < -30$ °C and $2 < S_i < 10$ ‰ the set $b_1 = 0.005$ K^{-1} and $b_2 = 0.06$ K‰$^{-1}$ gives L_{eff} within 1 % of its exact prediction. Moreover does this set imply errors of less than 0.5 %, when integrating the bulk $\overline{L_{eff}}$ of a winter ice slab with a linear temperature gradient.

A.2   Properties of NaCl brine and supercooled water

In the following the salinity and temperature dependencies of thermal properties are first discussed and approximated, before a liquidus relationship of each property is derived for NaCl brine and, if available, for seawater.

A.2.1   Mass fraction units

In the present work S is used for the weight fraction of salt in one kilogram solution in permille ‰ or g per kg. When thermodynamics and chemistry are involved often the number of moles of solute is a more suitable unit. The conversion from ‰ to *molality* M, the number of moles NaCl per kg solvent (here: water) is

$$M = \frac{S}{M_0(1 - 0.001\,S)},\tag{A.11}$$

with the molecular weight of $M_0 = 58.443$ grams per mole NaCl. In some applications of electrolyte theory (Falkenhagen, 1953; Robinson and Stokes, 1955), the *molarity* or concentration M_r (moles NaCl per liter solution) is of interest, which is given as $M_r = S\rho_b/M_0$, and hence requires the knowledge of the solution density ρ_b.

Seawater versus NaCl

The dominating salt in seawater is NaCl, yet it contains smaller fractions of many other ions and has a larger average molecular weight of $M_{sw} = 62.793$ g/mole, i. e., Millero and Leung (1976). The ionic strength I, which is relevant in the computation of some thermodynamic properties, equals M for NaCl solutions, while for standard seawater I=1.245 M (Millero and Leung, 1976). For practical reasons, the conventional salinity S_{psu} used by the oceanographic community (Millero and Leung, 1976; Fofonoff and Millard, 1983; Millero and Sohn, 1992) has been defined via the chlorinity and is given in 'practical salinity units' (psu), with the conversion $S(\text{g/kg}) = 1.004880\,S_{psu}(\text{psu})$. In most sea ice applications, like in the present work, the documented S is in g/kg.

A.2.2   Phase diagram of NaCl

The triple point of pure water at zero pressure is 273.16 K or 0.01 °C. The saturation of liquid water with respect to air and the atmospheric pressure of 1 bar lower the melting point by -0.0024 and -0.0075 K, respectively, giving a freezing point of pure surface water of ≈ 273.15 K (Dorsey, 1940). The presence of solutes lowers the freezing point of the solvent water to first order by $\Delta T_f \approx -1.86\,N_x\,M$, where -1.86 is the molar freezing-point depression or cryoscopic constant of the solvent water, M the molality of the salt in question and N_x the number of ions of the completely dissociated salt (Dorsey, 1940;

Sverdrup et al., 1946; Millero and Leung, 1976). In riverine water, with a typical salt content of 0.1 g/kg (Millero and Sohn, 1992), the freezing point depression is $\approx$ 0.004 to - 0.007 K, depending on the dominating salt. The thermodynamics of an exact equilibrium relationship $T_f(S)$ is more complex (Millero and Leung, 1976; Pitzer et al., 1984).

A critical evaluation of published experimental data on the liquidus-relationship of NaCl brine has been performed by Cohen-Adad and Lorimer (1991). For the range $17 < S < 235$ ‰ their reference values can be expressed by the simple polynom

$$T_f = -0.05558 \, S(1 + 1.12485 \times 10^{-3} \, S + 8.2982 \times 10^{-6} \, S^2 - 6.6644 \times 10^{-9} \, S^3), \quad (A.12)$$

with deviations of less than 0.004 K from the proposed form of Cohen-Adad and Lorimer (1991). At salinities below 50 ‰ the reference values from Cohen-Adad and Lorimer (1991) are better fitted by a polynom in $S^{1/2}$

$$T_f = -0.06473 \, S(1 - 0.04928 \, S^{1/2} + 0.006566 \, S - 0.00011464 \, S^{3/2}) \quad (A.13)$$

An alternative up-to-date source are the tabulated freezing point data published by Zaytsev and Aseyev (1992). These can be expressed in a similar form

$$T_f = -0.05818 \, S(1 + 6.5067 \times 10^{-4} \, S + 5.6015 \times 10^{-6} \, S^2 - 9.2265 \times 10^{-9} \, S^3) \quad (A.14)$$

for the tabulated values above 18 ‰, and as

$$T_f = -0.06286 \, S(1 - 0.03223 \, S^{1/2} + 0.005009 \, S - 0.0001549 \, S^{3/2}) \quad (A.15)$$

for the tabulated data between 1 and 50 ‰.

The equations (A.12) and (A.14) are, in connection with data above 50 ‰, shown in Figure A.3. Near 180 ‰ the difference between equations (A.12) and (A.14) reaches 0.4 K. A further verification has shown that the dataset from Oakes et al. (1990), also plotted in Figure A.3, thermodynamically founded derivations (Clarke and Glew, 1985; Akinfiev et al., 2000) and other tabulations (Kaufmann, 1960) agree more closely with the values given by Zaytsev and Aseyev (1992). Therefore equation (A.14) is suggested to be preferable over (A.12).

At salinities below 50 ‰, equations (A.13) and (A.15) differ only by 0.02 and 0.007 K, from the respective more approximate equations (A.12) and (A.14), but the polynomial expressions in $S^{1/2}$ are physically more consistent (Millero and Leung, 1976), and they correctly produces a minimum in the absolute value of the liquidus slope $m = \partial T_f / \partial S$. In Figure A.4 the liquidus slope m is shown. In this regime equations (A.13) and (A.15) differ by 2-3% in the freezing point depression. The minimum in m is situated near 10.6 and 8.4 ‰, respectively, in comparison to 8.9 ‰ for seawater. The theoretical liquidus slope at infinite dilution can be approximated by $m_0 \approx -1.86 \, N_{NaCl}/M_{NaCl} \approx -0.0637$, where $M_{NaCl} = 58.443$ g/mol and $N = 2$ is the number of ions of the completely dissociated

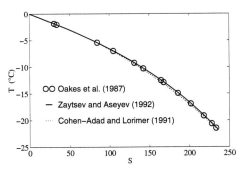
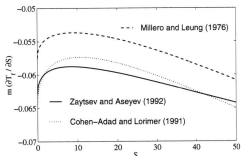

Fig. A.3: Phase relationship of aqueous NaCl, fitted to data above 18 ‰ tabulated by Cohen-Adad and Lorimer (1991), dotted line, and Zaytsev and Aseyev (1992), solid line. The circles are data points obtained by Oakes et al. (1990).

Fig. A.4: Liquidus slope m of aqueous NaCl and seawater at low salinity. Solid line is from (A.15) based on Zaytsev and Aseyev (1992), dotted line (A.13) based on Cohen-Adad and Lorimer (1991). The dashed line for sea water is from Millero and Leung (1976).

salt. The polynomial fits of both datasets yields consistent relations with values of $m = -0.0647$ and $m = -0.0629$ at infinite dilution.

In the present study (A.14) and (A.15) are applied as a general liquidus relationship above and below S=28.6 ‰, where the equations match. Their liquidus slopes match at 37.8 ‰. This has, for some calculations, to be taken into account to avoid a discontinuity in m. An estimate of the uncertainty in the freezing temperature from these equations has been obtained by comparison with the recent observations from Oakes et al. (1990) and other data cited therein. It is 0.02 K above a temperature of -10 °C, increasing to 0.1 K near -20 °C. The numbers and uncertainties can be compared to some historical work. Already Blagden (1788) obtained freezing point depressions of aqueous NaCl which were, for temperatures above -12 °C, always 0.4 K and typically 0.1 K within the modern predictions. The liquidus obtained by Ruedorff (1861), whose thermometer readings were limited to an accuracy of 0.1 K, agree with the present-day estimate always within 0.2 K and mostly within 0.1 K.

A.2.3   Relative volume of liquid brine in NaCl and sea ice

From the liquidus curves of both aqueous NaCl or seawater, the respective brine volumes can be computed as a function of the liquidus temperature from the equation

$$v_b = \left(1 + \left(\frac{S_b}{S_i} - 1\right)\frac{\rho_b}{\rho_i}\right)^{-1}, \tag{A.16}$$

where S_b and ρ_b are the liquidus brine salinity and density from paragraph A.2.2 and A.2.5, ρ_i is the density of pure ice, and S_i the bulk salinity of the ice brine mixture. Simple correlations of v_b with bulk ice salinity and temperature may only be found for limited regimes of S_b and S_i.

The computation of brine volumes for seawater differs from the NaCl system due to the slightly different liquidus relationship, density and the precipitation of salts. Following Cox and Weeks (1983) the main modification is to reduce the brine volume by the factor $1 + r_{pr}$, where r_{pr} is the precipitated salt fraction given by equations A.58 to A.61. For the density of seawater brine equations (A.27) or (A.28), which accounts for the small effect of precipitating mirabilite on the brine density, may be used. The precipitation implies an increase in the brine volume at -20 °C by about 0.4 %, compared to a decrease by ≈ 10 % due to the extraction of solid mirabilite from the brine.

A.2.4 Temperature of maximum density

Fresh water has a density maximum near 3.98 °C. With increasing salinity the density maximum shifts to lower temperatures. For seawater the shift is ≈ -0.215 K/‰ and approximately linear down to 40 ‰. Near 24.7‰ and -1.33°C the density maximum drops below the freezing point (Dorsey, 1940; Sverdrup et al., 1946; Dietrich et al., 1975).

The temperature of maximum density of NaCl solutions down to 80‰ can be expressed in the form (ICT, 1929; Kaufmann, 1960)

$$T_{\rho max} = 3.98 - 0.01482\, S^{1/2} - 0.1944\, S - 0.006961\, S^{3/2}, \qquad (A.17)$$

and is shown in Figure A.5 as a solid line. It crosses the liquidus curve near 22.8 ‰ and -1.31 °C, slightly different from seawater. The dashed line in Figure A.5 is the result of the approximate variation in β_{eff} for NaCl solutions given below (equation (A.21) and deviates slightly from the observations, presumably because no attempt has been made here to include a model of thermal expansion in the approximation of β_{eff}. The difference indicates the approximate character of equation (A.21 for the density, yet indicates that, down to its validation temperature of -20°C, the simple approach below implies a reasonable thermal expansion. For seawater brine, observations above 40 ‰ are lacking. Like for other thermal properties, the behaviour of the temperature of maximum density should closely follow NaCl brine.

A.2.5 Density

Pure supercooled water

Below its maximum at 3.98 °C the density of pure water decreases first very smoothly. Dorsey (1940) has summarised measurements down to -9 °C. More recent measurements

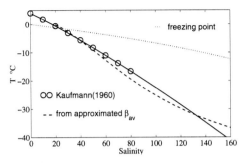

Fig. A.5: Temperature of maximum density of aqueous NaCl. The solid line is a fit to the data tabulated by Kaufmann (1960), the dashed line is implied by the approximate equation (A.21); the dotted line is the freezing temperature. The extrapolation below -20 °C is uncertain.

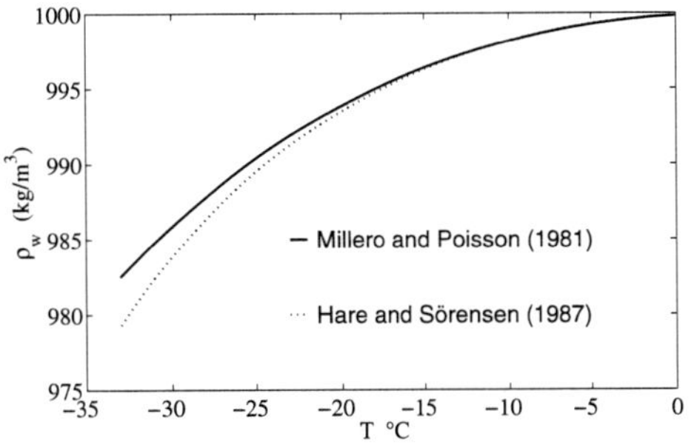

Fig. A.6: Density of supercooled water. Solid line is equation (A.19) from Millero and Poisson (1981). The dotted line represents the fit given by Hare and Sörensen (1987) to their data down to -33°C.

of the density of water supercooled down to -34 °C have been compared by Speedy (1987). Speedy's recommendation is based on the data from Hare and Sörensen (1987) and the fit given by the latter authors

$$
\begin{aligned}
\rho_w = {} & 999.86 + 6.690 \times 10^{-2}\, T - 8.486 \times 10^{-3}\, T^2 + 1.518 \times 10^{-4}\, T^3 \\
& -6.9884 \times 10^{-6}\, T^4 - 3.6449 \times 10^{-7}\, T^5 - 7.497 \times 10^{-9}\, T^6,
\end{aligned}
\tag{A.18}
$$

gives the density in k gm^{-3} and is shown in Figure A.6. For comparison, the equation of state of pure water at one atmosphere proposed by Millero and Poisson (1981) is

$$
\begin{aligned}
\rho_w = {} & 999.8426 + 6.7940 \times 10^{-2}\, T - 9.0953 \times 10^{-3}\, T^2 + 1.0017 \times 10^{-4}\, T^3 \\
& -1.1200 \times 10^{-6}\, T^4 + 6.5363 \times 10^{-9}\, T^5,
\end{aligned}
\tag{A.19}
$$

where the last two digits of each term of the original formula have been rounded, retaining predictions within 1 ppm. Equation (A.19) is also shown in Figure A.6. Although derived for the temperature range 0 to 40 °C, equation (A.19) performs well also for supercooled water down to -20 °C. At that temperature it gives 993.844 kgm^{-3} compared to 993.48 from equation A.18, while at -30°C the agreement is somewhat worse, the numerical values being 985.848 and 983.84, respectively.

NaCl brine

The density ρ_b of NaCl brine may be described in the form

$$
\rho_b = \rho_w + \beta_{eff}\, S,
\tag{A.20}
$$

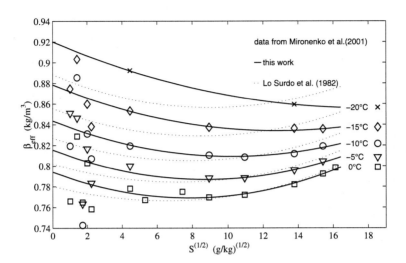

Fig. A.7: The effective haline density slope β_{eff} (kg m^{-3}‰$^{-1}$) of aqueous NaCl with respect to supercooled water. The observations (exact temperatures were -0.15, -5.15, -10.15,-15.15 and -20.15 °C) are from Mironenko et al. (2001), the solid lines from equation (A.21) at the corresponding temperatures. For comparison curves extrapolated from the equations given by Surdo et al. (1982) for the temperature range 0 to 50 °C are shown (dotted lines).

by defining an average haline density slope β_{eff} (kg m^{-3}‰$^{-1}$) based on the difference to the density ρ_w of supercooled pure water at the same temperature, divided by S. Recently, Mironenko et al. (2001) have published experimental results down to -20 °C and molalities between 0.0086 and 6 mol kg^{-1}. An approximative form to represent these measurements is

$$
\begin{aligned}
\rho_b &= \rho_w + (B_S + b_S\, B_T)\, S \\
B_S &= 0.7937 - 6.365 \times 10^{-3}\, S^{1/2} + 4.053 \times 10^{-4}\, S \\
b_S &= 0.1253 - 8.050 \times 10^{-4}\, S^{1/2} + 2.025 \times 10^{-4}\, S \\
B_T &= -2.841 \times 10^{-2}\, T + 1.074 \times 10^{-3}\, T^2.
\end{aligned}
\qquad \text{(A.21)}
$$

The coefficients were obtained as follows. First, at 0 °C a second-order polynom in $S^{1/2}$ was fitted to the observations of β_{eff}. Because at -20 °C only two data points were available these were supplemented by predictions from Figure 7 of Archer and Carter (2000), who simulated the data in questions by a thermodynamic model approach. Then a simple polynomial temperature dependence was estimated by using salinities above

$S = 20$, interpolating between the curves of β_{eff} for 0 and -20 °C. The relation is compared with the observations in Figure (A.7). Predictions which are obtained when applying the equations given by Surdo et al. (1982) for temperatures above 0 °C are also shown. The latter only underestimate β_{eff} slightly at 0 °C, yet an extrapolation to low temperatures may lead to errors of about 3 % in β_{eff}. Above 30 ‰ the equation (A.21) deviates by less than 0.5 % in β_{av} from the observations. At low salinities, the data show larger scatter due to limits in the accuracy of density measurements (Mironenko et al., 2001). Here the predictions of β_{eff} are likely correct within 1.5 %. When higher accuracies and other thermodynamic properties are needed, a more complex equation of state has to be used (Archer, 1992; Archer and Carter, 2000). For example, the depression of the temperature of maximum density with salinity, is only predicted by equation (A.21) within 10-15 %.

Density at the liquidus

Along the liquidus curve two properties, based on the density relationship, have been fitted. β_{eff} is distinguished from the non-dimensional haline contraction coefficient $\beta = \partial \rho_b / \partial S \rho_b^{-1}$, obtained by differentiating equation (A.21) with respect to salinity. Both β_{eff} and β were then determined in terms of liquidus T_f as polynomials in $(-T_f)^{1/2}$:

$$\beta_{eff,liq} = 0.7937 - 0.02722 \, (-T_f)^{1/2} + 0.01173 \, (-T_f)$$
$$- 0.0003487 \, (-T_f)^{3/2} - 0.0003487 \, (-T_f)^2 \tag{A.22}$$

$$\beta_{liq} = 0.7937 - 0.04054 \, (-T_f)^{1/2} + 0.008738 \, (-T_f)$$
$$- 0.0001550 \, (-T_f)^{3/2} - 0.0004609 \, (-T_f)^2. \tag{A.23}$$

β_{eff} has the unit kg m^{-3} ‰$^{-1}$, while β is nondimensional. The brine density ρ_b has been fitted along the liquidus as a polynomial in T_f

$$\rho_{liq} = 999.843 - 13.5249 \, T_f - 0.26213 \, T_f^2 - 0.0025719 \, T_f^3 \tag{A.24}$$

An alternative form based on the equilibrium brine salinity is

$$\rho_{liq} = 999.843 + 0.7640 \, S_b + 1.43 \times 10^{-4} \, S_b^2 + 6.82 \times 10^{-7} \, S_b^3. \tag{A.25}$$

The latter equation implies β_{eff} with an accuracy better than 0.1 % above $S_b = 40$, but underestimates it by about 3% at infinite dilution. A linearised form may be of practical interest for the salinity range $40 < S_b < 200$. It is

$$\rho_{liq} = 999.843(1 + 0.802 \times 10^{-3} \, S_b) \tag{A.26}$$

and over-(under) estimates the density at low (high) salinities by 4 (2) %. A less accurate, frequently used approximation to the density of NaCl brine in sea ice literature is $\rho_{b,liq} \approx$

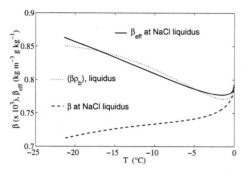

Fig. A.8: Average haline density slope β_{eff} (solid line), local density slope $(\rho_b\beta)$ (dotted line) and haline contraction coefficient β (dashed line) along the NaCl liquidus curve.

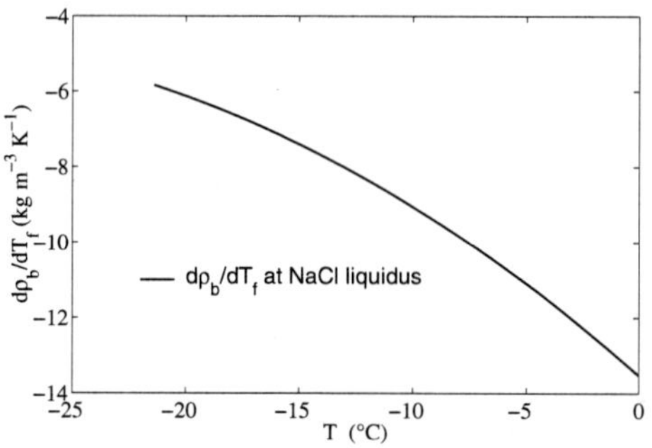

Fig. A.9: Density gradient $d\rho_b/dT_f$ along the liquidus of NaCl.

$1000 + 0.8\,Sb$ (Zubov, 1943; Assur, 1958; Ono, 1968; Richardson, 1976; Cox and Weeks, 1988).

From A.8 it is seen that both β_{eff} and β change considerably by 10 to 15 % between infinite dilution and the eutectic concentration, yet in the opposite direction. Due to the quasilinear density increase the local density slope $(\beta\rho_b)$ is similar to the bulk increase β_{eff} (also in kg m^{-3}‰$^{-1}$). For comparison, Maykut and Light (1995) have reported density observations of NaCl brine in the temperature range -4 to -18 °C. As indicated by the scatter in their figure, these determinations are probably less accurate than those from Mironenko et al. (2001). Comparing the second-order polynomial parametrisation proposed by Maykut and Light (1995) to equation (A.24), one finds that at -5 °C their predictions of β_{eff} are 5 % above, while at -20 °C they are 4 % below the present estimates. Between -10 and -15 °C they agree within 1 %.

Seawater brine

For seawater, both β_{eff} and β can be obtained from the above mentioned equation of state which is valid for salinities between 1 and 40 ‰ (Millero and Poisson, 1981). For this salinity range, and at 0 °C, the haline contraction coefficient β of sea water is 1.5 to 5.5 % larger than the NaCl value implied by equation (A.21). These numbers change to a more stable difference of 4.3-5.3 % when an improved equation of state from Feistel (2003) is considered. However, a use of these state equations at subzero temperatures and high salinities is not justified. Therefore the pragmatic approach

$$\beta_{sw} \approx 1.05\,\beta \quad and \quad \beta_{eff,sw} \approx 1.05\,\beta_{eff}, \tag{A.27}$$

is suggested, simply increasing β of aqueous NaCl by a factor 1.05 to obtain the corresponding seawater value. Due to the almost linear relation between density and salinity, one may assume the same ratio for β and β_{eff}. The value of 1.05 is consistent with a simple weighted summation over the three major salts in sea water using the correlations given by Surdo et al. (1982). Assuming that these ratios also hold at low temperatures, one would naively expect that the seawater/NaCl ratio of β drops from 1.05 at -8 °C to about 1.025 during complete precipitation of Na_2SO_4. The remaining density excess in seawater, compared to NaCl, is then mainly due to $MgCl_2$. Making this assumption it was found that equation (A.27) can be modified by approximating this effect in terms of a linear change in the solid fraction of Na_2SO_4

$$\beta_{sw} \approx (1.05 - 0.25 r_{pr})\,\beta \quad and \quad \beta_{eff,sw} \approx (1.05 - 0.25 r_{pr})\,\beta_{eff}, \tag{A.28}$$

where r_{pr} is the precipitated fraction of Na_2SO_4, with respect to the total salinity. The latter equation is valid down to -22.9 °C, when in turn NaCl starts to precipitate and $r_{pr} \approx 0.11$ according to section A.4. It is noted that equation (A.28) agrees within 2 to 4 % with $\beta_{eff,sw}$ from relations proposed by Maykut and Light (1995) based on less accurate density observations.

A.2.6 Isobaric specific heat

Pure supercooled water

The isobaric specific heat of supercooled water has been measured down to -37.1 °C (Angell et al., 1982; Archer and Carter, 2000). These data sets are shown in Figure (A.10) with the relation

$$c_{pw} = 5.9471 - 1.8245\,\epsilon_\theta^{-1/2} + 0.56944\,\epsilon_\theta^{-1} - 0.044219\,\epsilon_\theta^{-3/2}, \tag{A.29}$$

in units of $kJK^{-1}kg^{-1}$, where $\epsilon_\theta = (\frac{\theta}{\theta_c} - 1)$ is the reduced temperature introduced by Speedy (1987). It is defined via the absolute temperature θ and its critical value $\theta_c \approx 227.1$ K. The underlying thermodynamical model of supercooled water, and implications for other properties in the supercooled regime, have been discussed by Speedy (1987).

NaCl brine

The specific heat c_{pb} of NaCl brine decreases with increasing salinity. Archer and Carter (2000) presented accurate measurements at different molalities, up to the saturation for temperatures between 236 and 285 K. According to Archer and Carter it is likely that these observations are more reliable than values obtained by Thurmond and Brass (1987), differing by 2-3%. Early observations from Koch (1924) agree within 0.5 % with the data from Archer and Carter (2000). The dependence of c_{pb} on temperature and salinity is

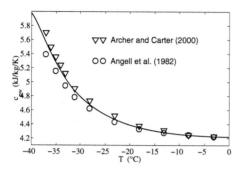

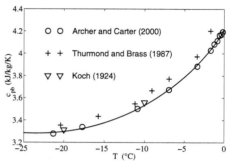

Fig. A.10: Isobaric specific heat of pure supercooled water. The solid line, equation (A.29) was fitted to the data from Angell et al. (1982) and Archer and Carter (2000).

Fig. A.11: Isobaric specific heat of NaCl brine at the liquidus temperature. The solid line, equation (A.30) was fitted to the data from Archer and Carter (2000), shown as circles. Plus signs are from Thurmond and Brass (1987) and the two triangles from Koch (1924).

complex and becomes anomalous below 0 °C. For example, Archer and Carter (2000) demonstrated that the temperature of the minimum in the specific heat, which is near 35 °C for pure water and shifts to about 0 °C for seawater of 35 ‰ salinity (Millero and Leung, 1976; Feistel, 2003), drops rapidly with salinity and reaches -20 °C at a salinity of 55 ‰. From Archer and Carter's tabulated data the following relation of c_{pb} for NaCl brine at the liquidus temperature was approximated

$$c_{pliq} = 4.2194 - 0.0246\,T_f - 0.1097\,T_f^2 + 0.01545\,T_f^3, \tag{A.30}$$

with c_{pliq} given in kJkg^{-1}. The equation was obtained by fixing the pure water value of 4.2194 kJ/kg at 0 °C. The agreement with the data from Thurmond and Brass (1987) and Koch (1924) is shown in Figure A.30.

Seawater

The specific heat of seawater has been investigated mainly in its natural salinity and temperature limits. Values once published by Krümmel (1907) at 17.5 °C and for salinities up to 40 ‰, agree within 2.5 % with more accurate tabulations from Millero and Leung (1976). The most recent thermodynamically consistent calibration for seawater by Feistel (2003) gives heat capacities for sea water which are, for the overlapping ranges of 0 to 10 °C and up to 0 to 40 ‰, ≈ 0.2 to 0.4 % lower than obtained by Archer and Carter (2000) for aqueous NaCl. In fact, the difference between the specific heat of pure water and aqueous $MgCl_2$ and Na_2SO_4 solutions near 0 °C is ≈ 10 − 20 % larger than the difference between pure water and aqueous NaCl (Zaytsev and Aseyev, 1992), which approximately holds also when comparing low temperature measurements of $MgCl_2$ and NaCl (Koch, 1924). With a simplified seawater composition of 24 parts NaCl, 5 parts $MgCl_2$ and 4 parts Na_2SO_4, one may, at 40 ‰, estimate a 0.4% smaller specific heat of seawater compared to NaCl, in agreement with the mentioned observations. By simple extrapolation one expects that this difference will not exceed 1.5 %, when focusing on temperatures above -22 °C. This estimate disregards the precipitation of Na_2SO_4 in the seawater system. Here it suffices to say that this implies an opposite effect and, near -22 °C, an increase by roughly 1 % in the specific heat capacity upon precipitation of Na_2SO_4. The heat capacities of aqueous NaCl and seawater brine are not expected to differ by more than 1 %, as long as temperatures above -22 °C are considered. More exact predictions for the multicomponent system seawater require low temperature data and extension of the thermodynamic evaluations from Millero (1974).

A.2.7 Thermal conductivity and diffusivity

Thermal conductivity

The thermal conductivity of NaCl solutions increases with temperature and decreases with salinity. From observations with saline solutions between 0 and -20 °C Riedel (1950) suggested the relations

$$K_w = K_{w0}(1 + 0.0032\,T) \tag{A.31}$$

$$K_b = K_{w0}(1 + 0.0032\,T)(1 - 0.00017\,S_b) \tag{A.32}$$

for the pure water conductivity K_w and brine conductivity K_b, with brine salinity S_b in ‰ and T the Celsius temperature. In Riedel's measurements and original formula $K_{w0} \approx$ 0.566 W m^{-1}K^{-1} was the proposed fresh water reference value at 0 °C. However, Riedel's data at higher temperatures are for all salinities systematically above the recommended value from IAPWS (1998), which is $K_{w0} = 0.562$ Wm^{-1}K^{-1}. A best fit of 0.562 $\pm$ 0.01 based on many recent data sources was also given by Ramires et al. (1994), fitting the salinity dependence of NaCl, and it is adopted in the present work. The above formula is intended to match the data below 0 °C, down to the eutectic concentration. According to Riedel (1950) and McLaughlin (1964) the pure water temperature dependence of K_w becomes approximately linear at temperatures near 0 °C, and only the linear term from Riedel (1950) was adopted in (A.32). It is noted that data for aqueous NaCl (Zaytsev and Aseyev, 1992; Ramires et al., 1994) and the pure water industrial standard (IAPWS, 1998) indicate a 10% larger temperature dependence than $(1 + 0.0032\,T)$, yet are mostly obtained by extrapolation from higher temperatures, while Riedels' results are truely obtained for cold brine. The lack in measurements in supercooled water was already pointed out by Powell (1958). It is notable that the six times stronger temperature dependence of K_b on salinity suggested by Schwerdtfeger (1963), frequently cited in the sea ice literature, is unreasonably large.

Thermal diffusivity

Using A.32 an approximation for the thermal diffusivity κ_w of supercooled water down to about -20 °C is

$$\kappa_w = \frac{K_w}{c_{pw}\,\rho_w} \approx 1.332 \times 10^{-3}(1 + 0.00311\,T - 0.0000782\,T^2), \qquad (A.33)$$

in units $cm^2 s^{-1}$. The validity at -20 °C is uncertain due to the linearised thermal conductivity dependence. κ_w computed from equation A.33 in the supercooled range is shown in figure (A.13).

Liquidus conductivity and diffusivity

Based on the above approximations, the liquidus thermal conductivity K_{liq} of NaCl brine is found to be almost linear and can be approximated as

$$K_{liq} = 0.562\,(1 + 0.00606\,T_f + 0.0000547\,T_f^2) \qquad (A.34)$$

in units $W\,m^{-1}K^{-1}$ with T_f the liquidus temperature in °C. One may then combine equations (A.34), (A.24) and (A.30) to compute the thermal diffusivity along the liquidus curve. It is best expressed as a polynomial in $(-T_f)^{1/2}$

$$\kappa_{liq} = 1.332 \times 10^{-3} + (5.46\,(-T_f)^{1/2} + 8.44\,(-T_f) + 2.57\,(-T_f)^{3/2}) \times 10^{-6} \qquad (A.35)$$

and gives κ_{liq} in $cm^2 s^{-1}$. Interestingly, κ_{liq} first increases to a maximum near -7 °C, 2% above the value of $\kappa_{w0} = 1.332 \times 10^{-3}$ $cm^2 s^{-1}$ at 0 °C, see figure A.13. Although there are some uncertainties in the properties and their relations, especially the linearisations used, this behaviour is mainly produced by the anomalies in density and specific heat and is likely a real feature of the liquidus diffusivity of aqueous NaCl.

Seawater

For seawater, one may again consider the difference compared to NaCl, due to the salts Na_2SO4 and $MgCl_2$. For the former the salinity dependence is similar to NaCl, for the latter it is about a factor of two larger (Riedel, 1950; Zaytsev and Aseyev, 1992). Assuming 24 parts NaCl, 4 parts Na_2SO4 and 5 parts $MgCl_2$, a 0.5% smaller conductivity can be estimated. This is in fact consistent with the limited measurements of seawater thermal conductivity by Jamieson and Tudhope (1970). When these measurements at 0

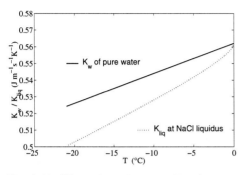

Fig. A.12: Thermal conductivity K_w of pure water (solid line) and K_{liq} of NaCl solutions at the liquidus temperature (dotted line)

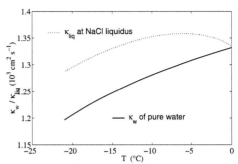

Fig. A.13: Thermal diffusivity κ_w of pure water (solid line) and κ_{liq} of NaCl solutions at the liquidus temperature (dotted line)

and 25 °C and salinities between 34 and 154 ‰ are normalised by the higher fresh water conductity obtained in the later study, the conductivities are on average $0.7 \pm 0.7\%$ smaller than from equation (A.32). Also the thermal diffusivity of seawater along the liquidus should not differ too much from an NaCl solution, when considering the slightly different values for c_{pliq} and ρ_{liq}.

A.2.8 Dynamic viscosity

Supercooled water

The dynamic viscosity of supercooled water increases with decreasing temperature, showing the same anomalous behaviour as the specific heat and the density (Angell, 1983). The following equation for the dynamic viscosity of supercooled water has been obtained by fixing $\mu_0 = 1.792$ mP s at 0 °C (IAPWS, 2003), and fitting a 4th-order polynomial to the data obtained by Hallett (1963) between 0 and -23 °C

$$\mu_w = 1.792 \, (1 - 0.03370 \, T + 0.0019855 \, T^2 + 0.00009472 \, T^3 + 0.00004415 \, T^4)) \qquad (A.36)$$

with μ_w in units mPa s. It is compared in Figure A.14 with curves representing the data from Osipov et al. (1977) and an equation that has been compiled by Pruppacher and Klett (1997) based on other data sources. Above -20 °C Hallett's values lie approximately between these references. The difference to each set is always less than 5 %. Below -20 °C, the agreement is worse. However, the reproducability of the experimental values given by Osipov et al. (1977) down to -35 °C has still to be verified.

NaCl brine

The viscosity of brine μ_b is often expressed as the product of a salinity-independent fresh water viscosity μ_0 and a factor accounting for the salt (Falkenhagen, 1953). The only study known to the author concerned with the NaCl brine viscosity below 0 °C is the comprehensive experimental work of Stakelbeck and Plank (1929). These data are shown as crosses in Figure A.15. Taking the work of Stakelbeck and Plank (1929) as a reference, several relations reviewed by Adams and Bachu (2002) were checked upon their possibility to be extrapolated to low temperatures, which was problematic for most relations. A reasonable agreement was found with the work from Korosi and Fabuss (1968), based on data in the temperature range 25 to 100 °C and salinities up to 170 ‰. Their high temperature relations can apparently be extrapolated to some degree, with deviations from the observations by Stakelbeck and Plank (1929) increasing at low temperatures and large salt fractions (dotted lines in Figure A.15). The difference seems to be related to the lack in a temperature dependence of μ_b/μ_w below 0 °C. Therefore the low temperature observations from Stakelbeck and Plank (1929) were correlated in the form

$$\mu_b = \mu_w \left(1 - 4.104 \times 10^{-5}\,S + 8.954 \times 10^{-6}\,S^2 + 2.037 \times 10^{-8}\,S^3\right) \qquad \text{(A.37)}$$

which is shown as solid lines in figure A.15. The deviation of equation (A.37) from any observation, with exception of the single measurement at -20 °C, is less than 1.6 %. The deviations might be related to the uncertainty in μ_w of supercooled water or to the calibration procedure applied by Stakelbeck and Plank (1929). It can be concluded that the measurements from Stakelbeck and Plank (1929) are consistent with Hallett's (1963) pure water viscosity measurements, yet more studies on the salinity dependence are desirable. It is noted that equation (A.37) is only valid below 0 °C. At 10 °C it performs worse than the correlations given by Korosi and Fabuss (1968), also seen in figure A.15.

Liquidus

Within 0.3 % of the above approximations the viscosity at the liquidus temperature can be represented by the relation

$$\mu_{liq} = \mu_w \left(1 - 0.0002001\,T_f + 0.002772\,T_f^2 + 0.0000552\,T_f^3\right), \qquad \text{(A.38)}$$

shown as a dashed line in Figure A.15). The ratio μ_b/μ_w increases to about 1.73 at the eutectic temperature of -21.3 °C.

Seawater

Already Krümmel (1907) reported the relative viscosity μ_b/μ_w of seawater to be 1.052 at 0 °C and 35 ‰ salinity, which compares to 1.053 at 5 °C obtained in a more recent study by Dexter (Millero, 1974). The corresponding NaCl solution values are: 1.01 measured by Stakelbeck and Plank (1929), 1.04 predicted by Korosi and Fabuss (1968) and 1.024 from Falkenhagen (1953). Na_2SO_4 shows an approximately 2 times larger concentration dependence of (μ_b/μ_w) compared to NaCl (Korosi and Fabuss, 1968), while the relative increase in (μ_b/μ_w) due to $MgCl_2$ is, at low concentrations, a factor 6-7 stronger than for NaCl (Stakelbeck and Plank, 1929). Applying again the simple weighted summation over the three major sea salts, one finds that 1.024 from Falkenhagen (1953) is most consistent with the seawater value of 1.05. However, at low concentrations this difference is within the uncertainty ranges of the experiments from Stakelbeck and Plank (1929), and therefore should not downgrade their observations at high concentrations and large viscosities. It is also possible that a slight temperature dependence of μ_b/μ_w observed by (Korosi and Fabuss, 1968) continues below 0 °C and has not been resolved by Stakelbeck and Plank (1929). It is noteworthy that $MgCl_2$ and NaCl display a different nonlinear temperature- and concentration viscosity dependence (Stakelbeck and Plank, 1929). Close to the eutectic temperature of NaCl, $\approx$ -22 °C, the relative (μ_b/μ_w) of aqueous $MgCl_2$ is only 4 times larger than for NaCl. More important is, however, that partial concentrations of $MgCl_2$ in seawater are still low near -22 °C, $\approx$ 35 ‰ , while the simultaneous effect due $\approx$ 180 ‰ NaCl becomes already nonlinear (figure A.15) and dominant. Hence, if a linear combination of the salts holds also at these low temperatures, then the seawater viscosity should always be within a few percent of an NaCl solution at the same liquidus temperature. However, accurate measurements of the viscosity of seawater brine or a theoretical multi-component model are clearly needed.

A.2.9 Solute diffusivity

The solute diffusivity D_s at the liquidus temperature is derived in consistency with the viscosity of brine in the separate appendix section A.3.

A.2.10 Solid-liquid interfacial free energy γ_{sl}

The solid-liquid interfacial energy γ_{sl} of the ice-water system has been determined by various methods. Due to its relevance in phase transition problems this property has been reviewed by several authors (Dufour and Defay, 1963; Jellinek, 1972; Jones, 1974; Hobbs, 1974; Hardy, 1977; Hillig, 1998). None of these authors has realised the importance and magnitude of the crystal anisotropy of γ_{sl} and a critical review of the different methods to obtain γ_{sl} with respect to crystal orientation is still lacking. A main goal of the

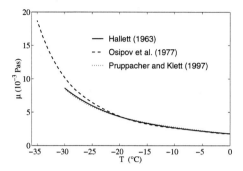

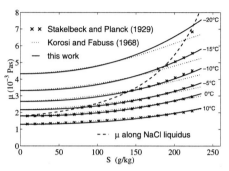

Fig. A.14: Dynamic viscosity μ_w of super-cooled water resembling the measurements from Hallett (1963), Osipov et al. (1977) and the compilation from Pruppacher and Klett (1997) shown as solid, dashed and dotted lines, respectively.

Fig. A.15: Dynamic viscosity μ_b of NaCl brine. Observations from Stakelbeck and Plank (1929) shown as crosses with equation (A.37) as solid lines, predictions based on high temperature data from Korosi and Fabuss (1968).

present section is to demonstrate, that most of the differences in the observations may be explained by account of the anisotropy in γ_{sl}.

The determination of γ_{sl} via the curvature measurement of an emerging grain boundary groove, performed by Hardy (1977), is often quoted as the most accurate method, e.g. Petrenko and Whitworth (1999). It yielded $\gamma_{sl} \approx 29.1 \pm 0.8$ mJ m^{-2} at a temperature of 0 °C. While Hardy assumed a pure ice conductivity of 2.2 W m^{-1}K^{-1}, the 'best' estimate of 2.14 W m^{-1}K^{-1} at 0 °C, adapted in the present work from Slack (1980), implies a slightly larger value 29.9 mJ m^{-2}. The value 33 ± 3 mJ m^{-2} obtained by (Ketcham and Hobbs, 1969) from a similar technique, yet based on solid-vapour grain-boundary groove angle measurements is another comparable and frequently cited estimate. Earlier applications of the grain-boundary method by Jones and Chadwick (1971) and Jones (1973a) resulted in a larger uncertainty bound and will be discussed below.

Homogeneous nucleation and temperature dependence of γ_{sl}

An estimation of γ_{sl} based on homogeneous nucleation theory of liquids was first proposed by Turnbull (1950). The advantage of Turnbull's approach is its generality. He noted that for a class of substances, including water, γ_{sl} is proportional to the latent heat of fusion L_f via

$$\gamma_{sl} = 0.32 \frac{L_f \rho_i^{2/3} M_w^{1/3}}{N_A^{1/3}}, \qquad (A.39)$$

where in case of water ρ_i is the ice density, M_w the molecular weight of water in kg mol^{-1} and N_A is Avogadros constant $(6.0221 \times 10^{23}\ mol^{-1})$. For a second class the prefactor was close to 0.45. According to this approach the interfacial energy is expected to display a similar temperature dependence as the latent heat of fusion in equation A.5. Its linear form between 0 and -36 °C implies

$$\gamma_{sl} \approx \gamma_{sl0}(1 + 0.0067\, T) \tag{A.40}$$

with T in °C. When inserting the molecular constants for water in equation (A.39), one obtains $\gamma_{sl0} = 31.3$ mJ m^{-2} at 0 °C, while Turnbull, using a different L_f, obtained $\gamma_{sl} \approx 32.1$ mJ m^{-2}.

Dufour and Defay (1963) discussed the application of homogeneous nucleation theory to ice and its solid-liquid surface tension in detail. They reviewed several studies and proposed a best estimate $\gamma_{sl} \approx 20.2$ mJ m^{-2} with an uncertainty of ≈ 0.2 mJ m^{-2}. This value was based on the assumption of ice freezing in form of hexagonal prisms and valid at -35 °C. The most plausible interpolation to 0 °C was suggested to result in $\gamma_{sl} \approx 22.4$ to 24.8 mJ m^{-2}. Ketcham and Hobbs (1969) reviewed several nucleation studies yielding a range 17 to 32 mJ m^{-2}, at a temperature of -40 °C. They concluded that, due to the large range, the results are inferior to grain-boundary measurements. Also Fletcher (1970) discussed several aspects and uncertainties of homogeneous nucleation theory in connection with minimum droplet diameters during freezing. His best estimate of $\gamma_{sl} \approx 22$ mJ m^{-2}, for the temperature range -30 to -40 °C, involved a graphical fit of a relation between droplet diameters and supercooling. While tentative, Fletcher's approach is not inconsistent with a recent discussion of curvature-dependent maximum supercooling (Bogdan, 1997). With A.40, the corresponding 0 °C value from Fletcher is 28.8 mJ m^{-2}. It thus appears that the surface tension at the in situ homogeneous nucleation temperature of ≈ -35 °C can be computed quite confidently from nucleation theory. However, the temperature dependence, derivable from the early scattered nucleation data, and thus the extrapolation to the melting point, is less confident.

An improved statistical method based on nucleation theory has been suggested by Wood and Walton (1969, 1970). They reported best estimates for γ_{sl} of 31.93 ± 0.44 and 27.53 ± 0.37 mJ m^{-2} for spherical and hexagonal nucleation clusters at 0 °C. These values were extrapolated by Wood and Walton (1970) from respective 'observed' values of 24.22 ± 0.02 and 20.95 ± 0.02 mJ m^{-2} at -36.55 °C. In their experiments the derived linear temperature dependence of γ_{sl} at low temperatures, was $d\gamma_{sl}/dT = 0.21$ mJ m^{-2} K^{-1} for the spherical and $d\gamma_{sl}/dT = 0.18$ mJ m^{-2} K^{-1} for the hexagonal cluster. Notably, γ_{sl} at -35 °C for a hexagonal cluster agreed within 1 mJ m^{-2} with the determination by Dufour and Defay (1963). The temperature dependence from this study was much less uncertain and implies factors 0.0065 and 0.0066 in equation (A.40), rather consistent with Turnbull's theory.

Equation A.39 can also be explored by applying Antonoff's rule (Pruppacher and Klett, 1997)

$$\gamma_{sl} \simeq \gamma_{sv} - \gamma_{lv}, \tag{A.41}$$

where γ_{sv} and γ_{lv} are the solid-vapour and liquid-vapour interfacial energies. For the former Pruppacher and Klett (1997) have estimated a temperature dependence $d\gamma_{sv}/dT \approx -0.05$ mJ m^{-2} K^{-1} while for the latter an approximately linear decrease $d\gamma_{lv}/dT \approx -0.14$ mJ m^{-2} K^{-1} is well-established near the melting point (Straub et al., 1980; IAPWS, 1994). Combining these dependencies implies $d\gamma_{sl}/dT \approx 0.19$ mJ m^{-2} K^{-1} for the solid-liquid interface energy. With γ_{slo} between 27 and 32 mJm^{-2} the normalised temperature dependence in equation (A.40) would become 0.006 to 0.007.

Other authors have proposed different temperature dependencies. However, the larger $\gamma_{sl} \sim \gamma_{sl0}(1+0.0089\,T)$, suggested by Pruppacher and Klett (1997), was obtained by mixing γ_{sl} from very different and uncertain methods (Hobbs, 1974; Hardy, 1977). Taborek (1985) estimated a weaker dependence of $\gamma_{sl} \cong \gamma_{sl0}(1 + 0.003\,T)$, based on nucleation of emulsified water. Huang and Bartell (1995) discussed the discrepancy to other observations in connection with additional nucleation data. They suggested a change in the preferred ice lattice from hexagonal (at the melting point) to cubic (at the nucleation temperature) as an explanation. Clearly, transitions between crystalline forms are complex and process-dependent (Fletcher, 1970; Hobbs, 1974; Petrenko and Whitworth, 1999) and need to be considered in connection with open questions in theories of supercooled water and homogeneous nucleation (Angell, 1983; Speedy, 1987; Koop, 2004; Holten et al., 2005). It seems likely that the mentioned variability in the temperature dependence of γ_{sl} obtained from nucleation measurements is related to these transitions in crystalline forms in the regime -35 to -40 °C. The low dependence $d\gamma_{sl}/dT \approx 0.03$ mJ m^{-2} K^{-1}, in normalized form $\sim (1 + 0.001\,T)$, suggested by (Huang and Bartell, 1995) refers to this transition regime and should not be interpolated to the melting point.

The above considerations indicate open questions concerning the classical homogeneous nucleation theory and the temperature dependence far from the melting point (Huang and Bartell, 1995). A recent discussion of the temperature dependence of $gamma_{sl}$ of supercooled water (Hruby and Holten, 2004), possibly offers an explanation for the different temperature dependencies derived from recent homogeneous nucleation studies (Wood and Walton, 1970; Taborek, 1985; Huang and Bartell, 1995) and earlier work (Dufour and Defay, 1963; Pruppacher and Klett, 1997). Theoretical model formulations of supercooled water make use of different approaches (Kiselev and Ely, 2001). Kiselev and Ely (2002) applied equation A.39 from Turnbull, while Jeffery and Austin (1997) introduced an additional temperature decrease, when considering Turnbull's model A.39 in their formulation of an equation of state. From the empirical basis for hexagonal ice it can, for temperatures above -35 °C, be stated that equations A.39 and A.40 are consistent with equation A.5 for L_f and with Antonoff's rule A.41.

Dynamic approaches

Another indirect and rather robust method to obtain the surface tension has been outlined by Coriell et al. (1971) and Hardy and Coriell (1973). These authors applied a nonlinear morphological stability analysis to the growth of vertical ice cylinders. Improving on earlier less successful linear theory, Coriell et al. (1971) obtained a value of 25 mJ m^{-2} for distilled water, for both freezing and melting. For 0.1 M HCl and NaCl solutions a larger value of 29 to 30 mJ m^{-2} was found only for freezing (Hardy and Coriell, 1973). Varying the pH of the solution, the melting value for 0.1 M NaCl solutions increased to 38 mJ m^{-2} in the presence of Cl$^-$-ions. This dependence of γ_{sl} on the pH-value was only found for melting. As the latter represents the true equilibrium, (Hardy and Coriell, 1973) concluded that adsorption during freezing might lower the true surface energy. Discussing some limitations of the approach and possible explanations, Hardy and Coriell (1973) suggested that complicated interfacial processes may be responsible for the complex results.

Vlahakis and Barduhn (1974) applied an approach suggested by Fernandez to describe dendritic growth under conditions, when strong forced fluid flow controls the heat flow. Their analysis yielded a value of $\gamma_{sl} \approx 33 \pm 6$ mJ m^{-2}. The main problem with this estimate, also noted by the authors, is that it was obtained by applying the maximum velocity hypothesis, yielding dendritic tip radii an order of magnitude less than observed. The maximum velocity hypothesis for free dendritic growth is now known to be incorrect, see Glicksman and Marsh (1993) and section 8.1. Alternatively, Kotler and Tarshis (1968) applied a modified Ivantsov approach of a freely growing nonisothermal dendrite, from which they derived 20 ± 2 mJ m^{-2} for the surface energy. Also this value would change considerably for different and more realistic dendritic growth assumptions. The method from Vlahakis and Barduhn (1974) may be thought to override the free growth selection of a dendrite due to the presence of forced convective heat transport. However, the agreement of their results with other values of γ_{sl} may be still fortuitous. In general, a dynamic approach can only be judged significant, if the theory also agrees with the observed microstructure, as in the studies by Coriell et al. (1971) and Hardy and Coriell (1973).

Pore size experiments

Kubelka and Prokscha (1944) reported results from a method which relates the critical undercooling of ice, growing through thin capillaries of silica gels, to the solid-liquid surface tension via the Gibbs-Thomson relation. They obtained a value of $\gamma_{sl} \approx 25.4 \pm 2.2$

mJ m^{-2} at -5 °C, which may be extrapolated to $\gamma_{sl} \approx 26.4 \pm 2.2$ mJ m^{-2} at the melting point of ice. Kubelka and Prokscha (1944) reported on careful studies already performed in 1933 and discussed several methods, leading to a spreading of 22 to 33 mJ m^{-2}. They noted that the final experimental setup gave the best reproducibility. The silica gels used by Kubelka and Prokscha (1944) had diameters in the range $10 - 20$ nm. For such small capillary radii r, it appears that a correction factor $1 + 2\delta_{tol}/r$ to the observed γ_{sl}, based on arguments once introduced by Tolman (1947), becomes relevant. With a Tolman length $\delta_{tol} \approx 0.42$ nm, suggested by Bogdan (1997) near the melting point, this correction of the experimental data from Kubelka and Prokscha (1944) yields $\gamma_{sl} \approx 29.3$ mJ m^{-2} at the melting point of ice. Indirect computations of δ_{tol} for the ice-vapour interface based on nucleation theory (Holten et al., 2005) are consistent with estimates from simplified concepts (Bogdan, 1997; Pruppacher and Klett, 1997). However, the Tolman approach is no validated physical theory (Pruppacher and Klett, 1997). The length scale $\delta_{tol} \approx 0.42$ nm is comparable to the intermolecular distance and ice lattice parameter 0.27 nm and 0.45 nm respectively see Petrenko and Whitworth (1999)), or to the capillary length itself ($l_\Gamma \approx 0.34$ nm, see paragraph A.44), indicating the plausibility of the correction. The suggested correction is heuristic and deserves, in view of the ongoing debate about the quasi-liquid layer on the surface of ice (Petrenko and Whitworth, 1999), further work.

Hillig (1998) devised a similar method, where ice was freezing through very fine pores. His results were obtained with filters of an order of magnitude larger pore sizes compared to those used by Kubelka and Prokscha (1944), making the Tolman correction neglegible. He reported an average value of $\gamma_{sl} \approx 31.7 \pm 2.7$ mJ m^{-2} with four sets of filter sizes. For Hillig's widest filter, the uncertainty was 5 %, being only $1 - 2$ % for the three fine-pore filters. Excluding the less certain run, the experiments give $\gamma_{sl} \approx 30.3 \pm 0.8$ mJ m^{-2} as best estimate.

Concentration dependence

The liquid-vapour interfacial energy of a NaCl solution is known to increase with molality at about 1.6 mJ m^{-2} per mole NaCl/ kg water, f.e. Kaufmann (1960); Pruppacher and Klett (1997). This may be converted to a liquidus temperature dependence of $d\gamma_{lv}/dT_f \approx -0.5 \ mJm^{-2}K^{-1}$ which should hold down to the eutectic point of -21.6 °C. If one assumes again that $\gamma_{sl} \simeq \gamma_{sv} - \gamma_{lv}$, then the solid-liquid interfacial energy γ_{sl} is expected to decrease with brine salinity. The liquidus temperature dependence would be almost a factor of three larger than the temperature dependence for pure water. On the other hand, equation A.9 suggests a weaker decrease in the isothermal latent heat of fusion of aqueous NaCl. Based on Turnbull's model this would imply only half the temperature dependence of supercooled water. However, Wood and Walton (1969), who also performed homogeneous nucleation experiments with aqueous NaCl solutions, found

a slightly larger γ_{sl} for 0.5 M aqueous NaCl solutions. Their observed weaker temperature dependence of γ_{sl} was consistent with Turnbull's model, considering the difference
in the latent heat of aqueous NaCl solutions. Although the observations point to the
applicability of Turnbull's model, some care is in order, as the saline solutions did not
nucleate at the same temperature as pure water, which necessitated extrapolation from
the local temperature dependence. Discrepancies and problems in the applied nucleation
theory in the presence of solutes, noted by Wood and Walton (1969), were also pointed
out by Rasmussen and MacKenzie (1972).

The morphological stability approach by Hardy and Coriell (1973) also yielded a
larger surface tension for a 0.1 M NaCl solution compared to pure water, similar to the
above mentioned pH-dependence. For melting, an increase from 25 mJ m^{-2} to 38 mJ m^{-2}
was found, for freezing an increase from 25 mJ m^{-2} to 30 mJ m^{-2}. Jones and Chadwick
(1971) applied a grain-boundary angle method to ice in a thin growth cell. They proposed
a method to extend pure water observations, which in their configuration yielded 44 $\pm$ 10
mJ m^{-2} (revised by Jones (1973a)), to NaCl solutions. The proposed linear increase of
γ_{sl} with solute concentration of approximately 13 mJ m^{-2} per mole NaCl, which was
qualitatively different from the change found by Hardy and Coriell (1973) at 0.1 M.
However, inspection of the data from Jones and Chadwick (1971) indicates that the
scatter in the observations is too large to prove a linear relation. Their results may also
be interpreted in terms of an approximately 15-25 % larger solid-liquid surface energy,
when solute is present, which would be consistent with the results from Hardy and Coriell
(1973). The average solutal γ_{sl} for the study by Jones and Chadwick (1971) may then,
in terms of the revised computation by Jones (1973a), be estimated as 56 $\pm$ 18 mJ m^{-2}.

Anisotropy

So far, nothing has been said about the anisotropy of surface tension. Applying equation
A.41 from Antonoff, in connection with simple bond considerations, one may arrive at a
pure water estimate of $\gamma_{sl} \approx 24$ mJ m^{-2} for the basal plane and $\gamma_{sl} \approx 33$ mJ m^{-2} for the
prismatic planes at 0 °C (McDonald, 1953; Pruppacher and Klett, 1997). The approach
is, however, heuristic and based on tentative considerations about elastic relaxation at
the solid-vapour interface (Pruppacher and Klett, 1997). It yields a solid-liquid interfacial
energy anisotropy of ± 0.15. Recent theories based on modern surface chemistry, in part
discussed by Petrenko and Whitworth (1999), pose some question-marks on the latter
simplified derivation. A larger value of the anisotropy, ± 0.3, has been derived by Koo
et al. (1991) from the equilibrium shape of water droplets in an ice matrix[1]. The latter
value would, if an average of $\gamma_{sl} \approx 30$ mJ m^{-2} is accepted, correspond to values of

---

[1] Koo et al. (1991)do not give a higher precision than 0.1, and from their photograph of an equilibrium
negative crystal a value of 0.35 $\pm$ 0.02 has been estimated within the limits of resolution

≈ 39 mJ m^{-2} for the prismatic and ≈ 21 mJ m^{-2} for the basal plane. A somewhat larger difference of ≈ 25 mJ m^{-2} between the solid-liquid surface energies for the basal and prismatic planes has been suggested by Yosida (1967) from considerations of grain boundary energies. Finally, Koo et al. (1991) suggested, from equilibrium shapes, that the anisotropy of γ_{sl} for all prismatic directions normal to the c-axis is rather small, 0.002 ± 0.001. How do these bounds compare to the experimental data discussed so far?

Jones and Chadwick (1971) and Jones (1973a) and report that the results in their experiments correspond to the energy of prismatic planes. This makes the larger value of 44 ± 10 mJ m^{-2} rather plausible. Also Ketcham and Hobbs (1969) reported that most of their grain boundaries did not expose a basal plane, implying that their result of 33 ± 3 mJ m^{-2} also corresponds to the prism surfaces. Jones (1974) has later, by pointing to an inaccuracy in the measurements from Ketcham and Hobbs (1969), suggested a refined value of 45 ± 15 mJ m^{-2}. Hardy (1977) did not report on the orientation of the crystals in his experiments. However, in contrast to the procedure by Jones and Chadwick (1971), who grew ice laterally in a horizontal Hele-Shaw cell, his ice was grown unidirectionally downward from a cold interface. His (here slightly modified) result $\gamma_{sl} \approx 29.8 \pm 0.8$ mJ m^{-2} therefore probably reflects mixed-grain boundaries with both surfaces exposed. This would at least be consistent with the two most plausible growth modes to be expected in his experiments, which are either(i) vertical c-axis orientation or (ii) horizontal alignment of the c-axis along the long axis the cell. Also Hardy's results are consistent with the above considerations on the anisotropy.

For the experiments by Hillig (1998), where ice grew through fine-pore filters, the revised $\gamma_{sl} \approx 30.3 \pm 0.8$ mJ m^{-2} has been estimated above. Hillig noted attempts to seed crystals with different initial crystal orientations. However, the ice always grew polycrystalline with no preferred orientation through the filters. The value given by Hillig (1998) should therefore be interpreted as an average value for different orientations. The same assumption may then also be valid for the similar method in the earlier study by Kubelka and Prokscha (1944). The corrected value from this study is $\gamma_{sl} \approx 29.3 \pm 2$ mJ m^{-2}. The gel and filter experiments are thus within 0.5 mJm^{-2} or 1.7% of the value from Hardy (1977).

Also the extrapolated γ_{sl} from homogeneous nucleation from Wood and Walton (1969, 1970), 31.93 ± 0.44 and 27.53 ± 0.37 mJ m^{-2} for spherical and hexagonal cluster shapes, may be interpreted as simple orientation averages. These values are consistent with other mixed orientation observations. Wood and Walton (1970) noted that it cannot be decided if the hexagonal cluster is the correct form for nucleation at low temperatures, yet the spherical value is interpretable as an upper estimate due to this method. The proper interpretation of homogeneous nucleation experiments with respect to anisotropy may require a more complex theoretical framework.

The morphological stability calculations from Coriell et al. (1971) relate to perturbations on a cylinder surface with its axis parallel to the c-axis. The surface energy obtained by these authors can be expected to fall between the prismatic and basal plane values, as both planes are exposed during the development of sinusoidal cellular pattern. The value 25 mJ m^{-2} obtained for pure water is indeed comparable to orientation averages from other studies. However, in the study perturbations were also found to interfere along and around the cylinder axis. It is noteworthy that, in dendritic stability theory, the anisotropy of the interfacial energy is enhanced by a factor $(n^2 - 1)$ for n-fold symmetry and that an effective γ_{sl} for a nonplanar interface is related to this enhancement via the dynamical evolution and shape of the perturbations. The treatment of an anisotropic correction of γ_{sl} likely requires a more complex analysis than performed by Coriell et al. (1971), with some general aspects discussed by Coriell and Sekerka (1976a).

There appear to be only two experiments eventually reflecting γ_{sl} in the basal plane. Hillig (1958) estimated, from growth kinetics of ice in capillaries and assuming a two-dimensional step-nucleation model, a value of 6.4 mJ m^{-2}. Michaels et al. (1966) have corrected a computational error and arrived, for their own and Hillig's experimental data, at a range of 1.8 to 2.2 mJ m^{-2}. These low values have been discussed by Hillig in terms of diffuse interface models. He conjectured that, during molecular step growth on the basal plane, a lower interfacial energy would be obtained, if the interface were several molecular spacings thick. These ideas were further discussed theoretically (Cahn et al., 1964) on the basis of a simple diffuse interface model. Such a view of the interface is principally consistent with more recent molecular dynamics simulations and indirect observations of the surface of ice (Petrenko and Whitworth, 1999). However, no determinations of γ_{sl} for the basal plane have appeared to date. The present understanding of the quasi-liquid surface layer and the nucleation process of ice (Pruppacher and Klett, 1997; Petrenko and Whitworth, 1999) seem still insufficient to derive a proper estimate of γ_{sl} from kinetic growth data.

Discussion

It appears that the account for anisotropy can explain much of the apparent scatter in the proposed values of γ_{sl}. A typical mixed orientation value for basal and prismatic planes of $25 < \gamma_{sl} < 32$ mJ m^{-2} may be allocated to one group of the studies (Kubelka and Prokscha, 1944; Wood and Walton, 1970; Hardy and Coriell, 1973; Hardy, 1977; Hillig, 1998), based on very different methods. If a best estimate is based on the three studies with lowest nominal uncertainties (Kubelka and Prokscha, 1944; Hardy, 1977; Hillig, 1998), it becomes $\gamma_{sl} \approx 30$ mJ m^{-2} (an exact average of the three methods yields $\gamma_{sl} \approx 29.8$ mJ m^{-2}), with a difference of less than ≈ 2 % between the methods and a standard observation of 3 % for each of them.

If the correction proposed by Jones (1974) to the data from Ketcham and Hobbs

(1969) is accepted, then the range of $44 < \gamma_{sl} < 46$ mJ m^{-2} becomes the most probable value for the prismatic planes (Ketcham and Hobbs, 1969; Jones and Chadwick, 1971; Jones, 1973a, 1974). Due to the lack in direct derivations of the surface energy of the basal plane, one may either use a shape-related anisotropy of 0.35 (slightly larger than 0.3 estimated by Koo et al. (1991)), in connection with the noted bulk value of 30 mJ m^{-2}, to estimate the typical γ_{sl} for the basal and prismatic planes to be close to 20 mJ m^{-2} and ≈ 41 mJ m^{-2}. Alternatively, a best estimate of 45 mJ m^{-2} from Jones (1974) for the prismatic planes can be used with the mixed value ≈ 30 mJ m^{-2} to estimate 15 mJ m^{-2} for the basal plane. The latter set implies an anisotropy of 0.5 and appears somewhat large, even if an uncertainty of 10 % is assumed for the equilibrium shape determination due to Koo et al. (1991). The determinations of the prismatic plane energies were subject to a nominal uncertainty of ≈ 10 mJ m^{-2}, while the mixed determinations appear correct to within ≈ 1 mJ m^{-2}. However, it is not known, to what proportion the mixed values represent prismatic and basal plane energies. While more experimental data are required to ascertain the exact anisotropy, it seems very likely that the latter is between 0.3 and 0.4. The conclusion reached here is that the anisotropy in the solid-liquid interface energy is much larger than suggested by Jones (1974) or Hobbs (1974).

Gibbs Thomson parameter Γ and capillary length d_0

In the solidification problems considered in the present work, not the solid-liquid surface tension alone, but the Gibbs-Thompson parameter Γ is the relevant term which enters the stability equations. It is given by

$$\Gamma = \theta \frac{\gamma_{sl}}{L_f \rho_i}, \tag{A.42}$$

where θ is the absolute temperature, L_f the latent heat of fusion and ρ_i the pure ice density. Γ has the dimension temperature $\times$ length. The main point made here is that, if the solid-liquid interfacial energy can, following Turnbull (1950), be predicted from nucleation theory, this implies no other temperature dependence than $\Gamma \sim \theta$. Inserting $\gamma_{sl0} \approx 29.8$ mJ m^{-2} at the melting point of water one obtains for $\theta \leq 273.15$ K

$$\Gamma = 2.68 \times 10^{-8} \frac{\theta}{273.15} \tag{A.43}$$

in units m K. Based on this model Γ is almost constant and only decreases slightly with the absolute temperature. This model has been adopted in the present work. A possible dependence on salinity needs, as discussed above, still to be established properly. The capillary length l_Γ for the pure ice-water interface is defined as

$$l_\Gamma = \Gamma \frac{C_{pw}}{L_f} \tag{A.44}$$

and takes the value $l_\Gamma \approx 3.39 \times 10^{-10}$ m at the melting point of pure water.

Conclusions

The most important result from the present review of studies of the solid-liquid interfacial energy of the ice-water system may be summed up as follows:

- Homogeneous nucleation experiments yield $\gamma_{sl} \approx 20 - 24$ mJ m^{-2} near -35 °C, in dependence on the cluster structure (Wood and Walton, 1970). The simplified model from Turnbull, equation (A.39), yields an extrapolated γ_{sl} near the melting point which agrees well with other studies.

- The temperature dependence of γ_{sl} between 0 °C and -40 °C is uncertain due to open questions in the theory of homogeneous nucleation of NaCl solutions and supercooled water. The model from Turnbull appears to be a reasonable approach as long as the nucleating ice form is hexagonal. For NaCl concentrations up to 200 ‰, and temperatures down to -20 °C considered in the present work, the possible error in γ_{sl} due to the uncertain temperature dependence is less than 7 %.

- Much of the reported discrepancies in observations of γ_{sl} near 0 °C are attributable to an anisotropy in γ_{sl} of 0.3 to 0.4. The effective γ_{sl} for a mixed grain-boundary, derived from rather different methods, covers the range 25 to 32 mJ m^{-2} and the best 'bulk' estimate is 30 mJ m^{-2}.

- Indirect estimates of γ_{sl} for the basal and prismatic planes, based on the shape-derived anisotropy, are 20 mJ m^{-2} and 41 mJ m^{-2}. Direct observations for the prismatic plane yield $44 < \gamma_{sl} < 46 \pm 10$ mJ m^{-2}.

- In dynamical experiments, γ_{sl} appeared to increase by 20 % when 0.1 M or more NaCl is added. The mechanism and the overall salinity dependence is less clear and awaits still accurate determination.

- Kinetic non-equilibrium values of γ_{sl} of pure water for freezing and melting are the same. For salt solutions γ_{sl} is larger during melting.

- For the analysis of unidirectional freezing of NaCl solutions, involving instabilities normal to the c-axis, $\gamma_{sl} \approx 30$ mJ m^{-2} is the most reasonable choice, in particular when morphological stability theory is applied.

- The model from Turnbull implies a Gibbs-Thompson parameter Γ which scales only with the absolute temperature.

A.2.11   KCl as an example for other aqueous solutions

For other electrolytes in principle all the derivations as performed for NaCl have to be repeated to derive a similar set of thermal and thermodynamical properties. First order estimates may be found by using these relative properties compared to NaCl. As information is mainly available for the temperature range 10 to 20 °C (Zaytsev and Aseyev, 1992), one has to be careful when extrapolating to subzero temperatures. When considering, for example, the solute diffusivity D_s at the liquidus, the activity coefficients need to be known.

In the present work, some validations of morphological stability theory have been done for aqueous solutions of potassium chloride, for which the liquidus slope m and the solute diffusivity D_s (relative to NaCl) were estimated from the compilation by Zaytsev and Aseyev (1992). The theoretical liquidus slope m_{KCl} at infinite dilution is expected to be smaller than m_{NaCl}, the ratio being given by the ratio of molecular weights $M_{NaCl}/M_{KCl} = 58.44/74.55 \approx 0.784$. At moderate concentrations, when water activity and ionic interactions become relevant, this value appears to stay relatively constant. It takes ≈ 0.77 between 30 and 40 ‰, but then decreases approximately linearly with concentration to 0.62 at a $C_\infty \approx 150$ ‰. The ratio of solute diffusivities, D_{KCL}/D_{NaCl}, is documented at 20 °C. It is 1.2 at low concentrations and 1.23 at 150 ‰.

Two other properties are relevant for directional solidification studies and morphological stability theory: The liquid thermal conductivity and the isothermal latent heat of fusion. The concentration dependence of the thermal conductivity K_b for KCl is approximately 30 % larger than for NaCl (Riedel, 1950). However, the total concentration-related decrease of K_b is, even at eutectic values, only of the order of 10 %. As the influence of the liquid thermal conductivity on morphological stability is small anyway, this modification was neglegbile in the present study. For the isothermal latent heat of fusion, also similar values as for NaCl have been assumed. Due to similar activity coefficients (Zaytsev and Aseyev, 1992) no L_f differing by more than a few percent is expected.

It is finally noted that the haline contraction coefficient β is typically 15 to 20% smaller for KCl than for NaCl, and that the dynamic viscosity of aqueous KCl decreases slightly with concentration, contrasting the behaviour for NaCl (Korosi and Fabuss, 1968). These properties are relevant at low growth velocities, when convection has to be included.

A.3 Diffusivity of aqueous NaCl solutions at the liquidus temperature

The solute diffusion coefficient D_s of aqueous solutions is of the order of 10^{-5} cm s^{-1}. A recent compilation of D_s for aqueous electrolytes can be found in Zaytsev and Aseyev (1992). Generally, D_s displays a strong temperature and a weaker salinity dependence. Its proper determination is critical for any process related to molecular diffusion in ice-brine mixtures. Observations have been mostly obtained between 20 and 50 °C (Zaytsev and Aseyev, 1992) and for aqueous NaCl only a few datasets extend down to 0 °C (Bruins, 1929; Kaufmann, 1960). The present section summarizes earlier approaches and observations and applies an extrapolation to subzero temperatures which is consistent with electrolyte theory (Robinson and Stokes, 1955; Harned and Owen, 1958).

A.3.1 Stokes-Einstein approach

Caldwell and Eide (1981) found, from experiments at a concentration of 28.5 ‰ NaCl and temperatures between 1.4 and 47.7 °C, good agreement with the Stokes-Einstein equation

$$D_s = \frac{k\theta}{6\pi\mu_w r_D} \qquad (A.45)$$

, where θ is the absolute temperature, k Boltzmann's constant, μ_w the viscosity of pure water and r_D an effective radius of the diffusing particle. This simple approach to relate the solute diffusivity inversely to the viscosity of pure water resulted in a standard error of only 4.4 %. The effective Stokes-Einstein radius r_D was estimated by Caldwell and Eide (1981) by combining their own diffusion measurements with an international standard of viscosity (IAPWS, 2003). It was found to be $r_D = 1.66 \pm 0.07 \times 10^{-10} m$ with no apparent trend with pressure or temperature. It can be compared to an alternative fit of the same data to a modified Arrhenius equation also given by Caldwell and Eide (1981) in the form

$$D_s = 0.14857 \, \theta \, exp\left(\frac{-\theta_B \left(1 - 0.00002414 \, P\right)}{\theta - \theta_C}\right), \qquad (A.46)$$

where θ is the absolute temperature, $\theta_B = 546.45$ K and $\theta_C = 139.09$ K are temperatures scales corresponding to an activation energy and free-volume temperature, respetively. P is the pressure in bars. The nonlinear least-squares fit to equation A.46 implied a slightly smaller standard error of 2.9 % than to equation A.45. The extrapolation of both equations to temperature below 0 °C is shown in Figure A.16.

Most of the data, used in the fitting procedure by Caldwell and Eide (1981), were from temperatures between 15 and 47 °C. The correct extrapolation to subzero temperatures is not trivial and likely requires a consistent theory of the anomalous regime of supercooled water (Angell, 1983; Speedy, 1987). The temperature scales θ_B and θ_C obtained by Caldwell and Eide (1981) are plausible results for 'normal' water, but in the supercooled regime an additional anomalous term is expected (Angell, 1983). On the other hand, the apparently temperature-independent Stokes-Einstein radius in equation (A.45) encourages the use of the latter formula, in combination with the actual viscosity of supercooled water, equation A.36, which empirically includes such an anomalous contribution. Therefore the Stokes-Einstein equation A.45 is suggested to be preferable over the Arrhenius model, at least if a temperature dependence of the Stoke-Einstein radius r_D can be excluded or quantified.

A.3.2   Temperature- and pressure-dependence of r_D

Caldwell and Eide (1981) did not note a temperature variability of r_D. Averaging their data for 1.2-4.2, 15.1, 31.3 and 47.7 °C gives r_D of 1.65, 1.65, 1.70 and 1.68 $\times 10^{-10}$ m, respectively, but the sample ensembles have standard errors of 5-8 % (Figure A.17). When the diffusivities, compiled by Zaytsev and Aseyev (1992) at a salinity of 20 ‰, are evaluated in the same manner to compute r_D, using standard values for the viscosity (IAPWS, 2003), the resulting r_D agrees well with Caldwell and Eide (1981) in the temperature range 15 to 30 °C, yet there is a discrepancy at higher temperatures (figure A.17). The tabulated data from Zaytsev and Aseyev (1992) imply increasing r_D with temperature. It is unclear, to what degree this difference can be attributed to the interpolation procedure used by Zaytsev and Aseyev (1992), as many different data sources were combined. Considering the method from Caldwell and Eide (1981) it has to be noted, that the data were obtained at different pressures up to 700 bars. One needs to consider the possibility of systematic bias due to the state equation of water: One of the many anomalies of pure water is that below 35 °C the viscosity decreases with increasing pressure, while above this temperature this behaviour inverts (Dorsey, 1940; IAPWS, 2003). However, already Krümmel (1907) has pointed out, that this behaviour changes in concentrated salt solutions for which the viscosity increases with pressure at any naturally accessible seawater temperature. As mentionend in the density and heat capacity sections, present theories and experimental data of supercooled water indicate that a lowering of the temperature of maximum density will also shift anomalies in the thermodynamic properties to lower temperatures (Angell, 1983; Speedy, 1987; Archer and Carter, 2000). Such a scenario would induce the strongest change in the pressure-viscosity relationship at low temperatures. The following quantitative estimate provides an idea of these changes in the experimental parameter range accessed by Caldwell and Eide (1981). At a concentration of 28.5 ‰ the temperature of maximum density is shifted by about 6 °C. From the typical average pressure of 600 bars in the low temperature runs at 1.2 to 4.2 °C, one may naively estimate a small correction, $r_D \approx 1.65 - 0.02 = 1.63 \times 10^{-10}$

m, for the lowest temperature ensemble, based on the expected change in the pressure-viscosity relation (Dorsey, 1940; IAPWS, 2003). For all higher temperature ensembles the corrections would be neglegible, due to lower average pressures involved, and due to the weaker expected shift in the pressure dependence. Hence, applying the correction as noted, r_D only changes within experimental scatter, and the lack in a temperature dependence in the data from Caldwell and Eide (1981) remains.

The data compiled by Zaytsev and Aseyev (1992) represent averages of many different sources. The latter authors have also tabulated diffusivities at infinite dilution for different temperatures. When the infinite dilution r_{D0} is plotted against temperature from 20 to 50 °C a somewhat smaller value of $r_{D0} = 1.50 \pm 0.015 \times 10^{-10}$ m (Figure A.17). A similar result is obtained with data for most other halides (not shown) tabulated by Zaytsev and Aseyev (1992): With a few exceptions, possibly related to the large amount of data compiled, r_{D0} at infinite dilution is found to be independent of temperature, while r_D at 20 to 40 ‰ was increasing with T. It further appears that, below 20 °C, also the temperature trend of r_D at 20-40 ‰ ceases or even slightly inverts. As the tables for some other halides only extend down to 10 °C and further evaluations are needed.

In Figure (A.17) also the observation for a 1 M NaCl solution from Clack at 18 °C (Kaufmann, 1960) has been included as a triangle. It differs from the 10 % larger values from Zaytsev and Aseyev (1992), indicating the scatter in the source data. Taking these additional aspects into account, it appears that below 20 °C, r_D depends only weakly on temperature. Hence, the compilation from Zaytsev and Aseyev (1992) and the data Caldwell and Eide (1981) are only inconsistent near 50 °C. The similarly large scatter in an earlier study from Caldwell (1973) indicates some limitations of the technique based on salt distribution equilibration times of several hours. On the other hand, a prediction based on activity coefficients, shown as a dashed line in Figure A.17, does not yield the strong temperature dependence from Zaytsev and Aseyev (1992). This prediction will be discussed below.

A.3.3 Concentration dependence

In Figure (A.18), the concentration dependence of D_s tabulated by Zaytsev and Aseyev (1992) is shown, D_s being normalized by D_{s0} at infinite dilution. A minimum in the diffusion coefficient is seen, at all temperatures and between between 20 and 40 ‰. The values from Zaytsev and Aseyev (1992) above 4 M or 190 ‰ have to be considered with some care. It appears that these have been partially extrapolated, no longer revealing a maximum in D_s near 4 M, as found by other authors at 25 °C (Vitagliano and Lyons, 1956; Harned and Owen, 1958; Tyrrell and Harris, 1984). The raw data from (Vitagliano and Lyons, 1956) are shown for comparison. They indicate a relative difference of only a few percent in D_s. The minimum tabulated by Zaytsev and Aseyev (1992) weakens with

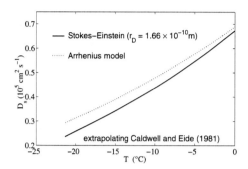

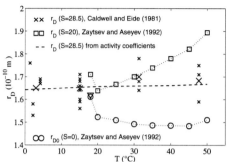

Fig. A.16: Simply extrapolated solute diffusivity of aqueous NaCl at a salinity of 28.5
‰ based on equations A.45 (solid line) and
A.46 (dashed line), both suggested by Caldwell and Eide (1981) from measurements between 1 and 47 °C.

Fig. A.17: Stokes-Einstein-radii given by
Caldwell and Eide (1981) for S=28.5
‰ (crosses) with large crosses showing the
ensemble mean at 4 temperatures. Radii
corresponding to the tabulations from Zaytsev and Aseyev (1992) are shown for 20
‰ as squares and at infinite dilution at circles. A triangle at 18 °C is from Clack at 55
‰ (Kaufmann, 1960)

decreasing temperature, while the increase of D_s at higher concentrations apparently
strengthens. Comparison with some limited observations at 5 °C (Bruins, 1929) and 18.5
°C (Gordon, 1937) indicates a very similar behaviour to the 25 and 30 °C curves in
Figure (A.18), supporting that there is no temperature dependence (not shown here).
The average value and observational range from Caldwell and Eide (1981), as mentioned
above independent of temperature and obtained close to the minimum in D_s, is also
shown in the figure and supports this view.

The infinite dilution value, obtained from the best estimate $r_{D0} = 1.50 \times 10^{-10}$ m, is
$D_{s0} = 0.74 \times 10^{-5}$ $cm^2 s^{-1}$ at 0 °C, and may be compared to $D_{s0} = 0.78 \times 10^{-5}$ $cm^2 s^{-1}$
reported by Gross (1968) based on ionic conductance measurements. Too little data are
available between 0 and 15 °C and an extrapolation to high salinities and temperatures
below 0 °C requires some further considerations.

Electrolyte theory

The Stokes-Einstein relation is a hydrodynamic principle based on rigid spheres. In the
present case of an electrolyte the Stokes-Einstein radius r_D or r_{D0} is a parameter in the
correlation between viscosity and diffusivity, but cannot be identified with a rigid radius.
In the theory of electrolytes, a comparable length scale comparable to the characteristic
radius of an ion from to the classical theory from Debye, Milchner and Hückel (Falken-

hagen, 1953; Robinson and Stokes, 1955). It is almost independent of temperature in the range 0 to 25 °C. Its value for NaCl is 4 to 4.2×10^{-10} m, and it is found from ionic activity coefficients (Falkenhagen, 1953; Robinson and Stokes, 1955). The ratio of this Debye length scale and r_D0 is approximately 2.7 to 2.8 for NaCl. A remarkably similar ratio 2.8 ± 0.2 is found (not shown here) for the infinite dilution r_{D0} for most other halides, tabulated by Zaytsev and Aseyev (1992). This similarity of diffusion and electrolytic conductivity allows reasonable predictions of the concentration dependence of D_s based on electrolytic conductivity data. This holds between infinite dilution and a molality of about 0.02 M, yet above this concentration an exact theoretical prediction is more difficult (Falkenhagen, 1953; Robinson and Stokes, 1955; Harned and Owen, 1958; Tyrrell and Harris, 1984). A detailed discussion of related problems and models has been given by (Tyrrell and Harris, 1984). For NaCl and many other halides the basic features of the concentration dependence of D_s/D_{s0} are reasonably represented by the following simplified equation after (Robinson and Stokes, 1955)

$$\frac{D_s}{D_{s0}} = (1 + f(M)) \left(1 + M\frac{d\,ln\gamma_a}{dM}\right) \frac{\mu_w}{\mu_b}, \tag{A.47}$$

where μ_w/μ_b is the viscosity ratio of pure water and the salt solution, M the molality, γ_a the ionic activity coefficient and

$$f(M) = 0.061\,M^{1/2} - 0.0537\,M + 0.018\,M^{3/2} - 0.0021\,M^2 \tag{A.48}$$

is a simplified function which accounts for ion interactions. f(M) has, for the sake of simplicity, been approximated from the computations by Robinson and Stokes (their table 11.8, p. 329) and is assumed valid at other temperatures. Representing the overall *electrophoretic* effect, it changes D_s by the order of 2% (Robinson and Stokes, 1955): the function f(M) increases rapidly to about 0.023 at 1 M and then decreases again slightly. It is noted that, in the original equation given by Robinson and Stokes (1955), the *electrophoretic* correction terms have been simplified, and a factor based on water diffusion and molality has been omitted in equation A.47. Both Robinson and Stokes (1955) and Harned and Owen (1958) noted the uncertainty in both factors, and introducing such terms did not give an improved agreememnt over the temperature range from Zaytsev and Aseyev (1992). Tyrrell and Harris (1984) and other contributors cited therein have discussed related problems, indicating that the theory of diffusion is incomplete. However, for the present temperature and salinity ranges, the unmodeled physics accounts likely for only a few percent in D_s.

D_s at subzero temperatures from activity coefficients

To compute D_s from equation A.47 at and below 0 °C, μ_w/μ_b has been computed from equation A.37. The the activity coefficients γ_a were taken from the study by Thurmond

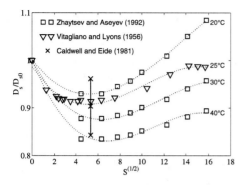 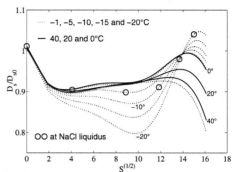

Fig. A.18: D_s/D_{s0}: ratio of NaCl diffusivity in water to its infinite dilution value taken from tables of Zaytsev and Aseyev (1992) and shown as squares. The temperature-independent average and scatter obtained by Caldwell and Eide (1981) at S=28.5 is also shown as a cross with a bar.

Fig. A.19: D_s/D_{s0} of NaCl in water based on equation (A.47) simplified after Robinson and Stokes (1955). Activity coefficients of NaCl above 0 °C have been taken from Clarke and Glew (1985), below 0 °C from Thurmond and Brass (1987).

and Brass (1987). For temperatures above 0 °C the equations given by Korosi and Fabuss (1968) were used for μ_w/μ_b and the activity coefficients are taken from Clarke and Glew (1985). The resulting D_s/D_{s0} is shown in Figure A.19 for different temperatures. A comparison with Figure A.18 shows that this procedure only yields a weak temperature dependence of D_s/D_{s0}. The prediction of equation (A.47), at a salinity of 28.5 ‰, using again $r_{D0} = 1.50 \pm 0.015 \times 10^{-10} m$, is shown as a dashed line in Figure A.17. It appears that equation (A.47) is more consistent with a lack in the temperature dependence of r_D . Below 30 °C, the agreement with the observations from Caldwell and Eide (1981) at 28.5 ‰ is almost perfect with differences of 1 % in D_s when using ensemble averages. Also the observations tabulated at 5 °C by Bruins (1929) are consistent with the graphs in Figure A.19 to within 2-3% over the whole salinity range.

A.3.4   Proposed diffusivity at the liquidus

In spite of some uncertainties in the observations and in the theories of diffusion, equation A.47 appears to predict D_s/D_{s0} within a few percent and has been used to extrapolate D_s into the regime below 0 °C, shown as dotted lines in Figure A.19. The predictions at the liquidus temperature appear as circles in Figure A.19. These points were used to obtain the polynomial

$$\frac{D_s}{D_{s0}} = 1 - 0.2480 \left(-T_f\right)^{1/2} + 0.2462 \left(-T_f\right) - 0.1191 \left(-T_f\right)^{3/2}$$

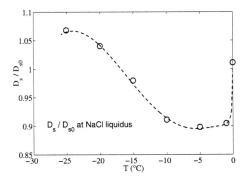

Fig. A.20: D_s/D_{s0}: ratio of NaCl diffusivity in water to its infinite dilution value taken computed via activity coefficients from equation (A.47).

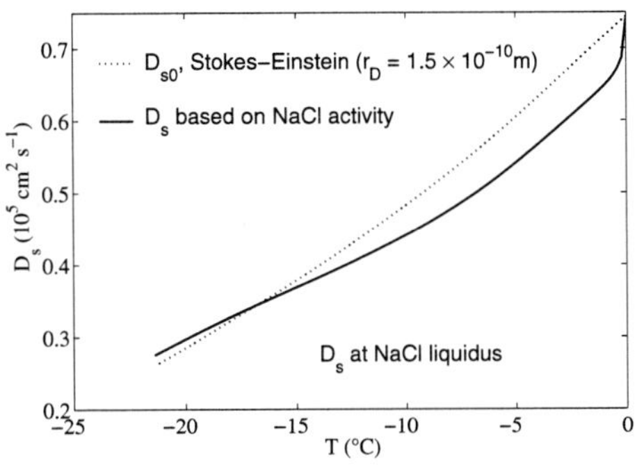

Fig. A.21: Diffusivity D_s of aqueous NaCl at the liquidus concentration estimated from activity coefficients and D_{s0} at infinite dilution (A.47) using $r_{D0} = 1.50 \times 10^{-10} m$

$$+0.02684 \left(-T_f\right)^2 - 0.002156 \left(-T_f\right)^{5/2} \tag{A.49}$$

shown in Figure A.20. Equation (A.49) is then used in connection with equation A.45 and $r_{D0} = 1.50 \pm 0.015 \times 10^{-10}$ m to compute D_s along the liquidus curve of aqueous NaCl. The result is compared in Figure A.21 to the relationships from Figure A.16 for infinite dilution.

A.3.5   Summary

The main results of the present analysis of the solute diffusivity of NaCl in aqueous solutions can be summarised as follows:

- At infinite dilution D_{s0} of NaCl in water may be described by the Stokes-Einstein equation (A.45) using $r_{D0} = 1.50 \pm 0.015 \times 10^{-10} m$. Here $D_{s0} = 0.74 \times 10^{-5}$ cm^{2}s^{-1} at 0 °C.

- There is some disagreement considering a possible temperature dependence of D_s/D_{s0}, the ratio of concentration-dependent diffusivity to the infinite dilution value, when comparing the experiments from Caldwell and Eide (1981) and the tabulations from Zaytsev and Aseyev (1992). Additional data sources (Bruins, 1929; Gordon, 1937) indicate that, below 25 °C, D_s/D_{s0} depends little on temperature. A simplified diffusivity model based on activity coefficients predicts a very weak temperature dependence.

- D_s/D_{s0} has been determined at the liquidus temperature of aqueous NaCl using activity coefficients at subzero temperatures. D_s/D_{s0} first decreases sharply from

1 to 0.9, then increases to values slightly above 1 at the eutectic concentration.

- At the typical freezing temperature of seawater, -1.9 °C, the present study predicts $D_s = 0.62 \times 10^{-5}$ cm^{2}s^{-1}, while the infinite dilution value of supercooled water at this temperature is $D_{s0} = 0.69 \times 10^{-5}$ cm^{2}s^{-1}. The uncertainty of these values may, on the basis of the available observations, be estimated as $\pm 0.02 \times 10^{-5}$ cm^{2}s^{-1}.

A.4   Seawater phase diagram

While dominated by NaCl, seawater contains several other salts that influence its phase relationship. The main elements dissolved in the world ocean are Chlorine, Sodium, Magnesium, Sulphur, Calcium and Potassium (Sverdrup et al., 1946; Dietrich et al., 1975; Millero and Sohn, 1992). The ratios of main ions in sea water are relatively constant. In standard ocean water these 6 ions account for $\approx$ 99.35 of the overall mass of salts, with individual percentages of 55.03 % Cl^-, 30.66 % Na^+, 3.65 % Mg^{2+}, 7.71 % SO_4^{2-}, 1.17 % Ca^{2+} and 1.13 % K^+ (Millero and Sohn, 1992). When sea water freezes, the ion ratios in the concentrating brine change due to the crystallisation of salts (Ringer, 1906; Nelson and Thompson, 1954; Assur, 1958). It also is useful to think of a recipe to produce standard ocean water in terms of relative salt fractions. The major salts in such a recipe make up 99.1 % by weight and may be given as 68.2 % NaCl, 14.3 % $MgCl_2$, 11.4 % Na_2SO_4, 3.2 % $CaCl_2$ and 2.0 % KCl, see f.e. Millero and Sohn (1992). This recipe is within 0.1% of each salt component the same as originally suggested by Lyman and Flemings (Sverdrup et al., 1946).

A.4.1   Data sources and previous interpretations

The multicomponent character of seawater implies that some salts start to crystallise close to, but not exactly at their respective eutectic concentrations. These eutectic transitions involve sharp changes in the liquidus relationship compared to binary solutions like, e. g., aqueous NaCl. A proper description of the phase relation of seawater requires the division into basic temperature regimes separated by the effective eutectic points of these salts. The first detailed observations of seawater ion concentrations below -2 °C and down to -50 °C have been performed by Ringer (1906). Nelson and Thompson (1954) extended Ringer's work in a more detailed study, measuring the major ion concentrations in natural seawater brine remaining at the bottom of small containers. Their work has become the standard reference for estimates of the liquidus relationship of seawater (Assur, 1958; Anderson, 1960; Richardson, 1976). A similar dataset of the six major ions, yet for artificial seawater, was published by Gitterman (1937) and reanalysed by Tsurikov and Tsurikova (1972). The following discussion is based on these three references (Ringer, 1906; Nelson and Thompson, 1954; Tsurikov and Tsurikova, 1972) and extends earlier interpretations of the latter (Assur, 1958; Anderson, 1960; Tsurikov and Tsurikova, 1972; Richardson, 1976; Marion et al., 1999). Here some corrections, mostly of systematic nature, are proposed. These result in a liquidus relation of seawater which is more consistent with oceanographic standard references (Millero and Leung, 1976) and recent thermodynamic simulations (Spencer et al., 1990; Marion and Farren, 1999; Marion et al., 1999).

A.4.2   Distinct temperature regimes

Discussing the phase diagram of seawater it is useful to distinguish the following temperature regimes:

$0 > T > $ -8 °C

Precipitation of $CaCO_3\ 6H_2O$
A small fraction of the Ca^+ ions starts to precipitate below -1.9 °C in form of $CaCO_36H_2O$ (Ringer, 1906; Tsurikov and Tsurikova, 1972). The precipitated fraction $r_{Ca(X)}$ of the calcium ion, with respect to its initial concentration in seawater, can be computed based on any non precipitating reference ion X from

$$r_{Ca(X)} = \frac{\left(\frac{C_{Ca0}}{C_{X0}} - \frac{C_{Ca}}{C_X}\right)}{\frac{C_{Ca0}}{C_{X0}}}, \tag{A.50}$$

where C_{Ca0} and C_{X0} denote the (initial) concentrations in normal seawater and C_{Ca} and C_X stand for the observed concentrations in the liquid brine. Equation A.50 has been applied by, for example, Tsurikov (1974) to describe ion fractionation in natural waters. In the following the notation Ca/X will be used to denote the ionic concentration ratio C_{Ca}/C_X. The same notation is used for other precipitating ions.

The observations from Nelson and Thompson (1954) indicate that, between -2 and -8 °C, approximately 10.35 % of the Calcium ion precipitates presumably as $CaCO_36H_2O$. Its precipitated fraction then stays rather constant down to -20 °C (Figure A.22). Assur (1958) arrived at a similar value of 10.0 %. Assuming standard seawater ionic ratios of $Ca/Cl=0.02127$, and $(CO_3+HCO_3)/Cl \approx 0.0065$ for the carbonate, and using the corresponding molecular weights (Millero and Sohn, 1992), one may derive that in the experiments from Nelson and Thompson (1954) 50.7% of the carbonate (CO_3+HCO_3) should have precipitated as $CaCO_36H_2O$. This is consistent with observations reported by (Tsurikov and Tsurikova, 1972) for sea ice and by Millero (1996) for high salinity brine in Mexican lagoons. Both authors report a maximum decrease in HCO_3^- to ≈ 50 % of the seawater reference. The CO_2-system is, of course, more complex than this simple view (Sverdrup et al., 1946; Millero, 1996), in particular when partitioning at the ice-water interface is considered (Papadimitriou et al., 2003). The given numbers are, however, consistent with available data and most representative. In warmer ice the precipiation of $CaCO_36H_2O$ implies only a 0.003 fractional decrease in the absolute brine salinity. This means that only at temperatures below -23 °C, when NaCl precipitates, the relative influence of carbonate relative to chlorinity increases to a few percent.

Precipitation of mirabilite, $Na_2SO_4\ 10H_2O$
Crystallisation of $Na_2SO_4\ 10H_2O$ starts between approximately -6 and -8°C   (Ringer,

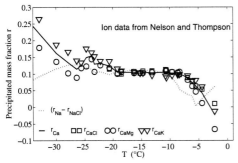

Fig. A.22: Precipitated fraction of Calcium $r_{Ca}(X)$ based on the reference ions X=Cl (squares, only above -23 °C), X=Mg (circles) and X=K (triangles). The average is indicated as a solid line. The precipitated fraction r_{Na} of Natrium (in form of $Na_2SO_4H_2O$) averaged from K and Mg-ratios is shown after subtraction of the Natrium fraction r_{Na} precipitated as hydrohalite of NaCl.

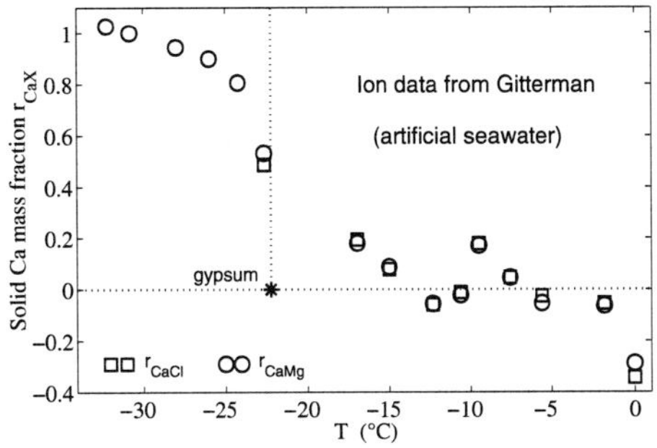

Fig. A.23: Precipitated fraction of Calcium $r_{Ca}(X)$ based on the reference ions Mg (circles) and K (triangles) for Gitterman's artificial seawater. The horizontal dotted line is theoretically expected due to the lack in HCO_3^-. The vertical dotted line marks the gypsum formation transition simulated by Marion et al. (1999).

1906; Nelson and Thompson, 1954). This transition is more relevant in terms of the overall brine composition, as Na_2SO_4 contributes more then 10 % by weight to the seawater salt content. Ringer (1906) concluded from observed SO_3-Cl-ratios, that slight precipitation of mirabilite likely begins between -5 and -8 °C, with heavy precipitation present near $\approx$ -8.2 °C. In the scattered ion data from Nelson and Thompson (1954) either the SO_4-Cl and Na-Cl-ratios (the latter shown in Figure A.24) or the precipitated fraction of Na_2SO_4 (shown as the dotted line in Figure A.22) may be taken as indicators for the onset of crystallisation in the range -6.2 and -8.3 °C. A frequently cited transition temperature is -8.2 °C(Assur, 1958; Weeks and Ackley, 1986; Eicken, 2003). It is worth a note that it was adopted by Nelson and Thompson (1954) from Ringer (1906), although it is neither manifest nor resolvable in the observations from Nelson and Thompson or Ringer. Also the ion data from Gitterman (Tsurikov and Tsurikova, 1972) show light precipitation between -5.6 and -7.6 °C, with heavy precipitation having started at -9.5 °C. Thermodynamic simulations from Marion et al. (1999) give -6.3 °C, as the theroretically expected eutectic temperature, while Gitterman himself proposed a value of -7.3 °C (Tsurikov and Tsurikova, 1972). All observations indicate that crystallisation of Na_2SO_4 $10H_2O$ sets in slowly from a slightly supersaturated state.

-8 > T > -23 °C

The precipitation of NaCl, presumably as hydrohalite $NaCl\, 2H_2O$, is observed near -23 °C (Ringer, 1906; Nelson and Thompson, 1954; Assur, 1958) and hence slightly below the eutectic point -21.3 °C of a binary NaCl solution (Cohen-Adad and Lorimer, 1991). Marion et al. (1999) found, in thermodynamic simulations, almost the same temperature of -22.9 °C, as given by Nelson and Thompson (1954). As close to this temperature heavy precipitation starts, a strong change in the liquidus slope takes place (Figure A.26).

-23 > T > -36 °C

The tabulated data from Nelson and Thompson (1954) imply a sudden change in Ca-Mg-ratio between -36.25 and -36.55 °C, which can be interpreted as the precipitation of the third major salt $MgCl_2\, 12H_2O$. Ringer (1906) had supposed a precipitation temperature of -36 °Cwhile thermodynamic simulations give -36.2 °C (Spencer et al., 1990; Marion et al., 1999). Notably, Assur (1958) has not correctly evaluated this transition. Another transition in this regime is the precipitation of potassium as anhydrous sylvite (KCl) at approximately -34 °C, which is evident from Nelson and Thompson's data and predicted by thermodynamic simulations (Spencer et al., 1990; Marion et al., 1999).

-36 > T > -54 °C

Assur (1958) first proposed brine salinity tables in this regime, which is only accessed by sea ice under extreme natural conditions. At about -54 °C antarctite (CaCl₂ 6H₂O) precipitates as the last minor salt in seawater. This transition is indicated in time-temperature plots during melting (Nelson and Thompson, 1954) and theoretically predicted by thermodynamic simulations (Marion et al., 1999). The only brine salinity observations in this regime are indirect estimates based on NMR measurements from Richardson (1976) whose quality is, as discussed below, difficult to assess. This temperature range is not considered in the present study, as also the thermo-physical properties of supercooled water become very uncertain. From a hydrodynamical standpoint, brine volumes in normal sea ice become so small that convection is no longer relevant.

A.4.3   Previously suggested relations: seawater freezing

In the high temperature regime down to -8 °C, the small initial influence of precipitating CaCO₃ 6H₂O is almost negligible, but the exact onset of the precipitation of mirabilite between -6.3 and -8 °C makes some difference when considering the freezing point depression. Most existing studies deal with freezing of natural seawater and thus with salinities

less then 40 ‰ and temperatures above -2 °C. Millero and Leung (1976) have compared
several data sources and, based on thermodynamic principles, derived the polynomial

$$T_f = -0.05722\,S + 1.69808 \times 10^{-3}\,S^{3/2} - 2.13412 \times 10^{-4}\,S^2, \qquad (A.51)$$

which gives, for $S < 40$ ‰, the freezing temperature of natural seawater in °C. It is noted
that the coefficients in this equation differ slightly from those provided by Millero and
Leung (1976), as S is here given in g/kg instead of the 1.00488 smaller practical salinity
used by the latter authors. Freezing points within 1 % of equation A.51 had already been
reported by Krümmel in 1907 (Krümmel, 1907; Dorsey, 1940).

Nelson and Thompson (1954) have given a relation between freezing temperature and
chlorinity. Modified for S in g/kg under the assumption of standard seawater it can be
written as

$$T_f = -0.05321\,S - 5.850 \times 10^{-7}\,S^3. \qquad (A.52)$$

Nelson and Thompson (1954) claimed its validity below $S<130$ ‰. However, equation
(A.52) is not founded on thermodynamic principles (Millero and Leung, 1976) and produces a difference of the order of 0.2 K in T_f at -8 °C, in comparison to equation A.51
and the results described below.

Pragmatic relations between S and T_f at lower temperatures have been proposed by
several authors (Assur, 1958; Pounder, 1964; Cox and Weeks, 1986). The relation most
widely used in sea ice modelling was suggested by Cox and Weeks (1986) on the basis of
Assurs (1958) phase relation table. It consists of two polynomial fits for the temperature
regimes below and above -22.9 °C. This neglects the change near -8 °C and implies an
inaccuracy which adds to the possible errors in Assur's interpretation of the original data
from mainly Nelson and Thompson (1954). In the following, additional data sources
mentioned above will be compared and critically evaluated to find a best estimate of the
liquidus relationship between -2 and -34 °C.

A.4.4 Critical comparison of reference data

The reference data proposed by different investigators from studies of the brine salinity
along the liquidus curve of seawater, are listed in table A.1. The main differences between
the experiments and interpretations are commented as follows.

Normal seawater

The freezing point relation of seawater given by Millero and Leung (1976),validated for
salinities up to 40 ‰ against many data sources, can likely be treated as the most
accurate reference value. The salinity by weight in the table, 36.53 at -2 °C, has been
obtained by increasing the practical salinity scale from these authors by 1.00488.

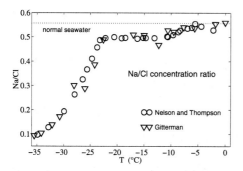

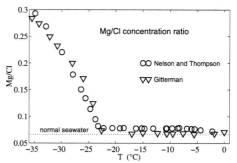

Fig. A.24: Na/Cl ionic concentration ratio for seawater brine from Nelson and Thompson (circles) and for artificial seawater from Gitterman (triangles). The dotted line is the ratio 0.5565 for normal seawater (Millero and Sohn, 1992).

Fig. A.25: Mg/Cl ionic concentration ratio for seawater brine from Nelson and Thompson (circles) and for artificial seawater from Gitterman (triangles). The dotted line is the ratio 0.06626 for normal seawater (Millero and Sohn, 1992), indicating a calibration error of 15% in the Mg dtermination by Nelson and Thompson.

Reference data from Nelson and Thompson (1954)

Nelson and Thompson (1954) obtained ion weight fraction data for the six major seawater ions Cl^-, Na^+, Mg^{2+}, SO_4^{2-}, Ca^{2+} and K^+. The covered temperature range was -2 to -43 °C, freezing seawater in small containers. The authors did not present their data as an integrated salinity. The values in the table are the sum of all measured ion concentrations with three further corrections: (i) The restriction to 6 analysed ions leaves a certain fraction of unresolved ions. With a ratio of 0.0065 of unresolved ions and chlorinity (Millero and Sohn, 1992) the overall salinity may be obtained by applying the factor $(1 + .0065 \times 1.81537) = 1.0118$, as long the ion ratios do not change. This is notably not the case. To account for this, the unresolved ion fraction of 0.0018 is assumed to be partitionable into a part which keeps its ratio to Cl constant (0.0053), and a carbonate fraction (0.0065). Appr5oximately 50 % of the latter precipitates between -1.9 and -8 °C as $CaCO_3 6H_2O$. The correction is most important at low temperatures when the dissolved carbonate increases relative to precipitating Cl. (ii) The concentration of SO_4^{2-}, not measured below -30 °C, was linearly interpolated between its -30 °C value and 0.3 ‰ obtained by Ringer (1906) for natural seawater at -40 °C. (iii) Figure A.25 shows that the reported ratios of Mg^{2+} to chlorinity and other ions are typically 15 % larger than in normal seawater. A thermodynamic simulation by Spencer et al. (1990) does not show such a change. This indicates a systematic overestimate of S due to incorrect Mg^{2+} determinations. This overestimate is expected to amount to 0.7% for the range -2 to -22 °C, increasing to 3 % at -34 °C, if the miscalibration of Mg^{2+} was 15 % through all runs. It was therefore decided to reduce the Mg^{2+} concentrations throughout by 15 %.

The initial ratios of Ca and K to chlorinity were 0.021 and 0.0197 as expected in normal seawater (Millero and Sohn, 1992). The initial ratio of Na to Cl in the data is scattered (Figure A.24). Taking a value of 0.545 ±0.01 it is approximately 0.01 smaller than the value of 0.556 for normal seawater (Millero and Sohn, 1992). This scatter creates, at higher temperatures, an uncertainty of ≈ 1 % in the salinity. Due to the apparent lack in precision no correction was applied to the Na concentrations.

The final salinity values, after correction, were then interpolated linearly to the temperatures at which they appear in the table A.1. The corrections bring the salinities into better agreement with the simulations from Spencer et al. (1990) at all temperatures, and with the data from Gitterman between -30 and -36 °C. The agreement with the reference value from Millero and Leung (1976) at -2 °C also improves accordingly. The discrepancy between the corrected values and the Gitterman data between -16 and -30 °C will be discussed below.

Interpretation by Assur (1958) and Anderson (1960)

Assur (1958) noted mainly having used the data from Nelson and Thompson (1954) to derive his proposed liquidus salinities. He discussed several possible methods of analysis, but did not give the final exact procedure to produce his table. It is possible that, as Assur starts from the Mg/Cl ratio of normal seawater, the bias in the Mg concentration from Nelson and Thompson (1954) was excluded. However, the origin of his tabulated values is not clearly explained. It appears, from the ratio of chlorinity to S (in g salt per kg water) in his table, that Assur did neither account for the difference between pratical salinity units and weight percent. Therefore Assur's initial values should be corrected by a factor 1.00488, which bhas been applied to all values given. As mentioned above, 50% of the unmeasured ions are carbonate ions, of which in turn the half appears to precipitate between -2 and -8 °C. Assuming that the other minor ions keep their ratio to chlorinity, this implies a change in the correction factor to 1.0037. However, at low temperatures this correction factor will increase again, as during precipitation of NaCl the relative contribution of HCO_3^- increases. For the sake of simplicity, Assur's values in table A.1 have been corrected by 1.00488. This facilitates a comparison above -20 °C but the values need to be viewed with care at lower temperatures.

Anderson (1960) also based his determinations on the major ions from Nelson and Thompson (1954). From his table 2, the brine salinities can be obtained down to -22 °C by substracting the weight fraction of Na_2SO_4, which is 0.4403 of the crystallizing salt $Na_2SO_4H_2O$ (Millero and Sohn (1992)). Down to -23 °C, the salinities proposed by Anderson agree to within 0.1 % with those obtained by Assur (1958) and are therefore not shown here. The uncorrected values from Assur (1958) and Anderson (1960) have also been listed by Cox and Weeks (1983), for use in sea ice modeling. At -2 °C they differ by 2 % from the reference point from Millero and Leung (1976), which indicates that the corrected Nelson and Thompson (1954) salinities are preferable over the valkues from Assur (1958). Between -10 and -24 °C, the agreement of Assur's liquidus values (increased by 1.00488) with the corrected data from Nelson and Thompson (1954) is better than 1 %. The assumption of precipitation of $MgCl_28H_2O$ near -24 °C, proposed by Assur (1958), is likely incorrect. However, in Assur's study this made little difference in the absolute salinity computation, because his assumed precipitation values for NaCl were correspondingly lowered by him. Below -30 °C, the salinities are 1 to 3 % below the corrected values from Nelson and Thompson (1954). This is probably related to the unaccounted increase in the HCO_3^- contribution.

Reference data from Ringer (1906)

To make reference data from Ringer (1906) comparable to other observations, they have been converted here from Ringer's chlorinities, by using the same S_b/Cl^--ratios as found in the observations from Nelson and Thompson (1954). Therefore only Ringer's results for natural seawater are given. The value given at -8 °C is an average of two samples, indicating the uncertainty. Ringers data agree better with the corrected values from Nelson and Thompson (1954) than with those given by Assur (1958) and Richardson (1976).

Interpretation using Nuclear Magnetique Resonance

Richardson and Keller (1966) and Richardson (1976) applied a Nuclear Magnetique Resonance (NMR) technique to estimate the total weight of liquid water. With additional information on eutectic concentrations, Richardson (1976) derived the liquidus salinity values in the last column of table A.1. These values hence depend on assumptions about the precipitation of salts. For example, in their first study, Richardson and Keller (1966) had borrowed the observed salinities from Nelson and Thompson (1954) to compute liquid volume fractions. Richardson (1976) has also compared his results with the salinities from Assur (1958) and Nelson and Thompson (1954), but several values of both authors are not correctly plotted by Richardson (1976) in his figure 7. Richardson repeated the inaccuracy from Assur (1958) interpreting standard salinity as weight fraction and not including the minor ions in seawater. Hence also Richardson's salinities have been increased by the factor 1.00488, the modified values being given in table A.1. At high temperatures above -8 °C, the liquidus salinities given by Richardson (1976) deviate by up to 7 % from the reference values from Nelson and Thompson (1954). In particular the

comparison with the reference value from Millero and Leung (1976) at -2 °C indicates
unaccounted problems with the devised NMR method.

Reference data from Gitterman

The original paper by Gitterman (1937) has not been available to the present author,
but all ion concentrations were tabulated by Tsurikov and Tsurikova (1972). These data
were first compared by the latter authors to the data from Nelson and Thompson (1954),
and have been further compared in later studies by Richardson (1976) and Marion et al.
(1999). The values given in the table have been obtained by linear interpolation of the
salinities tabulated by Tsurikov and Tsurikova (1972). No correction due to unaccounted
ions was applied, as Gitterman used artificial seawater. Above -16 °C Gitterman's results
are fully consistent with the corrected data from Nelson and Thompson (1954). A major
difference between the datasets is related to the precipitation of gypsum ($CaSO_42H_2O$),
which Gitterman proposed below -15 °C (Tsurikov and Tsurikova, 1972; Richardson, 1976;
Marion et al., 1999). It may be related to differences in the brine salinity of the order of
1 to 2 % and will be discussed below.

Discussion

Different attempts to derive the liquidus curve of seawater have been compared. From the
discussion so far it appears that the original data from Nelson and Thompson (1954) are
be preferable over the values tabulated by Assur (1958) and Anderson (1960), if they are
corrected in the described manner. Spencer et al. (1990) used a thermodynamical model
of the seawater multicomponent system based on a large set of empirical coefficients. This
model has been further developed and gives predictions of the phase equilibrium salinity
in fair agreement with Nelson and Thompson's ion data (Marion et al., 1999). The
corrected Mg^{2+} concentrations are also more consistent with simulations from Spencer
et al. (1990). The assumed overestimate of S_b, due to the likely miscalibration of the
Mg^{2+} measurements Nelson and Thompson (1954), starts only to exceed 0.7 % below -
22 °C. However, the correction also improves the agreement with the data from Gitterman
and Ringer (1906). Hence, with respect to the corrections applied here, all data sets are
consistent and the uncertainty in the derived equilibrium brine salinities is probably less
than 1 %. It is finally worth a note that first the corrections bring the salinity obtained
at -2 °C from the Nelson and Thomson data close to the most trustworthy standard
reference of seawater from Millero and Leung (1976).

 The reasonable agreement of the corrected Nelson and Thompson (1954) data with
the NMR derivations from Richardson (1976) between -10 and -30 °C is promising.

Liquidus salt content S_b of seawater

T °C	Millero	Nelson	Gitterman	Ringer	Assur[c]	Richardson[c]
-2	36.53	36.62	36.36		37.59	39.24
-4		70.05	67.93		70.59	67.08
-6		99.72	98.70		99.81	94.91
-8		123.5	123.7	117.5±3.0	126.5	122.7
-10		141.7	141.0	140.6	142.8	144.7
-12		156.4	155.5	148.0	157.6	158.5
-14		170.6	169.5		171.5	172.3
-16		184.4	186.5		184.4	185.7
-18		197.6	202.3		197.0	199.3
-20		210.7	214.5	213.4	209.9	212.7
-22		223.8	226.7		222.6	226.8
-24		231.1	227.9		230.5	232.8
-26		231.3	227.5		232.7	235.0
-28		232.6	233.5		234.1	235.7
-30		235.8	237.8	229.8	235.6	237.1
-32		241.1	241.0		237.5	246.9
-34		248.6	244.6		240.2	254.1

Tab. A.1: Liquidus salinity S_b of seawater obtained by several authors. The table includes the most accurate reference value from Millero and Leung (1976), the interpolated data from Nelson and Thompson (1954) with corrections applied as noted in the text; [c] salinities proposed by Assur (1958) and Richardson (1976) have been increased by the factor 1.00488. Some earlier determinations by Ringer (1906) are also listed.

However, the apparently bad performance of NMR above -10 °C calls for further NMR evaluations. Below -30 °C, the agreement decreases again. As already pointed out by Tsurikov and Tsurikova (1972), the NMR determination is not self-sustaining, but it requires additional informations about the ionic ratios during the precipitation of salts. The latter are uncertain in a multicomponent solution. One must rfecognise that the NMR results from Richardson (1976) were thus only validated with a few very scattered points from Ringer (1906) and Nelson and Thompson (1954). Later determinations using NMR have indicated complex responses to the rate of freezing and the initial salinity, indicating that this method must be used with care (Melnichenko, 1979; Cho et al., 2002).

For the temperature range -10 to -24 °Calso the values given originally by Assur (1958) and Anderson (1960) agree with the present corrected Nelson and Thompson estimates to within 1%. This is eventually a consequence of cancellation of the corrections not accounted by Assur, having underestimated the true salt mass due to the missing factor

1.00488, yet overestimated it due to the miscalibration of the Mg^{2+} data from Nelson and Thompson (1954). Because below -23 °C the relative contribution of Mg and HCO_3 to the brine salinity increases, the effect of the miscalibration will become more dominant and the values given by Assur, and used by, for example, Cox and Weeks (1983) are less confident in this regime .

A.4.5 Disequilibrium and gypsum precipitation

Before an optimal phase-relationship is suggested from the tabulations, it is useful to discuss the mentioned major difference between the Nelson-Thompson data and the observations from Gitterman. The main difference appears in the salinities near approximately -16 °C. As shown first by Tsurikov and Tsurikova (1972), their table 1, and discussed by Marion et al. (1999) and Marion and Farren (1999), this difference may be related to different pathways of freezing in the experiments from Nelson and Thompson (1954) and Gitterman. The latter author proposed that the precipitation of gypsum ($CaSO_4 2H_2O$) begins at a temperature of about -15 to -17 °C. The reaction

$$CaSO_4\, 2H_2O + 2(NaCl\, 2H_2O) + 10H_2O \rightleftharpoons CaCl_2\, 6H_2O + Na_2SO_4\, 10H_2O, \qquad (A.53)$$

noted by Tsurikov and Tsurikova (1972), implies a precipitation of gypsum and the re-dissolution of mirabilite. However, simulations show this precipitation to occur at a lower temperature of -22.2 °C (Marion et al., 1999). Figure 1 from Tsurikov and Tsurikova (1972) shows a crossing in the SO_4^{2-} concentrations from Nelson and Thompson (1954) and Gitterman at this temperature: below -22 °C, SO_4^{2-} appears to strongly redissolve in the experiments from Gitterman, while it decreases further according to the observations from Nelson and Thompson (1954). In Figure A.22 and A.23, the precipitated fractions of calcium in the data sets are compared. It appears that the data from Gitterman (A.23) are much more scattered above -20 °C than the results from Nelson and Thompson (1954), shown in Figure A.22. Because no samples were obtained by Gitterman in the temperature range -17.0 to -22.6 °C, his conclusion of a gypsum precipitation near -17 °C is actually not corroborated by his data. The simulated transition at -22.2 °C is quite plausible.

Thermodynamic modeling

Marion et al. (1999) suggested that the alternative Gitterman freezing pathway may have been a consequence of the much longer equilibration times in Gitterman's experiments, compared to those of Nelson and Thompson. From longterm experiments on artificial seawater Marion et al. (1999) reported a partly confirmation of their simulation. They suggested that the Gitterman phase relation is the more stable one. However, figure A.22 indicates that also in the experiments from Nelson and Thompson some gypsum was produced below -22 °C and accompanied by a re-dissolution of mirabilite. There are also indications of an oscillating character of the reaction A.53. As Gitterman used artificial seawater, while Nelson and Thompson (1954) froze natural samples, more experimental data for different equilibration times and corresponding simulations are needed to further elucidate the pathway questions, and the difference between different multicomponent solutions. The difference between the datasets from Gitterman and Nelson and Thompson (1954) had also been noted by Richardson (1976). The latter author, however, assumeing the precipitation of gypsum below -10 °Cin his calculations, did not apply a re-dissolution of $Na_2SO_410H_2O$. This re-dissolution is clearly seen in the Gitterman data Marion et al. (1999) and also indicated in the data from Nelson and Thompson (Figure A.22).

Precipitation and re-dissolution of gypsum

As both data and simulations are consistent with a gypsum precipitation below -22 °C, one would expect that the difference between the retrieved salinity from Gitterman's and Nelson and Thompson's data is small above -22 °C. Indeed, using the tabulated data from -2 to -22 °C, there is almost no bias, the difference being 0.03 ± 1.5 %. In the temperature range -16 to -22 °C, the estimated salinity based on the Gitterman data is 1-2 % larger than the corrected Nelson and Thompson values. This is likely related to the scatter in Gitterman's data in figures A.24 and A.23 and the lack of measurements between -17.0 to -22.6 °C. The second main difference between the datasets is found at -24 and -26 °C, close to which both Gitterman and Nelson and Thompson (1954) obtained samples. Now Gitterman's values are 1.7 % lower. The irregularities in the Na^+ measurements from Gitterman cannot explain this, as the larger Mg/Cl ratios in figure A.25 imply a smaller dissolved Cl fraction and thus stronger hydrohalite precipitation in the data from Gitterman. The range -24 to -26 °C also appears as the main regime of calcium precipitation shown in A.23. The enhanced hydrohalite precipitation is consistent with the reaction equation (A.53) for gypsum. If one assumes that $CaCl_2\,6H_2O$ on the right hand side of equation A.53 exists in dissolved form, then this transition implies the liberation of 4 water molecules per calcium ion. A 0.9 % decrease in the dissolved calcium concentration from 1 % (at -22 °C) to 0.1 % (at -26 °C) implies, due to its pure water content increasing, a decrease in the brine concentration which can be estimated from the molecular weights as $0.009(4M_{H2O}/M_{Ca}) \approx 0.018$. This is close to the difference between the Gitterman and Nelson/Thompson data.

A remaining question is, why the salinities from Nelson/Thompson and Gitterman approach each other again at -28 to -32 °C. One may suspect that the difference in supersaturation in hydrohalite near its eutectic point and the different equilibration times may play a role in this context. Figure A.26 and table A.1 are strongly indicative of some salinity overshoot and supersaturation between -23 and -26 °C in the Nelson/Thompson data, which appears less pronounced for Gitterman's values. One may speculate that a ceasing supersaturation at lower temperatures (and longer times) could have decreased the dissolved NaCl fraction with some delay in the Nelson/Thompson experiments. In addition, the Nelson/Thompson data in Figure A.22 indicate a weak delayed gypsum activity between -26 and -32 °C. The deliberation of water molecules according to equation A.53 would yield an ≈ 1 % decrease in the overall brine salinity. With respect to these plausible reactions both datasets appear consistent to within 1 %.

It appears that the simulations from Marion et al. (1999) fit the corrected data from Nelson and Thompson slightly better then the Gitterman values (their figure 1 and 2). However, the apparent presence of supersaturation near the eutectic points makes a detailed comparison difficult. The difference of 1.7 % between the Nelson/Thompson and Gitterman salinities is not very large, yet it needs to be understood when focusing on this supersaturation and the stability of chemical reactions. If effective phase-relationships depend on timescales involved (Marion et al., 1999), then hysteresis and slow changes in the chemical distribution may evetually also drive other processes, as for example convection. There is still need for more experimental work, in particular at low temperatures. If, as suggested by Marion et al. (1999), a longer equilibration time would favour the Gitterman pathway,– the precipitation of gypsum and redissolution of mirabilite below -22 °C –, then for shorter timescales the slightly lower salinities from Nelson and Thompson (1954) may be more appropriate in principle. They should then be relevant for the field of new ice formation.

A.4.6 Proposed empirical phase-relationship

Using the corrected data from Nelson and Thompson (1954) as the reference to derive the liquidus curve of seawater, the following equations have been fitted to the values in column 2 of table A.1:

$$T_f = -0.05535\,S + 5.520 \times 10^{-5}\,S^2 + 1.074 \times 10^{-6}\,S^3 \tag{A.54}$$

is suggested as long as $S < 131.92$.

$$T_f = 4.978 - 0.07942\,S - 1.875 \times 10^{-4}\,S^2 - 7.923 \times 10^{-9}\,S^3 \tag{A.55}$$

describes the range $131.92 < S < 228.72$ and

$$T_f = -22.90 - 1.18953(S - S_e) + 4.1557 \times 10^{-2}(S - S_e)^2 - 5.4271 \times 10^{-4}(S - S_e)^3 \tag{A.56}$$

where $S_e = 228.72$ is the seawater liquidus salinity at -22.9 °C and precipitation of NaCl 2H$_2$O sets in, is appropriate at temperatures between -22.9 and -36 °C.

The relations are shown with the corrected data from Nelson and Thompson (1954) in Figure A.26. Equation A.54 matches Millero's equation A.51 for normal seawater at

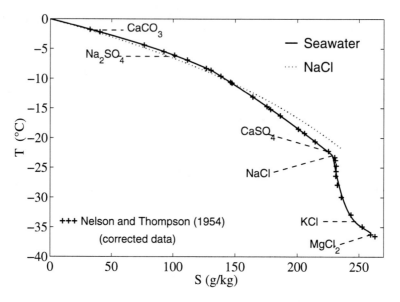

Fig. A.26: Phase relationship of seawater (solid line), fitted to the corrected ion data from Nelson and Thompson (1954) shown as crosses. The eutectic temperatures of the precipitating salts are indicated. The dotted line shows the liquidus curve of aqueous NaCl based on the data from Zaytsev and Aseyev (1992) derived in section A.2.2 .

36.1 ‰. If higher accuracies than 2 % in T_f are desired at salinities below 36 ‰, then equation A.51 should be used. Between 40 and 230 ‰, equations A.54 and A.55 deviate by less than 0.5 % from the corrected Nelson/Thompson salinities. Between -23 and -34 °C, corresponding to $230 < S_b < 247$ ‰, the inversion of equation A.56 gives a similar uncertainty in the seawater salinity. In connection with the computation of relative brine volumes at a given temperature the simplification due to equation (A.56) only leads to a mismatch of 0.5 %. However, it is noted that the data from Nelson/Thompson indicate an overshoot of the saturation value near the eutectic point of NaCl, followed by a slight decrease in S. Also shown in Figure A.26 is the corresponding phase-relationship for aqueous NaCl. At high temperatures the freezing temperatures are below those of seawater. The curves cross at a temperature of -10.5 °C. Below this temperature seawater freezes at a lower overall brine salinity by weight.

A.4.7 Precipitated ion fraction

The following assumptions are made to compute the overall precipitated ion fraction r_{prec} during freezing of seawater. Based on the calcium precipitation (figure A.22) it is found that solid $CaCO_3$ increases from 0 to 0.303 % weight fraction at -8 °C and stays constant below this temperature. Down to -34 °C, only natriumsulfate (Na_2SO_4 $10H_2O$) and hydrohalite $NaCl2H_2O$ are assumed to precipitate. An eventual precipitation of gypsum is assumed to be compensated by a redissolution of natriumsulfate as indicated in figure A.22 and neglected here. Above -34 °C the overall precipitated fraction may be computed using the concentrations of the reference ions Mg and K. As an example, the precipitated NaCl fraction $r_{NaCl(Cl,K)}$, with respect to total initial salinity, can be computed based on the reference ion K and the precipitating ion Cl via

$$r_{NaCl(Cl,K)} = \frac{\left(\frac{C_{Cl0}}{C_{K0}} - \frac{C_{Cl}}{C_K}\right)}{\frac{C_{Cl0}}{C_{K0}}} \frac{M_{0,NaCl}}{M_{0,Cl}} C_{Cl0},$$ (A.57)

where $M_{0,NaCl}$ and $M_{0,Cl}$ denote the molecular weights of Cl and NaCl, C_{Cl0} and C_{K0} the (initial) concentrations in normal seawater and C_{Cl} and C_K the observed concentrations in the liquid brine.

The overall precipitated salt content obtained from the Nelson/Thompson data is plotted in Figure A.27. r_{pr} may be approximated by the following equations. Below -1.9 °C the solid salt fraction is assumed to be zero, while

$$r_{pr} = -4.967 \times 10^{-4} \, (T + 1.9)$$ (A.58)

for the range -8<T<-1.9 °C , governed by precipitation of $CaCO_36H_2O$. Next,

$$r_{pr} = 0.00303 - 0.03144 \, (T + 8) - 4.099 \times 10^{-3} \, (T + 8)^2$$
$$-2.383 \times 10^{-4} \, (T + 8)^3 - 4.888 \times 10^{-6} \, (T + 8)^2$$ (A.59)

describes the regime for -22.9<T<-8 °C, when Na_2SO_4 $10H_2O$ precipitates, and

$$r_{pr} = 0.1088 - 0.1989(T + 22.9) - 2.052 \times 10^{-2}(T + 22.9)^2$$
$$-7.43 \times 10^{-4}(T + 22.9)^3$$ (A.60)

the range -36<T<-22.9 °C, when the precipitation of $NaCl2H_2O$ is dominant.

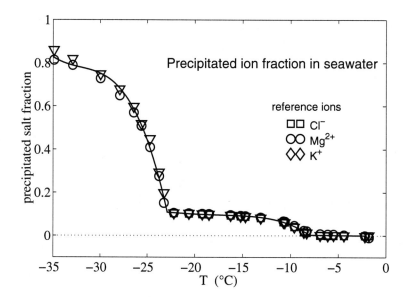

Fig. A.27: Overall precipitated ion fraction in seawater based on the reference ions Cl (squares, only above -23 ℃), Mg (circles) and K (triangles), computed from the corrected data from Nelson and Thompson (1954). The solid line is the prediction from the equations A.58 through A.61.

B. Unidirectional growth of ice

A first thorough discussion and theoretical analysis of the unidirectional growth of ice on seawater was given by Stefan (1889). The propagation of a phase boundary with latent heat release is since then known as the *Stefan Problem* and has received considerable interest (Carslaw and Jaeger, 1959; Muehlbauer and Sunderland, 1965; Rubinstein, 1971; Meirmanov, 1992; Alexiades and Solomon, 1993; Gupta, 2003). A first detailed application to the growth of freshwater ice and its dependence on boundary conditions like wind velocity and solar radiation was formulated by Devik (1931). Several aspects of the Stefan problem during the freezing of fresh natural waters were reviewed by Lock (1990).

As during freezing of seawater and saline solutions some mass fraction remains as brine in the liquid state, the effective latent heat to be released during freezing is less than for freshwater. A simple approach accounting for the reduction was given by Malmgren (1927), see appendix A.1.5. Nevertheless, sea ice has been found to grow as if it had the thermal properties of fresh ice (Johnson, 1943). This, as first glance surprising result, can be explained by the fact that not only the latent heat, but also the effective thermal conductivity K_{si} of sea ice decreases with increasing brine content. Quantitative determinations of K_{si} (Lewis, 1967; Trodahl et al., 2001) are close to the expected behaviour. The net effect of brine on the growth velocity of bulk sea ice is therefore often small compared to other factors like oceanic heat flux (Lewis, 1967), snow thickness and density (Nakawo and Sinha, 1981; Sinha and Nakawo, 1981) and wind-induced turbulent loss (Devik, 1931; Maykut, 1986). However, when brine volumes are high, with a large vertical gradient, then the ice-brine system may differ considerably from fresh ice growth. Such systems have been discussed in more recent years and approaches of different complexity have been suggested to model the growth rate (Huppert and Worster, 1985; Huppert, 1990; Worster, 1997). Applications to sea ice are, however, still semi-empirical (Wettlaufer et al., 2000).

In the following first the basic equations of one-dimensional pure ice growth are given in order to discuss some practical consequences for sea ice growth. To understand the more complicated freezing of an ice brine mixture is important in certain situations, like the early growth of a few centimeters of sea ice right before the onset of convective salt fluxes (Cox and Weeks, 1975; Wakatsuchi, 1983; Wettlaufer et al., 1997, 2000). Also at later stages this interfacial regime is present in form of the skeletal layer. A particular problem, relevant for the present work, is the change of the temperature gradient near the ice-water interface. It will be outlined on the basis of the discussion of early growth of saline ice. The main aspects of this appendix section are summed up in advance:

- The classical Neumann-Stefan solution for planar directional ice growth implies a curvature correction for the temperature profile which can be approximated by the Stefan Number (Stefan, 1889); its inclusion in ice growth models is simple, and its neglect may lead to an error of 5 % in the growth velocity, under natural conditions. This result is obtained when neglecting any temperature variation in the thermal properties of ice.

- The account of the temperature dependence in thermal properties changes the latter result. As the temperature dependence of the thermal conductivity of pure ice exceeds the specific heat release, the temperature curvature is theoretically expected to be inverted. This has not been noticed in earlier studies of the subject.

- During the growth of saline ice the delayed release of latent heat in a two-phase mixture dominates over the pure ice specific heat and conductivity changes. A practically simple approach to predict the early growth of saline ice is compared to an earlier simplification (Huppert and Worster, 1985; Wettlaufer et al., 2000). The present approach gives similar results but is more realistic in terms of the the the solid fraction.

- The temperature gradient near the cellular interface of ice growing from saline solutions is discussed. A proposed approximation for the interfacial temperature gradient is given. It is relevant in several sections of the present work, considering the average temperature gradient in the skeletal layer (section 5), and the modification of morphological stability (section 9) and subsequent derivation of relations between plate spacing and growth velocity.

B.1   The Neumann solution

The following idealized problem describes the one-dimensional propagation of the (ice) solid phase into its melt (water) and is often called the NEUMANN SOLUTION. First lectured by Franz Neumann during the 1860's, it has been communicated by Riemann in his lectures on partial differential equations, of which a revised elaboration was later published by Weber (1901). It is found in many standard works of heat conduction (Carslaw and Jaeger, 1959; Grigull et al., 1963; Lock, 1990). In the following the basic equations are therefore given without comments.

   Suppose that freezing proceeds into a water body of temperature $T_\infty > T_f$ from a cold boundary with temperature $T_b < T_f$, T_0 being the freezing temperature. Thermal conductivities K_i and K_w, diffusivities κ_i and κ_w and densities ρ_i and ρ_w for ice and water are assumed constant. The temperatures T_i in the ice and T_w in the water through which heat is flowing one-dimensionally in the z-direction obey

$$\frac{\partial T_w}{\partial t} = \kappa_w \frac{\partial^2 T_w}{\partial z^2} \tag{B.1}$$

$$\frac{\partial T_i}{\partial t} = \kappa_i \frac{\partial^2 T_i}{\partial z^2}$$

Assuming the boundary conditions

$$T_w(H_i, t) = T_i(H_i, t) = T_0 \quad \text{at the interface } z = H_i \tag{B.2}$$

$$T_w(\infty, t) = T_\infty \quad \text{far away in the water}$$

$$T_i(0, t) = T_u \quad \text{at the upper boundary } z = 0$$

the growth of ice of thickness $h_i(t)$ proceeds as

$$V = \frac{dh_i}{dt} = \frac{K_i}{L_f \rho_i} \left(\frac{\partial T}{\partial z} \right)_i - \frac{K_w}{L_f \rho_i} \left(\frac{\partial T}{\partial z} \right)_w \tag{B.3}$$

and may also be written in form of a diffusion equation

$$h_i = 2\eta (\kappa_i t)^{1/2} \tag{B.4}$$

where $\kappa_i = K_i/(\rho_i c_{pi})$ and c_{pi} is the the isobaric specific heat of ice. The solutions for the temperature fields in the solid and liquid are

$$T_i(z, t) - T_u = (T_0 - T_u) \frac{erf(\eta)}{erf \left(\frac{z}{2(\kappa_i t)^{1/2}} \right)} \tag{B.5}$$

$$T_w(z, t) - T_u = (T_\infty - T_u) \left(1 - \left(1 - \frac{T_0 - T_u}{T_\infty - T_u} \right) \frac{erfc \left(\frac{z}{2(\kappa_i t)^{1/2}} \right)}{erfc \left(\eta \left(\frac{\kappa_i}{\kappa_w} \right)^{1/2} \right)} \right) \tag{B.6}$$

wherein

$$erf(x) = \frac{2}{\sqrt{\pi}} \int_0^x e^{-u^2} du \tag{B.7}$$

$$erfc(x) = 1 - erf(x) \tag{B.8}$$

are the error function erf and complementary error function $erfc$.

The growth rate η is obtained combining (B.3), (B.5) and (B.6)

$$\eta \, Ste \, \pi^{1/2} = \frac{-\eta^2}{erf\eta} - \left(\frac{\rho_w c_{pw} \kappa_w^{1/2}}{\rho_i c_{pi} \kappa_i^{1/2}} \right) \left(\frac{T_\infty - T_0}{T_u - T_0} \right) \frac{exp \left(-\eta^2 \frac{\kappa_i}{\kappa_w} \right)}{erfc \left(\eta \left(\frac{\kappa_i}{\kappa_w} \right)^{1/2} \right)} \tag{B.9}$$

with the Stefan Number

$$Ste = \frac{L_f}{c_{pi}(T_0 - T_u)}. \tag{B.10}$$

B.2 Stefan's application to sea ice growth

If the water is at the melting point, $T_\infty = T_0$, equation (B.9) simplifies to

$$Ste^{-1} = \pi^{1/2}\, \eta\, exp(\eta^2)\, erf(\eta), \tag{B.11}$$

the solution discussed first in some detail by Stefan (1889) in his analysis of sea ice growth. The temperature gradient in the ice may be obtained by differentiating (B.5)

$$\frac{\partial T}{\partial z} = \left(\frac{T_0 - T_u}{h_i}\right)\left(\frac{2\eta}{\pi^{1/2}erf(\eta)}\right) exp\left(-\left(\frac{z}{h_i}\right)^2 \eta^2\right), \tag{B.12}$$

and hence the gradients at the top and bottom become

$$\left(\frac{\partial T}{\partial z}\right)_{z=0} = \left(\frac{T_0 - T_u}{h_i}\right)\frac{2}{\pi^{1/2}}\frac{\eta}{erf\eta} \tag{B.13}$$

$$\left(\frac{\partial T}{\partial z}\right)_{z=h_i} = \left(\frac{\partial T}{\partial z}\right)_{z=0} exp(-\eta^2), \tag{B.14}$$

demonstrating that the curvature in the temperature profile only depends on the growth rate η.

The exponential and error functions may be expended (Lock, 1990) in the form

$$Ste^{-1} = \pi^{1/2}\frac{2}{\pi^{1/2}}(1 + \eta^2 + ...)(\eta - \frac{\eta^3}{3} + ...) \approx 2\eta^2(1 + \frac{2}{3}\eta^2), \tag{B.15}$$

a solution which holds for most natural growth conditions. Stefan (1889) suggested a similar approximation to the growth equation (B.4). Combining the latter with equation (B.15) and setting $Ste^{-1} = 2\eta^2$ in the second order term on the right hand side of (B.15) one gets:

$$h_i\frac{dh_i}{dt} = 2\eta^2\, \kappa_i \approx \frac{\kappa_i}{Ste(1 + \frac{1}{3Ste})} = \frac{K_i(T_0 - T_u)}{L_f\rho_i}\frac{1}{(1 + \frac{1}{3Ste})} \tag{B.16}$$

Although Stefan gave this equation already in 1889, in ice growth models very often the approximation $Ste^{-1} = 2\eta^2$ is used, which assumes a strictly linear temperature gradient and leads to

$$h_i\frac{dh_i}{dt} \approx \frac{\kappa_i}{Ste} = \frac{K_i(T_0 - T_u)}{L_f\rho_i}. \tag{B.17}$$

The latter equation is strictly valid only for large Ste, and it is useful to evaluate the error implied by its use in many applications.

The change in the temperature gradient between the ice-water interface and the cold top boundary accounts for the specific heat necessary to cool the ice, conducted upwards

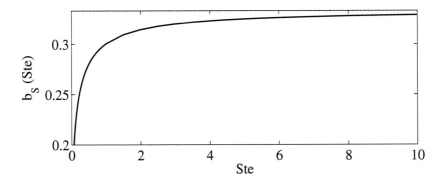

Fig. B.1: Parameter $b_S(Ste)$ in equation (B.18) for the growth velocity modification with Stefan number; the curve is asymptotic to $b_S(Ste) \approx 1/3$ at large Ste

in addition to the latent heat released at the interface. For a Stefan number of 8, corresponding to natural sea ice growth under cold conditions and a temperature difference of 20 °C between ice and water, $\eta \approx 0.245$. From equation B.14 one finds that the temperature gradient at the top is increased by $2\eta/\sqrt{\pi}\,erf(\eta) \approx 1.020$ compared to the bulk gradient $(T_u - T_i)/H$, while the bottom gradient is reduced by $1.020\,exp(-\eta^2) \approx 0.961$. Practically this means that the ice growth velocity, given by the bottom temperature gradient, is overestimated by about 4 % when the sensible heat is neglected. The thickness would then be overestimated by approximately 2 %.

The dependence can be illustrated by an equation of the form

$$h_i \frac{dh_i}{dt} = \frac{\kappa_i}{Ste(1 + \frac{b_S}{Ste})} \tag{B.18}$$

where $b_S(Ste)$ is a function of the Stefan number and plotted in Figure (B.1). The value $b_s = 1/3$, once given by Stefan, is optimal for infinite Ste and its inaccuracy increases at low Stefan numbers. In praxis an optimized constant, e.g. $b_s = 0.328$ for $Ste = 8$, may be used for the growth regime of interest. The latter value gives the growth velocity with an error of less than 0.6% as long the Stefan Number is larger than 2, being sufficient for ice growth in natural water. However, as we will be seen next, there is an important modification that changes this result under real conditions.

B.3 Variable thermal properties: Pure ice

As summarised in the appendix (A.1) the thermal conductivity of pure ice is temperature dependent. Above -50 °C it may be approximated as

$$K_i = K_{i0}\,(1 - 0.006\,T) \tag{B.19}$$

thus increasing with decreasing temperature from its value K_{i0} at 0 °C. On the other hand, assuming for simplicity no change in the temperature gradient, one may express the ratio of latent and sensible heat, to be conducted through the top boundary, to the latent heat of freezing at the advancing bottom boundary as

$$\frac{Q_0}{Q_{h_i}} \approx \frac{L_f - \frac{T_u - T_0}{2} c_{pi}}{L_f} \approx 1 - 0.0031\,(T_u - T_0) \tag{B.20}$$

where the temperature dependence of c_{pi} was neglected. The simple comparison with the change in K_i points to the interesting result, that the increase in thermal conductivity with decreasing temperature exceeds the additional specific heat by a factor of 2. Although a simple solution in the form of equation (B.5) is no longer possible for temperature-dependent diffusivities one may simplify the problem by assuming that only K_i varies with temperature (Carslaw and Jaeger, p. 11). Following Carslaw and Jaeger one may then proceed and replace $(T_u - T_0)$ by

$$(T_0 - T_u)' = \frac{\int_0^{T_0 - T_u} K_i d(T_0 - T)}{K_{i0}} = (T_0 - T_u)\,(1 - 0.003(T_u - T_0)) \tag{B.21}$$

in the above obtained solutions. To first order the Stefan Number is changed by the factor $(1 - 0.003(T_u - T_0))$. Although mathematically tractable, the second order terms make the equation (B.21) for the Stefan problem lengthy. The following ad-hoc approach is tentatively suggested as a growth velocity correction, based on the fact that the conductivity increase exceeds the specific heat release by a factor of two. Simply subtracting the temperature dependencies in equations (B.19) and (B.20) the effective $Ste \approx -\frac{1}{0.0029(T_0 - T_u)}$ may be inserted into equation (B.16) to obtain

$$h_i \frac{dh_i}{dt} \approx \frac{K_{i0}(T_0 - T_u)}{L_f \rho_i} \frac{1}{(1 - 10^{-3}(T_0 - T_u))} \tag{B.22}$$

It is seen that for a temperature difference $(T_0 - T_u) \approx 20$ K the ice growth velocity is expected to be approximately 2 % larger than with the assumption of a linear temperature gradient and a constant thermal conductivity. The correction is in the opposite direction of what constant conductivity models yield. In equation (B.22) the conductivity K_{i0} at the bottom has been used as a reference value. It is noted that, had one used a temperature-adjusted bulk value for K_i, at an average temperature of -10 °C and in connection with the bulk specific heat of the ice slab, almost no change would have been derived.

Equation (B.22), being an ad-hoc approximation for the growth of natural lake ice, shows that for most natural conditions the expected deviation from a linear temperature gradient growth law is small. Thickness corrections following from equation (B.22) are expected to be less then 1 % for natural growth of freshwater.

B.4 Effective thermal properties: Saline ice

Because sea ice incorporates liquid during growth its effective conductivity will depend on the conductivities of both phases which has to be accounted for when the volume fraction of brine is relatively large. Batchelor (1974) has given a detailed review on the bounds for random two-phase media. The upper and lower bounds of the effective conductivity are given by the well known parallel and serial conduction models

$$K_{par} = K_i \left((1 - v_b) + vb \frac{K_b}{K_i} \right) \tag{B.23}$$

and

$$K_{ser} = K_i \left((1 - v_b) + vb \frac{K_i}{K_b} \right)^{-1}, \tag{B.24}$$

where K_b and K_i are the thermal conductivities of brine and of pure ice, and v_b is the relative brine fraction by volume. The parallel and serial models refer, in fact, to non-random media, and therefore bound the random behaviour. Anderson (1960) has first discussed these models in connection with different sea ice types. For ice types where the phases are indeed randomly distributed, a weighted geometric mean conductivity

$$K_{geo} = K_i \left(\frac{K_b}{K_i} \right)^{v_b} \tag{B.25}$$

may be a useful approximation.

In practice, the conductivity of sea ice is, due to its columnar nature, expected much closer to the parallel bound. Field observations are not easy to perform and are scattered (Lewis, 1967; McGuiness et al., 1998; Trodahl et al., 2000). Many factors may influence local measurements of the thermal conductivity and render their interpretation difficult: most prominent are the deviation from vertical column orientation, the microscopic structure of the ice, its degree of granularity, horizontal layers of different texture, air inclusions, brine channel networks, and the transient nature of heat conduction.

In a model of the thermal conductivity that considers the basic principle microstructure of the ice, the following choices seem most appropriate:

- *Ice surface* - Because the orientation of the crystals is approximately random, it may be a reasonable choice to use an average of K_{par} and K_{ser} or apply the geometric mean model K_{geo}. Other models including effects of grain sizes have also been developed (Batchelor, 1974; Tavman, 1996). However, at present no attempt has been made to systematically study the effects of grain size and microstructure on sea ice conductivity.

- *Ice interior* - In columnar ice the crystals are parallel oriented. Brine layers and inclusions are not continuous, but much longer in the vertical, which is in favour of the parallel model. For granular ice types, like snow ice, the situation is similar as for the surface: the conductivity is always less than for columnar ice, but a particular model must be validated on the basis of the microstructure.

- *Skeletal bottom layer* - Here the heat is conducted through vertically oriented plates also favouring the parallel model. The parallel approach is frequently used in continuum models of two-phase materials. As we shall see below there appear some problems when approaching the ice-water interface.

B.5   Ice growth from saline water

When ice grows as a planar front into saline water only a small fraction of salt is incorporated in the solid. The rejected salt accumulates and depresses the freezing point in the interstitial liquid. In the absence of convection the growth rate is decreased and limited by the much slower diffusion of salt. Observations of the resulting salinity-dependence of growth rate of single crystals in supercooled solutions were first described by Tammann and Büchner (1935) and are now understood quantitatively, although for single crystals some open questions have remained, see section 8.1. The related problem of unidirectional solidification of a planar interface has been discussed by Weber (1977), who showed that the amount to which ice growth is limited by solute diffusion depends on the boundary conditions. Another analysis of the problem was given by Levin (1981). Huppert and Worster (1985) provided analytical solutions for this kind of solidification problem similar to the Neumann solution in section B.1 above. All approaches are restricted to planar freezing.

For a cellular ice interface the released solute can accumulate interstitially and the interface advance is much less restricted by solute diffusion at the macroscopic freezing front. In this modified Stefan problem only a fraction of the latent heat, conducted through the top of the ice, is released at the advancing cellular interface. The bulk is released through subsequent lateral freezing of cells. This delayed latent heat release requires the modification of the diffusion equations B.2 due to lateral molecular and eventually convective transport processes. A general theory for such heat and solute transport in a cellular array, based on the assumption of a continuum where effective thermal properties apply (paragraph B.4), was formulated by Hills et al. (1982). Due to the large heat/salt diffusivity ratio for saline solutions, it is often reasonable to assume that the brine salinity is in thermodynamic equilibrium and set by the local liquidus temperature. Under absence of convective heat flow, the growth rate of a two-phase medium is then determined by the thermal diffusion equation

$$c_{psi}\frac{\partial T}{\partial t} = \frac{\partial}{\partial z}\left(K_{si}\frac{\partial T}{\partial z}\right) + L_f\rho_i\frac{\partial(1-v_b)}{\partial t} \qquad (B.26)$$

where the subscript $_{si}$ denotes 'sea ice' and resembles the mass-averaged specific heat $\rho_{si}c_{psi} = \rho_i c_{pi}(1-v_b) + \rho_b c_{pb} v_b$. The effective conductivity K_{si} is based on a continuum model, for example the parallel conduction approach (equation B.3).

Worster (1986) extended the work by Huppert and Worster (1985) and solved the solidification of a two-phase *mushy layer* with vertically varying thermal properties numerically, applying the model from Hills et al. (1982) in the absence of fluid flow. He illustrated a complex dependence of the vertical distribution of liquid volume and freezing rate on the physical properties and boundary conditions. However, during solidification of aqueous NaNO$_3$ solutions, a rather simplified approach, using averaged bulk solid fraction and bulk thermal properties of the mushy layer, could predict the growth rates with similar accuracy (Huppert and Worster, 1985; Huppert, 1990). Also Kerr et al. (1990a) reported that the practically more feasible approach from Huppert and Worster (1985) gave predictions within 10 % of observed growth. This raises the question, how well it predicts the initial growth of sea and NaCl ice.

Linear temperature gradient approximations

Numerical solutions of equation (B.26) show that, for aqueous salt systems, the temperature gradient is rather linear (Kerr et al., 1989, 1990a). This assumption will be made here and eases the following discussion considerably. Assuming solute conservation, profiles of brine salinity and brine volume fraction can be obtained via the liquidus relationship. For ice grown from an NaCl-water solution corresponding to seawater concentrations ($S_0 \approx 35$ ‰) these are shown in figure (B.2).

From the definition of the growth rate η (equation B.4) the growth equations for pure ice, simplified for a linear temperature gradient, are:

$$\eta \approx \frac{1}{\kappa_i^{1/2}}\left(\frac{K_i(T_0 - T_u)}{2L_f\rho_i}\right)^{1/2} \qquad (B.27)$$

$$h_i = 2\eta(\kappa_i t)^{1/2} = \left(\frac{2K_i(T_0 - T_u)}{L_f\rho_i}\right)^{1/2} t^{1/2}, \qquad (B.28)$$

where T_0 is the melting point at salinity S_0 and T_u the temperature of the cold upper boundary. For a mixed-phase medium the model from Huppert and Worster (1985) simplifies to

$$\eta_{bulk} = \frac{1}{\kappa_i}^{1/2}\left(\frac{K_{par}(T_0 - T_u)}{2L_f\rho_i(1-v_{b,bulk})}\right) \qquad (B.29)$$

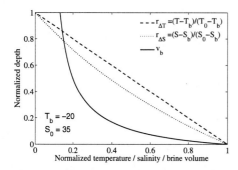
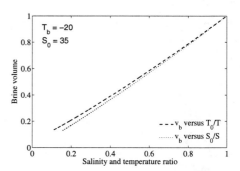

Fig. B.2: Normalised profiles of brine volume, salinity and temperature of ice growing from an aqueous solution with $S_0=35$ ‰ NaCl and a cold boundary at $-20\,°C$, the solution being at the melting point $T_0(S_0)$.

Fig. B.3: Relationship between brine volume, salinity and temperature ratios S_0/S and T_0/S for ice growing from an $S_0=35$‰ NaCl aqueous solution at the melting point, the cold boundary at $-20\,°C$.

$$v_{b,bulk} \;=\; 1 - \frac{S_u - S_0}{S_u + S_0} \tag{B.30}$$

where the bulk brine volume $v_{b,bulk}$ follows from conservation of solute in the mushy layer with brine salinity S_u at the cold boundary (Worster, 2000). The approach assumes (i) a constant solid fraction of the mushy layer, (ii) neglects specific heat, and (iii) computes $K_{si} = K_{par}$ via a parallel conduction model.

The constant solid fraction in this approach by (Worster, 2000) is unrealistic, because it implies that the bulk salinity strongly increases towards the upper cold boundary of the ice. The following alternative formulation is therefore suggested. The first, more realistic assumption, is a constant salinity throughout the medium. One may then, neglecting expansion, simply integrate the local brine volume $v_b \approx S_0/S$, shown in figure B.2, from S_0 to S_u, to obtain the average solid fraction

$$\overline{v_b} = \frac{S_0}{S_u - S_0} ln\frac{S_u}{S_0}. \tag{B.31}$$

Next, the overall latent heat to be conducted through the top boundary is found from this integration and related to the heat flux at the top of the ice. The thermal conductivity at this upper boundary is based on the local brine volume approximated by $v_{b,u} \approx S_0/S_u$, which is a good approximation for most cases, see figure (B.3). These assumptions are described by the set of equations

$$v_{b,u} \;=\; \frac{S_0}{S_u} \tag{B.32}$$

$$K_{si} \;=\; K_{par,geo,ser}\big(v_{b,u}\big) \tag{B.33}$$

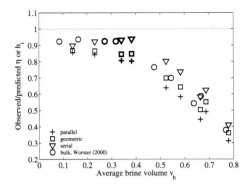

Fig. B.4: Observed growth rates in experiments from Wettlaufer et al. (1997) normalized by the growth rate from equation B.30 (Worster, 2000) and the predictions from equation B.35 with a parallel, geometric, or serial conduction model; the bold symbols denote experiments with zero superheat in the solution.

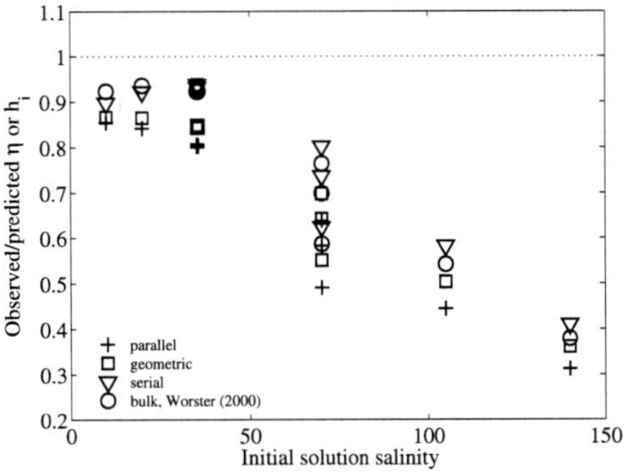

Fig. B.5: See caption of figure (B.4); here the normalized observed growth rates from Wettlaufer et al. (1997) are plotted against initial solution salinities. Note that all values are lower than 1 indicating that the models overestimate the growth.

$$\eta_{par,geo,ser} = \frac{1}{\kappa_i}^{1/2} \left(\frac{K_{si}(T_0 - T_u)}{2L_f \rho_i (1 - \overline{v_b})} \right)^{1/2} \tag{B.34}$$

$$h_{si} = \left(\frac{2K_{si}(T_0 - T_u)}{L_f \rho_i (1 - \overline{v_b})} \right)^{1/2} t^{1/2} \tag{B.35}$$

The subscripts $_{geo}$ and $_{par}$ and $_{ser}$ note the mode of conduction at the top. The geometric model may be realistic at the surface of natural ice, where crystals tend to be randomly oriented (Weeks and Ackley, 1986). However the surface in the laboratory may be different and at low solid fractions also a serial model may apply. The prediction B.35 is physically consistent with the actual solute distribution in ice. On physical grounds it is preferable over the approach from Worster (2000), and is still simple.

Figures B.4 and B.5 compare the described growth models with the experimental results of freezing NaCl solutions from Wettlaufer et al. (1997). The comparison refers to the early stage of growth after 100 to 200 hours, when Wettlaufer et al. (1997) report almost no salt loss from the ice-brine mixture. The integration of B.35 gives larger brine volumina than Worster's approach of constant v_b through the ice. The predictions do not differ much, but the relevance of the choice of the conduction model at the top boundary is emphasised by the present calculations. Considering that the neglected specific heat will decrease the predicted growth rates by approximately 7 to 8 % at low brine volumes and $T_u = -20$ °C, the serial conduction appears to represent the growth rates best, yet an overestimate of 5 to 10 % remains. At large brine volumes and solution salinities the overestimate becomes large. Apparently this cannot be explained by the mode in the surface heat conduction, as also the serial model overpredicts the growth.

Wettlaufer et al. (1997) outlined that the difference is related to thermal convection, as only during the runs at 35 ‰, indicated as bold symbols, the melt temperature was initially very close to the freezing point. At high salinities the freezing started at considerable superheat (see also section 11.6). The growth retardation by thermal convection was discussed by Wettlaufer et al. (1997) for the run at 70 ‰ and $-20°C$ (corresponding to $v_b \approx 0.53$ in figure B.4), where an approximate 20 % reduction in the growth rate was found. This compares roughly to the model overestimates in figures B.4 and B.5. At even higher brine salinities the superheat was largest and the observed increasing impact of thermal convection is as expected. From a more detailed discussion of the growth time series, Wettlaufer et al. (1997) suggested, that processes other than thermal convection might have been present. They invoked ideas from Kerr et al. (1989, 1990b), who proposed a kinetic undercooling growth law to improve numerical predictions, that always appeared to overestimate the growth rates slightly. [1] The physics behind the undercooling concept has not been clarified by the authors. The undercooling, eventually supposed by Wettlaufer et al. (1997) from time-temperature series in the liquid, amounts to more than 1 K during initial freezing. This appears rather large and has not been reported in experiments (Foster, 1969; Wintermantel, 1972; Cox and Weeks, 1975). Although Wettlaufer and colleages noted, that fluxes through container and bottom walls were suppressed by independent control of the freezing chamber, a simple estimate indicates that *vertical* heat conduction through the perspex tank walls may be relevant in the laboratory heat budget. Another mechanism affecting the growth rate may be expansion on freezing, which is not accounted in any of the above models. Depending on the nature of the phase transitions within the ice-brine mixture, brine must be expelled downwards. It may be retained in a thin boundary layer below the ice, where it may equilibrate and melt back the dendrites or stay supersaturated. The process will increase the average solid fraction within the interior, depending on brine density and solid fraction (appendix C.2.2).

From the runs at 35 ‰, convection being neglegible, it seems likely that the simplified equations B.31 through B.35, based on a linear temperature gradient approximation, are capable to predict the growth of NaCl ice in the absence of convection, if a bulk specific heat term is included. The proposed equations are more realistic than the constant brine volume approach from Huppert and Worster (1985) and Worster (2000). Coinci-

---

[1] The term *interfacial kinetics* utilized by Kerr et al. (1989, 1990b) is somewhat misleading in this context, because it is normally used in connection with molecular kinetics. For a macroscopic interface the undercooling mechanism encountered is likely related to the free growth of crystals and dendritic selection (chapter 8). Basically it is a problem of heat diffusion rather than kinetics.

dentally, as also pointed by Huppert and Worster (1985), the unrealistic approach gives similar results. The equations suggested here provide a simple means of computing the vertical distribution of the brine volume. Further theoretical improvements and their experimental validation should consider density changes upon freezing an the contact condition or conduction model at the cold boundary. In terms of the heat budget, interfacial supercooling is likely of less relevance. When morphological interfacial problems are considered, as in the present work, interfacial supercooling is more important. The next section B.6 suggests a simple approach to proceed in this direction by accounting for diffusion-limited interfacial effects. To keep the formulation macroscopically, a simple idea to replace the continuum approach and account for vertical and horizontal diffusion at the dendritic array interface, will be introduced. This allows a description of the nonlinearity in the temperature gradient at the interface.

B.6   Ice-water interfacial temperature gradient

The discussion from the previous chapter is of little relevance for the bulk growth of sea ice in nature. Due to the rapid loss of salt the latter has low brine volumes and will almost grow with the same velocity as freshwater ice. However, the bottom regime where this drop in the brine volume takes place, is very important, not only due to the desalination process itself, yet also because the sea ice microstructure is determined here. The following discussion attempts to provide a quantitative estimate of how the temperature gradient behaves in the regime of strongly changing liquid reaction, when comparing saline to fresh ice.

First, as in section B.5 above, basic assumptions are a linear temperature gradient and an equilibrium liquidus brine salinity, limitations to be discussed below. Expansion upon phase change is also neglected at the moment. Then the following equations for the local brine volume v_b and the average brine volume $\overline{v_b}$ are obtained:

$$v_b(S_b) = \frac{S_0}{S_b} \tag{B.36}$$

$$\overline{v_b}(S_b) = \frac{S_0}{S_b - S_0} ln \frac{S_b}{S_0} \tag{B.37}$$

The average brine volume $\overline{v_b}$ is obtained by integrating from the freezing interface where $S_b = S_0$ upwards to S_b. The local thermal conductivity $K_{si}(S_b)$ may then be computed from a parallel conduction model (B.23), based on the *local* v_b, while the average heat per unit volume, $L_{eff}(S_0, S_b)$, to be conducted upwards and released between S_0 and S_b, is based on the *average* $\overline{v_b}$:

$$K_{si}(S_b) = K_i(S_b)\left((1 - v_b) + v_b \frac{K_b(S_b)}{K_i(S_b)}\right) \tag{B.38}$$

$$L_{eff}(S_b, S_0) = (1 - \overline{v_b})L_{fv}(S_0) + (\rho_i c_{pi} + \overline{v_b}(\rho_b c_{pb} - \rho_i c_{pi}))\frac{T_b - T_0}{2} \tag{B.39}$$

These bulk equations are assumed to describe the steady heat flow at all levels in the ice with $S_b(z) > S_0$. The main contribution to the effective $L_{eff}(S_b, S_0)$ in equation B.39 is $L_{fv}(S_0)$, the volumetric isothermal latent heat of fusion of pure ice. The second term is an approximation and related to the changes in the overall specific heat, wherein c_{pb} and c_{pi} are reference values for pure ice and brine. Their temperature dependence is neglected in the present context. Being interested in mainly a region near the interface, it may be noted that the maximum relative contribution of the second term at high v_b is roughly $(\rho_b c_{pb} T_0)/L_{fv}(S_0)$, and ≈ 5 % for seawater salinities.

One may now define a virtual linear temperature gradient of *pure ice* growing with a planar interface velocity V as

$$\left(\frac{\partial T}{\partial z}\right)_{T_b} = \frac{V L_{fv}(T_b)}{K_i(T_b)}, \qquad T_b = T_{liq}(S_b) \tag{B.40}$$

where the properties on the right hand side are evaluated at the liquid temperature T_b. The temperature gradient in a saline ice-brine mixture which corresponds to the same interface velocity V and the flux balance from equations B.38 and B.39 would be

$$\left(\frac{\partial T}{\partial z}\right)_{S_b, S_0} = \frac{V L_{eff}(S_b, S_0)}{K_{si}(S_b)}. \tag{B.41}$$

The ratio of these gradients

$$r_{tz} = \frac{\left(\frac{\partial T}{\partial z}\right)_{S_b, 0}}{\left(\frac{\partial T}{\partial z}\right)_{S_0}} = \frac{L_{eff}(S_b, S_0)K_i(T_b)}{L_{fv}(S_b)K_{si}(S_b)} \tag{B.42}$$

depends only on S_0 and the liquidus salinity S_b, as these two variables also determine the brine volumes involved via equation B.37. The ratio r_{tz} may be understood as the reduction of $\partial T/\partial z$ due to the presence of liquid brine.

Boundary condition v_{int}

To compute $\overline{v_b}$ one might set $S_{int} = S_0$ and get $v_{int} = 1$ at the interface (from equation B.37). Within the concept of constitutional supercooling, however, v_{int} on the ice side may be defined differently. Let $S_{i,i}$ be the bulk ice salinity and S_{int} the brine salinity at the interface. By definition

$$k = \frac{S_{i,i}}{S_{int}} \tag{B.43}$$

$$S_{i,i} = \frac{S_{int}\rho_b v_{int}}{\rho_b v_{int} + \rho_i(1 - v_{int})} \tag{B.44}$$

which can be combined to give

$$v_{int} = \frac{k}{k + \frac{\rho_b}{\rho_i}(1 - k)} \tag{B.45}$$

If k is close to one, the approximation $v_{int} \approx k$ at the interface holds within a few percent and its maximum inaccuracy for small k is $\rho_b/\rho_i \approx 1.1$. Consequently, the simplified interface condition $v_{int} = k_{mi} = S_0/S_{int}$ with $S_{int} = (S_0 + \Delta S_{mi})$ from constitutional supercooling, is used to compute the brine volume in equation B.37, which is thus integrated from a non-zero value at the interface [2].

Temperature profiles

In figures B.6 and B.7 the normalized r_{tz} has been plotted for the conditions $T_b = -20$ °C, and $S_0 = 35$ ‰. These resemble the initial growth of seawater and correspond to the situation in laboratory experiments to be discussed below (Cox and Weeks, 1975; Wettlaufer et al., 1997). The relations of r_{tz} versus brine volume and temperature are shown as a solid curve in these figures. This curve is termed with 'no restriction' and will be discussed first. It is seen that r_{tz} at the warm side, the ice-water interface, drops to a value of almost one order of magnitude less than at the cold side, the top of the ice. This behaviour of the temperature gradient at the interface is related to the lower effective latent heat of fusion in the bottom fractions of ice, where the brine volume fraction is large.

It is recalled that the calculations are based on a linear temperature gradient approximation, whereas the resulting temperature gradient apparently is not linear but decreases towards the bottom. The following considerations serve as a first-order correction to this inconsistency. Integrating the gradient r_{tz} in figure B.7 one obtains the temperature distribution in figure B.8, again shown as a solid line for this case of unrestricted parallel conduction. Due to the curvature, with respect to the linear assumption, the integrated brine volume $\overline{v_b}$ decreases at any level in the ice, and so does the effective L_{eff}. The solution procedure therefore may be repeated by integrating equation B.42 stepwise, based on the new, nonlinear temperature profile, until convergence is obtained. It is noted that, if the curvature is not too large, as in figure B.8, these iterations only lead to minor changes in the relative temperature distribution in the figure. The latter already resembles closely the temperature profile in the steady state. However, the changes in the integrated $\overline{v_b}$ at a given temperature due to the curvature are more severe. This is seen in figure B.9, where the relative decrease in the integrated solid fraction, $\Delta\overline{v_b}/(1 - \overline{v_b})$, with respect to a linear gradient in figure B.7 is shown. While the difference is 8 % at the cold surface moderate, it peaks at values above 50 % near the ice-seawater interface. These numbers

---

[2] Surface tension is, for the sake of simplicity, neglected here to compute k_{mi}

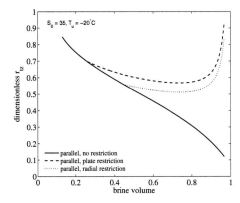
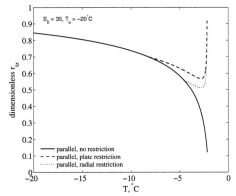

Fig. B.6: Normalised temperature gradient r_{tz} with respect to pure ice from equation B.42. Solid line: parallel conduction model. Dashed line: diffusive restriction of conduction near the tips due to equation B.47 for two-dimensional cells. Dotted line: diffusive restriction of conduction due to equation B.49 for three-dimensional cells.

Fig. B.7: As in figure B.6, but r_{tz} plotted against the temperature. Note that the *assumed* linear temperature profile would be a horizontal line. The curvature in the profile is shown in figure B.8.

indicate the systematic uncertainties due to the linear temperature approximation. It can be expected that they will underestimate the growth rate of seawater slightly. However, as will be seen next, it seems likely that other factors play a more important role, and therefore these corrections are not discussed further in terms of possible parametrisations.

Observed temperature gradients

Still looking only at the solid curves, it is outstanding in figures B.8 and B.7, that the temperature gradient at the bottom of the ice is fairly small. Before a comparison with observations is made, it is recalled that the above equations B.38 and B.39 describe the heat flow under the assumption of a steady state of heat conduction, frequently termed *frozen time* analysis. Upward heat conduction through any level in the ice treats the lower levels as a slab advancing with constant velocity and average effective phase transformation heat L_{eff}. Observed temperature profiles will deviate from the similarity solution due to time-dependence. As the heat is extracted from the top, a non-constant decreasing surface temperature, often observed in nature, will lead to increased curvature and larger surface temperature gradients. A comparison with observed profiles can be expected to be most reliable, if the layer of strongest brine volume change, for the conditions in figure B.7 the interface regime between -2 and -5 °C, is small compared

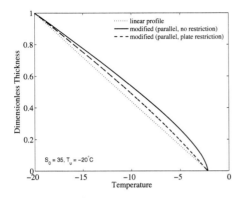

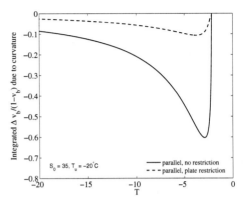

Fig. B.8: Deviation of the equilibrium temperature profiles from linearity (dotted line). Solid line: parallel conduction model. Dashed line: diffusive restriction of conduction near the tips due to equation B.47 for two-dimensional cells.

Fig. B.9: Relative increase $\Delta \overline{v}_b/(1 - \Delta \overline{v}_b)$ in the solid ice fraction, integrated upwards from the warm bottom to temperature T, with respect to a linear temperature profile. Solid line: parallel conduction model. Dashed line: diffusive restriction of conduction near the tips due to equation B.47 for two-dimensional cells.

to the total ice thickness.

For further reference the ratio of the temperature gradients in the 20 % bottom and 20 % upper fractions from figures B.8 and B.7 is approximately 0.39, while the ratio of upper to lower half is 0.62. Several studies indicate that near-bottom temperature gradients under comparable growth conditions are not that small. Wettlaufer et al. (1997) show temperature series from which the lower to upper half temperature gradients' ratio may be estimated as 0.7 to 0.8. The timeseries from a field study by Wettlaufer et al. (2000) indicate a variable ratio for the upper to lower 2 cm temperature gradients, being near 0.3 to 0.4, when the ice was 4 to 6 cm thick or when the surface temperature was decreasing, yet 0.5 to 0.7 under more steady conditions and when the ice had thickened to $\approx$ 15 cm. Although the results for the low thickness seem in agreement with the model predictions, they resemble nonstationary conditions with an increasing surface temperature and the problems noted above. This suggests the interpretation, that also in the study by Wettlaufer et al. (2000) the observed bottom temperature gradients were larger than predicted by the solid curves in figures B.6 and B.7. Another comparison can be made with the study from Cox and Weeks (1975) discussed in chapter 5. There the fresh ice gradient ratio r_{sk} was typically 0.65 to 0.7 in a 3 cm layer near the interface (figure 5.9). To be illustrated below, the standard prediction of r_{sk} for these growth conditions is also a factor of 2 too small. Finally, in a laboratory study by Niedrauer and Martin (1979) the interfacial gradients were smaller, amounting to a fraction of 0.3 to 0.5 of the ice surface gradients. The latter study, however, was performed in a 1.6 mm thin growth cell, the heat budget being strongly influenced by the lower conductivity of the plastic walls, which at least qualitatively can explain the stronger curvature in r_{tz}.

It is concluded that most mentioned studies, when carefully compared to the idealising assumptions, reveal an approximately 1.5 to 2 times larger near-bottom temperature gradient than predicted by the steady state heat conduction budget. In the following an explanation for this discrepancy is proposed.

B.6.1   Parallel conduction disequilibrium

The basic assumption in the above heat flow approach is that conduction can be calculated with a parallel thermal conductivity model. Such a model is likely appropriate for heat conduction in the direction of ice plates, yet its applicability not only requires a steady state in terms of the thickness of the skeletal layer, but also thermal homogeneity within the latter. Such a homogeneity is not expected near the interface, where the brine volume is strongly changing by lateral freezing. While the parallel conduction model assumes a liquid brine contribution of $v_b K_b$ to the effective conductivity, and associates the latter with the vertical advance of the ice interface, this 'exchange in specific heat' must break down near the interface, because the lateral diffusional distance in the liquid becomes comparable or larger than the widths of the growing ice plates. In this regime liquid vertical heat conduction cannot contribute to vertical ice growth. As a consequence, the parallel conductivity model overestimates the effective transport of latent heat.

Diffusional distance restriction

Quantitatively the problem can be addressed by introducing a restriction for the brine volume fraction available for conduction, modifying the term $v_b K_b$ in equation B.38 and replacing it by $(v_{bcon} K_b)$. This concept is based on the requirement of lateral freezing normal to the plates in the skeletal layer. For a parallel conduction model to be valid the released latent heat may be imagined to be diffused away from the lateral boundary into (i) the liquid brine and (ii) into the solid ice. Only if the typical lateral diffusional distances are similar, the application of a parallel continuum model is justified. The following scalings for v_{bcon} are simple and physically reasonable:

- 1. *Plate like growth.* For a parallel conduction model to be valid, it is simply assumed that the lateral diffusional distance in the liquid, $(\kappa_b t)^{1/2}$ times v_b, must be equal or less the diffusional distance in the ice plates $(\kappa_i t)^{1/2}$ times the solid fraction $(1 - v_b)$. This one-dimensional concept leads to

$$
\begin{aligned}
v_{bcon} &= (1 - v_b)\left(\frac{\kappa_b}{\kappa_i}\right)^{1/2} \qquad \text{for} \qquad v_{bcon} < v_b \\
v_{bcon} &= v_b \qquad \text{for} \qquad v_{bcon} < v_b
\end{aligned}
\tag{B.46}
$$

This condition is denoted as 'plate restriction' in figures B.6 through B.8.

- 1. *Growth of circular cells.* An approximate validity condition for a parallel conduction model to be valid may be formulated by assuming an array of cylinders of
  spacing R and radius r. Now it is assumed that the lateral diffusional distance in
  the liquid, $(\kappa_b t)^{1/2}$ times $(R-r)$ must equal the diffusional distance in the ice plates
  $(\kappa_i t)^{1/2}$ times r. If this two-dimensional concept is combined with an approximate
  relation

$$
\left(\frac{R}{r}\right)^2 - 1 \approx \frac{v_b}{1 - v_b}
\tag{B.47}
$$

between R, r and v_b, one may obtain

$$
\begin{aligned}
v_{bcon} &= (1 - v_b)\left(2\left(\frac{\kappa_b}{\kappa_i}\right)^{1/2} + \frac{\kappa_b}{\kappa_i}\right) \qquad \text{for} \qquad v_{bcon} < v_b \\
v_{bcon} &= v_b \qquad \text{for} \qquad v_{bcon} < v_b
\end{aligned}
\tag{B.48}
$$

This condition is denoted as 'radial restriction' in figures B.6 through B.8.

Applying these rescalings to the $v_b K_b$ term in equation B.38, the resulting temperature distributions and gradients change fundamentally, to be discussed on the basis of
figures B.6 through B.8. Although near the interface the cells may be more irregular,
the approach most consistent with the sea ice microstructure is probably the 'plate restriction'. The meaning of the results, shown as dashed curves, will be emphasised, now
noting that they are slightly modified in the 'radial restriction' scaling.

Restricted latent heat flow

First, the temperature distribution of the 'restricted conduction', shown in figure B.8 as
a dashed line, is now much less curved and close to linear. Therefore all results based on
the linear temperature approximation apply within a few percent. This may be deduced
from the small solid fraction correction shown by the dashed line in figure B.9. Forming
again the ratio of the average r_{tz} in the 20 % bottom and 20 % upper parts of the profile,
one obtains a value of 0.74, which is now in good agreement with the experimental studies
noted above (Wettlaufer et al., 1997, 2000). The temperature gradient in figure B.8 now

shows the interesting behaviour to increase towards the interface, which will be discussed below.

One may ask, what these results, which are based on an ad-hoc correction, imply for the mode of freezing and the microstructure near the interface. By using a smaller v_{bcon} in the $(v_b K_b)$-term in equation B.38, one has actually 'stolen' a certain local fraction of the parallel conductive heat flux. The conduction models are seen in figures B.6 and B.7 to match near $v_b \approx 0.25$ and $T \approx -8$ °C. One may then speculate that the 'stolen' heat, as it does not contribute to the local lateral ice growth, creates supercooling in the inter-cellular liquid brine. At some level this supercooling will be large enough to drive the lateral growth. One may also argue, that the mode of freezing is lateral, and that therefore a serial conduction mode should be associated with this heat flow component. The solutions for the latter have not been shown in figures B.6 and B.7, yet it suffices to note that, near the interface, they approach $r_{tz} \approx 4$ due to the ratio K_i/K_b, serial conduction being rather small. Therefore, serial conduction may be neglected with respect to the ice front propagation, and a parallel model with a restricted conducting brine volume fraction is a reasonable approximation.

An important note is that the solutions, here obtained for the specific conditions $S_0 = 35$ ‰ and $T_b = -20$ °C, are almost general in terms of the relationship between the brine volume v_b and r_{tz}. This is so because the local and average brine volumina, related to S_0/S_b via equations B.37, dominate the heat flux balance, while the temperature and salinity dependence of the thermal properties is comparatively small. The relation in figure B.6 can therefore, in principle, be transfered to other salinities, with one important difference, given by the boundary condition B.45 to be discussed next.

Conjecture of minimum r_{tz}

From the simplified approach the increase in r_{tz} near the interface is predicted to take place in a very thin layer. For 35 ‰ it corresponds to a temperature drop of ≈ 0.6 K, but is more generally given by a corresponding change in the brine volume. The temperature resolution in most studies (Cox and Weeks, 1975; Wettlaufer et al., 1997, 2000) is too coarse to test the proposed r_{tz} properly. However, there are some temperature profiles which indicate that the gradient indeed steepens at the interface (Niedrauer and Martin, 1979; Haas, 1999). It is suggested that, under directional temperature forcing from a cold boundary, an increasing r_{tz} towards the ice-water interface may be interpreted in two ways. In terms of the vertical heat flux in the solid ice dendrites, this gradient would have the tendency to accumulate released latent heat locally, instead of accelerating its transport towards the cold interface. In terms of the equilibrium temperature in the interdendritic liquid, the larger r_{tz}, compared to parallel conduction, implies that the liquid becomes supercooled near the interlace. How the temperature evolves in detail

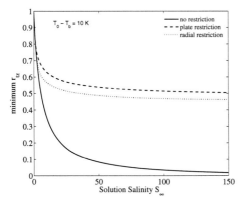

Fig. B.10: Minimum r_{tz} from figure B.6 in dependence on the solution salinity and for a fixed temperature difference 10 K between bottom and top of the ice. Solid line: parallel conduction model. Dashed line: diffusive restriction of conduction near the tips due to equation B.47 for two-dimensional cells. Dotted line: diffusive restriction of conduction due to equation B.49 for three-dimensional cells.

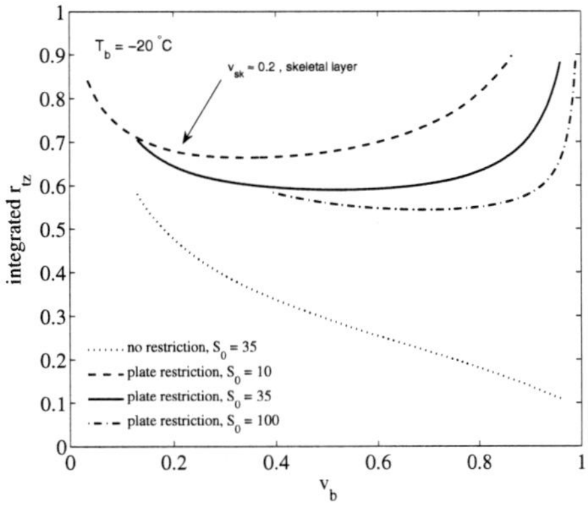

Fig. B.11: Normalised temperature gradient $\overline{r_{tz}}$ integrated from the ice-water interface to a certain brine volume level. The integration is based on the curved temperature profiles. Dotted line: parallel conduction model, $S_\infty = 35$ ‰. Two-dimensional diffusive restriction for $S_\infty = 35$ ‰ (solid line), $S_\infty = 10$ ‰ (dash-dotted line) and $S_\infty = 100$ ‰ (dashed line). The brine volume at the top of the skeletal layer may be associated with $r_{sk} \approx \overline{r_{tz}}$.

cannot be shown by the present approach, but the basic principles are well illustrated,

The considerations lead to the following heuristic approach to model the interface behaviour. It is conjectured, that the vertical interfacial temperature gradient near the ice dendrites is controlled by the minimum r_{tz}, which is the driving force for dendritic advance. In figure B.10 the computations of this minimum r_{tz} are shown for the selected difference between cold boundary and solution freezing temperatures, $T_b - T_0 = 10$ K, and in dependence on the solution salinity S_∞. It is seen that, above a certain concentration, the results depend little on S_0. As already noted, this dependence mainly comes from the marginal stability boundary condition B.45, implying that growth from low salinity solutions requires a smaller interfacial brine volume. Therefore the values of minimum r_{tz} in figure B.10 are plausible interfacial boundary conditions for morphological stability theory.

Interfacial undercooling

The suggested approach may serve as a simple perspective on the formation of undercooling near a macroscopic cellular freezing interface. As the liquid fraction contribution to the parallel heat conduction requires lateral heat diffusion, one expects that the system adjusts by allowing liquid to supercool, which than yields the necessary driving force to transport the latent heat. This mechanism may conceptually explain the interfacial undercoooling discussed by Kerr et al. (1989, 1990b) for isopropanol solutions. For aqueous solutions of NaCl and other salts, Wintermantel (1972) has documented interfacial undercoolings in dependence on the growth velocity and S_0. For example, his maximum undercoolings, asymptotically approached at highest growth velocities, are in reasonable agreement with the difference between the temperature of the minimum in r_{tz} in graphs like figure B.7, and the respective freezing temperatures.

It must be stressed that the present approach is heuristic, absorbing several unmodeled processes, to predict the effective temperature gradient necessary to drive the advance of the cellular ice interface. The difference between this approach and the parallel model, see figures B.6 and B.7, may therefore eventually be interpreted as the supercooling potential near the interface. The applicability of the formulation likely depends on the question, how much supercooling is possible near the tip, before sidearms form. For ice the approach is probably justified under natural growth conditions: Due to the large anisotropies in the free growth rate and interfacial energy, the plates are smooth and allow some supercooling between them. It is likely that in other systems, like NH_4Cl-H_2O , the interfacial and interdendritic undercooling near the tip of cells will favour dendritic growth and formation of side arms, which increases the effective conductivity and decreases r_{tz} near the interface. Sidebranching is also expected to strongly effect the effective permeability. This may, for example, explain the apparent relation between convective thresholds and undercoolings reported by Worster and Kerr (1994). The derivation of a practically simple matching condition for sidebranching arrays most likely requires different approximations, accounting for microstructural pattern formation as discussed in chapter 8.

Hence, without considering any geometric details, it may be conjectured from figures B.6 and B.7, that the heat flow r_{tz} implies undercooling near the interface. The comparison of parallel and lateral heat flow thus provides an idea, how interfacial undercooling is produced for a macroscopic interface. The detailed real temperature profiles and possible supercoolings cannot be predicted in this way.

Skeletal temperature gradient r_{sk}

The calculations so far have always assumed that the ice contains the same salinity as the solution from which it freezes. This is not the case under normal growth conditions once convection has set in. Instead, one typically finds a bulk ice salinity decreasing from the solution value S_0 at the interface to a fraction of 0.2 to 0.5. However, the principle dependence of r_{tz} on the brine volume, shown in figure B.6, is only little affected by such changes, in particular if considered in connection with the conjecture of minimum r_{tz}. This stable behaviour of the system follows from the almost plateau-like r_{tz} considering brine volumes between 0.2 and 0.9 in figure B.6. Hence, if not considerable superheat sources from the liquid, or massive change in the microstructure are induced by convection, then the noted results should also apply to naturally growing ice, provided that an evaluation is made in terms of v_b.

The results can then be presented in a somewhat different form in figure B.11, showing the average r_{tz} when integrated upwards from the interface to a specific brine volume. Again, computations are done for the seawater reference $S_0 = 35$ ‰ and $T_b = -20$ °C , yet in addition the relations are shown for $S_0 = 10$ and $S_0 = 100$ ‰. The higher the salinity, the higher the interfacial brine volume, and the curvature in the temperature profile, resulting in a smaller integrated r_{tz}, similar to the behaviour in figure B.10. There is a weak salinity dependence, yet it is not large for the range 10 to 100 ‰. This form can be compared with finite temperature gradient measurements near the interface, if the brine volume of the upper level is known, as it was the case in the study by Cox and Weeks (1975), extensively discussed in chapter 5, for example in figure 5.9. The analysis of these data had, in connection with thermal convection corrections, resulted in a best value of $r_{sk} \approx 0.66$ for the average r_{sk} in a 3 cm interfacial layer, limited by a brine volume of mostly 0.15 to 0.20. The present predictions result in $0.64 < \overline{r_{tz}} < 0.68$ for this brine volume, in rather good agreement with the observations.

In all calculations so far, the effect of expansion on freezing has been neglected. As discussed in chapter C the latter implies in general a loss of salt and thus an increase in the solid fraction, yet its exact magnitude depends on the detailed nature of the expansion flow. An upper bound for the reduction in the salinity and increase in the solid fraction is $(1 - \rho_i/\overline{\rho_b})$, yet a more plausible value, accounting for the redistribution of salt and downward displacement of the whole intracellular brine, is a reduction of this scale by $\overline{v_b}$ (see section C.2.2). For the bottom regime of interest and $S_0 = 35$ ‰, relevant values are $\overline{v_b} \approx 0.4$ and $\overline{S_b} \approx 90$ ‰ to compute $\overline{\rho_b}$. One may then estimate a relative increase of 6 % in the solid fraction, which to first order should translate to r_{tz}.

B.6.2 Growth rate of saline ice

The above considerations also imply a relatively simple approximation to predict the growth rate of saline ice before it starts to loose salt. The growth rate must only be augmented by the factor r_{tz}^{-1}, at the cold boundary, which is obtained from equation B.42. The enhancement factor may be seen in figure B.7, where it is approximately

1/0.85 for a solution of 35 ‰ and a cold boundary of -20 °C, if a parallel conduction model is assumed. A critical question in freezing experiments and during natural freezing, is probably the conductivity mode near the ice surface. Here the crystals are frequently not parallel but randomly oriented, which gives rise to a serial or geometrical conduction model. Saline ice may then grow at a smaller velocity than fresh ice.

B.6.3   Conclusion

In conclusion, a simplified approach to describe the early growth of saline ice has bee formulated. It is in agreement with observations, in particular the temperature distribution near the macroscopic cellular freezing interface of saline NaCl solutions and seawater. An important modification of earlier models is a restriction of thermal conduction by finite lateral diffusion of latent heat. A simplified approximate analysis indicates that the ice-seawater interfacial region or *skeletal layer* can, when compared to freshwater ice advancing at the same velocity, be characterised by a 0.6 to 0.7 smaller interfacial temperature gradient. The approach offers the possibility to include reasonable first order corrections that are relevant for the stability in the boundary layers near the interface. It may also be helpful to understand phenomena like interfacial supercooling. While the diffusive corrections B.47 and B.49 seem sufficient to solve some problems concerned with the macroscopic freezing interface of saline solutions, detailed numerical integrations must provide more significant results and restrictions of the present estimate.

C. Review: Basic mechanisms of sea ice desalination

Aged bulk sea ice may be characterised by two main structural features (Hamberg, 1895; Bennington, 1963, 1967; Cherepanov, 1973; Wakatsuchi, 1983; Weeks and Ackley, 1986; Eicken et al., 2000; Light et al., 2003). These are

- Inclusions and pores of submillimeter size that have formed by transition of the original brine layers.

- Vertically oriented brine channels of widths of a millimeter or more, extending through considerable fractions of the ice and frequently being organised in networks.

The present section first gives an overview over the two basic desalination mechanisms with respect to these features: *expansion upon freezing* and the *migration of brine pockets* by diffusion. In connection with the process of expansion some new ideas are suggested, concerned with the question, if the expelled brine may leave the ice or not.

Gravity drainage or *free convection* is important near the bottom or during warming of sea ice. Its rigorous form in the skeletal layer has been discussed in chapter 11. A general theory for thicker bulk ice is not available yet, as the evolution of the microstructure is complex, difficult to quantify, and as the interaction of possible processes is not understood. Some empirical work and heuristic models are mentioned at the end of this appendix.

C.1 Brine pocket migration

C.1.1 Theory

While the solute diffusion coefficient in pure ice is many orders of magnitude less than in water (Hobbs, 1974; Petrenko and Whitworth, 1999), this situation is different for solute inclusions, when the ice is exposed to a temperature gradient. In a steady state a phase boundary will migrate in the direction of the brine salinity gradient with the velocity

$$V_{prop} = \frac{\rho_b}{\rho_i} \frac{D_s}{S} \frac{dS}{dz} = -\frac{\rho_b}{\rho_i} \frac{D_s}{m S} \frac{dT}{dz} \approx -\frac{\rho_b}{\rho_i} \frac{D_s}{T} \frac{dT}{dz}, \qquad (C.1)$$

where $m = dT/dS$ is the liquidus slope, ρ_b and ρ_i are densities of brine and ice and dT/dz is the imposed temperature gradient. The approximate form on the right hand side is obtained by assuming a linear liquidus relationship. Hoekstra et al. (1965) applied equation to the migration of brine pockets of NaCl and KCl aqueous solutions and found that observations and predictions disagreed by a factor of 2 for KCl and 3 for NaCl. Seidensticker (1966) analysed their results critically. He suggested that, for spherical pockets, the gradient with in a brine sphere should differ from the gradient in the solid. Based on a solution of the Laplace equation (Carslaw and Jaeger, p. 426), Seidensticker proposed

$$V_{mig} \approx - \left(\frac{3K_i}{2K_i + K_b} \right) \frac{\rho_b}{\rho_i} \frac{D_s}{T} \frac{dT}{dz}, \tag{C.2}$$

with a prefactor depending on the difference in conductivities of ice and brine. A value of 1.46 was used by Seidensticker (1966) and assumed constant. Using more realistic diffusion coefficients and liquidus slopes than Hoekstra et al. (1965), he found that, for KCL, the predicted V_{mig} was only 10 to 20 % above the observations. For NaCl the predictions were still a factor 1.5 to 1.7 too large. If the conductivity-related prefactor is based on the property equations from the present appendix, it is found to increase from 1.45 to 1.66 while T is lowered to -12 °C, compared to the constant 1.46 from Seidensticker (1966). The mismatch with the observations then increases to a factor 1.6 to 1.8, on average. Neveretheless, Seidensticker's approach has been adopted by Weeks and Ackley (1986) as a reasonable correction to the original work from Hoekstra et al. (1965).

The following explanations may be suggested for the mismatch. First, D_s evaluated in the appendix A.3 gives values which are 10 to 20 % smaller than those suggested by Seidensticker. Second, equation C.2 from Carslaw and Jaeger (1959) is valid for a spherical pocket in the steady state. In the pocket migration case the movement towards the warmer side implies an acceleration that requires the warming of the liquid, which will slow down the movement. A second retarding effect is that the increase in inclusion size releases latent heat. Furthermore, melting implies, if no overpressure exists, the formation of a vapour bubble. The latter has been described to retard the migration of liquid inclusions (Nakaya, 1956). Finally, what appears to be most important seems, that pocket migration is associated with intermittent latent heat release and consumption. This defines a Stefan problem at a phase boundary, for which the temperature gradient in the liquid, for the given conditions, does not exceeds the temperature gradient in the solid, see section B. Without analysing this problem in detail, no definite conclusions can be drawn. It appears inuitively more reasonable that, as the liquidus concentration steadily adjusts to the temperature of the surrounding ice, the temperature gradient in the solid is a better approximation to the actual process chain. This would mean that the velocity of the pockets should be better determined by the original equation C.1. In fact, dismissing the conductivity-related factor introduced by Seidensticker (1966), its application then predicts the migration velocities of NaCl brine pockets from Hoekstra et al. (1965) to within 10 %. The agreement for KCl brine becomes, however, worse. The results for KCl need to be reevaluated by proper computation of molecular property data at the liquidus. The difference may also indicate the relevance of other, unaccounted, aspects of the phase transition.

It is concluded that, for NaCl brine, the results from the experimental study from Hoekstra et al. (1965) are fully consistent with equation C.1, in case that the temperature dependence of D_s from appendix A.3 is applied. Equation C.2, suggested by Seidensticker (1966) and adopted by Weeks and Ackley (1986), may be viewed with caution, when a confined volume with phase changes in considered. Migration velocities measured by Harrison (1965a) are also in reasonable agreement with equation C.1, but left unexplained deviations. The possible role of kinetics and defects in ice were considered by Jones (1973b). Some other noteworthy aspects, considering brine pocket migration, are that Hoekstra et al. (1965) did not find a dependence of V_{mig} on crystal orientation. Neither did V_{mig} depend on the direction with respect to gravity, which indicates the absence of free convection effects that would introduce a pocket size dependence. Pockets investigated had diameters less than 0.07 mm. Harrison (1965a) confirmed such a lack in size dependence. He noted some occurrences of pocket elongation and diagonal migration which were, however, not general. On the other hand, Hoekstra et al. (1965) reported that all larger inclusions, migrating in a temperature gradient, broke up into smaller pockets after a few days. While one important factor in the problem appears to be the pocket shape, the basic thermodynamical aspects of the problem still need to be clarified. The subject requires more work.

C.1.2 Effects in natural sea ice

Untersteiner (1968) discussed the implications of brine pocket migration for the desalination of natural Arctic sea ice. He concluded that brine pocket migration is unimportant when the change in salinity profiles or desalination of sea ice are considered. For example, at a temperature of -10 °C, and a temperature gradient of 0.1 K/cm, corresponding to growth velocity of ≈ 0.65 cm d^{-1}, the downward migration of a brine inclusion would be about 1.1 mm per month, roughly 0.5 % of the growth velocity. Cherepanov (1973) reported comparable average migration rate estimates of 8 to 15 mm/year, based on optical analysis of Arctic multi-year ice cores. It is worth a note, that the ratio of pocket migration velocity to growth velocity simply depends inversely on temperature, because both velocities scale with the temperature gradient. Assuming that the latter is linear through the ice, this ratio can be expressed as

$$\frac{V_{mig}}{V} = \frac{\rho_b}{\rho_i} \frac{D_s L_f \rho_i}{K_i} \frac{1}{T}. \tag{C.3}$$

With $T = -5\,°C$, typical for thin ice or close to the ice-water interface, one has $V_{mig}/V \approx$ 0.014. In praxis it may be of interest to integrate equation (C.3) from the top of the skeletal layer, (where $T \approx -3\,°C$ is typical, depending on growth conditions), to the desired temperature. Integration from -3 to -5 °C gives $\overline{V_{mig}/V} \approx 0.019$. Considering a growth velocity of 3 cm d^{-1} for young ice the pockets would migrate 1.7 cm per month. Although the pocket velocity is small compared to the growth velocity, this length scale may be sufficient in redistributing brine between inclined feeder systems of channel networks, for which this is a realistic length scale (Lake and Lewis, 1970; Eide and Martin, 1975; Niedrauer and Martin, 1979; Kawamura, 1988) or to horizontal banding features (Bennington, 1963; Cole and Shapiro, 1998). The possibility of the latter process might imply differential migration of small and large pores, and indicating the importance of understanding the thermodynamic details of the process.

The considerations indicate that, despite the low velocities, brine pocket migration may play a role in reconnecting brine pockets and creating certain features of brine channel networks. This process is still poorly understood. Stander (1985) has described different stages of microstructure development and also pointed out the observation of grain boundary migration in first-year sea ice based on optical analysis of cores (Stander, 1984). In the field of metallurgy the migration process has been called 'zone melting' and equation (C.1) was applied to derive solute diffusion coefficients (Wernick, 1956). Recently, Nuclear Magnetic Resonance (NMR) investigations have been applied to determine diffusion coefficients and their anisotropy from such principles (Callaghan et al., 1999; Menzel et al., 2000). Furthermore, preliminary conclusions about the the size and behaviour of smallest brine inclusions were drawn. However, in the latter studies the temperature dependence of solute diffusivity has not been considered. A more fundamental analysis of such results must not only involve a temperature-dependent D_s and the above noted time-dependent aspects of brine-pocket migration. It will likely also have to consider aspects of molecular interface physics that become relevant when the micrometer scale is approached (Hobbs, 1974; Petrenko and Whitworth, 1999).

C.2   Expansion and expulsion

When ice is cooled, the equilibrium concentration of the brine within the inclusions increases and more liquid freezes. The overall fractional brine volume decreases. Due to the ice-liquid density difference the internal freezing leads to expansion. For disconnected inclusions this will, due to the incompressibility of water, rapidly lead to the formation of small cracks, allowing some brine to move away from the partially solidifying volume element. For connected inclusions or larger channels the *expansion flow* will not involve cracking.

As, under normal winter conditions, sea ice is colder and less permeable at the top, the expansion flow may therefore be thought to be directed downward towards the more open ice-water interface. In dependence on its pathway, and the accompanying thermodynamic interaction with the solid ice, this brine may either be expelled from the sea ice, or it may remelt some ice at the locations to which it flows, turning its freezing out process to some degree backwards. In the first case the bulk salinity is definitely decreased. In the second case brine is redistributed within the ice, but there will still be some net effect. The basic concepts of these concepts will be outlined next.

C.2.1 Complete expulsion of brine

An approximation of the salinity change due to expulsion during cooling of sea ice was first proposed by Untersteiner (1968). He suggested

$$\frac{S_{i2}}{S_{i1}} = \left(\frac{T_2}{T_1}\right)^{\frac{\Delta\rho_*}{\Delta\rho_*-1}} \tag{C.4}$$

to estimate the decrease in salinity during cooling of ice from temperature T_1 to T_2. This assumes that all locally expelled brine is lost from the ice. Untersteiner (1968) applied $\Delta\rho_* = 0.1$ as a typical constant normalised density difference between ice and water.

Implicit to the derivation of equation C.4 is a linear relation between the brine salinity S_b and the temperature, which is assumed in the following: $T_1/T_2 = S_1/S_2$. Cox and Weeks (1986) have derived more accurate equations to describe the expulsion process. Following Cox and Weeks (1986), and neglecting the small thermal expansion of pure ice, the change in the volumetric fraction v_b during cooling can be expressed as

$$\frac{dv_b}{dT} = -\frac{\rho_b v_b}{\rho_i S_b}\frac{dS_b}{dT}. \tag{C.5}$$

Assuming the brine density ρ_b to be constant, equation C.5 has the solution

$$\frac{v_{b2}}{v_{b1}} = \left(\frac{S_{b1}}{S_{b2}}\right)^{\left(\frac{\rho_b}{\rho_i}\right)} \tag{C.6}$$

As the brine density increases with brine salinity, the change in brine volume may be obtained by stepwise integration of equation C.6 for changing ρ_b. Cox and Weeks (1986) simplified the integration as follows by introducing a linear salinity dependence of the brine density ρ_b as

$$\rho_b \approx \rho_{b*} + \beta_* S_b \tag{C.7}$$

where $\rho_{b*} = 1000$ kg m^{-3} is a reference water density and $\beta_* = 0.8$ an effective linear density slope. Assuming a constant fresh ice density $\rho_{i*} = 917$ kg m^{-3}, and integrating equation (C.5 from T_0 to $T < T_0$, gives

$$\frac{v_b}{v_{b0}} = \left(\frac{S_{b0}}{S_b}\right)^{\left(\frac{\rho_{b*}}{\rho_{i*}}\right)} exp\left(\frac{\beta_*}{\rho_{i*}}(S_{b0} - S_b)\right). \tag{C.8}$$

Equation C.8, derived by Cox and Weeks (1986), has been used by the same authors to compute expulsion in a sea ice growth model (Cox and Weeks, 1988), but unfortunately was not given correctly in the latter publication. To obtain the salinity loss, Cox and Weeks (1986) proceeded as follows. The bulk sea ice salinity is defined as

$$S_i = \frac{v_b \rho_b S_b}{\rho_{si}}. \tag{C.9}$$

and the bulk sea ice density given by

$$\rho_{si} = v_b \rho_b + (1 - vb)\rho_i. \tag{C.10}$$

Differentiating S_i with respect to temperature yields

$$\frac{dS_i}{dT} = S_i \left(\frac{1}{S_b}\frac{dS_b}{dT} + \frac{1}{\rho_b}\frac{d\rho_b}{dT} + \frac{1}{v_b}\frac{dv_b}{dT} - \frac{1}{\rho_i}\frac{d\rho_i}{dT} \right). \tag{C.11}$$

The last term on the right hand side is related to the sea ice density change. Cox and Weeks (1986) neglected this term and found from integration from T_0 to T

$$\frac{S_i}{S_{i0}} = \frac{S_b}{S_{b0}}\frac{\rho_b}{\rho_{b0}}\frac{v_b}{v_{b0}}, \tag{C.12}$$

which by inserting equation C.8 gives

$$\frac{S_i}{S_{i0}} = \left(\frac{S_b}{S_{b0}} \right)^{\left(1 - \frac{\rho_{b*}}{\rho_{i*}}\right)} \left(\frac{\rho_{b*} + \beta_* S_b}{\rho_{b*} + \beta_* S_{b0}} \right) exp\left(\frac{\beta_*}{\rho_{i*}}(S_{b0} - S_b) \right). \tag{C.13}$$

As S_b is an unique function of temperature, and all other properties in equation C.13 are constant, it can be used to predict the change in salinity from an initial value S_{i0}, while the ice is cooled from T_0 to T and the brine salinity changes from S_{b0} to S_b.

The analytic solution C.13 is possible due to the linearisation of brine density and the assumption of constant β_* and bulk sea ice density ρ_{i*}. However, the sea ice density change neglected by Cox and Weeks (1986), is relevant at large brine volumes. Accounting for the change in ρ_{si} during cooling, makes a numerical solution necessary. Combining equations C.9 and C.6 gives

$$\frac{S_{i,k+1}}{S_{i,k}} = \left(\frac{S_{b,k+1}}{S_{b,k}} \right)^{\left(1 - \frac{\overline{\rho_{b(k,k+1)}}}{\overline{\rho_{i(k,k+1)}}}\right)} \frac{\rho_{si,k}}{\rho_{si,k+1}} \tag{C.14}$$

where indices k and k+1 are used to indicate its differential character and numerical scheme to solve the differential equation C.5. In the present section a stepwise computation with average values $\rho_{b(k,k+1)}$ and $\rho_{i(k,k+1)}$ has been used to solve for the change in S_i when the temperature changes from T_0 to T. This includes the nonlinear changes in β, ρ_b, pure ice density ρ_i and bulk sea ice density ρ_{si}. Most important is that the bulk sea

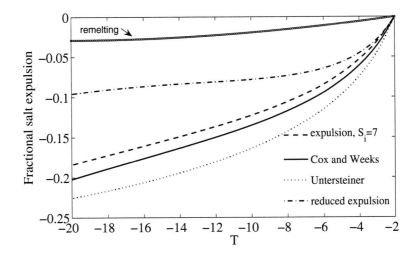

Fig. C.1: Fractional average desalination of sea ice due to brine expulsion during cooling. The solid line corresponds to equation C.13 from Cox and Weeks (1986), the dotted to C.4 from Untersteiner (1968). The dashed line is the numerical integration of equation C.14 accounting for all property changes. The curve termed 'reduced expulsion' assumes that the brine at the bottom of the ice is expelled. The uppermost curve for 'remelting' allows only for brine density changes.

ice density, equation C.10, depends on the brine volume v_b and requires the specification of the initial ice salinity S_{i0}. Hence, expulsion is salinity-dependent, which is not retained in the simplification from Cox and Weeks (1986).

In Figure C.1 the fractional salinity change obtained by numerical integration is compared to the simplified approach from Untersteiner (equation C.4) and the approximation from Cox and Weeks (equation C.13). An initial ice salinity of 7 ‰ and temperature of -2 °C are used as reasonable natural conditions. It is found, that both approximate formulas overestimate the decrease in sea ice salinity salinity. Two other curves with even lower desalination are discussed in the following, in connection with the question, under what conditions the described total expulsion may be expected.

C.2.2 Reduced expulsion concepts

The expulsion equations from Cox and Weeks (1986) had once been formulated to describe the change in salinity during cooling of stored samples. In this case it may be reasonable to assume that brine is expelled laterally from the sample. However, the ap-

plication to floating sea ice by Untersteiner (1968) and Cox and Weeks (1988) requires
further considerations, because the ice has no lateral boundaries. A realistic conceptional
model might be (i) formation of cracks through which (ii) the expelled brine can be communicated to preferential paths. Associating such paths with the larger brine drainage
channels and networks, it is difficult to imagine that the brine will leave unmixed from the
level where it was expelled. To formulate such a mixing model requires detailed knowledge of the preferential paths. Moreover would *horizontal* gradients and boundary layers
within the pores and channels be expected to become relevant. Such a system can no
longer be treated macroscopically. However, a plausible approach might be to assume,
that the brine in a channel is displaced instantaneously downwards by the expansion.
Thus, only brine at the salinity near the bottom of the ice will compensate the internal
density change. This can be accounted for by always using ρ_{b0}, the bulk solution density
at the bottom of the ice, instead of the brine salinity at the expulsion level, in equation
C.14. The integration for this *reduced expulsion* concept is also shown in Figure (C.1)
for an initial ice salinity of 7 ‰. It is approximately 50 % of the maximum possible
expulsion.

However, there is another factor that needs to be considered. When the brine is
pushed downwards it will be exposed to the equilibration process discussed earlier. The
downward displaced brine equilibrates to the local temperature keeping its higher salinity
and some ice will remelt. This in turn leads to a contraction of the bulk sea ice due to the
inverse phase transition. One would expect an upward migration of seawater, after it had
been pushed out. The overall result from this simple picture is a downward redistribution
of salt, but only a small change in the average salinity. The salinity decrease is due to
the fact that salt has been redistributed to levels of less dense brine, hence due to the
haline contraction of the solution. A first-order estimate can be made. One may assume
for simplicity, that the momentary downward displacement implies, via remelting and
return flow, an average brine density decrease of $(\rho_b - \rho_{b0})/2$. The average ice brine
density decrease, caused by a volume element cooled from T_0 to T_b and thus densified
from ρ_{b0} to ρ_b, may again be estimated by assuming a linear increase of ρ_b with depth.
Integrating ρ_b from ρ_{b0} gives a value of the order of $(\rho_b/\rho_{b0} - 1)/4$ as the fractional
decrease in the average ice salinity. This curve is termed *remelting* in Figure C.1 and
implies a much smaller decrease in the ice salinity of only a few percent.

It must be noted, that the three curves for the maximum expulsion (from Untersteiner
(1968), Cox and Weeks (1986) and the present integration) describe the local change in
absence of internal salinity redistribution. However, the curves *remelting* and *reduced
expulsion* are no longer locally valid, but are integrated estimates. They indicate the
bulk ice salinity decrease, implied by expansion of a bulk ice volume element, while it is
cooled from the interface temperature to T at some level in the ice. As the first order
approximations did not account for the increase in the ice salinity at lower levels due to

this downward displacement, they overestimate the expulsion slightly. Thus, averaging these curves for ice with a linear temperature gradient, gives an upper bound of the overall fractional salt loss by expulsion: For an ice sheet with surface temperature of -20 °C it is -0.019 and -0.073 for *remelting* and *reduced expulsion*, respectively. These values can be compared to a fractional average salinity loss of -0.120 for the maximum expulsion.

The approach should be understood to yield approximate upper and lower bounds of expansion effects. These appear more realistic than the classical expulsion concept from Cox and Weeks (1986). Exact simulation with appropriate boundary conditions would be needed to obtain better estimates, in particular with respect to the question of the detailed redistribution of the expelled brine. One needs then also to consider the question, if the expelled brine might create supersaturation in front of the dendrites, or leaves mainly through the larger brine channel network. The mechanism is likely an important aspect of brine channel network formation.

C.2.3 Simple models of brine redistribution

Mass and salt continuity

The simple averaging averaging approach does not yield the corresponding redistribution profile. As noted by Cox and Weeks (1975), the redistribution may be computed by using the continuity equations for solute (C.15) and mass (C.16) to obtain an equation (C.17), which relates the temporal changes in brine volume $\partial v_b / \partial t$ and brine salinity $\partial S_b / \partial t$ to the vertical fluid velocity W:

$$\frac{\partial(v_b \rho_b S_b)}{\partial t} + \frac{\partial(v_b \rho_b S_b W)}{\partial z} = 0 \tag{C.15}$$

$$\frac{\partial(v_b \rho_b)}{\partial t} - \rho_i \frac{\partial v_b}{\partial t} + \frac{\partial(v_b \rho_b W)}{\partial z} = 0 \tag{C.16}$$

$$\frac{\partial v_b}{\partial t} + \frac{v_b \rho_b}{S_b \rho_i} \left(\frac{\partial S_b}{\partial t} + W \frac{\partial S_b}{\partial z} \right) = 0 \tag{C.17}$$

Assuming thermal equilibrium, S_b can be expressed in terms of the local temperature change due to the moving interface and one may compute $\partial v_b / \partial t$ based on the vertical velocity. This was done by Cox and Weeks (1975) letting velocity signals propagate downwards from a rigid, constant temperature cold upper boundary, where $W = 0$ was assumed. Cox and Weeks' computations (in their figure 29) show a smaller decrease in salinity compared to the classical expulsion model, being approximately 50% of the latter, and comparable to the 'reduced expulsion' above. A validation by measurements was impossible, because the salinity loss due to the influence of gravity and convection, was much larger than the expulsion change.

Equations C.15 to C.17 describe the flow due to expansion and may be a useful formulation for thick sea ice, where convective processes are much less rigorous than in the bottom skeletal layer. Its applicability is, however, not completely clear. Cox and Weeks (1975) integrated these equations from the cold end downwards, letting a displaced liquid volume first equilibrate and dissolve ice in cold upper layers. However, due to the incompressibility of water, it would appear more logical to assume an instantaneous downward displacement of brine, followed by, as discussed above, a dissolution at all levels. Chiareli and Worster (1992) have applied a similar set of equations to predict the effect of expansion on the evolution of the solid fraction of mushy layer. In their approach they use the boundary condition of no normal flow at the mush-liquid interface. As will be seen in the next section, this is appropriate if the solute equilibrates infinitely rapid to thermal equilibrium by remelting or freezing, but it neglects the possibility of instantaneous outflow from the ice. Meaningful applications of equations C.15 through C.17 will need to make assumptions about the flow paths or brine channel networks and the boundary conditions.

The local solute redistribution equation

The question of reduced or complete expulsion may be considered on the basis of a variant equation C.17. In the metallurgical literature it is called the 'local solute redistribution equation' (Flemings and Nereo, 1967; Flemings, 1974) and often written in the form

$$\frac{\partial v_b}{\partial S_b}\frac{S_b}{v_b} = -\frac{\rho_b}{\rho_i}\left(1 + \frac{W\frac{\partial T}{\partial z}}{\frac{\partial T}{\partial t}}\right). \tag{C.18}$$

It is based on the assumption that the brine salinity instantaneously locally equilibrates to the temperature. This equation is valid locally, but may conceptually be used for the average change in a volume element.

The left hand side of equation C.18 represents the relative change in the bulk salinity S_i due to changing v_b and S_b. W on the right hand side is the fluid velocity and appears as an additional term compared to equation C.5. It can be imagined as the ratio of heat advection in the z-direction, $W\partial T/\partial z$, to the local rate of temperature change $\partial T/\partial t$. For a linear temperature gradient, neglecting the effect of advection on the rate of temperature change, one has

$$\frac{\partial T}{\partial t} \approx -\frac{\partial T}{\partial z}V\frac{z}{H_i}, \tag{C.19}$$

where V is the growth velocity and z/H_i the normalized distance from the top cold boundary. Inserting $\partial T/\partial t$ into equation C.18 gives

$$\frac{\partial v_b}{\partial S_b}\frac{S_b}{v_b} = -\frac{\rho_b}{\rho_i}\left(1 - \frac{W}{V}\frac{H_i}{z}\right). \tag{C.20}$$

If now cooling is considered, in the normal case v_b decreases while S_b increases, and the right left hand side is negative, if the bulk salinity decreases. The condition for a decreasing ice salinity is thus

$$\left(1 - \frac{W}{V}\frac{H_i}{z}\right) > 0, \tag{C.21}$$

which under absence of fluid flow is fulfilled. Condition C.21 is also fulfilled if the brine moves upwards, in the opposite direction of growth $\frac{W}{V} < 0$. However, if the brine however moves downward, it must pass the ice bottom $z = H_i$ moving at growth velocity V. Therefore the loss of salt requires $W > V$. With these conditions equation C.21 is no longer fulfilled. Hence, the conceptual consideration of equation C.20 demonstrates that vertical fluid flow, induced by expansion, may not lead to desalination of ice but to redistribution of salt by remelting. This conclusion, however, requires thermal equilibrium.

Equation C.20 implies a second important information. If there is fluid flow out of the ice, $W > V$, then it follows that $\partial v_b / \partial S_b > 0$. Now the local solid fraction changes, although the brine salinity increases upon cooling. This is apparently a condition, where wider channels can form at any level and the flow can become unstable. The concept provides a simple understanding for the formation of brine channels (Flemings, 1974; Beckermann, 2002). Some aspects of brine channels in sea ice will be considered next.

C.3 Drainage through brine channels

Expulsion and brine channels

The local redistribution equation C.18, which describes the basic coupling of local brine volume changes with fluid flow, gives also a simple indication for the formation of brine channels. Channels will likely form when the fluid velocity exceeds $W > -\frac{\partial T}{\partial t}\frac{\partial T}{\partial z}$ because in this case the brine volume v_b does no longer decrease locally. The formation of a brine channel is nonlinear: Flow towards the warm end causes remelting which increases the liquid fraction. A larger liquid fraction reduces the resistance to flow which increases the flow velocity and induces further remelting. However, as summarised in section 11.3.1, the concise theoretical prediction of conditions that lead to such channel formations is difficult. Here only empirical studies of brine channels are summarised and some simple considerations with respect to redistribution of brine are offered.

Identification of brine channels

Many optical observations of brine channels have been reported. The existence of brine channels may, to some degree, be indirectly validated from comparative analysis of sea ice salinity cores. For first generation brine channels in thin ice, spacings of 1 to 4 cm have been reported (Wakatsuchi, 1977; Niedrauer and Martin, 1979; Wakatsuchi and Saito, 1985; Wettlaufer et al., 1997). When brine channel networks are present in thicker ice, they appear with a typical distance of 5-20 cm (Bennington, 1963; Wakatsuchi and Saito, 1985; Shapiro and Weeks, 1993; Cottier et al., 1999). The salinity obtained from conventional 7.5 cm core samples will therefore depend on the existence of a brine channel in the sample. Its standard deviation may, due to the comparable distance of larger feeder systems, be a measure of the contribution of brine channels to the overall salinity. For thin ice of 10-15 cm thickness the normalized standard deviation of conventional samples is typically 5-10 % of the ice salinity. A systematic study of a large laboratory dataset with crude vertical resolution has been performed by Weeks and Lee (1962). Some related observations from the present study are the first three profiles given in the appendix (figure D.1). For this ice, standard deviations of the vertical average k_{eff} were obtained from 5 to 7 cores at each date, and are shown in figure C.3. After a day of high temperatures of -5 to 0 °C, the normalised (by the average k_{eff} of the profile) standard deviation increased to approximately 20 %. During a period of stable temperatures it decreased again. The results may be interpreted in the way that, when the ice warms, brine channels become activated. While the ice thickens and ages, they increase in general in their volume fraction, presumably due to different processes: expansion and lateral expulsion of brine, slow pocket migration and merging processes, complex convective-diffusional redistribution due to melting and refreezing by weak thermal cycling.

The absolute value of the standard deviation of k_{eff}, also seen in figure C.3 shows a similar increase after the strong warming event, when it reaches 0.04 in k_{eff} or 1.4 ‰ in terms of bulk salinity. Over longer times the standard deviation of k_{eff} remains relatively stable in the range 0.02 to 0.03, corresponding to an ice salinity standard deviation of 0.7 to 1 ‰. This range reflects the contribution of brine channel systems and appears to be rather constant, while the average k_{eff} is decreasing (leading to a relative increase of the standard deviation). A possible interpretation is that the large brine channels keep their state at some convective limit, while the thinner feeder systems are the locations from where brine is lost by feeding the large channels. There may be other interpretations, involving complex redistribution process chains, and a more detailed discussion requires refined resolution in time and space. Similar magnitudes of the ice salinity variability are indicated by the profiles from Bennington (1967) for ice of 60 to 120 cm thickness, yet by comparing only two different cores. In addition, Bennington (1967) cut his 7.5 cm diameter cores into two halves to illustrate the variability.

The normalised standard deviation of 20 % after warming contrasts a decrease in standard deviation after a warming poeriod observed by (Cottier et al., 1999). Before the warming, the range was 20 to 30 % in the latter study, comparable to the present samples. There are, however many differences between the studies. Cottier et al. (1999) obtained horizontal standard deviations of single cores analysed on a small scale-grid, the ice was

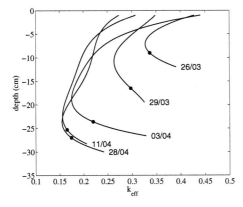

Fig. C.2: Normalized salinity profiles during the growth of thin ice described in the appendix (D). Sampling dates for the profiles are denoted. The dots indicate the position of the 3 cm distance from the macroscopic ice-water interface.

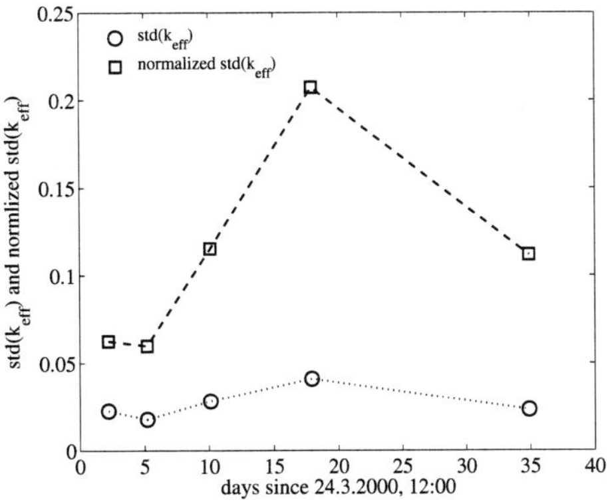

Fig. C.3: Standard deviations of k_{eff} with respect to the average profiles in figure (C.2) plotted against time since the onset of growth. Ice thickness increases from 12 to 30 cm during the period (see appendix D). Standard deviations normalized by the vertically averaged k_{eff} are also shown.

grown in a tank, and the timescale of the warming was different. Clearly, more systematic data for different ice thicknesses and thermal histories are required, to understand the processes of desalination and microstructure evolution. However, even bulk ice salinity samples may provide an indirect means of understanding the redistribution of solute into brine channels.

C.4   Gravity drainage

One may principally distinguish between two gravity drainage processes. The first one is driven by a pressure head imposed at the upper boundary of ice, for example due to melt water, and thus related to the density difference between water and air. The second one is driven by the internal gradient in the brine salinity within the ice.

C.4.1   Flushing and freeboard drainage

The most pronounced mode of gravity drainage becomes active during summer when melt water forms on the ice. This hydraulic head then gives rise to a vertical pressure gradient, which drives a net flow in the direction of gravity. The mass flow may be described by Darcy's equation 11.4 and depends on the permeability, viscosity and brine volume of the ice. The process has been termed *flushing* in an early discussion by Untersteiner (1968). Flushing may also be initiated by warming of the freeboard of the ice and has long been suggested to be responsible for the very low freeboard salinities of multiyear ice (see section 1.2). During the growth of cold first-year ice, upper brine volumes appear to be sufficiently small to prevent freeboard drainage (Nakawo and Sinha, 1981). However, also here the freeboard effect could play some role under temperature cycling.

C.4.2   Internal gravity drainage

The second gravity drainage process is related to internal density gradients, given by the upward increase of brine salinity with temperature towards the top of the ice, where the temperature is lower. As shown in section 11, the simple permeability law in the lamellar skeletal layer allows a concise prediction of the onset of salt release from thin ice. Above the skeletal layer, both the microstructure transforms into more complex pattern, including disconnected cavities and larger brine channels with feeder systems. A description of this system is much more complicated and at present only very little observations exist. The only dataset that gives detailed quantitative results of local salinity changes appears to be again the study by Cox and Weeks (1975).

The experiments from Cox and Weeks (1975) discussed in Chapter 5, indicated that internal gravity drainage above the skeletal layer is still important for the desalination of thin sea ice. Based on the evolution of nondestructive salinity profiles, the authors later proposed a crude parametrization of the desalination rate (Cox and Weeks, 1988). The local change in salinity was proposed to be proportional to the temperature gradient and the local brine volume by the equation

$$\frac{dS_i}{dt} = 3.37 \times 10^{-4} \left(v_b - 0.05\right) \frac{dT}{dz}. \tag{C.22}$$

It gives the local salinity change in ‰ per second, if the temperature gradient is given in K/cm and the local brine volume fraction exceeds 0.05. It must be noted that, although equation C.22 implies a brine volume of 0.05, below which no desalination is assumed, the data shown by Cox and Weeks (1975) do not reveal such a cutoff (their figure 31). Also, as the above equation was fitted to the data from Cox and Weeks (1975) after subtraction of the expulsion term C.13, the latter must be added to give the total effective salt loss, even if expulsion is, as discussed above, less pronounced. Also, an evaluation of figure 31 from Cox and Weeks (1975) indicates, that equation C.22 is an absolute upper bound, with average desalination being less than half this value. The desalination rate dataset supplied by Cox and Weeks (1975) is very scattered and should be reanalysed with respect to the changing experimental conditions discussed in Chapter 5. Petrich et al. (2006) have recently discussed these data by adopting a similar form as C.22 to constrain the ice permeability. However, equation C.22 has not been based by Cox and Weeks (1988) on a physical model. The critical point can be seen by the following approximation. The solute conservation equation C.15 may, assuming a constant brine volume v_b and brine density ρ_b, and setting $S_i = v_b S_b$, put into the simple form

$$\frac{dS_i}{dt} = W v_b \frac{\partial S_b}{\partial z} + S_i \frac{\partial W}{\partial z} \qquad (C.23)$$

By dropping the second term on the right hand side this equation can be compared to Cox and Weeks' (C.22). As $\partial S_b/\partial z$ is related to $\partial T/\partial z$ via the liquidus slope, it follows that Cox and Weeks' parametrization implies a constant velocity W, in connection with a reduced brine volume fraction $(v_b - 0.05)$ available for flow. The assumption that W depends neither on the pressure or brine salinity gradient, nor on the brine volume, is contradictory to Darcy's law discussed in Chapter 11. From physical principles, one would expect a stronger dependence of desalination rates on v_b. More concise parameterisation and theoretical treatments are needed.

C.4.3 Profile analysis by Untersteiner (1968)

Untersteiner (1968) studied the desalination of multiyear ice. He related its longterm salt and temperature profiles to a simple empirical approach to estimate the strength of flushing. Untersteiner was concerned with the equilibrium profile of multiyear ice evolving during several seasonal cycles. In his concept the bottom level of the ice is rising at annual average velocity V_{abl} due to surface ablation and reaches the top during a decade or so. This implies a persistent shift to a state of lower temperature during winter and higher temperature during summer. The parameter $V_{abl}\delta S_i/\delta z$, based on the bulk salinity profile, was found to be proportional to v_b^3, which is consistent with general permeability scalings discussed in Chapter 11. Untersteiner also investigated an expulsion model. The upward movement of each level creates not only higher temperature during summer, but also a temperature lowering during winter. This drives intermittent expansion and contraction. The comparison of the equilibrium profiles with actual profiles indicated, that permeability and flushing are more important than expulsion. As Untersteiner's approach was based on the local evaluation of v_b and S_i, as well as average annual ablation rates, it was a heuristic approach, meant to demonstrate the contribution of different processes to brine redistribution. While it also indicated the potential to predict the evolution of ice salinity profiles, the necessary theoretical and observational ground to do so have still not been laid.

C.4.4 Threshold brine volume for percolation of sea ice

Malmgren (1927) reported that sea ice at salinities of less than 5 ‰ kept a constant salt content for a rather long time, as long as the temperature remained below -4 to -5 °C.

These properties correspond to a limiting brine volume of 0.05 to 0.06. Since Malmgren's work similar comparable stable salinities have been documented and this has raised the question, under what conditions the sea ice salinity may become stable and how this may be related to the macroscopic permeability of sea ice. Recently several authors (Golden et al., 1998; Golden, 2003; Petrich et al., 2006) have discussed this question in terms of simplistic theoretical models of percolation. Also Cox and Weeks (1988) assumed, proposing equation C.22 for sea ice desalination, a threshold brine volume of $v_{b0} = 0.05$, below which no desalination can take place. However, it must be stated that the data from Cox and Weeks (1975), also shown by Weeks and Ackley (1986) are not consistent with a threshold at $v_{b0} = 0.05$. The observations still reveal desalination rates of 0.1 ± 0.05 ‰/day at this brine volume.

Another example for the lack in a cutoff brine volume of 0.05 are vertical property profiles of Baltic Sea ice presented by Fransson et al. (1990). This ice has generally a much lower brine volume fraction, and its permeability would also be smaller due to lack in brine channels. Looking at average profiles of many cores, the salinity apparently decreases from a value of 0.65 ‰, at a brine volume of 0.05 (35 cm depth), to 0.54 at a brine volume of 0.025 (25 cm depth). This 20 % change in salinity cannot be accounted by expulsion alone and indicates that brine drainage continued at low v_b. The ice may, of course have been exposed to thermal cycling.

The brine volume estimate 0.05 for the stable salinity and temperature values reported by Malmgren (1927) is only a crude one. However, also the much better resolved salinity profile time-series from Nakawo and Sinha (1981) appear too scattered, to preclude the presence of a slow desalination in the shown series, at levels in the ice where the brine volume was below 0.05. Similar time series of profiles of ice temperature and ice salinity from *Syowa Station*, Antartica, have been reported by Wakatsuchi (1974b) in form of contour plots. These indicate more clearly, that seasonal ice salinities may reach a stable state (over the course of two months), when $S_i \leq 4.5$ ‰ and $T_i \leq -10$ to -12 °C. This corresponds to a brine volume of 0.02 to 0.025. The latter range is probably a more realistic threshold for the desalination. It is consistent with in *in situ* studies of Arctic ice by Freitag (1999), showing that at such low brine porosities the pores are essentially disconnected.

It is unlikely that a general percolation threshold can be defined for sea ice. Its value will very likely depend on growth conditions, the initial microstructure and its evolution, but also on the distribution of inclusion sizes. Its definition in terms of desalination rates is another matter of debate. At low brine volumes of the order 0.025, it seems more likely that only slow diffusion processes and expulsion lead to further desalination. Fluid permeability is then no longer a useful definition at all, and one needs again a synthesis of different mechanisms to describe dthe system's metamorphosis.

In this sense it is thought that the approaches to restrict sea ice percolation by a *three-*

dimensional percolation threshold (Cox and Weeks, 1988; Golden et al., 1998; Golden, 2003; Petrich et al., 2006) must be viewed with caution. However, in the present work a *two-dimensional* percolation formulation is applied to the transition of brine layers into sheets, see section 12.3. In the relevant regime right above the skeletal layer of sea ice, this formulation is a theoretically concise approach, which is fully consistent with observations.

C.5 Conclusions

The main processes that may contribute to desalination of bulk ice have been summarised. As pointed out earlier by many other authors, gravity drainage through brine channel networks is doubtless the most important mechanism. The present chapter has focused on some aspects, how slower salt transport mechanisms may be related to the formation of brine channels.

- The role of brine pocket migration in the framework of brine channel formation is unclear. Its exploration requires further monitoring and modeling of the migration process. The effect of shapes on migration velocity of inclusions seems important. The process must be re-investigated in general.

- Figure C.1 and the above discussion indicate, that expansion alone will likely lead to redistribution of salt in sea ice. This does not mean that brine expulsion is not relevant for desalination. It must, however, be considered and modeled in connection with other mechanisms. An illustrative example is the evolution of multiyear ice considered in a heuristic approach by Untersteiner (1968). Any level in the ice shifts to a state of lower temperature during winter and higher temperature during summer. The former temperature lowering drives the expulsion process, while the latter may provide the necessary permeability and pathway for laterally redistributed brine through convection in brine channels. Pocket migration could, as noted, be viewed as a process that creates the necessary feeder systems of brine channels. Hence, expansion can be very effective when combined to thermal cycling.

- Observations of desalination of young ice and parameterisations of the latter are not confident at the moment. Models suggested so far are rather heuristic and imply parametric forms that neither physically, nor empirically convince. The important aspect of brine channel network evolution has not been addressed yet.

- Desalination above the rigorously convecting skeletal layer depends likely on the redistribution of solute to large brine channel features. Modeling their evolution is a challenging task. Even their initial formation process is not fully understood in a quantitative manner.

D. Ice observations from Adventfjorden

During the winter season 2000, a series of salinity profiles has been obtained by the author and his colleague Cecilia Bennet[1] during the growth of thin ice in the inner part of Adventbay (Svalbard), 500 meters from the UNIS (University Courses on Svalbard) building. The width of the bay is approximately 2 kilometers, with water depths between 5-20 m at the sampling site. Observations were obtained 50 to 100 m from the shore line.

Sample analyses and weather data

Samples were taken during March/April 2000 with a standard sea ice coring instrument of 7.5 cm diameter. Collected subsamples of each core, approximately 10 cm in length, were put in plastic bags. Some were immediately (1 to 2 hours after sampling) analysed, others were stored typically 1 to 4 weeks in a cold room at UNIS at a temperature $-25 < T < -20$ °C. Samples were then cut into 3 to 4 cm slices which were melted in small closed plastic boxes. The salinity of the melt was measured with a standard conductivity instrument, calibrated repeatedly by a more exact standard salinometer for seawater salinity determinations. The accuracy of the measured ice salinities given below is better then 0.1 ‰, resulting in an accuracy in the effective distribution coefficient k_{eff} of typically 1 to 2 %. Air temperatures and wind velocity were read from graphical screen output from a weather station at the UNIS building, before and after each sampling trip. Unfortunately, it was later realised, that the UNIS weather station data were not stored routinously. However, continuous weather station data from a station in Adventdalen, 10 km apart, were found to deviate by less than 1 to 2 °C from the notated UNIS observations. These are documented in the following description as 10 meter air temperature T_{10} and wind velocity U_{10}.

During a period of strong easterly winds and spring tide the young winter ice, which has reached a thickness of 50 to 60 cm, had completely disappeared from Adventbay on 18.03.2000, followed by a warm period with temperatures not too much below 0 °C. When temperatures dropped again, growth of a solid ice cover was reinitialised during the 24.03. Approximately 30 to 40 hours after the onset of growth (25.03., 22:00), a thin ice cover of thickness 7 cm had formed under slightly decreasing (from -15 to -20 °C) air temperatures and light winds ($U_{10} \approx 2 - 3$ m/s). This initial thin ice was covered moderately with frost

---

[1] The help of Magnus Larsson in ice core sampling and analysis is gratefully acknowledged

Ice observations Adventfjord (Svalbard)

date	hours	cores	H, cm	H_0, cm	V_0, cm d^{-1}	$\left(\frac{dT}{dz}\right)_0$, K cm^{-1}	k_0
25/03	≈ 35	1	7				
26/03	53	7	12.0	4.9	5.88	0.624	0.328
29/03	125	7	19.5	8.7	4.39	0.466	0.264
03/04	243	5	26.6	7.8	2.26	0.240	0.175
11/04	430	4	28.3	5.9	1.54	0.163	0.156
28/04	838	4	30.0	6.4	1.31	0.139	0.155

Tab. D.1: Results from ice core sampling during March/April 2000. The columns from the left give the day of sampling, hours since the onset of growth, number of profiles and ice thickness H (averaged from the profiles), and the distance H_0 between the ice-water interface and the level of minimum k_{eff}. V_0 at each level was estimated from a polynomial fit of the growth history. For the conversion between V_0 and the bottom temperature gradient $(dT/dz)_0$ a reduction of the latter by a factor 0.66 with respect to pure ice growth has been assumed. It is noted that in figure 5.14 the observations of k_0' at 6 cm are shown, while the tabulated k_0 refers to the salinity minimum of each profile, which differs slightly from $6cm$.

flowers, approximately corresponding to 10 to 15 % coverage classification according to the description by Martin et al. (1996). Scraping off the ice surface with a plastic edge, two 1 m^2 samples of the combined frost flowers/slush layer were found to have salinities of 55 and 60 ‰, their water equivalents being 0.35 and 0.14 mm, respectively. These values agree with the results from Martin et al. (1995) during early stages of frost flower development for similar growth conditions in a laboratory.

Over the coarse of the following weeks, ice cores were taken regularly at the site, 50 to 100 m from the shore line. At each sampling date between 4 and 7 cores were obtained, separated by several meters in the field. The observations are plotted in figure D.1. During the first 9 days of growth air temperatures were rather constant near -20 °C and the ice reached a thickness of 26 cm (profile from 03.04). During the next week the temperature was higher, reaching peak values between 0 to -5 °C during the whole 09.04. During this period also 3 cm of snow fell on the ice. The next samples, taken on 11.04., revealed only a further increase of ≈ 2 cm in thickness. During the next two weeks air temperatures were stable and mostly between -10 and -15 °C, but there was little further growth until 28.04., when the final samples were taken.

In figure D.1 the average profiles are represented by individual 4th-order polynomials. Based on these polynomials the minimum k_{sk} level denoted as H_0 was found. The values fall between 4.9 and 8.7 cm and are given together with the growth conditions in table D.1. These data are discussed in sections 2.4 and 5.5.6. It is noted that the standard deviation of k_{eff} was typically 5 to 10 % during early growth and stable low temperatures,

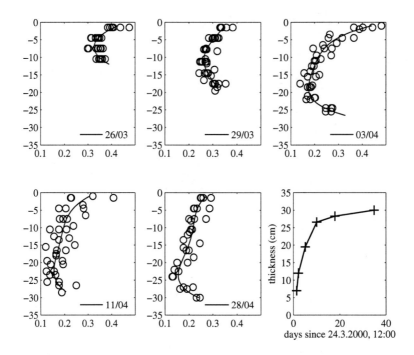

Fig. D.1: Salinity measurements based on ice core samples during March/April 2000. Y-axis: position from the ice surface in cm; x-axis: k_{eff} or the ice salinity normalized by the seawater salinity $S_\infty \approx 34.5$. Each circle represents the average salinity of a 2-4 cm vertical slice. The average profiles are indicated by fits to individual 4th-order polynomials.

but increased to 20 % for the two profiles following from 11.04 and 28.04. This is likely a consequence of intermediate warming and redistribution of brine to larger channel features (Tucker et al., 1984; Cottier et al., 1999). The numbers indicate the scatter to be expected when only single profiles are obtained in a growth series. Growth velocity estimates based on single thickness measurements may imply an even larger uncertainty. For the small thickness increments between the samples, the latter varied by ±50 %. By averaging the present averaging over several samples more confident estimates were obtained. The final uncertainty of the values given in table D.1 are estimated as 10 % for k_0 and 20 % for the growth velocity V_0.

Finally, it is noted that heat budget estimates indicate, that the reduced growth during early spring may be explained by either increasing shortwave radiation absorbed in the ice, or by oceanic heat fluxes driven by pronounced tidal currents. The reduction in growth is reasonable due to the potential of a considerable fraction of the incoming shortwave radiation to penetrate through thin, saline ice into the water below, e.g., Maykut (1986); Perovich and Richter-Menge (2000). An accurate determination of these fluxes is beyond the scope of the present study. However, the expected (based on a ice surface energy balance) conductive fluxes through the ice have been compared to the observed growth rates for the last two profiles from 11.04. and 28.04. Doing so it was estimated that the growth retarding processes are likely equivalent to temperature gradients $(dT/dz)_0$ in the range ≈ 0.04 to 0.05 K cm^{-1}. Comparison with table D.1 indicates that $(dT/dz)_0$, based on growth velocity estimates neglecting an oceanic heat flux, might be underestimated by 20 to 30 % for the lowest growth velocities. Considering the relationship between k_0 and dT/dz, discussed for example in section 5.5.6, such a change is relevant, in particular at lower growth velocities.

E. Reanalysis of the dataset from Cox and Weeks (1975)

The general description of the dataset from Cox and Weeks (1975) that is considered here in detail has been given in section 5.1 from the main part. Here a detailed discussion of the observations is given. It will focus on the following questions:

- The laboratory protocol suffers from a number of caveats that need to be discussed quantitatively to evaluate, if this dataset resembles natural growth conditions.

- The simple *skeletal model* for solute distribution, equation 5.9 from section 5.3 will be discussed in connection with the dataset. It is based on the assumption of a growth-velocity-independent brine volume, here set to $v_{sk} = 0.17$, at the top of a skeletal layer of constant thickness $H_{sk} = 3$ cm. Although it appears from figure 5.10 that the constancy of $v_{sk} = 0.17$ is not fulfilled below a growth velocity of 3 cm d^{-1}, it looks reasonable at larger V. It will be used as a first approach to discuss some particular aspects of the Cox and Weeks (1975) laboratory protocol.

- A description of the observations in terms of desalination near the 3 cm distance from the interface, the nominal skeletal layer, is given.

E.1 Original k_{sk} observations

The observations of k_{sk} given by Cox and Weeks (1975) are plotted in figure E.1 against the ice growth velocity for the two freezing runs. Values not based on the $H_{sk} = 3$ cm level, obtained by Cox and Weeks from the upper portions of the first profile of the runs, are emphasised with crosses. It is seen that the dotted line, the δ_{bps}-fit given by Cox and Weeks and used in many later studies (Weeks and Ackley, 1986; Cox and Weeks, 1988; Weeks, 1998a; Eicken, 2003), apparently matches these points properly. Also shown are the relation between k_{sk} and V, predicted by the simple skeletal scaling concept from section 5.3, using $v_{sk} = 0.17$ and $H_{sk} = 3$ cm. These predictions are shown for two solution salinities $S_\infty = 35$ and 60 ‰, approximately corresponding to the initial and final experimental conditions in the experiments from Cox and Weeks (1975), as respective solid and dashed lines. In both runs S_∞ increased from 35 to $40 - 45$ ‰ while

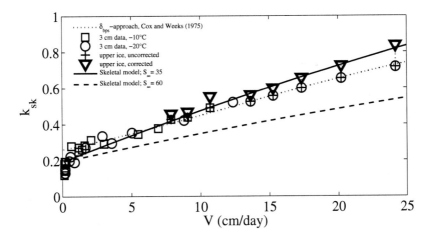

Fig. E.1: Experimental data from freezing NaCl solutions obtained by Cox and Weeks (1975) shown as squares and circles; the plus signs denote data points not evaluated instantaneously at a distance of 3 cm from the interface, and influenced by later desalination. The triangles are these data after corrections due to brine drainage and expulsion. The dotted line is the δ_{bps}-fit from Cox and Weeks (1975). The solid and dashed lines correspond to the constant skeletal layer model (5.4) with $S_\infty = 35$ and 60, respectively, and $H_{sk} = 3$ cm. In the experiment S_∞ was initially 35 ‰, reaching typically $40 - 45$ ‰ at $V = 2 - 3$ cm d^{-1}, with final values $60 - 65$ ‰ at lowest V.

the growth velocity V decreased to 2 to 3 cm d^{-1}. The observations are consistently closer to the dotted line for 35 ‰. In figure E.2, where the ratio of observed k_{sk} to predicted k_{sk} based on the actual S_∞ during the experiment is shown, this is illustrated in more detail. From the original data, shown as squares and circles, it is apparent that the simple skeletal prediction overestimates the values of k_{sk} at high growth velocities.

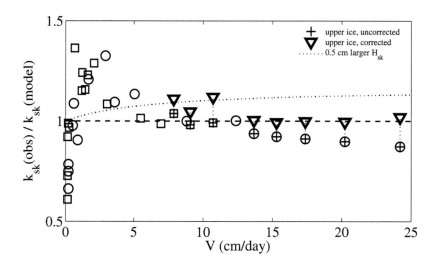

Fig. E.2: Experimental data from Cox and Weeks (1975) with symbols corresponding to figure (E.1). Shown is the ratio of observed to predicted k_{sk},
with predictions based on the observed, temporarily changing S_∞ in equation (5.4). The additional plus signs denote the uncorrected ice values from
the first profile of each of the two freezing runs, the bold triangles are the
corrected data due to brine drainage and expulsion.

E.2 Brine drainage corrections of k_{sk}

So far it appears that the δ_{bps}-fit matches the observations better than the physically more
consistent skeletal model. However, as already noted, the values denoted with plus-signs
were not obtained 3 cm from the interface, yet from the upper parts of the first salinity-
profile of each run. It is evident from figures (5.8) and (5.7) that the upper salinities
were still decreasing subsequent to the first profile. Hence, all upper ice k_{sk}-values, not
evaluated at the 3 cm level, must, for consistency, be corrected for desalination with
respect to the time when they were 3 cm from the interface. This correction has been
performed with an equation suggested by Cox and Weeks (1988), based on an analysis
of the actual dataset. It is discussed in a later section C.4.2 and takes the form

$$\frac{dS}{dt} = c_1 \left(v_b - 0.05\right) \frac{dT}{dz}, \tag{E.1}$$

for the local salinity change in ‰ per second, if the temperature gradient is given in
$K cm^{-1}$. Cox and Weeks (1988) noted it to be valid for local brine volume fractions
exceeding a threshold 0.05. Here this equation is used as an empirical means to correct the

upper ice k_{sk} values given by Cox and Weeks. However, instead of using $c_1 = 3.37 \times 10^{-4}$ ‰ cm s^{-1} K^{-1} suggested by Cox and Weeks (1988), an evaluation of the data from Cox and Weeks (1975), in particular their table VIII, showed that at brine volumes $v_b \approx 0.15$, which is the typical scale of interest here, a constant $c_1 \approx 1.7 \pm 0.7 \times 10^{-4}$ ‰ cm s^{-1} K^{-1} is a better description of the desalination. The correction was then estimated as follows: an average brine volume, an average temperature gradient, and the elapsed time Δt since a given level from the first profile was situated 3 cm from the interface were estimated. The average brine volume was simply estimated as an average of the actual v_b in the first profile and $v_{sk} = 0.17$, which always is assumed as the initial H_{sk}-value, consistent with the above simplified distribution concept. An average temperature gradient and the time of desalination have been estimated in the same simple manner, assuming linear growth during the first hours. The resulting correction in k_{sk} is to first order proportional to $\Delta k_{eff} \approx \Delta t v_b dT/dz$. As during diffusion-dominated growth $V \sim dT/dz \sim t^{-1/2}$ and all variables were changing by less then a factor two, it is expected that the nonlinear deviation from this simplification is less then 10%, clearly neglegible compared to the uncertainty of ± 50 % in the gravity drainage parametrization E.1 from Cox and Weeks (1988). A second minor correction was assumed due to brine expulsion. It was calculated as $\Delta k_{eff} = k_{eff} \left((v_b/v_{sk})^{(1-\rho_{b0}/\rho_i)} \right)$, based on the 'reduced expulsion' concept from section (C.2.1), also assuming an initial $v_{sk} = 0.17$.

The 'reduced expulsion' gives approximately half of the conventional expulsion discussed in section (C.2.1). It corresponds quantitatively to the numerically simulated reduced expulsion contribution which Cox and Weeks (1975) simulated and subtracted before computing the desalination due to gravity drainage alone, their table VIII, on which the actually used $c_1 \approx 1.7 \pm 0.7 \times 10^{-4}$ ‰ cm s^{-1} K^{-1} is based. Hence, let the 'reduced expulsion' concept be correct or not, the corrections applied here reflect the observed desalination rates in the data $CW75$. The corrections increase k_{sk} at the upper levels by 10 to 15% with respect to the values of k_{eff} given by Cox and Weeks. Clearly these corrections should be viewed as estimates with an uncertainty of eventually 50 %, considering the spread in the data leading to equation C.22. They are, of course, quantitatively also consistent with the desalination between the first and second profiles, obvious in figures 5.3 and 5.4.

The corrected values of k_{sk} are shown as bold triangles in figures E.1 and E.2. The data appearance clearly changes when gravity drainage is accounted. Looking at figure E.1 the skeletal model prediction now performs well. This is more clear in figure E.2 where the ratio of predicted and observed k_{sk} is shown, also accounting for the increase in C_∞ during the experiments via equation 5.9. For velocities between 5 and 25 cm d^{-1}, the prediction with constant $v_{sk} = 0.17$ and $H_{sk} = 3$ cm deviates only by a few percent from the observed k_{sk}, with three exceptional points near $V \approx 10$ cm d^{-1}. These exceptions all stem from the upper portion of the freezing run at 10 °C. The three noted exceptional points may admit some systematic explanation. They might, for example be influenced by non-equilibrium effects during the first centimeters of ice growth. Another possibility is that the respective growth velocities are underestimated. Figure 22 from Cox and Weeks (1975) indicates that during the 10 °C run the first thickness observation was obtained after a few hours when the ice was 2 cm thick (not shown here). Hence, an underestimate in the freezing interface position by just 0.4 cm, implying an initial freezing rate underestimate by 20%, would be sufficient to explain the differences between observations and predictions. Also indicated is, as a dotted curve in figure E.2, how a 0.5 cm increase in H_{sk} would affect k_{sk} in the skeletal prediction. For the three points it differs slightly from the uncertainty due to a growth velocity error, the latter being more severe when the ice is very thin.

Due to the simplicity of the approach and the noted error sources it does not seem suitable to press for any better agreement between predicted and observed k_{sk}. Vice versa, considering the coarse resolution of observed temperature gradients and the noted difficulty to measure the cellular freezing interface position to within 0.5 cm, the good agreement down to 5 cm d^{-1} is, at first glance, surprising. It is worth a note that the corrected upper ice values were not related to in situ observations of the interface, but to ice thickness estimates based on polynomial fits of the whole early growth time series (Cox and Weeks, 1975). The quality of such a fit during the first centimeters of growth is probably somewhat fortuitous, which might also explain why all early points for the -20 °C-run are matched by the predictions, while those for the -10 °C-run are not. The close agreement of the predictions with the k_{sk}-values obtained from profiles down to 5 cm d^{-1} is, on the other hand, expected, as the constant $v_{sk} = 0.17$ used in equation 5.9 in thus range closely resembles the actually observed values, see figure 5.10. Moreover, as an experimental error enters equation 5.9 via $v_{sk} \times r_{sk}$, a comparison of figures 5.9 and 5.10 above indicates that for growth velocities down to $3 - 5$ cm d^{-1}the deviations from the assumed constant $v_{sk} \times r_{sk}$ partially cancel each other.

E.3   Variability in k_{sk} due to v_{sk}

The simple $v_{sk} = const.$ approach reasonably predicts k_{sk} above a growth velocity of 3 to 5 cm d^{-1}, but performs less well at low growth velocities. A first comparison of figures E.2 and 5.10 indicates that it is mainly the deviation in v_{sk} from a constant 0.17 which resembles the deviation in k_{sk} from the predictions. It is therefore natural to proceed in asking for the expected dependence of v_{sk} on growth velocity.

The concept of a structural transition at the top of the skeletal layer was once suggested by Anderson and Weeks (1958) in connection with sea ice strength studies. At a critical brine layer width d_0 a transition to inclusions was suggested as an explanation for the mechanical behaviour of sea ice. If one identifies a_0 and the critical d_0 reported by Anderson and Weeks (1958) with plate spacing a_0 and brine layer width d_{sk} at the skeletal transition, one must have $v_{sk} \approx d_{sk}/a_{sk}$ for large crystals. While Anderson and Weeks (1958) reported $a_0 \approx 0.46$ mm and $d_{sk} \approx 0.07$ mm (and thus $v_{sk} \approx 0.15$), it has been documented in subsequent studies that a_0 depends on growth conditions. While all investigators agree in terms of an increase of a_0 with growth velocity, the form of the relationship has been less clear (Tabata and Ono, 1962; Weeks and Assur, 1963; Rohatgi and Adams, 1967b; Lofgren and Weeks, 1969a; Nakawo and Sinha, 1984), but has now been reasonably established by the present work.

The brine layer width d_{sk} has so far been suggested to be associated with a process of minimisation of the ice-water interfacial energy (Anderson and Weeks, 1958; Weeks and Assur, 1963; Tsurikov, 1965) and has been assumed constant. A physical model or scaling of d_{sk} has never been derived. This task was addressed in Appendix F and Chapter 12, where some evidence for the conjecture $d_{sk} = const.$ is presented. Here it suffice to say that the transitional v_{sk} is expected to decrease with increasing a_{sk} and decreasing growth velocity, provided that d_{sk} remains constant or shows a weaker velocity dependence. Then, under the assumption $d_{sk} \approx const.$, the maximum in v_{sk} and k_{sk} near $2 - 3$ cm d^{-1} in figures E.2 and 5.10 is the principal unexplained feature of the observations. To investigate its origin two particular aspects of the laboratory experiment are addressed. These are

- The influence of thermal convection on v_{sk}.

- The intrinsic skeletal layer thickness or: the relevance of the choice of the level H_{sk} for the apparent relations between k_{sk}, v_{sk} and the growth velocity.

E.4 Skeletal temperature gradient $(dT/dz)_{sk}$ and thermal convection

The approximation $r_{sk} \approx 0.7$ reasonably represents the relation between V and the skeletal layer temperature gradient as long $V \approx 3$ cm d^{-1} (figure 5.9). Below this velocity an increase in r_{sk} to a ten-fold value near ≈ 0.15 cm d^{-1} is found. This increase in r_{sk} was accompanied by a complete halt in the ice growth when the thickness reached 30 to 40 cm (figures 5.1 and 5.1). A possible explanation for this behaviour is *thermal convection* in the liquid below the ice, implied by the laboratory setup. The convective heat flux Q_{con} in a container of water depth H_w and temperature difference ΔT_w between the bounding surfaces of the liquid is expected to scale as $Q_{con} \sim H_w^3 \Delta T_w$ (e g., Turner (1973)). During the $CW75$ experiments Q_{con} is expected to have decreased slightly with decreasing H_w (ice thickness increasing). It may have increased with C_{infty} due to $\Delta T_w = T_{infty} - T_{bot}$, if a constant container bottom and room temperature T_{bot} are assumed. Hence, Q_{con}

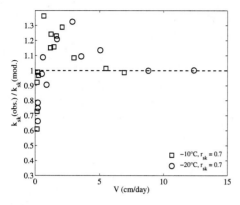
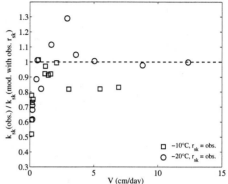

Fig. E.3: Experimental k_{sk} from Cox and Weeks (1975) normalized by the predictions with the skeletal model setting $r_{sk} = 0.7$; data points as in E.1, but lacking the upper ice values.

Fig. E.4: Experimental k_{sk} from Cox and Weeks (1975) normalized by the predictions with the skeletal model using the observed temperature gradient r_{sk} from figure (5.9).

depends on several experimental conditions like the room temperature and the isolation, thermal properties and geometry of the growth apparatus. Lacking detailed information it is most important, that the influence of Q_{con} *relative to the conduction through the growing ice* should have increased with thickness, becoming dominant at the end of the experiment, when growth ceased. Accordingly, figure 5.9 may also be interpreted as the relative strength of thermal convection.

For present goal it is most important that thermal convection implies a heat flux which changes the relation between growth velocity and the temperature gradient at the bottom of the ice. As thermal convection results in a larger skeletal layer salinity gradient this implies an increase in k_{sk}. The effect can be evaluated by inserting the observed r_{sk} from figure 5.9 into equations 5.4 and 5.8. The result is shown in figure E.4, and contrasted with the model predictions with constant $r_{sk} = 0.7$ in figure E.3. The data points correspond to figure E.2, lacking the upper ice k_{sk} for which, as discussed, no temperature profiles exist. k_{sk} has decreased but still shows a maximum in the ratio of observed to modeled salinity. The comparison of figures E.4 and E.3 indicates, that thermal convection is responsible for an increase of k_{sk} by $10 - 30$ % at lower growth velocities, always with respect to the same V without convection present.

The principle effect is more clearly quantified as follows. The assumption of a constant heat flux due to thermal convection in the liquid implies that the growth velocity takes the general form $V = c_1 (dT/dz)_{sk} - c_2$, where c_2 represents the thermal convection forcing. This simplification assumes that the thermal fields in the ice and the liquid are

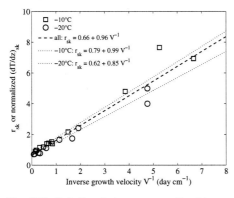

Fig. E.5: Relation between normalized temperature gradient r_{sk} and the growth velocity V in the experiments from Cox and Weeks (1975). The least squares fit to all data is shown as a dashed line, the separate fits as dotted lines.

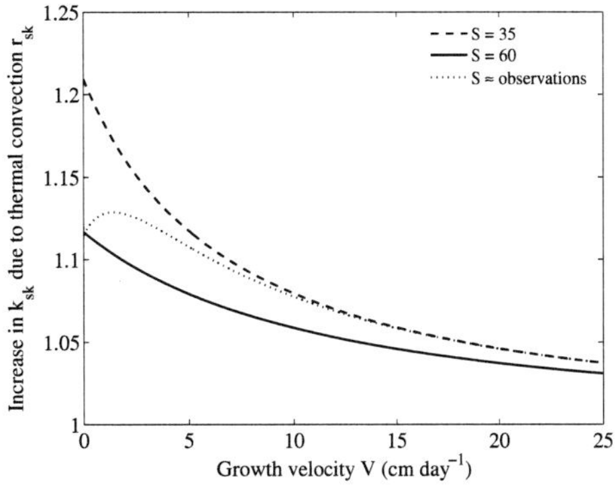

Fig. E.6: Increase in k_{sk} from the skeletal model by using the variable $r_{sk}(V)$ due to thermal convection, equation (E.2), and $r_{sk} = 0.66 = const.$, shown for two solution salinities. The dotted line takes the approximate salinity increase into account.

not interacting and implies

$$r_{sk} = r_{sk0} + \frac{V_{th}^*}{V} \tag{E.2}$$

as the relation between r_{sk} and V. In the absence of thermal convection one finds $r_{sk} = r_{sk0}$, whereas the second term may be interpreted as a critical velocity scale. For example, $2V_{th}^*/r_{sk0}$ indicates the growth velocity where thermal convection increases r_{sk0} by a factor of 2. A correlation with all $CW75$ data shown in figure E.5 gives $r_{sk0} = 0.66$ and $V_{th}^* = 0.96$ cm d^{-1} which appears reasonable in connection $r_{sk} = 0.7$ assume above due to figure 5.9. The observations are consistent with a quasi-constant thermal convection strength. The largest deviations at high V^{-1} could be related to the coarse temperature resolution. As will be discussed below, the fits for the individual runs are different, with $r_{sk0} = 0.62$ and $V_{th}^* = 0.85$ cm d^{-1} for -20 °C and $r_{sk0} = 0.79$ and $V_{th}^* = 0.99$ cm d^{-1} for the freezing run at -10 °C.

The principal effect of thermal convection on k_{sk} is obtained by introducing relation E.2 for r_{sk} into equation 5.8 for k_{sk}. The curves in figure E.6 show the result for $r_{sk0} = 0.66$ and $V_{th}^* = 0.96$ cm d^{-1}, the lumped fit for the dataset. The relative increase of k_{sk} at a given velocity depends on C_∞ and is shown for initial and final solution salinities (dashed and solid lines). When the actual (approximated) change of solution salinity with V is included in the computations one obtains the dotted curve in figure E.6. The calculation demonstrates that at low growth velocities the effect of thermal convection should have been a rather constant increase in k_{sk} by 10 to 15 %. At high V the contribution the

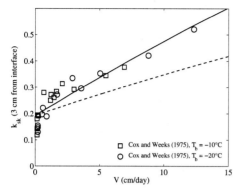

Fig. E.7: Experimental k_{sk} (3 cm) from Cox and Weeks (1975) plotted against the growth velocity. Only profile data from figure (E.1) are shown. The solid and dashed lines show the skeletal model for $C_\infty = 35$ and 60 ‰.

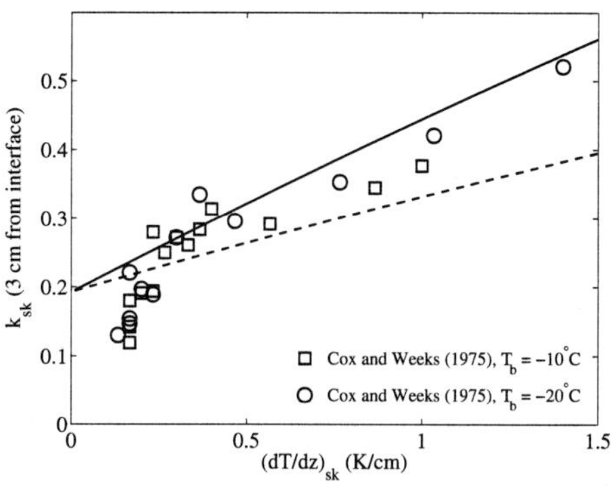

Fig. E.8: Experimental k_{sk} (3 cm) from Cox and Weeks (1975) plotted against the actual skeletal temperature gradient $(dT/dz)_{sk}$, instead of V as in figure (E.7). The solid and dashed lines show the skeletal model for $C_\infty = 35$ and 60 ‰.

change in the interfacial heat balance may still have increased k_{sk} by 5 %. These numbers are, of course, only valid for the particular conditions in the experiments from Cox and Weeks (1975).

In figure E.5 also the separate fits for the two freezing runs are shown. In the run at -10 °C the temperature gradients at low V^{-1} are slightly larger and the data appear to approach a 15 to 20 % larger $r_{sk} \approx 0.79$ at high V. This explains that in figures E.4 and E.3, where $r_{sk} = 0.7$ was assumed constant, the computed difference based on the predictions eventually reached 30 % for this run. The intrinsic effect of thermal convection is less and coresponds to figure E.6. It appears that the thermal convection in the -10 °C run was slightly stronger, which might be related to thinner ice and a deeper water colum H_w. Considering the scatter in figure E.5, the noted coarse temperature gradient resolution, and the unknown boundary conditions, a more detailed discussion seems not feasable here. The effect of thermal convection on k_{sk} has been quantitatively estimated in a consistent manner and is significant.

E.5   Changing variables from V to $(dT/dz)_{sk}$

It is useful to visualize a consequence of the preceding paragraph by plotting k_{sk} in figure E.8 in dependence on the actual temperature gradient, contrasting the growth velocity plot, figure E.7. The main difference between the data appearance is that in the plot of $(dT/dz)_{sk}$ versus k_{sk} the convection-biased k_{sk} values at low V are moved to the right.

Principally one may argue that for an experiment influenced by thermal convection, and thus the $CW75$ data, the relation between k_{sk} and $(dT/dz)_{sk}$ is preferable to describe the variation of k_{sk} with growth conditions.

If figure E.7 resembles natural growth without convection is, however not certain. The figure apparently emphasises a strong deviation from the assumption $v_{sk} = constant$, leading to the question, whatever the exact dependence might be, if it is related to $(dT/dz)_{sk}$ or V. This question has been addressed in figure E.9 by plotting v_{sk} against the growth velocity for two narrow actual temperature gradient ranges $0.17 < (dT/dz)_{sk} < 0.23$ and $0.24 < (dT/dz)_{sk} < 0.34$ K cm^{-1}. It seems that v_{sk}, and thus k_{sk}, during slow growth increases with the interface velocity V. For the regime $0.24 < (dT/dz)_{sk} < 0.34$ K cm^{-1} such a dependence is not apparent, but there are only a few data points and the two freezing runs also appear to differ. The scattered data and possible correlation of several variables do not allow a decision, if V and $(dT/dz)_{sk}$ influence k_{sk} independently. One may further look for an indication that external thermal convection influences k_{sk} by modifying v_{sk} internal to the ice. In figure E.10, v_{sk} has been drawn versus the normalized temperature gradient r_{sk} for three growth velocity regimes. Considering the regimes separately, no systematic variation of v_{sk} with r_{sk} is apparent. This has been ascertained by comparing the first v_{sk} maxima in figures 5.9 and 5.10 above. These maxima are not associated with an exceptional increase in r_{sk}. The change in v_{sk} between the three growth velocity regimes likely reflects the principle effect of the absolute V or $(dT/dz)_{sk}$ on v_{sk}. Recalling that not the strength of the convective flow but rather its relative contribution to the interfacial heat balance changes with r_{sk}, figure (E.10) does not reveal a systematic influence of this relative heat flux strength on v_{sk}.

The above analysis thus suggests that the observations obtained by Cox and Weeks (1975) overestimate salt entrapment under those natural growth conditions where thermal convection is absent, and that this overestimate is largest at low growth velocities. [1]. Thermal convection changes the relation between the temperature gradient in the skeletal layer and the growth velocity, leading to an apparent increase in k_{sk}, when compared to the situation at the same V without convection. Therefore the laboratory data should be better presented in terms of $(dT/dz)_{sk}$. A direct effect of thermal convection on v_{sk}, eventually by the triggering of fluid motion within the ice, is not evident in the observations, although the data are too sparse to draw a definite conclusion. The overall effect of thermal convection in the experiments of Cox and Weeks (1975) appears to have been an increase in k_{sk} by 5-15%, becoming more pronounced at low growth velocities.

---

[1] While thermal Rayleigh-convection is a laboratory mode and not expected during natural sea ice growth, turbulent oceanic heat fluxes may play a similar role in the field (Wettlaufer et al., 2000). As these are often connected to time-dependent unidirectional under-ice currents, it is not clear if the present results are transferable

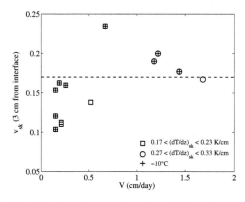

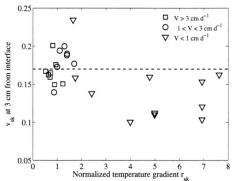

Fig. E.9: Experimental v_{sk} (3 cm) from Cox and Weeks (1975) at two low temperature gradient regimes versus the growth velocity. The run at -10 °C is marked with plus signs.

Fig. E.10: Experimental v_{sk} (3 cm) from Cox and Weeks (1975) plotted against the normalized temperature gradient r_{sk} for three growth velocity regimes.

E.6 H_{sk}, v_{sk} and the skeletal layer conjecture

When the thermal convection bias is accounted for in the evaluation of k_{sk} from Cox and Weeks (1975), by presenting the data in the form of figure E.8, it emerges that the $v_{sk} = const.$ skeletal model is not valid. The correspondence of the deviation of k_{sk} from the $v_{sk} = const.$ approach to the actual variation in v_{sk} is emphasised by comparing in figures E.11 and E.12. The maxima in k_{sk} and v_{sk} are still unexplained.

Evaluation level H_{sk} versus intrinsic skeletal thickness H_{sk}^*

For a further discussion of the maximum in v_{sk}, first the main principle of the proposed skeletal layer conjecture is recalled. The justification to define k_{sk} at the top of a skeletal layer is, that the microstructure changes at its upper boundary, a distance H_{sk}^* from the interface. The transition is ideally thought to lead to an abrupt drop in the permeability, strongly suppressing further desalination. The selected profiles from Cox and Weeks (1975), shown in figures 5.3 and 5.4 above, support such a transitional behaviour, arguable from the strong change in the salinity gradient, approximately 3 cm from the interface. Leaving the detailed physics of this sharp transition untouched at the moment, there is empirical evidence that it exists.

While the profiles from $CW75$ indicate, that also above the skeletal layer desalination continues slowly, the first focus is on the sharp transition near $H_{sk} \approx 3$ cm. It is natural to ask, how the evaluation of k_{eff} is influenced, if the intrinsic transition level H_{sk}^* would change with growth conditions. From figures 5.3 and 5.4 the principle consequence of

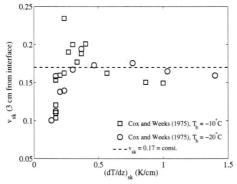
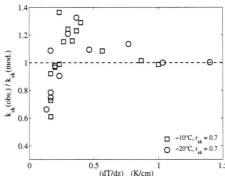

Fig. E.11: Brine volume v_{sk} at 3 cm distance from the ice-water interface; see also the corresponding relation versus V in figure (5.10).

Fig. E.12: Experimental k_{sk} from Cox and Weeks (1975) normalized by the skeletal model with $r_{sk} = 0.7$, in dependence on the skeletal layer temperature gradient.

a larger intrinsic $H_{sk}^* > 3$ cm is that, due to the larger salinity gradient in the skeletal layer, the evaluation at $H_{sk} = 3$ cm would overestimate k_{sk} considerably. This effect is shown in figures E.13 and E.14, in terms of the relative change $\delta k_{sk}/k_{sk}$ for a small variation $H_{sk} = 3 \pm 0.5$ cm[2]. There is little systematic dependence of $\delta k_{sk}/k_{sk}$ on the solution salinity S_∞, when compared with $H_{sk} = 2.5$ cm (figure E.14). When considering the dependence on the temperature gradient, $\delta k_{sk}/k_{sk}$ is larger at low values of $(dT/dz)_{sk}$ where k_{sk} is smaller, whereas absolute changes δk_{sk} are approximately the same. Setting the evaluation to a larger $H_{sk} = 3.5$ cm, generally decreases k_{sk}, with a few exceptions where the change is negligible. The positive shifts of $\delta k_{sk}/k_{sk}$ for $H_{sk} = 2.5$ cm are, due to the steeper salinity gradients, a factor of 3 to 4 larger. This demonstrates the importance to either know the intrinsic H_{sk}^* or to select a sufficiently large H_{sk} to avoid this uncertainty.

The asymmetry in the uncertainty of k_{sk} is emphasised in figure E.15. The large upper uncertainties clearly show that the relationship discussed so far, in particular the maximum at temperature gradients 0.3 to 0.5 K cm^{-1} noted earlier (figures E.11 and E.12), must be viewed with caution. It could easily be related to a change in H_{sk}^* by 0.5 cm, implying $\delta k_{sk}/k_{sk}$ of 30 to 40 %.

However, as the noted maximum in v_{sk} and k_{sk} in figures E.11 and E.12 appears to be quite stable, a systematic explanation in terms of H_{sk}^* seems appropriate. In figure E.16 some limited data on the sea ice skeletal layer thickness H_{sk}^*, summarized

---

[2] This range also resembles the noted problem to measure the position of the ice-water interface at an accuracy of better than ± 0.5 cm

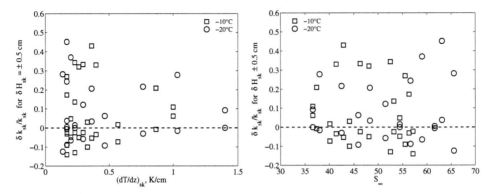

Fig. E.13: Relative change in k_{sk} in the profiles from Cox and Weeks (1975) associated with a ± 0.5 cm change in the H_{sk}-level versus the actual temperature gradient $(dT/dz)_{sk}$. Almost all negative δk_{sk} correspond to assuming $H_{sk} = 3.5$ cm, positive ones to $H_{sk} = 2.5$ cm.

Fig. E.14: Relative change in k_{sk} in the profiles from Cox and Weeks (1975) associated with a ± 0.5 cm change in the H_{sk}-level versus the solution salinity C_∞. Almost all negative δk_{sk} correspond to assuming $H_{sk} = 3.5$ cm, positive ones to $H_{sk} = 2.5$ cm.

in a later section 3.1, are shown. These indicate that the skeletal layer increases with thickness and eventually saturates at a constant value, once the latter exceeds 20 to 30 cm^3. In the $CW75$ experiments the thickness of 20 cm coincides with the fifth profile and $(dT/dz)_{sk} \approx 0.35$ K cm^{-1} for the run at $-10\,°$C, and with the third profile and $(dT/dz)_{sk} \approx 0.8$ K cm^{-1} for the run at $-20\,°$C (see figure 5.1. The apparent initial increase of v_{sk} may thus consistently be associated with a change in the intrinsic skeletal layer thickness H_{sk}^*. It is suggested that for the thinnest samples k_{sk} and v_{sk} were evaluated well above the intrinsic skeletal layer and therefore are slightly underestimated. Once the thickness reached 20 to 30 cm, the skeletal layer thickness H_{sk}^* exceeded the $H_{sk} = 3$ cm level, which would overestimate k_{sk} considerably, even for a difference $H_{sk}^* - H_{sk}$ of only 0.5 cm. As Cox and Weeks (1975) only reported the skeletal layer thickness at 1 cm resolution, such a variation must remain undetected and has likely created considerable bias in their derived k_{sk}.

Towards the end of the experiment, when growth ceased, Cox and Weeks (1975) reported several profiles with a skeletal layer thickness of only 2 cm. It seems that, at these low growth velocities, the skeletal layer thickness decreases again in connection with an increase in thermal convection. This is speculative, but appears plausible in connection with the rather steep thermal gradients in the liquid, and the generally expected effect

---

[3] No studies on the salinity dependence of H_{sk}^* are available

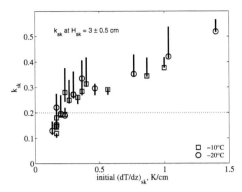

Fig. E.15: Variability in the experimental k_{sk} resulting from varying $H_{sk} = 3 \pm 0.5$ cm, obtained from the profiles of Cox and Weeks (1975).

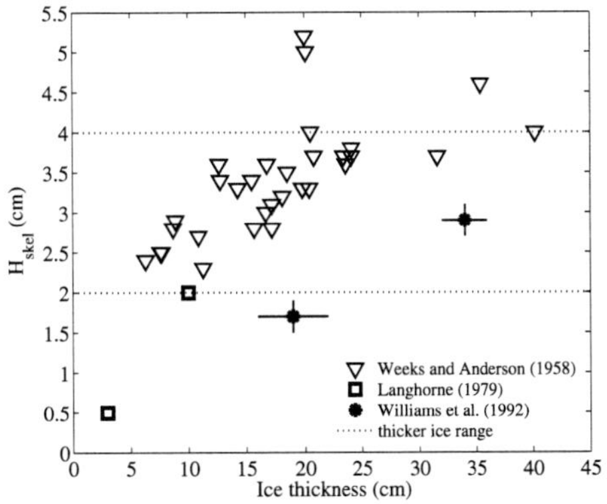

Fig. E.16: Field data of skeletal layer thickness of thin ice from different sources discussed in chapter 3. The dotted lines span the typical range for thicker sea ice.

of fluid motion in the liquid, for which a few examples were noted in section 3.1.

E.7 Delayed desalination

The conjecture that the intrinsic skeletal layer H_{sk}^* exceeded the evaluation level $H_{sk} = 3$ cm in the course of the $CW75$ experiments is consistent with the analysis so far. A difference $H_{sk}^* - H_{sk}$ of only 0.5 cm may lead to an overestimate of 40 % in k_{sk} (figure E.13) and is thus capable to explain the anomalous behaviour in, for example, figure E.12, eventually also the noted maximum in v_{sk}. The negative values of $\delta k_{sk}/k_{sk}$ in figure E.13 show that, also above the skeletal layer H_{sk}^*, some desalination continues. This raises the question of the most appropriate level to determine a quasistable k_{eff}. This is further investigated by considering how the initial k_{sk} at a fixed level changes when the latter distance towards the interface increases above H_{sk}. The fractional desalination, defined as $\Delta k_{sk} = (1 - \overline{k_{eff}}/k_{sk})$, will be used as a measure of the relative decrease of k_{sk} from the 3 cm level. For the lack of longer time series this is simply based on an average $\overline{k_{eff}}$ for all following profiles. It thus illustrates the principal changes, noting that Δk_{sk} at low temperature gradients, towards the end of the experiment, may be underestimated.

The relations between Δk_{sk} and several system variables are plotted in figures E.17 to E.20. It is seen that the maximum in k_{sk}, at temperature gradients 0.3 to 0.5 K cm^{-1} (see figure E.12), is comparable to a maximum in fractional desalination in figure E.17. As note above, the largest Δk_{sk} may correspond to an evaluation of k_{sk} within the skeletal layer and a corresponding overestimate of k_{sk}. For the normalized temperature gradient or thermal convection strength r_{sk} no simple dependence emerges from the plot of δk_{sk} versus r_{sk}, as shown in figure E.18. It seems, however, that at large enough r_{sk} desalination Δk_{sk} is limited to 20 to 30 %.

Different desalination for freezing runs at −20 and −10 °C

A so far unrecognised aspect emerges most clearly from figure E.19, where δk_{sk} is plotted versus the ice thickness. The runs at −20 and −10 °C show a very different delayed desalination. The former indicates a relatively constant δk_{sk} in the range 20 to 30 %, increasing eventually slightly with thickness and supporting an increase in the intrinsic H_{sk}. For the latter run, δk_{sk} is much larger and reaches values of 40 to 50% throughout the experiment, decreasing only at the end, when the growth rates were small. In contrast, the maxima k_{sk} (in figure E.12) were comparable for the runs. The result that, while the initial entrapment is similar, the delayed desalination is larger for the run at −10 °C, indicates the difficulties in merging the datasets, and the necessity to understand the factors affecting the intrinsic H_{sk}^*. A closer look at figure E.11 shows that the maximum in v_{sk}, contrasting k_{sk} in figure E.12, mostly contains observations from the run at −10 °C. The difference between the delayed desalination Δk_{sk} for the runs would therefore be reduced, if for the warmer run a slightly larger H_{sk} would have been used. The possible

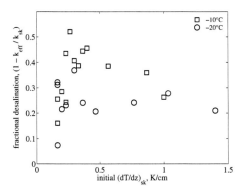
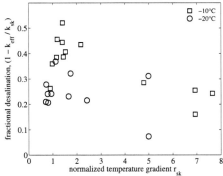

Fig. E.17: Fractional desalination $\Delta k_{sk} =$ $(1 - \overline{k_{eff}}/k_{sk})$ at a fixed level versus temperature gradient $(dT/dz)_{sk}$.

Fig. E.18: Fractional desalination $\Delta k_{sk} =$ $(1 - \overline{k_{eff}}/k_{sk})$ at a fixed level versus normalized temperature gradient r_{sk}.

effect of C_∞ on the intrinsic H_{sk}^* has not been touched so far. The relation between δk_{sk} and C_∞ (not shown) is almost identical to the appearance in figure E.19, because C_∞ in the experiments increased almost linearly with thickness (see figure 5.2). Therefore it is unclear to what degree C_∞ may influence H_{sk}^* and v_{sk}, or if the suspected increase of the H_{sk}^* is mainly related to the increasing thickness, figure E.16.

While the growth conditions considered so far, $(dT/dz)_{sk}$, r_{sk}, C_∞ and ice thickness, only identify but do not explain the difference between the two runs, figure E.20 emphasises another aspect. It is seen that the fractional desalination increases with skeletal brine volume v_{sk}. Despite the scatter, one may tentatively suggest from the graph that Δk_{sk} might vanish somewhere in the range $0.05 < v_{sk} < 0.1$. As already pointed out, one finds that the observations for the two runs are clustered around different v_{sk}. It further appears that Δk_{sk} was almost a factor of 2 larger for the freezing run at -10 °C, because the initial brine volume v_{sk} at 3 cm distance was systematically larger in this case. It further supports the conjecture, that during the run at -10 °C the level $H_{sk} = 3$ cm fell within the skeletal layer.

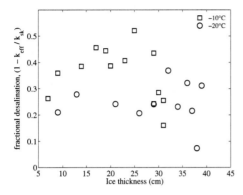
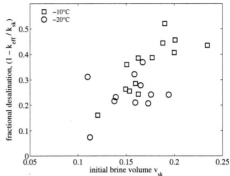

Fig. E.19: Fractional desalination $\Delta k_{sk} = (1 - \overline{k_{eff}}/k_{sk})$ at a fixed level versus ice thickness.

Fig. E.20: Fractional desalination $\Delta k_{sk} = (1 - \overline{k_{eff}}/k_{sk})$ versus brine volume v_{sk} at $H_{sk} = 3$ cm.

E.8　Discrepancies at the ice-water interface

The different desalination during the runs at -20 and -10 °C has emerged figures E.17 to E.20. The difference between the runs is further illustrated in figure E.21, showing the ratio k_{sk}/v_{sk} which, considering equation 5.4, closely resembles the brine salinity enhancement factor S_{sk}/S_∞ within the skeletal layer. The ratio approaches, at low temperature gradients, the expansion parameter $r_\rho \approx 1.14$ for the solution salinity in question (equation 5.4). Also in figure E.21, the runs at -20 °C and -10 °C appear systematically different. However, in the latter relation between k_{sk}/v_{sk} and $(dT/dz)_{sk}$, this is rather unexpected, as the brine salinity gradient should be given by the temperature gradient via the liquidus slope, which is very similar at the brine salinities of the two runs. Figure E.21 may then be interpreted in two ways. Either the skeletal temperature gradients, derived by Cox and Weeks (1975) in one of the experiments, are subject to a systematic error, or the position of the ice-water interface in the run at -20 °C must have been underestimated, eventually being the consequence of the coarse temperature resolution.

Further elucidation comes from figure E.22. It shows the ratio of the interfacial temperature T_{int} from $CW75$ and the melting temperature of the underlying solution, T_∞. The brine near the interface is expected to be colder and more saline than the solution. Taking a constitutional interfacial supercooling of 0.1 K as a conservative estimate [4], T_{int}/T_{infty} should be slightly above 1 for the solute concentrations in question. However, figure E.22 shows that it is clearly below 1 during almost the whole run at -10 °C, while for the -20 °C run it is considerably larger. Hence, in the former case of too

---

[4] This value has been observed in experiments (Rohatgi and Adams, 1967b; Foster, 1969; Terwilliger and Dizio, 1970) and is theoretically founded by the present work's interface analysis.

small values, the estimated interface is not at the freezing point. The latter range of 1.1 to 1.2 would imply supercoolings of 0.3 to 0.6 K which appear unrealistic. Although Cox and Weeks (1975) reported some differences between observed solution salinities and those obtained from a solutal budget of the ice, these were of the order of 1 % for most parts of the experiments and always less than 5 %, being insufficient to explain the apparent interfacial temperature difference. This suggests again that, at least in the run at -10 °C, the reported interface temperatures are subject to some systematic error, most likely related to the coarse resolution of the temperature sensors, in particular by the end of the experiment. Alternatively, the ice thickness and interface position might be overestimated systematically. It is recalled that the normalised temperature gradient r_{sk}, according to equation E.2 and figure E.5 was systematically larger for the run at -10 °C. Requiring that the interface temperature should not be above the melting point, also these temperature gradients would be reduced. To conclude, it seems most likely that, for some reason, the interface temperature interpolations by Cox and Weeks did not match the actual temperature profile in the same way for the two freezing runs. One may also suspect, that the observational procedure was changed between the runs, or that the visual appearance of the interface was different with a different temperature forcing and adjustment of thermal insulation. Other effects related to experimental boundary conditions and their control might be at work.

In conclusion, it appears that the reported ice-water interface is 'too warm' in the freezing run at -10 °C. This error fundamentally effects the interpretation of the solute distribution in the interfacial regime. The deviations from an expected T_{int} are qualitatively consistent with the earlier noted conjecture, that a constant $H_{sk} = 3$ cm overestimated the skeletal layer for the run at -20 °C, yet underestimated it for -10 °C. The former conditions implies a slight underestimate of k_{sk}, while in the latter case the overestimate is considerable, as one evaluates k_{sk} still within a regime of a large bulk salinity gradient. Although also a difference in the intrinsic H_{sk}^{*} between the runs cannot be excluded, it seems unlikely that it would have pertained over the observed range in thickness and growth velocity. The variation in H_{sk}^{*} with thickness and growth conditions is a different aspect in connection with the possibility to formulate a simple model. Figure E.22, on the other hand, shows the problems to discuss such a model concisely based on the $CW75$ dataset, because the interfacial temperature field is not properly resolved.

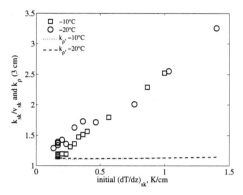

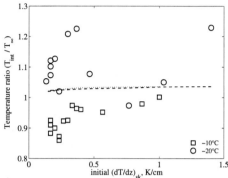

Fig. E.21: Brine salinity enhancement k_{sk}/v_{sk} across the skeletal layer $H_{sk} = 3$ cm and density factor r_ρ versus temperature gradient $(dT/dz)_{sk}$, for the two runs at -20 °C and -10 °C.

Fig. E.22: Ratio of interfacial temperature to melt temperature T_{int}/T_∞ versus temperature gradient $(dT/dz)_{sk}$. Dotted $(-10$ °C$)$ and dashed lines $(-20$ °C$)$ show expected ratios for an interfacial undercooling of 0.08 K.

E.9   Conclusions: Changing the evaluation level of k_{sk}

At present two non-destructive timeseries of sea ice salinity and temperature profiles from Cox and Weeks (1975) comprise the most relevant dataset on brine entrapment during freezing of NaCl or seawater solutions. In the present appendix to Chapter 5, the potential of the data to understand fundamental desalination processes near the ice-water interface has been investigated. The detailed reanalysis of these data, with respect to a simplified interfacial layer concept, section 5.3, has yielded the following main results:

- A consistent relation between growth velocity and k_{sk} at a fixed level H_{sk} from the interface requires brine drainage corrections for observations in the upper levels of the first profile of each run. This step changes the data appearance considerably (figure E.1).

- The data from $CW75$ are shown to be biased by thermal convection and/or lateral heat flow (figure 5.9 and E.6). Therefore an evaluation of the relation between k_{sk} and the skeletal layer temperature gradient $(dT/dz)_{sk}$ is preferable (figure E.8) over the velocity relation (figure E.7).

- It is likely that the evaluation level of H_{sk} at later stages of the experiments was situated within the intrinsic skeletal layer of width H_{sk}^*, leading to a considerable overestimate of k_{sk}^* at the top of the skeletal layer. This interpretation is consistent with profiles (figure E.13) and longterm desalination (figure E.17).

- The skeletal layer H_{sk}^* generally varies with ice thickness (E.16), but may also be influenced by fluid flow and salinity in the bulk solution, for which some indirect indications are found (figures E.18 and E.19).

- The two freezing runs indicate systematic differences in fundamental aspects of the skeletal layer appearance (figures E.21 and E.22. These seem to be related to uncontrolled or unaccounted experimental conditions, a systematic change in the determination of the interface position, or a systematic difference in the intrinsic H_{sk}^*.

Due to the unknown contributions of the different error and variability sources, the present discussion of the interfacial processes within the skeletal layer must remain speculative. The dataset may therefore be better presented in a form, which largely avoids the uncertainties in the interface dynamics and intrinsic width of the skeletal layer. Instead of investigating k_{eff} near the skeletal transition level $H_{sk}^* \approx 3$ cm, where gradients are large, a larger distance can be expected to be a better evaluation level. This level has, with respect to the discussion of the 5 to 10 cm desalination boundary layer in thin ice in section 2.4, been set to 6 cm. Here this distance is termed H_0' to distinguish it from the level H_0, where desalination is expected to be halted based on physical grounds, to be discussed in chapters 3 and 12.

Figure E.23 shows k_0' in relation to the temperature gradient in the skeletal layer at time of its origin, but only when it has reached a distance $H_0' = 6$ cm from the interface, normally realised in one of the following salinity profiles. In a few cases, when this level was overshoot in the subsequent profile, linear interpolation was applied to find k_0'. The resulting solute distribution for a distance of 6 cm from the interface should be compared to figure E.15 for the skeletal level 3 cm. The uncertainties associated with a change by 0.5 cm, are now, as H_0' is safely above the skeletal layer, strongly reduced in the latter figure. Although the asymmetry in the bars resembles that desalination is still ongoing, the relationship between k_0' and growth conditions is now much less affected by uncertainties in the interface position and the intrinsic skeletal layer thickness H_{sk}^*.

Desalination beyond $H_0' = 6$ cm is emphasised in figure E.24. It shows the final k_{min}, reached by the end of the growth phase at each level (after 25 and 40 days for the runs at at -20 and -10 °C, respectively). Due to observational scatter the final two profiles of each run have been averaged to obtain k_{min}. The bars denote the difference to k_0' from figure E.23. Evidently, the desalination has continued beyond a distance of 6 cm from the interface and reduced k_{eff} to final fractions amounting to 70 to 90 % of k_0'. The final k_{min} during the higher temperature run at -10 °C is seen to be considerably lower. This is only partially related to the longer durance of this run. It is likely a consequence of the larger brine volumes associated with the warmer ice and has been touched at some points of the present appendix, like the brine volume dependence indicated by equation E.1. It

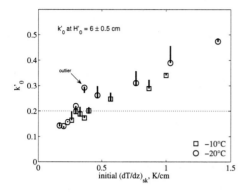
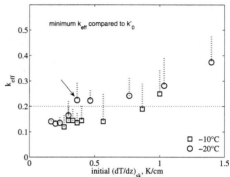

Fig. E.23: k'_0 at $H'_0 = 6 \pm 0.5$ cm with uncertainties indicated, versus the initial 3 cm temperature gradient $(dT/dz)_{sk}$. Reconstructed from those $CW75$ profiles which yield $(dT/dz)_{sk}$ initially followed by at least one profile with $H'_0 = 6$ cm.

Fig. E.24: Minimum k_{eff} from $CW75$ shown as symbols, bars denoting the difference towards k'_0 at $H'_0 = 6$ cm from figure E.23, versus the initial 3 cm temperature gradient $(dT/dz)_{sk}$.

is also notable, that the final minimum values of k_{eff} are reduced to approximately 50 and 70 % of k_{sk} at 3 cm, during the runs at -20 and -10 °C, respectively.

If, as it seems, already k'_0 at some 6 cm from the interface is different for the runs at -20 and -10 °C, this would mean that it is impossible to define a level where k_{eff} is only dependent on the temperature gradient $(dT/dz)_{sk}$ or growth velocity. However, on average the difference is less at 3 cm from the interface (figure E.15), which indicates that the *model concept* might be fruitful, if one allows for variability in the intrinsic skeletal layer boundary H^*_{sk}. As already noted, the latter is an important, apparently not constant, parameter, and cannot be discussed properly due to the coarse temperature resolution and the 'too warm' interfacial temperature estimates for the run at -10 °C (figure E.22). As long as this problem is unsolved, a level of 6 cm is a reasonable compromise, on the one hand avoiding the uncertainties near the interface, on the other hand minimizing the temperature-dependent ongoing desalination. The difference between the runs in figure E.23 is not too severe at 6 cm distance.

Uncertainties and practical consequences

The change in the evaluation level of k_{eff} in the $CW75$ data, and in other observations in general, reduces the problems of delayed desalination and uncontrolled skeletal layer dynamics. It provides a reasonable lower bound on k_{sk} at the skeletal transition. This approach cannot remove certain problems faced, when constructing a relation between

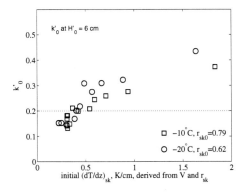

Fig. E.25: k_0' at $H_0' = 6 \pm 0.5$ versus the initial $(dT/dz)_{sk}$ derived from the growth velocity and r_{sk} using equation E.2. To compute $(dT/dz)_{sk}$ from V for both runs the same $r_{sk0} = 0.66$ and $V_{th}^* = 0.96$ cm d^{-1} were used.

Fig. E.26: k_0' at $H_0' = 6 \pm 0.5$ versus the initial $(dT/dz)_{sk}$ derived from the growth velocity and r_{sk} using equation E.2. To compute $(dT/dz)_{sk}$ from V the individual fits for -20 and -10 °C to r_{sk0} and V_{th}^*, see figure E.5 were applied.

k_0' and the temperature gradient. As in the $CW75$ experiments the relation between V and the temperature gradient $(dT/dz)_{sk}$ is biased by convection, an expression of k_0' in terms of $(dT/dz)_{sk}$ is preferable, yet the latter was only coarsely resolved. How this can influence the data appearance and what uncertainties are created may be illustrated with figures E.23 and E.24 wherein one observational point, near $(dT/dz)_{sk} \approx 0.4$ K cm^{-1} and denoted with an arrow, is outstanding from the otherwise more smooth data. Resembling observations from the fifth profile during the -20 °C run, this point is anomalous in many other graphs of which figure 5.8 is emphasised. There this anomaly appears as a rather low interfacial brine volume $v_{int} \approx 0.63$ compared to expected values 0.8 to 1. In view of the discussion of uncertainties this is most likely be interpretable as an error in the interface position or temperature. For the given conditions it could be explained by an underestimate of the temperature gradient by 30 to 40 %. Inspection of figures E.23 and E.24 shows that such a correction would bring the observation in agreememnt with the rest of the data from $CW75$. Figures 5.8 and 5.7 therefore may be taken as an indication of typical uncertainties of 10 to 20 % in $(dT/dz)_{sk}$.

Under certain growth conditions it can be most fruitful to fit a smooth time-thickness relationship to the observations and determine the growth velocity by differentiation at each level for which k_0' is to be determined. In the absence of thermal convection or other heat sources, $(dT/dz)_{sk}$ may then simply obtained from V and $r_{sk0} \approx 0.7$. In the $CW75$ experiments with thermal convection and lateral heat conduction this method has been shown to be inaccurate. The correction of the relation between V

and $(dT/dz)_{sk}$, accounting for the convection parameter r_{sk} as shown in figure E.6, was therefore performed by applying the least square-fitted relation E.2 between r_{sk} and V. This has been done, first by inserting the individual fits for each run and second, by inserting the bulk relation. The results are shown in figures E.25 and E.26, respectively. Slight differences with respect to figure E.23 appear, because in the latter figure k_{sk} had sometimes to be interpolated to the 6 cm level of the profile for which $(dT/dz)_{sk}$ was observed, and because some high temperature gradient data become available now from the first profiles. The most notable difference between figures E.25 and E.26 is that, applying individual convective corrections and different r_{sk0}, the runs appear more different than they do when an average correction is applied. As the individual corrections manifest the systematic differences and errors in interfacial temperatures noted above, one may interpret figure E.26 as some sort of correction for this effect. The latter figure is therefore also taken as an indication that the intrinsic difference of k_0' between the runs is smaller than figures E.23 and E.25 indicate, and that the skeletal transition value k_{sk} may indeed be only a function of the temperature gradient near the interface.

The approach does not completely remove the apparent outlier for the -20 °C run. In addition to the error bars from figures E.23, which also apply in figures E.25 and E.26, denoting the uncertainty of k_0' due to the determination of the ice-water interface , the resolution of the growth velocity by the end of the experiment was rather limited. Other deficiencies in the accurate control of the growth were pointed out in section 5.4, and may contribute to the scatter. In view of these limitations, figures E.26 and also E.23, are thought to be the best interpretation of the dataset from Cox and Weeks (1975), when considered in terms of its generality to characterise natural ice growth. The values of k_0', 6 cm from the interface, reflect the dependence of the interfacial solute distribution on growth conditions reasonably. The better reference is probably the run at -20 °C, where drainage between ≈ 3 and 6 cm is least, and of the order of 10%. Figure E.24 then highlights the differences in delayed desalination when comparing warm and cold ice.

F. Sea ice microstructure: Bridging of the skeletal layer

In the present appendix, the processes near the advancing ice-water interface are considered, supplementing the discussion of the plate spacing from the main part of this work. The length scale of interest here is the width d_{sk} of brine layers at the onset of bridging between the plates. In situ data of d_{sk}, the brine layer width at the skeletal transition, have not been reported so far. It has been considered, to indirectly compute $d_{sk} \approx v_{sk}a_{sk}$, utilising the nondestructive observations of v_{sk} from Cox and Weeks (1975). This would, however, require the proper knowledge of the level H_{sk}, where brine layer transform to sheets. However, as shown in chapter 5 and appendix E, the assumption of a constant H_{sk} is likely not justified. An evaluation is uncertain, simply because the vertical gradients in the brine volume v_{sk} and k_{sk} still are large near H_{sk}. In the particular dataset of Cox and Weeks (1975), the only systematic nondestructive study available to date, the coarse resolution of the temperature gradient near the interface, and the interface position itself, make a detailed study even more problematic.

The purpose of the present chapter is to discuss microstructure statistics of older sea ice and samples far away from the freezing interface with respect to the possibility to reconstruct the scale d_{sk}, at which they once started to transform into sheets. This makes it necessary to describe quantitatively, how inclusions change in shape during cooling and warming and subsequent changes in the brine volume. After reviewing the observational methods and their caveats and potentials, first the threedimensional change in inclusion scales is discussed in F.2. Then an attempt is made to interpret statistical analyses of sea ice microstructure by projecting them to the temperature close to the skeletal transition.

To understand the following statistical analysis, and avoid some confusion with previous work, the microstructural length scales are denoted and distinguished as follows:

- The *brine layer width* is generally denoted as d, and d_{sk} is defined as its value at the level H_{sk}, the top of the skeletal layer already discussed in section 5.6. Presumably, at this level the brine layers start to loose their rectangular shape, with the first appearance of inclusions and disconnections.

- l_g and l_d will be discussed in this section as the principal *major and minor axis lengths* of liquid inclusions in a *horizontal* cross-section. The minor length l_d of brine sheets corresponds to the original width direction of brine layers d which may have changed during cooling.

- The vertical inclusion length, normally the major length one measured in a vertical thin section is termed l_v, while the *average* horizontal length is l_h.

- *Inclusion lengths* are *projected* from any position and /or temperature to a brine volume that resembles the upper skeletal layer (here $v_b \approx 0.2$ is assumed). The projection is based on empirical relationships between aspect ratios and brine volumina. These lengths are given the subscript $_{sk}$. Hence, l_{gsk} becomes the major length in a horizontal thin section and may maximally take the length L_{gra} of a grain. If the projection is correct, than all inclusions merge by definition at this level, and the minor length $l_{dsk} = d_{sk}$.

- The intrinsic horizontal wavelength of perturbations during the bridging process above the skeletal layer is called l_{g*}. Assuming the onset of bridging with small amplitudes one must have, of course, $l_{d*} = d_{sk}$ for the width.

- The *percolation transition* level H_0 is the level, where bridging has made the brine layers so weakly permeable that desalination stops (the salinity minimum in a profile). Here the width of brine sheets (in a horizontal cross-section) is therefore d_0. Notably, in the earlier concept suggested by Assur (1958), d_{sk} does not change between H_{sk} and H_0 and thus $d_0 = d_{sk}$.

Because their work is frequently cited, it is worth a note, that Anderson and Weeks (1958) described the smallest observed brine layer width in samples before the appearance of predominantly circular vertical tubes. It was discussed by these authors irrespective of position. It is clear that, the smoother the transition from layers to cylinders takes place, the more subjective is its optical determination. It will, where it appears in the text, be distinguished by terming it d_0'. Two other scales introduced by Anderson and Weeks are: d_{cy} is the diameter of cylinders after the splitting of brine sheets or layers of width d_0', as it appears in a horizontal cross-section with predominantly circular inclusions; d_{cp} is the diameter of predominantly spherical inclusions which eventually by splitting of cylinders of diameter d_{cy}.

F.1   Sea ice microstructure observations

Analysis techniques have, in comparison to the heuristic work by Weeks and Anderson (1958), considerably improved during the past decade, mainly due to the possibility to digitize images. A principle problem in interpreting core sections is still, how to construct a threedimensional picture from horizontal or vertical thin sections. It is best illustrated by looking at composite construction from Cole and Shapiro (1998), here

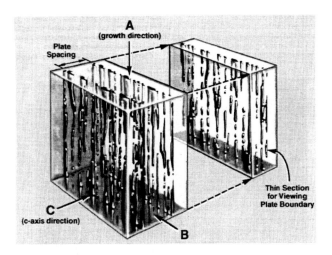

Fig. F.1: Sketch of typical brine inclusions distribution in sea ice drawn by
Cole and Shapiro (1998). The width d or l_d of brine layers is measured in the
c-axis direction normal to the plate boundaries. Thin sections are normally
obtained by viewing in the A direction or some mixture of B and C.

shown as figure F.1. It is clear, that the information on the pore structure, which can be
obtained from a cross-section, depends on the viewing direction. Figure F.2 from Cole
(2001) is a comparison of micrographs in the principal directions. Even the derivable
average porosities can be biased due to the finite size of sections. Furthermore, if one
is interested in the three-dimensional network and distribution of inclusions, one needs
to obtain several cross-sections, which is a laborious task. It would be ideal, also in
order to describe the morphological transitions of brine layers near the skeletal layer, to
obtain information about the brine layer planes by viewing them in the c-axis direction.
So far, only Cole and Shapiro (1998) and Cole et al. (2002) have reported some limited
observations and statistics for such sections. One is shown by the lower micrograph of
figure F.2 from Cole (2001). One may imagine, that it is particularly difficult to monitor
the brine structures between the plates without destroying or modifying them.

Published data on bottom samples are sparse, because the liquid nature of the brine
makes sampling more difficult. As it is difficult to realise sampling of bottom fractions
without loss of brine, one has to investigate levels further up in the ice, where the microstructure has changed with respect to the interface, or the transition level, where
the brine is closed off. Freezing the sample, and analysing it at a temperature different
from the in situ value, imposes a similar problem, concerning the change in the volume

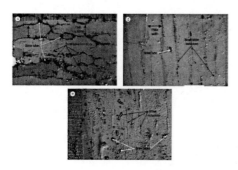

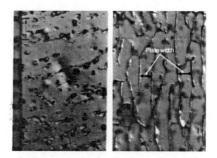

Fig. F.2: Thin section micrographs of first-year ice from Cole (2001). Standard horizontal cross-section (c) with brine layers and inclusions; (d) vertical cross-section normal to the c-axis; (e) vertical cross-section of an original brine layer viewing in the c-axis direction.

Fig. F.3: Horizontal thin section micrographs of first-year ice from Cole (2001). Right: obtained in situ within several hours after sampling; left: stored below -20 °C for several months and warmed up to -10°C.

fraction, size and aspect of inclusions. Finally, longterm changes during storage under isothermal conditions may be different from the situation in the field. These problems are particularly serious for bottom ice, where changes due to the high liquid fractions and temperature are largest. Before datasets obtained so far are discussed, it is useful to summarise the techniques applied to circumvent or solve the noted problems.

F.1.1 Observational techniques

Sample storage is essential for the analysis and interpretation of microstructure statistics. This is very clearly demonstrated in figure F.3 from Cole (2001), showing horizontal cross-sections of first-year ice. The right hand is a micrograph, obtained in the field within several hours after sampling, and clearly shows the intact brine layers. The left hand micrograph has been obtained after storage below -20 °C for several months. Although figure F.3 is an extreme example, while normally the plate spacing can be still identified also after storage, it indicates the difficulties to obtain information about the in situ microstructure from stored and thermally modified samples.

Temperature cycling

The standard procedure in most studies to date is to sample an ice core of diameter 7 to 8 cm, ship it to a cold room and perform a microstructure analysis at later times (Nakawo and Sinha, 1984; Perovich and Gow, 1996; Eicken et al., 2000). As seen in figure F.3, this procedure may involve, even if the brine is made immobile by storing the samples at temperatures below -23 °C, temperature hysteresis effects due to inclusion shape changes. Another modification due to cooling prior to the analysis is an expulsion or redistribution of brine due to expansion upon freezing (Cox and Weeks, 1986). It may typically decrease the bulk salinity, and hence the brine volume, by 10 to 20 %, see section C.2. These too processes, bulk salinity loss due to expansion and inclusion shape hysteresis during thermal cycling, are serious disadvantages of standard sampling and analysis procedures.

To give an example, Light et al. (2003) analysed ice from a field program (Cole et al., 1995) for which also Cole and Shapiro (1998) described the microstructure on a broader scale. The samples analysed by Light et al. (2003) came from the depth levels 60 to 80 cm, and were obtained by the end of the winter at an in situ temperature of ≈ -5 °C. At that time the ice was already in a stage of warming, originally having been cooled to ≈ -12 °C in mid-winter at the levels in question. Micrographs of this ice were taken at a temperature of -15 °C, but only after storage at -20 °C for 3 years. With respect to temperature, natural thermal cycling in the upper parts of sea ice may be similar to such artificial laboratory procedures. With respect to time they are not. To what degree the temperature-related changes in the microstructure, as derived by Light et al. (2003) and other studies of similar procedures (Perovich and Gow, 1996; Eicken et al., 2000), may represent real sea ice transitions, has not been demonstrated yet.

If storage is performed at -25 °C, then subeutectic transitions may occur, which are unrealistic for the major part of the ice. The difference increases towards the bottom fractions which, in terms of interfacial desalination, are most important to be retained. For relatively warm bottom ice, the natural temperature cycles are much reduced, and it is likely that the transitions produced during storage and cooling may be very different from the in situ evolution.

Processing in the field

In some studies thin section analysis has been performed in a cold room during field work (Stander, 1985; Sinha and Zhan, 1996; Cole and Shapiro, 1998). Cole and Shapiro (1998) prepared their thin sections at -14 °C, within several hours after sampling. Stander (pers. comm.) performed the preparation of cross-sections mostly within several minutes, and at relatively high temperatures of -10 °C. However, even under very rapid sampling and analysis conditions, many precautions have to be taken in order to monitor an undisturbed in situ microstructure. The following estimate may demonstrate this. After a core has been removed from its warmer environment, the length scale of propagation of a temperature signal from the ambient is $(\kappa_i t)^{1/2}$, from which one obtains 2.6 cm for a time of 10 minutes. This indicates a rapid response of an ice core sample which normally

has a radius of 3.5 to 4 cm. Moreover, once a very thin section is cut, the equilibration
to the environmental temperature may be even faster. Therefore this mode of sample
preparation in most cases can be expected to reflect the temperature of the cold room
rather than the in situ values.

Alternative approaches have been applied to conserve the in situ pore structure. Bennington (1963) noted the application of a procedure, where dye and brine were used to
seal a thin section, and then the ice was drained for some hours at −25 °C, to finally elucidate the microstructure. The sensitivity of the in-situ microstructure to this method
has not been investigated. Eicken et al. (2000) have, for ice samples grown in an ice
tank (Eicken et al., 1998), applied a technique to minimize the degree of brine inclusion
metamorphosis during storage. Samples were centrifuged at *in situ* temperatures. At
temperatures of −3 to −4 °C, most relevant for the skeletal transition at the bottom of
sea ice, this method has been shown to remove 70 to 90 % of the brine from the ice
(Weissenberger et al., 1992). This technique is elaborate but particularily suited for bottom samples with an open pore structure. However, at lower temperatures the residual
fractions in the latter study were more like 50 to 70 %. Considering the centrifugation
technique, it might be most consistent to store these samples at in situ temperatures for
further analysis, while so far low storage temperatures have been used also in this case
(Eicken et al., 2000). It is neither clear, if not a strong centrifugating force may create
changes in the microstructure, because also in a thin sample gradients in the equilibrium
temperature are present. The forced flow might also create new drainage pattern by
remelting, as well as hydrodynamic stability pattern, which might translate to the pore
structure.

Coarsening and aging

Hence, whatever method is used, one must be aware that the removal of an ice core
induces rapid changes in the physical conditions, which may immediately change the
microstructure of a reactive porous medium like sea ice. It appears that studies focusing
on such rapid changes are completely lacking.

Considering very long storage times, neither studies of slow equilibrium processes have
been performed on sea ice. It has, for example, been shown that freshwater ice accreted
in wind tunnels, with subgrain sizes comparable to sea ice brine inclusions, changes its
structure considerably during annealing at low temperatures of ≈ −20 °C (Prodi and
Levi, 1980). Although in these ice samples the liquid fractions between the subgrains are
extremely small and and the physical mechanisms are likely related to phase transition
kinetics on the molecular scale (Dash et al., 1995; Petrenko and Whitworth, 1999), the
observation indicates the importance of slow surface energy processes to drive the merging
of inclusions. These processes have certainly to be considered in case of storage times of

three years like in the study by Light et al. (2003).

In addition to the irreversible behaviour of inclusion shapes during cooling and warming, hysteresis has also been observed for the phase transition process itself, implying for example a delay in equilibration, when crossing the eutectic temperature during cooling (Koop et al., 2000; Cho et al., 2002). It appears that very small inclusions may become supercooled until they change phase spontaneously. The formation mechanism of smallest inclusions reported in sea ice (Cole and Shapiro, 1998; Light et al., 2003) and pressure-related 'corrosion features' (Bennington, 1963; Stander, 1985; Cole et al., 2002) is presently not understood. The formation of these features might be more probable during rapid cooling of a sampled core than it is under slowly changing natural conditions. Also the role of molecular nucleation processes under strong disequilibrium eventually deserves attention.

Due to the lack in understanding of morphological transitions, it is important to properly control and document storage times and temperatures, cooling rates and the details of the thin section preparation process, in connection with microstructural analyses. Only this may facilitate a later reevaluation when comparative studies of influencing factors become available. Sinha (1977) reported that even the choice of the light source was observed to influence the nature of thin sections.

F.1.2 Resolution

Modern techniques of micrography analysis allow the preparation of thin sections, wherein brine inclusions of 0.01 mm or even smaller in size are clearly distinguishable (Sinha, 1978; Light et al., 2003), see also figure F.3. However, the sample preparation process is elaborative and susceptible to a reaction of brine inclusions to the procedure (Sinha, 1977). A trade-off between resolution, sample size and sample preparation time must be coordinated with the main goals of a study. The pixel resolution during digital statistical analysis has been another limiting factor influencing the effectivity of a statistical analysis, but should no longer be a limitation with present day computing and image processing capabilities.

The pixel size of 0.03 mm limited the size of resolvable inclusions in the studies by Perovich and Gow (1991, 1996) to a value approximately two times this scale. A useful approach, applied by Perovich and Gow (1996) during digital analysis, is to complement such a method by a combination of grey-shading pixel allocation and definition of an ice-brine threshold grey scale. If the latter is determined via the actual phase relation brine volume, errors may eventually be minimised. Pixel sizes in the NMR study by Eicken et al. (2000) were larger, 0.09 mm. The above scales, in particular the NMR values, are a clear limitation at low temperatures, when inclusion sizes are small. However, somewhat counter-intuitively, the NMR-derived brine volume at −21 °C was larger than the phase-equilibrium value by more than 50 (Eicken et al., 2000). The latter authors interpreted the differences as a consequence of the resolution limit, a conclusion which also was drawn by Perovich and Gow (1996), who found an increasing number of inclusions during warming. In a later NMR study, Bock and Eicken (2005) found even larger discrepancies, the result being under- and overestimates of brine volumes for columnar and fine-grained granular ice, respectively.

Hence, interpretation of microstructural statistic obtained to date, is not straightforward. On the one hand, it seems that the temperature dependence of size and number distributions may be influenced in a complex manner by several factors like (i) the merging process of large inclusions during warming, (ii) the growth and merging of small inclusions during warming, exceeding the pixel threshold, (iii) tube splitting and size increase during cooling, (iv) time allowed to reach equilibrium shapes. On the other hand, as discussed by Eicken et al. (2000), there are still several systematic problems with the NMR technique. In addition, a proper study of the methodology of automatic pixel allocation, in dependence on resolution and size distribution, is lacking for sea ice. Therefore the detailed results of the NMR studies performed so far have to viewed with care, as the method is not fully developed and appears to require complementary observations. The potential of a gray-shading method and use of brine fractions per pixel is certainly worth future investigations. A much better resolution of ≈ 0.01 mm, based on pixel sizes 0.003 mm, was recently obtained in a study by Light et al. (2003) and provided some insight on single inclusion behaviour discussed above. It was limited to vertical cross sections and has so far not been extended to a study of the temperature dependence. Recognizing, that an automated analysis can only be successful if the transition mechanisms are better understood, there is a need for further studies, in particular the acquirement of larger statistical data sets.

Structure of brine layers

Considering the three-dimensional monitoring of pore networks, it is fundamental to have good informations on the brine layers between the crystals and their transformation to inclusions. Here, of course, the resolution limitations become most severe. Examples of these layers are the c-axis viewing direction in figure F.1 and the lower thin section image in figure F.2. The information published so far on these features is rather limited (Cole and Shapiro, 1998; Cole et al., 2002). It is possible that sampling and analysis techniques, which definitely avoid the destruction upon sampling, have still to be developed. Alternatively to thin sections, the method of multi-slice NMR has been suggested (Eicken et al., 2000; Menzel, 2002), but to obtain high resolution one must use very thin slices thereby loosing the advantages of a nondestructive method. NMR was shown to perform well when brine is centrifuged and replaced by a contrast agent (Eicken et al., 2000),

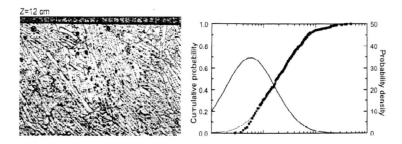

Fig. F.4: Left: Digitized horizontal thin section of at 12 cm depth of 20 cm thick lead ice from Perovich and Gow (1996); right: corresponding observed (dots) and fitted cumulative size distribution as well as probability density function.

but the results obtained from unmodified ice samples to date are less convincing (Bock and Eicken, 2005). At present, it seems unclear how NMR can be utilized in an optimal and effective sense to obtain detailed relevant information on brine layer metamorphosis.

Statistical distribution functions

Perovich and Gow (1996) to some degree circumvented the problem of limited resolution, by representing the observed distribution of brine inclusion sizes by a cumulative distribution function (CDF), from which mean and median were computed (figure F.4). The approach however assumes a distribution, which is not validated at small inclusion sizes. The analysis from Light et al. (2003) indicates the presence of a large number of very small pockets. However, inclusions of width $l_d < 0.05$ mm, presumably unresolved by Perovich and Gow (1996), probably contribute little to the average brine volume. The vertical sections studied by Light et al. (2003), yielded a less than 2 % volumetric contribution of inclusions of vertical lengths. Considering typical aspect ratios, a similar percentage seems reasonable for a horizontal threshold of 0.05 mm. It seems more probelmatic, that the samples from Light et al. (2003) had been stored for three years at -20 °C, compared to -25 °C and a few months reported by Perovich and Gow (1996). Without knowledge about shape changes due to slow diffusion and splitting mechanisms, it is thus unclear, to what degree these results resemble sea ice in nature. For example, the maximum in the inclusion area reported by Perovich and Gow (1996) near 0.006 mm^2, is not revealed in the study of vertical sections by Light et al. (2003).

Clearly more high resolution studies are necessary to clarify the distribution function of inclusions and their metamorphosis. As will be discussed below, the assumption of a *lognormal* distribution is not necessarily justified. If inclusion sizes are determined by an instability process, like the one that creates the plate spacing, then one would rather expect a *normal* distribution. If the ice, however, is processed through temperature cycles or stored for a very long time, then normal distributions might become be more relevant, but may be an artifact of the analysis procedure.

F.2 Temperature dependence of inclusion shapes

The brine volume change of a sea ice volume element is to first order proportional to the relative change in the Celsius temperature. This has to be taken into account, when interpreting thin section analysis in terms of their absolute length scales. Here it will determined, how length scales of inclusions may be best projected to a length scale at a different brine volume. The main purpose is to project them to the top of the skeletal layer. In the following discussion of published observations it will, due to the lack in better knowledge, be assumed that the microstructure of sea ice reacts without hysteresis on temperature changes, as long these do not lead to a considerable loss of brine.

The temperature dependence of sea ice microstructure statistics has to some degree been investigated in a number of studies (Perovich and Gow, 1991, 1996; Eicken et al., 2000), for which several differences in the procedures need to be mentioned. In both warming experiments from Perovich and Gow, the sample in question had been obtained close to the bottom of 20 cm thick new ice (temperature $\approx -3\,°\mathrm{C}$), stored at $-25\,°\mathrm{C}$, and finally the microstructural changes were documented during warming from $-20\,°\mathrm{C}$ to -1 °C. Both cooling and warming procedures were therefore artificial. The samples from Eicken et al. (2000) were taken from the 10 to 13 cm top level of a 110 cm thick first-year ice core. The storage temperature of $\approx -20\,°\mathrm{C}$ corresponded to the *in situ* value and the cooling had thus taken place in nature, before the analysis of the microstructure was performed during a warming sequence in the laboratory. Samples analyzed by Light et al. (2003) had been stored at a higher temperature of $\approx -20\,°\mathrm{C}$ for a rather long time of three years, prior to the analysis also to be mentioned below.

The present section focuses on the question, how the width and length of inclusion changes with brine volume v_b, the latter being related to temperature and salinity. To compare the results from Eicken et al. (2000) and Perovich and Gow (1996) the following procedure has been applied. The temperature dependence of l_d is assumed to have the form

$$\frac{l_d}{l_{d0}} \sim \left(\frac{v_b}{v_{b0}}\right)^{b_1},\tag{F.1}$$

where v_{b0} and l_{d0} are a reference brine volume and reference minor elliptic diameter taken from the dataset in question. Brine volumes have been computed from salinities and temperatures given in each study. In the following, these reference values are chosen at a temperature of $-2\,°C$ for the datasets from Perovich and Gow (1991, 1996), to resemble a value closest to the bottom of sea ice *in situ*. For the data from Eicken et al. (2000), the highest processing temperature of $-6\,°C$ is used as reference.

The following limits are expected: If all brine inclusions keep their shape during changes in the brine volume, one must have $b_1 = 1/3$. If the brine inclusion width does not change at all and always reflects the value at its fixation, then $b_1 = 0$. For the skeletal layer, with lateral freezing, the present simplified concept implies a shrinking of brine layers in the c-axis direction and hence $b_1 = 1$.

The compilation of published statistical data is shown in figures (F.5) and (F.6). For the studies from Perovich and Gow both mean and median values are reported. The slope based on the statistics from Perovich and Gow (1991, 1996) is $b_1 \approx 0.11$ with a corresponding estimate of $b_1 \approx 0.13$ based on the NMR statistics from Eicken et al. (2000). The mean value at $-1\,°C$ in the study study by Perovich and Gow (1996) appears to be out of range. This can be explained by the fact that the large brine volumes of ≈ 0.4 imply a stage or level in the skeletal layer, where mainly the average brine layer width must increase. The median from Perovich and Gow (1996), and the correlation length from Perovich and Gow (1991), show a similar slope as for lower brine volumes and these values were included in the least square fit.

If the lowest temperature samples (-20 and $-21\,°C$) are excluded from both fits, recalling the noted resolution problems, the slopes increase to $b_1 \approx 0.14$ and $b_1 \approx 0.18$, respectively. The larger value corresponds to the data from Eicken et al. (2000) but is then based on only two temperature populations. The datasets are too limited to discuss the significance of the slopes, but the agreement of three different studies in the range $0.11 < b1 < 0.18$ is promising. The slope is also principally consistent with a later NMR analysis by Bock and Eicken (2005). However, due to large discrepancies between the NMR-derived and expected brine volumes, the results of the latter study are omitted here.

It is possible to indirectly compare these results to the study of vertical thin sections by Light et al. (2003). The vertical-horizontal aspect ratio is in the present notation expressed as

$$\frac{l_v}{l_h} \sim l_v^{b_2}, \tag{F.2}$$

where l_h is now the average horizontal length of inclusions seen in a vertical section. Recognising that $v_b \sim l_h^2 l_v$, equation F.2 leads to the proportionality

$$l_h \sim v_b^{\left(\frac{1-b_2}{3-2b_2}\right)} = v_b^{b_3}. \tag{F.3}$$

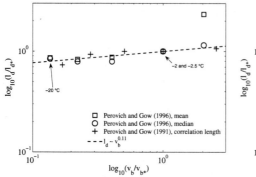

Fig. F.5: Normalized minor diameter l_d of horizontal brine inclusions versus normalized brine volume, base on a warming sequence of thin section analysis from Perovich and Gow (1996); data from a correlation length analysis from Perovich and Gow (1991) are also included. The mean l_d at -1 °C has not been included in the least square fit, see text.

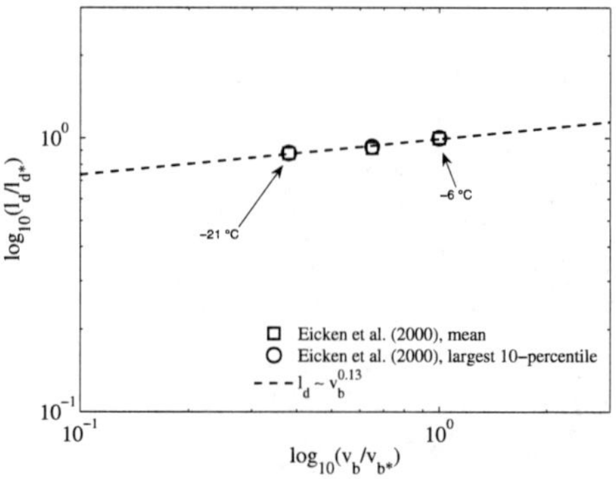

Fig. F.6: Normalized minor diameter l_d of horizontal brine inclusions versus normalized brine volume based on a warming sequence of NMR thin section analysis by Eicken et al. (2000). The sample temperature were -21, -10 and -6 °C. For the upper 10-percentile the thinnest inclusions are excluded.

As one may write the average horizontal inclusion size as $l_h \sim l_d(l_g/l_d)^{1/2}$, equation (F.3) can be used to estimate the dependence of the minor horizontal diameter l_d on v_b, if the relation between the horizontal aspect ratio $\epsilon_h = l_g/l_d$ and v_b,

$$\epsilon_h = \frac{l_g}{l_d} \sim v_b^{b_4} \tag{F.4}$$

is known. The datasets from Perovich and Gow do not show a significant dependence, while from Eicken et al. (2000) a weak increase of the aspect with brine volume, $b_4 \approx 0.05$, was estimated in the same way as for l_d (not shown). Then, with equation (F.3), one has

$$l_d \sim v_b^{(b_3 - b_4/2)} \tag{F.5}$$

where the exponent $(b_3 - b_4/2)$ is equivalent to b_1 in equation (F.1). Light et al. (2003) obtained $b_2 \approx 0.67$ by least square fitting. This implies $b_3 \approx 0.2$. The range $0.17 < b_1 < 0.2$ depends on the question, if the assumed aspect ratio slope is negligible ($b_4 = 0$) or weak ($b_4 = 0.05$). It is noted that a somewhat larger $b_2 \approx 0.73$ is also consistent with the data from Light et al. (2003) and eventually represents the smaller brine pockets with widths ≈ 0.1 mm better. This would, assuming the bounds $0 < b_4 < 0.05$ imply a smaller $0.15 < b_1 < 0.18$ for the study of Light et al. (2003). Finally, the dependence of the vertical major axis may be written as

$$l_v \sim v_b^{b_5}, \tag{F.6}$$

where $b_5 = (1 - 2b_3)$. This gives a range $0.60 < b_5 < 0.64$ by considering either $b_3 \approx 0.2$ from Light et al. (2003), or the slightly smaller $b_3 \approx 0.18$ suggested here. This much larger sensitivity of the vertical major axis l_v on brine volume is supported by fitting the observations from Eicken et al. (2000) in the same manner as above for l_d, which gives a consistent $b_5 \approx 0.53$ (not shown). The latter value is, however more uncertain. The studies from Perovich and Gow (1996), Eicken et al. (2000) and Light et al. (2003) are thus reasonably consistent with each other, considering the change in inclusion shapes in both the horizontal and vertical directions with v_b. It is recalled, that the resolution of inclusions and hence the average values were very different in the three studies (Perovich and Gow, 1996; Eicken et al., 2000; Light et al., 2003). It is particularly remarkable, that similar dependencies for the shape changes with brine volume are derived. The *size distribution* at a *fixed temperature*, obtained by Light et al. (2003), apparently corresponds to the *variation of an average or median length scale* with *changing temperature*.

Lacking more detailed observations of the temperature dependence of the relationships between brine volume and pore scales, the following compromise is made here. First, constant exponents of the dependence are assumed. Second, a somewhat stronger dependence than derived in figures F.5 and F.6 is suggested, because the interest in the present study is the temperature regime above $-15\,°C$. The best estimate is tentatively chosen by giving less weight to the data points at the lowest temperatures due to the above noted uncertainties. An exponent $b_2 = 0.15$ seems then most plausible and, after all, it is suggested to take

$$l_d \sim v_b^{0.15}, \qquad l_g \sim v_b^{0.2}, \qquad l_v \sim v_b^{0.65}, \qquad \epsilon_h \sim v_b^{0.05} \tag{F.7}$$

for the three principal axes of inclusions and the horizontal aspect ratio ϵ_h. These exponents are most consistent with the studies discussed above. The uncertainty of each exponent, to be viewed as an average between -2 and $-15\,°C$, is probably 0.03, and may be larger when particular temperature ranges are considered. These scalings are applied in the next section F.3, to compare microstructural data obtained at different temperatures. The uncertainty in these relations will be seen to be not too large, leading to length scale errors of less than 6 %. For future investigations of sea ice microstructure evolutions, and the validation of more advanced models, it is desirable to obtain more detailed observations of inclusion transitions, in dependence of warming and cooling rates andthe ice's history.

F.3 Bridging length scales: Derivation from inclusion statistics

F.3.1 Physical process scales

As a variety of metamorphosis mechanism exists for sea ice, a statistical microstructure analysis will be most effective in terms of its generality, if it is combined to a physical model. Of course, studies of larger brine channel features require a different resolution than studies focusing on brine layers. These features are, however also governed by different physics and distribution functions. Brine channel networks evolve in time and may be described by a lognormal distribution. If a physical length scale can be expected to be determined by an instability process, as it is supposed for the bridging of brine layers, this is likely different. At some scale the mechanisms may overlap. This must be known to decide, if the process in question may be more closely linked to the mean or to the median of a variable. So far, there is a lack in investigations, which link the physical process of the formation of an inclusion length scale, to an appropriate statistical analysis. The present section is an attempt to proceed in this direction. Its goal is to derive the stability length scale or wave length l_{g*} at the skeletal transition of brine layers of width $d_{sk} = l_{d*}$. Due to the lack of direct informations this must be done on the basis of microstructural data from horizontal cross section far away from the interface.

The observations to be discussed are from Perovich and Gow (1996) and Cole and Shapiro (1998). Both teams provided profile statistics of inclusions for different ice types. The study from Perovich and Gow (1996) is most appropriate for a detailed analysis, because a large number of statistics was presented for horizontal cross sections. In addition to the warming sequence, discussed in section F.2, these authors reported the mean and median brine inclusion areas, their perimeter and ellipticity, as well as the mean and median major elliptic diameters for several ice types. The analysis was performed on stored samples warmed to the original in situ temperatures. However, some remarks concerning the reported statistics, which are not perfectly suited for the purpose of the present discussion, are necessary in advance.

F.3.2 Distribution functions and process length scales

Small inclusions

In the study from Perovich and Gow (1996), the resolution of inclusions was limited to $0.003 \, \text{mm}^2$ in area. This corresponds to a diameter larger than $\approx 0.05 \, \text{mm}$. The statistics supplied by Perovich and Gow (1996) are based on fitting a lognormal distribution to the digitized observations and thereby, as may be seen in figure F.4, include typically 10 to 20 % of very small inclusions which actually have not been measured. For example, the maximum in the probability density, seen in the latter figure is no observed feature, but an outcome of this tentative fit. While Light et al. (2003) observed even smaller inclusions at larger fractions, their observations were obtained at lower temperatures of $-15 \, °\text{C}$, compared to typically $-5 \, °\text{C}$ in many samples from Perovich and Gow (1996). They can therefore not be taken as a support for the existence of the small inclusions in first-year ice, which remains as a fundamental open question.

Temporal distribution changes

Grain size distributions may deviate considerably from a lognormal behaviour, in particular when coarsening processes are considered (Thompson, 1990; Rohrer, 2005). For example, for grain growth in freshwater ice, the work by Jellinek and Gouda (1969) has shown that initially normal distributions have a tendency to become lognormal with time. However, the case of inclusions confined to brine layers is likely different and eventually comparable to the structuring of grain boundaries, which can develop very different distributions in case of anisotropic interfacial energy (Rohrer, 2005). Hence, there is no justification to assume that microstructural parameters of sea ice follow a lognormal distribution. Moreover, provided that the mentioned unresolved inclusions exist in the samples, one must ask, if they would have existed under natural conditions. They might be formed during storage at $-25\ ^{\circ}\mathrm{C}$, under subeutectic conditions, recalling the splitting of inclusions at the tentatively suggested threshold near 0.03 mm, discussed in Chapter 3. Upon warming, they might have survived and thereby represented an artificial mode in the statistics.

The interpretation of microstructural data from mature ice, in terms of structural conditions at the skeletal layer, is further complicated by the expected possible distribution change, due to the cooling-storage-warming procedure on the one hand, and due to aging in the field on the other hand. This may effect the length scales l_d, l_g and l_v and differently. For example, the *pore area* distributions from Perovich and Gow (1996) and the NMR study by Eicken et al. (2000) at lower resolution actually bear characteristics of a lognormal distribution. The *major pore length*, obtained by Eicken et al. (2000) via a nondestructive method, does rather indicate a normal peak. No investigators have so far reported distributions of minor inclusion length l_d, probably due to the limited resolution. Such distributions are likely to be more normally distributed, if the multiplicative splitting or joining of inclusions in the brine layer planes is the main lognormal driving force. In any case, distributions are likely to be multi-modal and their proper interpretation will involve the account of several physical processes.

Mean inclusion length and bridging transition

In a volumetric sense, the small inclusions only account for a few percent of the brine volume of sea ice (Light et al., 2003). The practical consequence is that, let the smallest inclusions from Perovich and Gow (1996) be real or not, including them in the statistics is to lower the median and mean length scales. A critical bridging brine layer width d_{sk}, or inclusion width l_{dsk}, proposed in the present work, is more likely to to imply a non-multiplicative physical process. It would lead to a normal rather than a lognormal distribution for the brine layer width and l_{gsk}. While in a volumetric weighted average the small inclusions would not play a role, their counting will bias an estimate of length scales. The bias will be largest for the major diameter l_g and it will be more pronounced when computing mean values. On the other hand, there are some large inclusions, the brine channels, which are also included in the statistics from Perovich and Gow (1996). These large features should neither be included when estimating the brine layer stability length scales. As shown by alternative studies (Eicken et al., 1998, 2000; Light et al., 2003), only a few brine channels can considerably affect the mean inclusion size, while they will leave the median almost unchanged. These large inclusions will most strongly bias the mean minor axis l_d to larger values.

Tentative conversions of Perovich and Gow's statistics

Due to the noted aspects only median values from the work of Perovich and Gow (1996) will be considered in the following discussion. In addition, some tentative conversion of the reported scales, denoted in the following with primes, are made. Recall that Perovich and Gow (1996) computed mean and median values from lognormal distributions via unweighted counting. It is suggested that the ratio of the reported median cross-sectional area A' and inclusion length l'_g, is the most significant length scale, as the bias from small inclusions can be expected to cancel to some degree. How close this length scale will be to the intrinsic average l_d in a normal distribution, cannot be known. This treatment of the statistics provided by Perovich and Gow (1996) is justified, if a non-multiplicative process controls it.

The brine inclusion width may then be estimated as $l_d = s_h A'/l'_g$, where s_h depends on the shape of inclusions, being $4/\pi$ for an ellipse and 1 for a rectangle. A realistic shape, with respect to the original brine layers, is perhaps a rectangle with circular edges and thus $s_h = (1 + (\pi/4 - 1)/\epsilon_h)^{-1}$. The median aspect ratios from the statistics of Perovich and Gow (1996) are typically 4 to 6, implying a range in s_h between 1.04 and 1.06. In the present section thus simply $s_h = 1.05$ has been used. Perovich and Gow (1996) provided ϵ'_h indirectly in terms of the ellipticity $E = P^2/A$, P being the perimeter of an inclusion. As the width l_d of brine inclusions in their study was always close to or only a few times the resolution limit, the ellipticity is not a well-suited parameter. For moderate aspects, it depends on the conversion of the digitized edge pixels into a perimeter. Perovich and Gow (1996) did not report how the commercial software used by them treats this problem. Here it is assumed, that it will in principle classify inclusions as being rectangles, and that only in an average sense the area will be less by the typical shape factor $s_h = 1.05$, as sometimes the edge pixels will not be classified as brine. The aspect ratio is therefore derived from the reported ellipticity E by solving $E = s_h(l_g + l_d)^2/(l_g l_d)$ for $l_g/l_d = \epsilon_h$.

Due to the the fact, that the bias of l'_g by small inclusions is opposite to the primary bias of l_d by large inclusions, the aspect of the median values of l_d and l'_g can be expected to be less than the median of the aspect ratio of these variables, the ϵ'_h provided by Perovich and Gow. The ratio of ϵ'_h and l_g/l_d in the data is between 1.2 and 1.8 for the median. The best estimate of the characteristic major inclusion length is therefore suggested to be the product $l_d\epsilon_h$ (also using median values), in contrast to the presumably biased median l'_g given by Perovich and Gow (1996). In conclusion, the converted properties

$$l_d = 1.05A'/l'_g, \qquad \epsilon_h = \epsilon'_h, \qquad l_g = 1.05A'\epsilon'_h/l'_g \qquad \text{(F.8)}$$

are expected to be closest to the process length scales thought to be most relevant here. As will be seen below, these choices also make a comparison with values of l_d, reported by by Cole and Shapiro (1998), most feasible.

F.3.3 Inclusion width l_d: projection to the skeletal transition

Figure F.8 shows the median values of the minor inclusion length l_d converted from area and major length given by Perovich and Gow (1996). Results are plotted versus thickness for several ice types and represent the in situ values. Next, the above derived temperature dependence from equation (F.7) can be applied to project the brine inclusion widths to their value at the top of the skeletal layer, l_{dsk}. This is the value relevant for the present discussion of brine layer transition. The projection is performed by using the temperature and salinity information from Perovich and Gow (1996), to compute the actual brine volume of the sample, and a typical skeletal layer brine volume of $v_{sk} = 0.20$, see chapter 5. Possible deviations from this constant v_{sk} will be discussed below. The results for l_{dsk}, obtained in this way, are shown in figure (F.8).

The main interest here is in ice samples classified by Perovich and Gow (1996) as *columnar*. Two granular near-surface samples, marked with stars in figure F.8, are shown for completeness, while one pancake sample of considerable air content is omitted. Giving similar length scales, they deviate from the clear picture that emerges in the following for all other measurements. The increase in the in situ l_d, by projection to l_{dsk}, is not very large, as the in situ temperatures were relatively high. It appears however, that the projection makes the data somewhat more consistent. For lead I, for example, the temperature correction creates a decreasing l_{dsk} as it is observed for all other samples. In the following l_{dsk} will be discussed. In figure F.8 the following aspects are noted:

- The median values of l_{dsk} are between 0.05 and 0.11 mm.

- For each profile l_{dsk} decreases with depth.

- No difference between young lead ice and first-year ice at similar depth levels is evident.

The limited precision in the statistics reported by Perovich and Gow (1996) implies an uncertainty in the values of l_{dsk} of approximately 5 %. The variability and uncertainty,

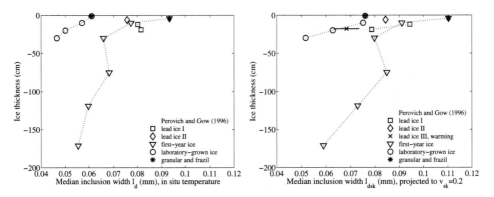

Fig. F.7: Median minor inclusion width l_d estimated from the statistics by Perovich and Gow (1996) for different ice types and vertical profile locations. Two granular samples at the surface are denoted with a star.

Fig. F.8: Median minor inclusion width l_{dsk} estimated from the statistics by Perovich and Gow (1996). Estimates of l_d in figure F.7 are projected to the typical brine volume $v_{sk} = 0.20$ at the top of the skeletal layer.

due to the temperature normalisation procedure, may be indicated in the standard deviation of the repeated measurements during warming for lead III (the cross in figure F.8). It appears to be ≈ 10 %. The depth dependence is thus a siginificant finding. At first glance, two explainations for the decreasing l_{dsk} with depth may be suggested. These are are (i) a dependence of l_d on the age of the ice and (ii) on the growth velocity during its origin.

F.3.4   Possible factors affecting l_d

Growth velocity

The growth velocity V decreases under most natural conditions with thickness, and this indicates the possibility of a dependence between l_{dsk} and the growth velocity. It is notable that the laboratory ice, shown as circles in figure F.8, while being thin, grew at low velocities, probably close to the lower levels of thick first-year ice. This also supports an increase in l_{dsk} with growth velocity.

Within the present framework, the only parameter which unambiguously can be related to the growth velocity is the brine volume v_{sk} at the top of the skeletal layer. In figure F.8 all observations have been projected to a value of 0.20. From chapter 5 and figure 5.10, it can be suggested that at low growth velocities a critical v_{sk} of half this value may be plausible. The temperature projection of l_d would then, in accordance with the brine volume dependence $l_d \sim v_b^{0.15}$, create a 11 % *decrease* in l_{dsk} for the lower first-year ice values in figure F.8. This effect should enhance the depth dependence. An increase of l_{dsk} with growth velocity normally corresponds to an increase with the brine salinity gradient. As the latter increases the Rayleigh number, one would intuitivly expect a dependence opposite to the observed one. Hence, if l_{dsk} depends on growth velocity, then the involved processes must be different from what has been discussed in, for example, Chapter 11.

Aging

Ice in the upper levels has aged for a longer time. However, the thin lead ice samples from Perovich and Gow (1996) show similar values of l_d as the thick first-year ice at the same level. Aging therefore does not serve as an easy explanation of the observed variability. The less detailed statistical study by Cole and Shapiro (1998) is to date the only report on observations of the change in brine inclusion distribution at a fixed vertical level, over the coarse of the winter. These authors reported that at 20 cm distance from the surface of growing first-year ice the average l_d increased from 0.080 mm in November via 0.112mm in March to 0.138 mm in May. While similar, for a quantitative comparison with values from Perovich and Gow (1996) some adjustments are necessary, which is discussed below. Here the focus is on the aspect of aging. Cole and Shapiro (1998) provided additional information on the aspect ratio distributions for the three stages. These are reproduced here in figure F.9. It is seen that (i) almost all inclusions with an aspect larger than 5 disappeared during winter and (ii) most inclusions with an aspect of 1 to 2 disappeared from March to May. A first plausible interpretation is that the loss of the small aspect inclusions probably contributes little to the change in l_d, as it was not observed between November to March when a 30 % increase in l_d occurred. As pointed out by Cole and Shapiro (1998), the main factor of changing the average l_d appears to be, that the longest inclusions break up into shorter ones during winter. If these large aspect inclusions were originally thicker than the average, the increasing number of wider inclusions implies an increase in the average l_d. If this is the main effect reflected by figure F.9, it is likely that the volume-averaged l_d has not changed that much, its principal value being retained during winter. How this can be explored further, is described and discussed more quantitativly in figure F.11 below.

Hence, the observations from Cole and Shapiro (1998) do not necessarily support the idea that aging is responsible for the vertical variability of l_{dsk} in figure F.8, as aging mostly leads to a change in the distribution of l_g. On the basis of this finding and the formula $l_d = s_h A'/l'_g$, used to estimate l_d from the statistics given by Perovich and Gow (1996), one would not expect a change in l_d due to changes in l_g. This makes it plausible, that a mechanism different from aging creates the vertical dependence in figure F.8. The ideas can be traced further by considering the relation between l_d and the aspect ratio ϵ_h.

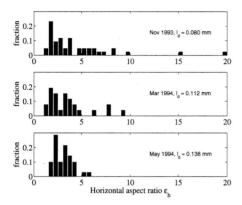

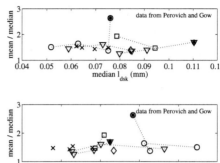

Fig. F.9: Distribution of horizontal inclusion aspect ratios reported by Cole and Shapiro (1998) at 20 cm distance from the surface, for early winter to early spring. The average inclusion width at each date is noted.

Fig. F.10: Ratio of mean to median inclusion width l_{dsk} estimated for the samples from Perovich and Gow (1996). Upper: in dependence of l_{dsk}; lower: in dependence on median aspect ϵ_{hsk}. Symbols are as in figure F.8.

F.3.5   In situ observations by Cole and Shapiro (1998)

For a quantitative comparison with the study of Cole and Shapiro (1998) discussed below, two adjustments of the latter observations are necessary. In the latter study the processing temperature was $-14\,°C$. The salinity variations, observed over the coarse of the winter at the 20 cm level, are likely related to normal sampling variability and a constant value of 5.5 ‰ is assumed (Cole et al., 1995; Cole and Shapiro, 1998). This leads to a phase-relation brine volume of 0.024 during processing and the correction factors $(0.2/0.024)^{0.15} \approx 1.37$ and $(0.2/0.024)^{0.05} \approx 1.11$, for l'_d and ϵ'_h respectively, to project them to $v_b = 0.20$. Next it has to be considered that Cole and Shapiro (1998) computed mean inclusion widths. To facilitate a correction, the ratio of mean and median l_d derived from the Perovich and Gow data is shown in figure F.10. It does not reveal a depth or ice-type dependence. The columnar ice observations in the figure yield 1.49 ± 0.16. The non systematic scatter, also indicated in some CDF plots from Perovich and Gow (1996), indicates the variable influence of a few large inclusions on the mean. The larger mean-median ratio of the few granular samples, also reported by Bock and Eicken (2005), points to a dominance of larger inclusions in disordered samples. The mean to median ratio of 1.49 for the columnar samples implies that larger inclusions contribute by about one third to the overall porosity. A similar ratio is indicated by statistics of vertical sections from Eicken et al. (2000), when comparing the distribution maxima in the PDF to averages of l_g. Consequently, the mean values of l_d from Cole and Shapiro (1998) need to be corrected by the factor $1.37/1.49 \approx 0.92$, to make them comparable to the median

l_{dsk} derived for the Perovich and Gow data.

In the statistics from Perovich and Gow (1996), the mean ϵ_h' exceeded the median by a remarkably constant factor ≈ 1.25 which varied by less than ± 10 %. For the aspect distributions from Cole and Shapiro (1998) in figure F.9, the mean to median ratio decreased from 1.34 to 0.94 as the distribution becomes almost normal by the end of the season. The median aspect is 3 to 3.5 in the original data. Due to the limited statistical analysis and eventually different procedure from Cole and Shapiro (1998) it is difficult to compare these estimates. In the following, l_{dsk} derived from the observations of Cole and Shapiro (1998), is therefore compared to both the mean and the median aspect ratio.

F.3.6 Co-variability of l_{dsk} and aspect ratio ϵ_{hsk}

In figures F.11 and F.12 the data from Perovich and Gow (1996) and Cole and Shapiro (1998) discussed so far have been plotted in dependence on the aspect ratio ϵ_{hsk} of inclusions. As discussed above, the proposed relevant length scale $l_{gsk} = l_{dsk}\epsilon_{hsk}$ to be discussed here is shown here. All symbols for the different ice types from Perovich and Gow (1996) are as shown earlier and appear in figure F.11. The additional temporal evolution data from Cole and Shapiro (1998) are shown as pentagrams. The adjusted l_{dsk} from these authors is plotted against the mean and the median aspect ratios. The following analysis focuses first on the scales derived from the Perovich and Gow (1996) report.

The results for two granular samples are only shown for completion. The focus is on the other columnar samples. The interpretation of the apparently different warming sequence, shown as crosses, is delayed at the moment. For all other observations from Perovich and Gow (1996), it is seen in both figures that the data can be in principal divided into two groups.

- One group is denoted with additional solid dots and contains the four surface ice samples from a depth 6 to 10 cm below the ice surface. It represents the early growth regime. This group is described by the curves $l_{gsk} = const. = 0.46$ mm in figure F.12 which appears as $l_{dsk}\epsilon_{hsk} = const. = 0.46$ mm in figure F.11.

- The second group comprises all other ice samples, from levels below at and below 30 cm in first-year ice and the laboratory tank study. This group is approximated by the curve $l_{gsk} = const. = 0.46$ mm and represents lower growth rates.

Figures F.11 and F.12 are not independent of each other and rather show the relation in a different coordinate frame. A fundamental result is that the major length scale of inclusions for a certain ice growth regime may be described by $l_{gsk} = l_{dsk}\epsilon_{hsk} = const.$, even if the aspect ratio varies over a relatively wide range. The relatively small overall range of $0.35 < l_{gsk} < 0.47$ mm indicates only a slight dependence of l_{gsk} on growth

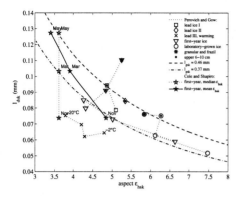

Fig. F.11: Median inclusion width l_{dsk} estimated after Perovich and Gow (1996) ver-
sus median aspect ϵ_{hsk}.   Additional dots
mark the surface samples, which are closely
matched by the dashed curve $l_{gsk} = 0.46$
mm. The corresponding dash-dotted curve
$l_{gsk} = 0.37$ mm is a lower bound for the
data.   The in situ observations from Cole
and Shapiro (1998) are shown as pentagons
against the median (dotted connecting line)
and mean (solid connecting line) ϵ_{hsk}. Symbols of ice types from Perovich and Gow
(1996) are as in figure F.8.

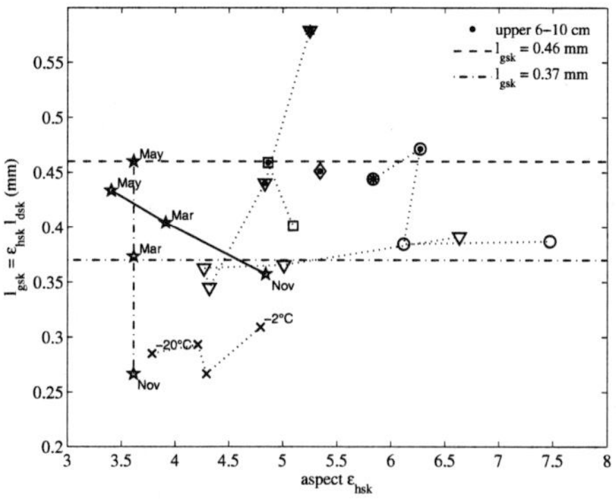

Fig. F.12: The data in figure F.11 yet plot-
ted as $l_{dsk}\epsilon_{hsk} = l_{gsk}$ against the aspect ratio. The data from Perovich and Gow (1996)
fall between $l_{gsk} = 0.46$ mm (upper,fast
growth) and $l_{gsk} = 0.36$ mm. An exception
is the warming sequence (crosses) discussed
in the text. The temporal in situ estimates
based on the data from Cole and Shapiro
(1998), shown as pentagons, approach the
upper limit of $l_{gsk} = 0.46$ mm in May, when
the distribution in figure F.9 becomes normal
and is no longer based on large inclusions.

conditions. It appears that l_{gsk} increases slightly with aspect ratio for the high and the
low growth velocity regimes, but this indication is certainly not significant, considering
the scatter and the approximate treatment.

    The field observations from Cole and Shapiro (1998) supplement the results in two
interesting manners. First, the data appear different when plotted either against the
median or the mean ϵ_{hsk}. This indicates the importance of using a proper distribution
of inclusions to evaluate critical length scales. Second, by the end of the season the
observations approach l_{gsk} for the first high velocity group from Perovich and Gow. This
supports again, that temporal changes in the data from Cole and Shapiro (1998) have not
to do with real changes in inclusion width l_d, but need to be interpreted as a break-up of
wider inclusions, increasing their number density, and thus, the mean of an unweighted
average.

F.3.7 Discussion of dominant length scales

It is still necessary to discuss the meaning of the apparently constant microstructure scale. At first glance, it appears that l_{gsk}, and thus the major length scale of inclusions, is the characteristic process length scale of the microstructure. However, l_{gsk} has been computed from the data of Perovich and Gow by a combination of median length scales, see equation F.8. It must be viewed as a somewhat tentative estimate of the dominant length scale. One can argue that l_{dsk}, as derived from Perovich and Gow (1996), should, due to the apparent importance of volumetric averaging, be corrected in this sense. However, the discussed statistical data do not provide sufficient information for such a correction.

Process length scale l_{gsk} versus l_{dsk}

Due to the lack in information, it has to be further argued tentatively. Light et al. (2003) found, in vertical thin sections, a tendency for the major length to increase with aspect ratio. In principle one may also suspect such an evolution when considering the bridging process: if bridging is favoured by somewhat thinner brine layers, these will be split into smaller inclusions. Hence, the wider brine layers in a spectrum may survive with the larger aspect ratios. It must then, in a volumetric sense, be given more weight to the larger inclusions when computing the average width l_{dsk}. The appearance of the dataset from Cole and Shapiro (1998) in figure F.11, may be taken as an indication, that an approximate volumetric correction of l_{dsk} is simply proportional to the aspect ϵ_{hsk}. In this sense the values of l_{gsk} in figure F.12 could then be interpreted as an almost constant effective, volume-averaged l_{dsk}, for each of the two growth velocity groups. The apparently smaller l_{dsk} could be explained as the consequence of an inappropriate representation of the process length scale by an unweighted median. This needs to be shown in future analysis. Here this is taken as support for the plausibility of a constant inclusion or brine layer width l_{dsk} at the skeletal transition.

Volumetric brine layer estimates: l_{g*} and d_{sk}

At the moment the available data are limited and the corresponding interpretations are still tentative. In the following therefore an approach is made which supplements the above contention by turning to the comparison of the two groups, the high growth velocity group of surface ice with $l_{gsk} \approx 0.46$ mm, and the bulk ice group with $l_{gsk} \approx 0.36$ mm. With approximate knowledge of the growth conditions for these groups, it is possible to invoke an independent simple test of the derived scales. The brine layer width at the considered level $v_{sk} = 0.2$ is, by definition of the skeletal transition, given by $d_{sk} = v_{sk}a_0$. Assuming, for simplicity, perfectly rectangular inclusions, such that the derived l_{dsk} matches d_{sk}, one may define a basic length scale l_{g*} of the inclusions given as

$$l_{g*} = \frac{v_{sk}a_0}{l_{dsk}}l_{gsk} = v_{sk}a_0\epsilon_{hsk}. \tag{F.9}$$

It is important to distinguish it from the scales used so far: l_{g*} is defined as the *intrinsic* average length of brine inclusions at the onset of bridging. As at this stage the inclusions still touch each other, l_{g*} may be interpreted as the wave length at which they separate. In contrast, l_{gsk} is an estimate from the statistics of ice which has been cooled and warmed and therefore related to a width l_{dsk} that may differ from d_{sk} at the original transition. From equation F.9 it is seen that, if one knows $v_{sk}a_0$, one may test the consistency of the statistical estimates. First, l_{g*} may be either compared to the observed ϵ_{hsk} which under some conditions might be a more relevant paramter. Second, as a correct statistical estimate should, by definition, give $l_{gsk} = l_{g*}$ at the skeletal transition, one should expect that $v_{sk}a_0 = d_{sk} \approx l_{dsk}$, if the scales obtained from the statistics are consistent. In the following this comparison of d_{sk} with l_{dsk} is illustrated on the basis of the two growth velocity regimes:

- *High growth velocities.* For the upper ice, the high growth velocity group, one may, based on the discussion in earlier chapters, suggest $a_0 \approx 0.45$ mm, which is the theoretical and empirically expected value for growth velocities near 5 cm d^{-1}, a reasonable freezing rate for 6 to 10 cm thick ice. Furthermore, recalling the result from chapter 5, that $v_{sk} \approx 0.20$ is a reasonable brine volume at the skeletal layer of thin ice, one obtains $d_{sk} \approx 0.2 \times 0.45 = 0.09$ mm. This value is consistent with the values of l_{dsk} in figure F.11, which is further shown in figure F.14, where the inclusion width has been plotted against the bulk salinity of the samples. Three points of the group are within 5 % of $l_{dsk} = 0.09$ mm. It is notable that these three points correspond to salinities between 9 and 16 ‰, which serves as a further indirect confirmation of a growth velocity of the order 5 cm d^{-1}. It is not the detailed agreement which is emphasised here. Most important is that, for the surface ice sample group, the estimated l_{dsk} is consistent with an independent calculation based on the brine volume and the brine layer spacing a_0.

- *Low growth velocities.* For the low velocity group a typical range of $0.6 < a_0 < 0.8$ mm is realistic for growth velocities between 0.5 and 1 cm d^{-1}, the range reported for the ice tank growth by Perovich and Gow (1996) and generally expected for average first-year ice growth. This range of a_0 is approximately confirmed by estimates based on micrographs shown by Perovich and Gow (1996). For this group one therefore obtains $0.12 < d_{sk} < 0.16$ mm, assuming the same $v_{sk} = 0.20$. This value is considerably larger than the estimated range $0.06 < l_{dsk} < 0.08$ mm, within which most of the low growth velocity samples fall. Therefore one must conclude that the estimate of l_{dsk} and l_{gsk} is not consistent with a cross-check of l_{g*} via the brine volume and a_0 due to equation F.9. To make d_{sk} consistent with the estimated l_{dsk}, one has two possibilities. Either the skeletal brine volume needs to be reduced to ≈ 0.1, reducing in turn d_{sk} by a factor of two, or one has to assume that the simplifications have led to an underestimate of l_{dsk}. A smaller required skeletal transition brine volume of $v_{sk} \approx 0.10$ is, in fact, not inconsistent with the results from Chapter 5 for low growth velocities. The scatter in these observations appears, however, too large to obtain a limiting estimate of v_{sk}. Nevertheless, it seems justified to adopt $0.12 < k_{eff} < 0.15$ for slowly growing first-year ice (chapter 5), which implies $0.11 < v_{sk} < 0.14$. With $a_0 = 0.7$ mm this leads to a revised range $0.08 < d_{sk} < 0.10$ mm in agreement with the high velocity data.

In summary, the volumetric brine layer estimates eventually support the scaling

$$d_{sk} = v_{sk} a_0 \approx 0.09 \pm 0.02 \qquad \text{(F.10)}$$

for the width d_{sk} of brine layers at the skeletal transition, in millimeters.

F.3.8 Effects of the laboratory protocol on derived scalings

The difference between the derived $l_{dsk} \approx 0.11 \pm 0.02$ mm and $\epsilon_{hsk} \approx 4$ from Cole and Shapiro (1998), and the smaller l_{dsk} deduced from the observations from Perovich and Gow, appears to be correlated with larger ϵ_{hsk}. This has been tentatively suggested to be a consequence of the laboratory protocol. So far the question, what mechanism might be capable to create the differences in ϵ_{hsk} and l_{dsk}, has remained untouched. Next a possible explanation for the appearance of the the data from Perovich and Gow (1996) is suggested. The focus is on the effect of cooling and warming of the sample from Perovich and Gow.

Aspect change due to merging inclusions

Figure F.13 shows the same data as figure F.12, yet for clarity the observations from Cole and Shapiro (1998) have been omitted. In addition the lines $l_{dsk} = 0.10$ and 0.06 mm, which are the approximate limits of the columnar ice data, have been drawn. In the coordinate frame in figure F.13, two aspects are highlighted. First, $l_{dsk} = 0.10$ mm is chosen as an approximate upper bound of the data and shown as a solid line. This illustrates its intersection with the dashed line $l_{gsk} = 0.46$ mm near $\epsilon_{hsk} \approx 4.5$. In a volumetric sense one can state that, the higher the l_{dsk} of inclusions, the shorter they must be and the smaller the aspect ratio. One may than view $\epsilon_{hsk} \approx 4.5$ as a bound for the critical aspect ratio implied by the data: All larger values of ϵ_{hsk} are interpreted as a consequence of merging of some inclusions of this basic aspect. Such an interpretation

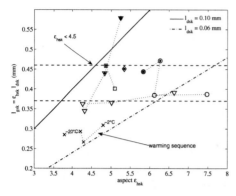

Fig. F.13: This is figure F.12 with the observations from Cole and Shapiro (1998) omitted. Instead upper and lower bounds of $l_{dsk} = const.$ have been drawn. The warming sequence, shown as crosses, approximately follows a line $l_{dsk} = const.$, indicating the merging of inclusions, see the discussion in the text.

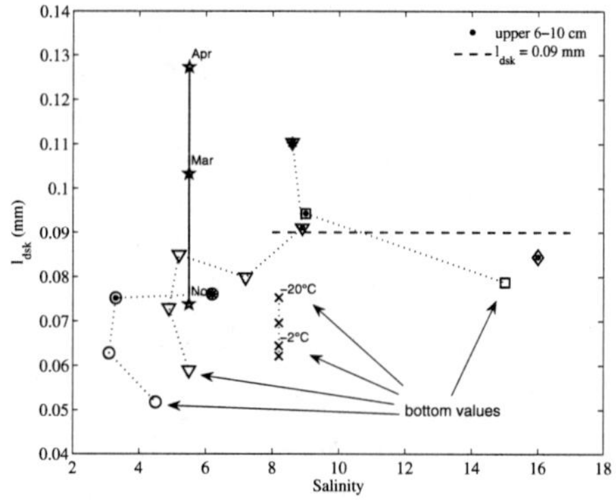

Fig. F.14: Inclusion width l_{dsk} versus bulk salinity for different samples. The dashed line indicates the typical range of upper ice, high growth velocity samples marked with a dot. The bottom samples of all ice cores show lower l_{dsk}. Symbols are the same as in previous figures.

is consistent with the fact, that remelting inclusions will preferentially elongate in the basal planes.

The warming sequence documented by Perovich and Gow (1996), shown as crosses in figure F.13, indicates the mentioned merging behaviour more directly. Noting that these data have already be normalized by the expected average changes in ϵ_h with brine volume, the warming sequence indicates the principal increase in ϵ_{hsk} along a $l_{dsk} \approx const.$ curve in a reasonable manner. It is also notable that, even while in the warming sequence the *number* of inclusions was increasing with temperature, there is a tendency of inclusions to merge and increase their average aspect ratio.

Comparing the warming sequence from Perovich and Gow (1996) with the change in l_{gsk} from Cole and Shapiro (1998) between March and May (earlier figures), when the ice was also warming under real conditions, gives further support to these ideas. The change in $l_{gsk} - \epsilon_{hsk}$-space are very different and almost orthogonal to each other. This is another indication of artificial changes due to the cooling and warming procedure in the laboratory.

The difficulties in interpreting the statistics from Perovich and Gow (1996) are therefore suggested to be a consequence of the cooling-warming procedure. To first order the following hysteresis effect may be a reasonable explanation: During cooling, the brine inclusions are, within the limits of equation F.7, almost shape-preserving when viewed in the horizontal plane. During warming, their growth is limited by Tyndall figure dynamics (Nakaya, 1956; Kaess and Magun, 1961; Knight and Knight, 1972; Mae, 1976). It implies a tendency for producing longer and thinner inclusions than originally had existed at the skeletal layer transition level. This idea can be extended further. Figure F.14 indicates an increase in l_{dsk} with bulk salinity. The following explanation is based on the contention that the salinity only indirectly affects l_{dsk}, as it varies inversely with the plate width. It is suggested, that the larger the plate width, the easier it is for new thin brine layers to form during warming *within* the plates. Clearly, if the brine layers are not restored during warming at their original sites, then this must happen and the average brine layer or inclusion becomes thinner, while the aspect increases. The warming process thus creates additional brine inclusions. Indeed, such a formation of inclusions within the plates is indicated in figure F.4, and can clearly be identified in a number of micrographs shown by Perovich and Gow (1996), in particular for their first-year ice samples. In micrographs obtained without temperature cycling such features appear to be much less frequent, as seen in figure F.3 or other micrographs from Cole et al. (2004).

Bottom samples

Hence, now the range $0.06 < l_{dsk} < 0.08$ mm from the statistics needs to be increased by only 20 to 30 % to reach volumetric consistency with d_{sk}. Noting that the ratio l_{dsk}/d_{sk} is almost the same as l_{gsk}/l_{g*} for the low velocity group, it can be argued that, even without knowing *why* l_{dsk} derived from the statistics might be an underestimate of d_{sk}, the data may be either interpreted in terms of a single constant l_{g*} or d_{sk}.

Finally another aspect, which likely biased the observations from Perovich and Gow (1996) towards smaller l_{dsk}, may be deduced from figure F.14. It is emphasised that l_{dsk} of all bottom samples is systematically lower by an amount of $\approx$ 20 %. As these samples come from levels 1 to 2 cm from the bottom of the ice, it is likely that during sampling some brine has been lost, or that brine expulsion during the cooling-melting cycle is most effective here, where boundaries are open. In any case, one may expect that the brine inclusions will, during later analysis, contain some air. The potential to restore the original brine layers upon remelting will be reduced. It is therefore consistent that smaller l_{dsk} (and l_{gsk} in figure F.12) are found for bottom samples.

F.3.9 Conclusion: Process length scales during bridging

Having discussed the limitations which emerge from considerations of an unmodified l_{dsk} from Perovich and Gow (1996), it is now easier to draw conclusions and best estimates of process length scales from the data. It has been hypothesised that the skeletal layer starts to form bridges at a limiting width d_{sk}. The most relevant dataset that may be used to reconstruct this scale, by projecting observed inclusion scales to a brine volume of ≈ 0.2, has been obtained in the field by Cole and Shapiro (1998). As in the latter study micrographs were taken shortly after sampling, and by cooling the samples slightly below in situ conditions, these observations can be rated as unbiased estimates. By the end of the winter, the statistical distribution in these observations becomes normal and the averages reported by Cole and Shapiro (1998) can likely be interpreted in terms of volumetric mean values. These are $l_{dsk} \approx 0.127$ mm, an aspect $\epsilon_{h*} \approx 3.6$ and an inclusion length of $l_{gsk} \approx 0.46$ mm with a standard deviation of approximately 20 %. The binning resolution of 0.5 in the aspect ratio may also indicate the uncertainty of these values.

The microstructure statistics published by Perovich and Gow (1996) were discussed on the basis of the procedure

- 1. Compute a median l_{dsk} for the dataset.

- 2. Derive a length scale $l_{gsk} = l_{dsk}\epsilon_{hsk}$ from the data

- 3. Interpret l_{gsk} in terms of the true process length scales l_{g*}, ϵ_{h*} and $l_{d*} = d_{sk}$ at the skeletal transition.

The laboratory protocol from Perovich and Gow (1996) is, due to thermal cycling, suggested to produce thinner inclusions of larger aspect ratio than originally existed during freezing. This is supported by the micrographs shown by these authors. However, the major length scale of those inclusions, which most likely were least affected by desalination effects, was

$$l_{gsk} = 0.46 \pm 0.02 \qquad (F.11)$$

mm, and in agreement with the work from Cole and Shapiro (1998). The lowest aspect ratio in the dataset from Perovich and Gow (1996) takes a value between 3.5 and 4.5, also close to the observations from Cole and Shapiro. A reasonable condition, which appears to be justified by all the observations considered here, is thus an unbiased aspect ratio $\epsilon_{h*} \approx 4$ at the skeletal transition. As l_{gsk} appears to show a very slight aspect dependence, the most plausible intrinsic transition scales, suggested from the statistics by Perovich and Gow (1996) projected to $\epsilon_{h*} = 4$, are

$$\epsilon_{h*} \approx 4, \qquad l_{g*} \approx 0.44 \quad \text{mm}, \qquad l_{d*} = d_{sk} \approx 0.11 \quad \text{mm} \qquad (F.12)$$

Significance bounds of these estimates cannot be given here but, after all, it seems that an uncertainty of 10 % is realistic, based on the discussed data sources. Also the very similar l_{g*} supports the overall consistency with the scales in F.12.

It was mentioned that NMR studies (Eicken et al., 2000; Bock and Eicken, 2005) have a too coarse resolution to derive l_{dsk}, yet l_{gsk} seems comparable to the above values. Eicken et al. (2000) obtained average major length scales between 340 and 380 μm for natural sea ice samples and 500 μm for an artificially grown sample, with little dependence on test temperature and a standard variation of approximately ± 200 μm. However, also in these statistics the length scales were no volume averages. The comparison with them must therefore be rated as approximate. At least they are not in disagreement with the process length scales derived here.

An important implication is that now

$$v_{sk} = \frac{d_{sk}}{a_0} \quad \text{with} \quad d_{sk} \approx const. \approx 0.11 \quad \text{mm} \tag{F.13}$$

is a condition for the brine volume at the skeletal transition. Hence, although the observations are limited, the microstructural sea ice statistics may be interpreted in terms of a constant width d_{sk} of brine layers, at the skeletal transition to the quasi-solid ice matrix. It will be shown next, that these results can be tested by a second indirect approach, the analysis of ice strength test data.

F.4 Bridging layer width: Estimate from sea ice strength tests

The predictability of sea ice strength was one of the main motivations which led Anderson and Weeks (1958) to formulate the cylinder splitting model for the ice microstructure discussed above. The basic concept may be understood in terms of early work on sea ice strength (Tsurikov, 1939; Anderson and Weeks, 1958; Assur, 1958; Weeks and Assur, 1967) which indicated that a square-root dependence of the form

$$\sigma(v_b) = \sigma_\infty \left(1 - \left(\frac{v_b}{v_{b*}}\right)^{1/2}\right) \tag{F.14}$$

often described the relationship between the brine volume v_b and a yet unspecified sea ice strength index σ reasonable well. Two examples of strength tests are shown in figures (F.15) and (F.16). The constant v_{b*} in equation F.14 has, in simple terms, been interpreted as the brine volume, at which the strength vanishes and the bridges between the plates disappear. It has been suggested to estimate $d_{sk} = a_0 v_{b*}$ via a fit of strength test data (Anderson and Weeks, 1958; Weeks and Assur, 1964). An estimate of d_{sk} is, of course, only feasible when a_0 has been measured. This has been done by Weeks and Assur (1964) at 7 levels of ≈ 30 cm thick laboratory ice samples. They utilised the extensive strength ring-tensile test data from Weeks (1961) to compute v_{b*} at each level in the ice. Using a linear strength reduction model with $(v_b/v_{b*})^1$, they estimated $d_{sk} = a_0 v_{b*} \approx 0.112$ mm, a value which varied little with depth. This result seems to be in perfect agreement with the result F.12 from the previous section. However, the correlation found by Weeks and Assur (1964) with the square-root equation F.14 was actually better. Had v_{b*} therefore consistently been derived from F.14, then Weeks and

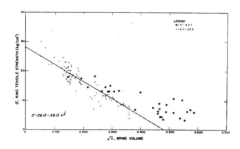

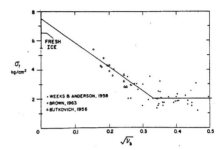

Fig. F.15: Ring tensile strength test data versus brine volume. Note the flattening of the relation near $v_b^{1/2} \approx 0.4$. From Frankenstein (1969).

Fig. F.16: Flexural strength test data compiled by Weeks and Assur (1967) from different studies. Note the flattening of the relation near $v_b^{1/2} \approx 0.35$.

Assur (1964) had obtained a 30 % larger value of $d_{sk} = \approx 0.15$ mm. In the following it will be shown, that some modifications to the approach from Weeks and Assur (1964) are necessary and lead to a reevaluation of these results.

Residual strength

First it is, on the basis of figures F.15 and F.16, and other comparable relations (Weeks and Assur, 1967; Weeks and Ackley, 1986), concluded that $\sigma(v_b)$ does not drop to zero at low brine volumes. In fact, it approaches a constant $\sigma_0' \approx (1 - 0.85)\sigma_\infty$, or ≈ 15 % of the solid ice strength. The corresponding brine volume at this threshold is $v_b \approx 0.85^2 v_{b*}$ and thus a factor ≈ 0.7 smaller than the zero-intercept v_{b*}. The argument, that this reduced v_{b*} should represent the strength transition better, is empirically confirmed by observations. The suggested equation to estimate d_{sk} then becomes

$$d_{sk} = r_{gra}0.85^2 v_{b*}a_0 \qquad\qquad (F.15)$$

and includes also a structural correction factor r_{gra}, computed from equation 5.13 and accounting for the finite plate lengths. The grain length variation has been approximated by $L_{gra} = 5 + z/15$ mm, where z is the level in the ice sheet (in mm), and L_{gra} is the horizontal length of a grain in the direction of the brine layers. It is consistent with grain sizes observed in other studies and discussed in section 5.6, yet should not be extrapolated above the upper thickness of 60 cm on which it is based. A minor geometrical aspect, with respect to the original approach from Weeks and Assur (1964), may also be tested. The factor $cos\phi$, applied by Weeks and Assur (1964) to correct plate spacings a_0 measured in horizontal cross sections, ϕ being the vertical inclination of the plates, has been reversed to compute also $a_0 = a_0'/cos\phi$. This may be more consistent with the fact, that in the

table title

$\overline{a_0}$ (mm)	$\overline{a'_0}$ (mm)	$\overline{z}$ (cm)	$\overline{V}$ cm d^{-1}	$\overline{L_{gra}}$ (mm)	v_{b*}	r_{gra}	d_{sk} (mm)	d'_{sk} (mm)
0.468	0.407	3.7	6.7	3.9	0.302	1.048	0.107	0.095
0.470	0.441	7.3	5.5	5.7	0.304	1.038	0.112	0.112
0.483	0.482	11.6	4.8	7.8	0.277	1.033	0.110	0.110
0.504	0.503	17.2	4.1	10.6	0.261	1.028	0.109	0.109
0.546	0.546	22.0	3.5	13.0	0.276	1.024	0.119	0.119
0.589	0.589	26.3	3.1	15.2	0.273	1.021	0.120	0.120
0.610	0.610	31.1	2.7	17.6	0.233	1.020	0.105	0.105

Tab. F.1: Observations of plate spacing a_0, inclined spacing a'_0 and strength test parameters from Weeks and Assur (1964). Calculations of the critical brine layer width d_{sk} are described in the text.

ring-tensile strength tests applied by Weeks (1961) the failure plane was normal to the horizontal. The following results are presented in terms of the minimum brine layer width d'_{sk}, and its value d_{sk} parallel to the failure plane.

Reanalysis of observations from Weeks and Assur (1964)

To calculate d_{sk}, first v_{b*} and a'_0 are obtained from Weeks and Assur (1964). Alternatively, using a_0 in the failure plane leads to slightly different results, using the inclination angles also reported in the original study. The grain size parameter r_{gra} is computed from the assumed grain-size-depth relationship. As the strength tests were obtained with average samples of 5 to 6 cm thickness, the values of a_0 and r_{gra} have been linearly interpolated to the average levels $\overline{z}$ given by Weeks and Assur (1964). The final results for d_{sk} in the failure plane and the minimum width d'_{sk} along with all values in the calculations are given in table (F.1).

In table F.1 also the estimates d'_{sk}, based on the corrected a'_0, are included. It is seen that these differ from d_{sk} only in the upper levels, where the vertical inclinement of the plates is significant. Averaging the results gives $d_{sk} \approx 0.112 \pm 0.006$ mm and $d'_{sk} \approx 0.110 \pm 0.008$ mm as best estimates. Hence, the modified analysis of the strength test data seems to confirm the results from the microstructure statistics rather well. Again, no significant dependence on growth conditions is found for d_{sk}, indicating the consistency of the approach and an eventually constant d_{sk} as the flattening criterion for the strength curve.

Reanalysis of observations from Dykins (1967)

Another laboratory study, where simultaneous tensile strength tests and observations of a_0 have been obtained, is the work by Dykins (1967, 1969). Ice was also grown in the laboratory. For those runs initiated at a tank salinity of 32.5 ‰ NaCl, the plate spacings increased from 0.35 to 0.45 mm during the experiment. The change of grain size with depth in this study was very similar to the dependence assumed for the ice from Weeks and Assur (1964). However, strength data from different depth levels have not been distinguished and, from a fit of tensile strength test results to equation F.14, Dykins (1969) obtained $v_{b*} \approx 0.291$ for horizontal samples and $v_{b*} \approx 0.251$ for vertical samples. Assuming average values $0.35 < a_0 < 0.45$ mm and a grain length of $L_{gra} \approx 10$ mm, resulting in a small correction $r_{gra} \approx 1.04$, one obtains, from equation (F.15), a range $0.066 < d_{sk} < 0.099$ mm. The detailed growth history has not been published, yet Dykins (1967) reported that growth took place in a cold room at -29 °C, and was enhanced by fans above the ice surface. It therefore can be expected that growth was more rapid than in the experiments from Weeks and Assur (1964). However, due to the following argument it is unclear, if the smaller estimated d_{sk} may be related to faster growth conditions. The solution salinity increased during the experiments and high brine volumes corresponded to ice frozen from highest solution salinities. These were the bottom subsamples in the tests, where growth was slowest and a_0 largest. This should, in principle, imply that the larger values of a_0 dictates the critical v_{b*} in the strength relation. Hence, the upper bound $d_{sk} \approx < 0.10$ mm appears most plausible for this dataset.

Reliability of strength approach

In most other strength experiments known to the author, a_0 has not been reported at all or with insufficient detail. Considering, however, that for thick first-year ice the plate width range is typically $0.6 < a_0 < 0.9$ mm, one may perform some bulk estimate. The first-year ice study by Frankenstein (1969), for example, see figure F.15, gave a value of $v_{b*} \approx 0.23$. Assur's data (Assur, 1958) for saline first-year ice and most other first-year ice strength studies reported by (Weeks and Assur, 1967) indicate $v_{b*} \approx 0.20$. Using the latter range in v_{b*}, with the mentioned a_0 range for thick, more slowly growing ice, an approximate range $0.09 < d_{sk} < 0.13$ mm can be estimated from equation F.15. Within the noted uncertainty limits this range agrees with the result from the young ice strength tests. It is another indication that d_{sk} does not depend on growth velocity, thickness, or plate spacing.

It must be finally noted, that the procedure to estimate d_{sk} from strength tests, has to be viewed with caution. The ring-tensile strength test utilized by Weeks and Assur (1964) is, due to relatively large scatter and critics in its theoretical basis, no longer in use when testing sea ice (Weeks, 1998a). It has further emerged, that saline ice failure is more complicated than assumed in the early simple structural models leading to equation F.14. It depends on microstructural anisotropy, other geometric factors and the mode of failure (Cole, 1998; Schulson, 2001; Cole, 2001; Kachanov and Sevostianov, 2005). Also the least-square fits to obtain v_{b*} may be influenced by effects due to changing inclusion shapes as discussed above (section F.3). The residual strength, assumed as 15 % in equation (F.14), is certainly variable. The earlier noted hysteresis effects upon cooling and warming may imply a lack in reproducability of the same microstructure upon warming of stored ice. Clearly, a large number of strength tests is needed for a reliable estimate. Therefore the quantitative agreement of the indirect ice strength estimates with the microstructural d_{sk} should not be overrated. Nevertheless, whatever the exact strength dependence might be, the flattening of the strength-brine volume is consistent with the principle, that the residual strength becomes limited by *intergranular* failure, when brine volumes larger than the connectivity limit of brine layers are encountered. At lower brine volumes *intragranular* brine layer connectivity strengthens the ice. Therefore the rather simplified idea of a failure transition at a certain brine volume, indicated in figures F.15 and F.16, might still be promising, making the strength-brine volume relationship a proxy for d_{sk}, at least in tests, where failure occurs in flexure or tension.

Bibliography

Adams, J., Bachu, S., 2002. Equations of state for geofluids: algorithm review and inter-comparison of brines. Geofluids 2, 257–271.

Ahlers, G., Grossmann, S., Lohse, D., 2002. Hochpräzision im Kochtopf. Phys. J. 1 (2), 31–37.

Akamatsu, S., Faivre, G., 1998. Anisotropy-driven dynamics of cellular fronts in directional soldification in thin samples. Phys. Rev. E 58 (3), 3302–3314.

Akinfiev, N. N., Mironenko, M. V., Duneva, A. N., Grant, S. A., 2000. Thermodynamic properties of NaCa solutions at temperatures lower than 0°C. Geochem. Int. 38 (7), 689–697.

Alexiades, V., Solomon, A. D., 1993. Mathematical Modeling of Melting and Freezing Processes. Taylor & Francis. Hemisphere Publishing Corporation, 323 pp.

Amberg, G., Homsy, G. M., 1993. Nonlinear analysis of buoyant convection in binary solidification with application to channel formation. J. Fluid Mech. 252, 79–98.

Anderson, D., 1966. Heat of freezing and melting of sea ice. Research Report 202, Cold Regions Research and Engineering Laboratory, Hanover, New Hampshire.

Anderson, D. L., 1960. Physical constants of sea ice. Research 13 (8), 310–318.

Anderson, D. L., Weeks, W. F., 1958. A theoretical study of sea ice strength. Trans. Amer. Geophys. Union 39, 632–640.

Andersson, O., Suga, H., 1994. Thermal conductivity of the Ih and XI phases of ice. Phys. Rev. B 50 (10), 6583–6588.

Angell, C., M.Oguni, Sichina, W., 1982. Heat capacity of water at extremes of supercooling and superheating. Journal of Physical Chemistry 86, 998–1002.

Angell, C. A., 1983. Supercooled water. Ann. Rev. Phys. Chem. 34 (34), 593–630.

Arakawa, K., 1954. Studies on the freezing of water (ii): formation of disc crystals. J. Faculty of Science, Ser. 2 5, 311–339.

Archer, D. G., 1992. Thermodynamic properties of the NaCl + H₂O system: II. thermodynamic properties of NaCl(aq) and NaClH₂O(cr), and phase equilibria. J. Phys. Chem. Ref. Data 21 (4), 793–829.

Archer, D. G., Carter, R. W., 2000. Thermodynamic properties of the NaCl + H₂O system. 4. heat capacities of H₂O and NaCl(aq) in cold-stable and supercooled water. Journal of Physical Chemistry B 104, 8563–8584.

Arcone, S. A., Gow, A. J., McGrew, S., 1986. Microwave dielectric, structural, and salinity properties of simulated sea ice. IEEE Trans. Geosc. Rem. Sensing GE-24 (6), 832–839.

Assmann, K., Helmer, H. H., Beckmann, A., 2003. Seasonal variation in circulation and water mass distribution on the Ross Sea continental shelf. Antarct. Sci. 15 (1), 3–11.

Assur, A., 1958. Composition of sea ice and its tensile strength. In: Arctic Sea Ice. Vol. 598. Proc. Conf.on Arctic Sea Ice, Natl. Acad. Sci., pp. 106–138.

Assur, A., Weeks, W. F., 1963. Growth, structure, and strength of sea ice. Int. Association of Scientific Hydrology 63, 95–108.

Baker, D. R., Paul, G., Sreenivasan, S., Stanley, H. E., 2002. Continuum percolation threshold for interpenetrating spheres. Phys. Rev. E 66, 046136.

Baker, I., Cullen, D., Illiescu, D., 2003. The microstructural location of impurities in ice. Can. J. Phys. 81, 1–9.

Barbieri, A., Langer, J. S., 1989. Predictions of dendritic growth rates in the linearized solvability theory. Phys. Rev. A 39, 5314–5325.

Barduhn, A. J., Huang, J. S., 1987. Why does ice grow faster from salt water than from fresh water ? Desalination 67, 99–106.

Barnes, H. T., 1928. Ice Engineering. Renouf Publishing, Montreal, 353 pp.

Bassom, A. P., Blennerhassett, P. J., 2002. Impusively generated convection in a semi-infinite fluid layer above a heated flat plate. Q. J. Mech. Appl. Math. 55 (4), 573–595.

Batchelor, G. K., 1967. An introduction to fluid mechanics. Cambridge University Press, 615 pp.

Batchelor, G. K., 1974. Transport properties of two-phase materials with random structure. Ann. Rev. Fluid Mech. 6, 227–255.

Bechhoefer, J., Libchaber, A., 1987. Testing shape selection in directional soldification. Phys. Rev. B 35, 1393–1396.

Beckermann, C., 2002. Modeling of macrosegregation: applications and future needs. Int. Materials Reviews 47 (5), 243–261.

Beckermann, C., Gu, J. P., Boettinger, W. J., 2000. Development of a freckle predictor via Rayleigh number method for single-crystal nickel-base superalloy castings. Metall. Mater. Trans. A 31 A, 2545–2556.

Beckermann, C., Wang, C. Y., 1995. Multiphase/scale modeling of alloy solidification. Ann. Rev. Heat Transf. 6, 115–198.

Ben Amar, M., Pelce, P., 1989. Impurity effect on dendritic growth. Phys. Rev. A 39 (8), 4263–4269.

Bennett, M. J., Brown, R. A., 1989. Cellular dynamics during directional soldification: interaction of multiple cells. Phys. Rev. B 39 (16), 11705–11723.

Bennington, K. O., 1963. Some crystal growth features of sea ice. J. Glaciol. 4, 669–688.

Bennington, K. O., 1967. Desalination features in natural sea ice. J. Glaciol. 6 (48), 845–857.

Bensimon, D., Kadanoff, L. P., Liang, S., Shraiman, B. I., Tang, C., 1986. Viscous flows in two dimensions. Rev. Mod. Phys. 58 (4), 977–999.

Billia, B., Jamgotchian, H., Thi, H. N., 1996. Influence of sample thickness on cellular branches and cell-dendrite transtion in directional solidfication of binary alloys. J. Cryst. Growth 167, 265–276.

Billia, B., Trivedi, R., 1993. Handbook of Crystal Growth. Vol. I. Elsevier, Ch. 14: Pattern formation in crystal growth, ed. D. T. J. Hurle, pp. 901–1071.

Blagden, C., 1788. Experiments on the effect of various substances in lowering the point of congelation in water. Philos. Trans. Royal Soc. London 78, 277–312.

Blair, L. M., Quinn, J. A., 1969. The onset of cellular convection in a fluid layer with time-dependent density gradients. J. Fluid Mech. 36, 385–400.

Bock, C., Eicken, H., 2005. A magnetic resonance study of temperature-dependent microstructural evolution and self-diffusion of water in arctic first-year sea ice. Ann. Glaciol. 40, 179–184.

Boettinger, W. J., Coriell, S. R., Greer, A. L., Karma, A., Kurz, W., Rappaz, M., Trivedi, R., 2000. Solidification microstructures: recent developments, future directions. Acta mater. 48, 43–70.

Boettinger, W. J., Warren, J. A., 1999. Simulation of the cell to plane front transition during directional solidification at high velocity. J. Crystal Growth 200, 583–591.

Bogdan, A., 1997. Thermodynamics of the curvature effect on the ice surface tension and nucleation theory. J. Chem. Phys. 106 (5), 1921–1929.

Bolling, G. F., Tiller, W. A., 1960. Growth from the melt: I. influence of surface intersections in pure metals. J. Appl. Phys. 31 (8), 1345–1350.

Bouchard, D., Kirkaldy, J. S., 1997. Prediction of dendrite arm spacings in unsteady- and steady-state heat flow of unidirectionally solidified binary alloys. Metall. Mater. Trans. B 28 B, 651–663.

Bower, T. F., Brody, H. D., Flemings, M. C., 1966. Measurements of solute redistribution in dendritic soldification. Trans. Metall. Soc. AIME 236, 624–634.

Brattkus, K., 1989. Capillary instabilities in deep cells during directional solidification. J. Phys. France 50, 2999–3006.

Brattkus, K., Misbah, C., 1990. Phase dynamics in directional soldification. Phys. Rev. Letters 64 (16), 1935–1938.

Bruins, H. R., 1929. International Critical Tables. Vol. 5. McGraw-Hill, New York, Ch. Coefficients of diffusion in liquids, pp. 63–76.

Buchanan, J. Y., 1887. On ice and brines. Proc. Royal Soc. Edinburgh 14, 129–149.

Burden, M. H., Hunt, J. D., 1974. Cellular and dendritic growth. II. J. Cryst. Growth 22, 109–116.

Burton, J. A., Prim, R. C., Slichter, W. P., 1953a. The distribution of solute in crystals grown from the melt. Part I. Theoretical. J. Chem. Phys. 21, 1987–1991.

Burton, J. A., Prim, R. C., Slichter, W. P., 1953b. The distribution of solute in crystals grown from the melt. Part II. Experimental. J. Chem. Phys. 21, 1991–1996.

Busse, F. H., 1981. Hydrodynamic instabilities and the transition to turbulence. Springer, Ch. 5. Transition to turbulence in Rayleigh-Benard convection, pp. 97–137.

Cadirli, E., Karaca, I., Kaya, H., Marasli, N., 2003. Effect of growth rate and composition on the primary spacing, dendrite tip radius and mushy zone depth in the directionally solidified succinonitrile-salol alloys. J. Cryst. Growth 255, 190–203.

Cahn, J. W., Hillig, W. B., Sears, G. W., 1964. The molecular mechanism of solidification. Acta Metall. 12, 1421–1439.

Caldwell, D. R., 1973. Thermal and fickian diffusion of sodium chloride in a solution of oceanic concentration. Deep-Sea Res. 20A, 1029–1039.

Caldwell, D. R., Eide, S. A., 1981. Soret coefficient and isothermal diffusivity of aqueous solutions of five principal salt constitutents of seawater. Deep-Sea Res. 28A, 1605–1618.

Callaghan, P. T., Dykstra, R., Eccles, C. D., Haskell, T. G., Seymour, J. D., 1999. A nuclear magnetic resonance study of Antarctic sea ice brine. Cold Regions Science and Technology 29, 153–171.

Canuto, V. M., Goldman, I., 1985. Analytical model for large-scale turbulence. Phys. Rev. Lts. 54 (5), 430–433.

Carman, P. C., 1956. Flow of Gases Through Porous Media. Butterworths, London, 182 pp.

Caroli, B., Caroli, C., Misbah, C., Roulet, B., 1985a. Solutal convection and morphological instability in directional solidification. J. Phys. Paris 46, 401–413.

Caroli, B., Caroli, C., Misbah, C., Roulet, B., 1985b. Solutal convection and morphological instability in directional solidification. II. Effect of the density difference between the phases. J. Phys. Paris 46, 1657–1665.

Caroli, B., Müller-Krumbhaar, H., 1995. Recent advances in the theory of free dendritic growth. ISIJ Int. 35 (12), 1541–1550.

Carsey, F. D., 1992. Microwave Remote Sensing of Sea Ice. Vol. 68 of Geophysical Monograph. American Geophysical Union, Washington.

Carslaw, H. S., Jaeger, J. C., 1959. Conduction of Heat in Solids, 2nd Edition. Clarendon Press, Oxford, 510 pp.

Chalmers, B., 1964. Principles of solidification. John Wiley, 319 pp.

Chan, S. K., 1971. Infinite Prandtl number convection. Studies Appl. Math. 1 (1), 13–49.

Chandrasekhar, S., 1961. Hydrodynamic and Hydromagnetic Stability. Clarendon Press, 652 pp.

Chen, C. F., 1995. Experimental study of convection in a mushy layer during directional soldification. J. Fluid Mech. 293, 81–98.

Chen, C. F., Chen, F., 1991. Experimental study of directional solidfication of aqeous ammonium chloride solution. J. Fluid Mech. 227, 567–586.

Chen, F., 2001. Stability analysis on convection in directional soldification of binary sokutions. Proc. Natl. Sci. Counc. ROC 25 (2), 71–83.

Chen, F., Chen, C. F., 1988. Onset of finger convection in a horizontal porous layer underlying a fluid layer. ASME J. Heat Transfer 110, 403–409.

Chen, F., Chen, C. F., Pearlstein, A. J., 1991. Convective stability in superposed fluid and anisotropic porous layers. Phys. Fluids 3 (4), 556–565.

Chen, F., Lu, J. W., Yang, T., 1994. Convective stability in ammonium chloride solution directionally solidified from below. J. Fluid Mech. 276, 163–187.

Chen, P., Chen, X. D., Free, K. W., 1998. Solute inclusions in ice formed from sucrose solutions on a sub-cooled surface - an experimental study. J. Food Engin. 38, 1–13.

Cherepanov, N. V., 1964. Structure of sea ice of great thickness (in russian). Trudy Arkt. Antarkt. N.-I. Institut Leningrad 267, 13–18.

Cherepanov, N. V., 1973. Main results of an investigation of the crystal structure of sea ice. Problemy Arktiki i Antartiki 41, 43–54.

Cherepanov, N. V., Kamyshnikova, A. V., 1971. Sizes and shapes of congealed ice crystals. In: Yakovlev, G. N. (Ed.), Studies in Ice physics and ice engineering (fiziko-tekhnicheskie issledovaniya l'da). Israel Program for Sceintific Translations, Gidrometeorologichevskoe Izdatel'stvo (Transalted from Russian), pp. 170–176.

Chiareli, A. O. P., Worster, M. G., 1992. On measurement and prediction of the solid fraction within mushy layers. J. Cryst. Growth 125, 487–494.

Cho, H., Shepson, P. B., Barrie, L. A., Cowin, J. P., Zaveri, R., 2002. NMR investigation of quasi-brine layer in ice/brine mixtures. J. Phys. Chem. B 106, 11226–11232.

Chung, C. A., Chen, F., 2000. Onset of plume convection in mushy layers. J. Fluid Mech. 408, 53–82.

Chung, C. A., Worster, M. G., 2002. Steady-state chimneys in a mushy layer. J. Fluid Mech. 455, 387–411.

Clarke, E. C., Glew, D. E., 1985. Evaluation of the thermodynamic functions for aqueous sodium chloride from equilibrium and calorimetric measurements below 154°C. J. Phys. Chem. Ref. Data 14 (2), 489–610.

Cohen-Adad, R., Lorimer, J. W., 1991. Alkali Metal and ammonium chlorides in water and heavy water (binary systems). Vol. 47 of Solubility Data Series. Pergamon Press, Ch. 2, Sodium Cloride, pp. 64–209.

Cole, D. M., 1998. Modeling the cyclic loading response of sea ice. Int. J. Solids and Structures 35, 4067–4075.

Cole, D. M., 2001. The microstructure of ice and its influence on mechanical properties. Eng. fracture mech. 68, 1797–1822.

Cole, D. M., Dempsey, J. P., Shapiro, L. H., Petrenko, V. F., 1995. In-situ and laboratory measurments of the physical and mechanical properties of first-year sea ice. In: Dempsey, J. P., Rajapakse, Y. D. S. (Eds.), Ice Mechanics. Vol. 207. Am Soc. Mech. Engin., pp. 161–178.

Cole, D. M., Eicken, H., Frey, K., Shapiro, L. H., 2002. Some observations of high porosity bands and brine drainage features in first-year sea ice. In: 16th International Symposium on ice. IAHR, Dunedin, New Zealand, pp. 179–186.

Cole, D. M., Eicken, H., Frey, K., Shapiro, L. H., 2004. Observations of banding in first-year arctic sea ice. J. Geophys. Res. 109 (C08), C08012.

Cole, D. M., Shapiro, L. H., 1998. Observations of brine drainage networks and microstructure of first-year sea ice. J. Geophys. Res. 103 (C10), 21739–21750.

Conti, M., Marconi, U. M. B., 2000. Groove instability in cellular solidification. Phys. Rev. E 63, 011502.

Coriell, S. R., Boisvert, R. F., MacFadden, G. B., Brush, L. N., Favier, J. J., 1994. Morphological stability of a binary alloy during directional soldification: initial transient. J. Cryst. Growth 140 (1-2), 139–147.

Coriell, S. R., Cordes, M. R., Boettinger, W. J., 1981. Convective and interfacial instabilities during unidirectional solidification of a binary alloy. J. Cryst. Growth 49, 13–28.

Coriell, S. R., Hardy, S. C., 1969. Morphology of unstable ice cylinders. J. Appl. Phys. 4 (40), 1652–1655.

Coriell, S. R., Hardy, S. C., Sekerka, R. F., 1971. A non-linear analysis of experiments on the morphological stability of ice cylinders freezing from aqueous solutions. J. Cryst. Growth 11, 53–67.

Coriell, S. R., MacFadden, G. B., 2002. Applications of morphological stability theory. J. Crystal Growth 237-239, 8–13.

Coriell, S. R., MacFadden, G. B., Sekerka, R. F., 1985. Cellular growth during directional solidification. Ann. Rev. Mater. Sci. 15, 119–145.

Coriell, S. R., McFadden, G. B., 1993. Handbook of Crystal Growth. Vol. I. Elsevier, Ch. 14: Morphological stability, ed. D. T. J. Hurle, pp. 785–857.

Coriell, S. R., Sekerka, R. F., 1976a. The effect of the anisotropy of surface tension and interface kinetics on morphological stability. Journal of Crystal Growth 34, 157–163.

Coriell, S. R., Sekerka, R. F., 1976b. Interface stability during crystal growth: the effect of stirring. Journal of Crystal Growth 32, 1–7.

Cottier, F., Eicken, H., Wadhams, P., 1999. Linkages between salinity and brine channel distribution in young sea ice. J. Geophys. Res. 104 (C7), 15859–15871.

Cox, G. F. N., Weeks, W. F., 1973. Salinity variations in sea ice. Research Report 310, U.S. Army Cold Regions Research and Engineering Laboratory.

Cox, G. F. N., Weeks, W. F., 1974. Salinity variations in sea ice. J., Glaciol. 13 (67), 109–120.

Cox, G. F. N., Weeks, W. F., 1975. Brine drainage and initial salt entrapment in sodium chloride ice. Research Report 345, U.S. Army Cold Regions Research and Engineering Laboratory, this research report consists of the dr.-philos. thesis *Brine drainage in sodium chloride ice* (1974) by G. F. N. Cox, Dartsmouth College, 179 pp.

Cox, G. F. N., Weeks, W. F., 1983. Equations for determining the gas and brine volumes of sea ice samples during sampling and storage. J. Glaciol. 32, 371–375.

Cox, G. F. N., Weeks, W. F., 1986. Changes in the salinity and porosity of sea ice samples during shipping and storage. J. Glaciol. 32 (112), 371–375.

Cox, G. F. N., Weeks, W. F., 1988. Numerical simulations of the profile properties of undeformed first-year ice during the growth season. J. Geophys. Res. 93 (C10), 12449–12460.

Cragin, J. H., 1995. Exclusion of sodium chloride from ice during freezing. In: 52nd Eastern Snow Conference. Toronto, Ontario, Canada 1995, pp. 259–262.

Cullen, D., Baker, I., 2002. Observation of impurities in ice. Micr. Res. Techn. 55, 198–207.

Currie, I. G., 1967. The effect of heating rate on the stability of stationary fluids. J. Fluid Mech. 29 (2), 337–347.

Dash, J. G., Fu, H., Wettlaufer, J. S., 1995. The premelting of ice and its environmental consequences. Rep. Prog. Phys. 58, 115–167.

Davenport, I. F., King, C. J., 1974. The onset of natural convection from time-dependent profiles. Int. J. Heat Mass Transfer 17 (17), 69–76.

Davis, S. H., 1993. Handbook of crystal growth. Vol. I. Elsevier, Ch. 13: Effects of flow on morphological stability, ed. D. T. J. Hurle, pp. 861–897.

DeCheveigne, S., Guthmann, C., Kurowski, P., 1988. Directional soldification of metallic alloys: the nature of the bifurcation from planar to cellular freezing. J. Cryst. Growth 92, 616–628.

Dee, G., Mathur, R., 1983. Cellular patterns produced by the directional solidification of a binary-alloy. Phys. Rev. B 27 (12), 7073–7092.

Degan, G., Vasseur, P., 2003. Influence of anisotropy on convection in porous media with nonuniform thermal gradient. Int. J. Heat and Mass Transf. 46, 781–789.

Denton, R. A., Wood, I. R., 1979. Turbulent convection between two horizontal plates. Int. J. Heat Mass Transfer 22, 1339–1346.

Devik, O., 1931. Thermische und dynamisch bedingungen der Eisbildung in Wasserläufen, auf norwegische Verhältnisse angewandt. Geof. Publ. 9 (1).

Dietrich, G., K.Kalle, Krauss, W., Siedler, G., 1975. Allgemeine Meereskunde, 3rd Edition. Gebrüder Borntraeger, Berlin, 593 pp.

Ding, G., Huang, W., Zhou, Y., 1997. Prediction of average spacing for constrained cellular/dendritic growth. J. Cryst. Growth 177, 281–288.

Dombre, T., Hakim, V., 1987. Saffman-Taylor fingers and directional solidification at low velocity. Phys. Rev. A 36 (6), 2811–2817.

Doronin, Y. P., Kheisin, D. E., 1975. Morskoi Led (Sea Ice). Gidrometeoizdat, Leningrad, english translation 1977 by Amerind Publishing, New Delhi, 318 pp.

Dorsey, E. N., 1940. Properties of ordinary water-substance. Monograph Series, Nr. 81. Reinhold Publishing, American Chemical Society, New York, 673 pp.

Drost-Hansen, W., 1967. The water-ice interface as seen from the liquid side. J. Coll. Interf. Sci. 25, 131–160.

Drygalski, E., 1897. Grönlands Eis und sein Vorland. Vol. 1 of Grönland-Expedition der Gesellschaft für Erdkunde zu Berlin 1891-1893. Berlin, Kühl, 555 pp.

Dschu, I.-L., 1968. Ausfrieren von Eis aus einer turbulent strömenden Salzlösung an gekühlten Oberflächen. Thesis, Eidgenössische Technische Hochschule Zürich, juris Druck Zürich.

Dufour, L., Defay, R., 1963. Thermodynamics of clouds. International Geophysics Series. Acadenic Press, London.

Dullien, F. A. L., 1979. Porous Media: Fluid Transport and Pore Structure. Academic Press, 396 pp.

Dykins, J. E., 1967. Tensile properties of sea ice grown in a confined system. In: Physics of Snow and ice. Vol. 1. Hokkaido University, Sapporo, Japan, pp. 523–537.

Dykins, J. E., 1969. Tensile and flexure properties of saline ice. In: Physics of ice. 3rd International Symposium on the Physics of Ice, Plenum Press, New York, pp. 251–270.

Eicken, H., 1992. Salinity profiles of Antarctic sea ice: field data and model results. J. Geophys. Res. 97 (C10), 15545–15557.

Eicken, H., 1998. Deriving modes and rates of ice growth in the Weddell Sea from microstructural, salinity and stable-isotope data. Vol. 74 of Antarctic Research Series. pp. 89–122.

Eicken, H., 2003. From the microscopic to the macroscopic to the regional scale: growth, microstructure and properties of sea ice, thomas, d. n. and dieckmann, g. s. (eds.) Edition. Blackwells Scientific Ltd., London, pp. 22–81, 402 pp.

Eicken, H., Bock, C., Wittig, R., Miller, H., Poertner, H.-O., 2000. Magnetic resonance imaging of sea-ice pore fluids: methods and thermal evolution of pore microstructure. Cold Reg. Sci. and Technology 31, 207–225.

Eicken, H., Weissenberger, J., Bussmann, I., Freitag, J., Schuster, W., Dlegado, F. V., Evers, K. U., Jochmann, P., Krembs, C., Gradinger, R., Lindemann, F., Cottier, F., Hall, R., Wadhams, P., Reisemann, M., Kousa, H., Ikävalko, J., Leonard, G. H., Shen, H., Ackley, S. F., Smedrud, L. H., 1998. Ice tank studies of physical and biological sea ice processes. In: Shen, H. T. (Ed.), Ice in Surface Waters, Vol.1. Vol. 1. A. A. Balkema, Rotterdam, Netherlands, pp. 363–370.

Eide, L. I., Martin, S., 1975. The formation of brine drainage features in young sea ice. J. Glaciol. 14 (70), 137–154.

Elder, J. W., 1967. Steady free convection in a porous medium heated from below. J. Fluid Mech. 27 (1), 29–48.

Emms, P., Fowler, A. C., 1994. Compositional convection in the solidification of binary alloys. J. Fluid Mech. 262, 111–139.

Emms, P. W., 1998. Freckle formation in a solidifying binary alloy. J. Engineer. Math. 33 (33), 175–200.

Falkenhagen, H., 1953. Elektrolyte. S. Hirzel Verlag, Leipzig, 263 pp.

Farhadieh, R., Tankin, R. S., 1975. A study of the freezing of sea water. J. Fluid Mech. 71 (2), 293–304.

Feistel, R., 2003. A new extended Gibbs thermodynamic potential of seawater. Progr. in Oceanogr. 58, 43–114.

Feistel, R., Hagen, E., 1998. A Gibbs thermodynamic potential of sea ice. Cold Reg. Sci. Technol. 28, 83–142.

Fenech, E. J., Tobias, C. W., 1960. Mass transfer by free convection at horizontal electrodes. Electrochimica Acta 2, 311–325.

Feng, J., Huang, W. D., Lin, X., Pan, Q. Y., 1999. Primary cellular/dendrite spacing selction of Al-Zn alloy during unidirectional solidification. J. Crystal Growth 197, 393–395.

Fernandez, J., Kurowski, P., Limat, L., Petitjeans, P., 2001. Wavelength selection of fingering instability inside Hele-Shaw-cells. Phys. Fluids 13, 3120–3125.

Ferrick, M. G., Calkins, D. J., Perron, N. M., 1998. Stable environmental isotopes in lake and river ice cores. In: Shen, H. T. (Ed.), Ice in Surface Waters, Vol.1. Vol. 1. A. A. Balkema, Rotterdam, Netherlands, pp. 207–214.

Ferrick, M. G., Calkins, D. J., Perron, N. M., Cragin, J. H., Kendall, C., 2002. Diddusion model validation and interpretation of stable isotopes in river and lake ice. Hydrol. Proc. 116, 851–872.

Fertuck, L., Spyker, J. W., Husband, W. H., 1972. Computing salinity profiles in ice. Can. J. Physics 50, 264–267.

Flemings, M. C., 1974. Solidification Processing. McGraw-Hill, Inc., 364 pp.

Flemings, M. C., Nereo, G. E., 1967. Macrosegregation: Part I. Trans. Metall. Soc. AIME 239, 1449–1461.

Fletcher, N. H., 1970. The Chemical Physics of Ice. Cambridge University Press, Cambridge, 271 pp.

Fofonoff, P., Millard, R. C., 1983. Algorithms for computation of fundamental properties of seawater. Unesco Tech. Pap. in Mar. Sci. 44, 53 pp.

Foster, T. D., 1968a. Effect of boundary conditions on the onset of convection. Phys. Fluids 11 (6), 1257–1262.

Foster, T. D., 1968b. Haline convection induced by the freezing of sea water. J. Geophys. Res. 73 (28), 1933–1938.

Foster, T. D., 1969. Experiments on haline convection induced by the freezing of sea water. J. Geophys. Res. 74 (28), 6967–6974.

Foster, T. D., 1971. Intermittent convection. Geophys. Fluid Dyn. 2, 201–217.

Foster, T. D., 1972. Haline convection in polynyas and leads. J. Phys. Oceanogr. 2, 462–469.

Fowler, A. C., 1985. The formation of freckles in binary alloys. IMA J. Appl. Math. 35, 159–174.

Frankenstein, G. E., 1969. Ring tensile strength studies of ice. Crrel research report 172, Cold Regions Research and Engineering Laboratory.

Franks, F., 1982. Water: A comprehensive treatise. Plenum, volumes 1-7, 1972 to 1982.

Fransson, L., Hakansson, B., Omstedt, B., Stehn, L., 1990. Sea ice properties studied from the icebreaker Tor during BEPERS-88. Tech. Rep. RO 10, SMHI.

Freitag, J., 1999. Untersuchungen zur Hydrologie des arktischen Meereises - Konsequenzen für den Sofftransport. Ph.D. thesis, Universität Bremen, Fachbereich Physik/Elektrotechnik, auch: Ber. z. Polarforsch., Bd. 325, 1999.

Frick, H., Clever, R. M., 1980. Einfluss der Seitenwände auf das Einsetzen der Konvection in einer horizontalen Flüssigkeitsschicht. ZAMP 31, 502–513.

Fujioka, T., Sekerka, R. F., 1974. Morphological stability of disc crystals. J. Cryst. Growth 24/25, 84–93.

Fukusako, S., 1990. Thermophysical properties of ice, snow and sea ice. Int. J. Thermophys. 11 (2), 353–372.

Fukutomi, T., Saito, M., Kudo, Y., 1952. On the structure of ice rind, especially on the structure of thin ice sheet and ice-sheet block. Low Temp.Sci. 9, 113–123.

Furukawa, Y., Shimada, W., 1993. Three-dimensional pattern formation during growth of ice dendrites - its relation to universal law of dendritic growth. J. Crystal Growth 128 (128), 234–239.

Garrison, G. R., Francois, R. E., 1991. Acoustic reflections from arctic ice at 15-300 kHz. J. Acoust. Soc. Am. 90 (2), 973–984.

Georgelin, M., Pocheau, A., 1997. Onset of sidebranching in directional soldification. Phys. Rev. E 57 (3), 3189–3203.

Georgelin, M., Pocheau, A., 2004. Characterization of cell tip curvature in directional soldification. J. Cryst. Growth 268, 272–283.

Gershuni, G. Z., Zhukovitskii, E., 1976. Convective stability of incompressible fluids. Israel Program for Scientific Translations, Keter Publishing House, in russian 1972 by Izdatel'stvo Nauka, Jerusalem, 329 pp.

Giauque, W. F., Stout, M. F., 1936. The entropy of water and the third law of thermodynamics. The heat capacity of ice from 15 to 273 k. Journal of the American Chemical Society 58, 1144–1150.

Gitterman, K. E., 1937. Thermal analysis of seawater (termichesky analiz morskoy vody). Trudy Solyanoy Laboratorii Akad. nauk SSSR 15 (1), 5–23.

Glicksman, M., Lupulescu, A., 2004. Dendritic crystal growth in pure materials. J. Cryst. Growth 264, 541–549.

Glicksman, M. E., Koss, M. B., 1994. Dendrititic growth velocities in microgravity. Phys. Rev. Letters 73, 573–576.

Glicksman, M. E., Marsh, S. P., 1993. Handbook of Crystal Growth. Vol. I. Elsevier, Ch. 15: The dendrite, ed. D. T. J. Hurle, pp. 1075–1122.

Golden, K. M., 2003. Critical behaviour of transport in sea ice. Physica B 338, 274–283.

Golden, K. M., Ackley, S. F., Lytle, V. I., 1998. The percolation phase transition in sea ice. Science 282, 2238–2241.

Goldstein, R. J., Chang, H., See, D. L., 1990. High-Rayleigh-number convection in a horizontal enclosure. J. Fluid Mech. 213, 111–126.

Goldstein, R. J., Volino, R. J., 1995. Onset and development of natural convection above a suddenly heated horizontal surface. Trans. ASME 117, 808–821.

Gordon, A. R., 1937. The diffusion constant of an electrolyte, and its relation to concentration. J. Chem. Phys. 5, 522.526.

Gow, A., 1986. Orientation textures in ice sheets of quietly frozen lakes. J. Cryst. Growth 74 (2), 247–258.

Gow, A. J., Arcone, S. A., McGrew, S. G., 1987. Microwave and structural properties of saline ice. CRREL REPORT 87-20, Cold Regions Research and Engineering Laboratory, Hanover.

Gow, A. J., Meese, D. A., Perovich, D. K., III, W. B. T., 1990. The anatomy of a freezing lead. J. Geophys. Res. 95 (C5), 18221–18232.

Gow, A. J., Weeks, W. F., Kosloff, P., Carsey, S., 1992. Petrographic and salinity characteristics of brackish water ice in the Bay of Bothnia. CRREL REPORT 92-13, Cold Regions Research and Engineering Laboratory, Hanover.

Goyeau, B., Benihaddadene, T., Gobin, D., Quintard, M., 1999. Numerical calculation of the permeability in a dendritic mushy zone. Metall. Mater. Trans. B 30, 613–622.

Graf, F., Meiburg, E., Härtel, C., 2002. Density-driven instabilities of miscible fluids in a Hele-Saw-cell: linear stabilty analysis of the three-dimensional Stokes equations. J. Fluid Mech. 451, 261–282.

Granskog, M. A., Leppäranta, M., Kawamura, T., Ehn, J., Shirasawa, K., 2004a. Seasonal development of the properties and composition of landfast sea ice in the Gulf of Finland, the Baltic Sea. J. Geophys. Res. 109, C02020.

Granskog, M. A., Virkkunen, K., Thomas, D. N., Kola, H., Martma, T., 2004b. Chemical properties of brackish water ice in the Bothnian Bay, the Baltic Sea. J. Glciol. 50 (169), 292–302.

Gresho, P. M., Sani, R. L., 1971. The stability of a fluid layer subjected to a step change in temperature: transient vs. frozen time analysis. Int. J. Heat Mass Transf. 14, 207–221.

Grigull, U., Gröber, H., Erk, S., 1963. Die Grundgesetze der Wärmeübertragung, 3rd Edition. Springer-Verlag, Berlin, 436 pp.

Gross, G. W., 1967. Ion distribution and phase boundary potentials during the freezing of very dilute ionic solutions at uniform rates. J. Colloid Interface Science 25, 270–279.

Gross, G. W., 1968. Some effects of trace inorganics on the ice/water system. Vol. 73 of Advances in Chemistry. Amer. Chem. Soc., Ch. 3, pp. 27–97.

Gross, G. W., Gutjahr, A., Caylor, K., 1987. Recent experimental work on solute redistribution at the ice/water interface. Implications for electrical properties and interface processes. J. de Phys. C1 (3), 527–533.

Gross, G. W., McKee, C., Wu, C., 1975. Concentration dependent solute redistribution at the ice-water phase boundary. I. Analysis. J. Chem. Phys. 62 (11), 3080–3084.

Gross, G. W., Wong, P. M., Humes, K., 1977. Concentration dependent solute redistribution at the ice-water phase boundary. III. Spontaneous convection, chloride solutions. J. Chem. Phys. 67 (11), 5264–5274.

Grossmann, S., Lohse, D., 2000. Scaling in thermal convection: a unifying theory. J. Fluid Mech. 407, 27–56.

Gündüz, M., Cadirli, E., 2001. Directional solidfication of aluminium-copper alloys. Mater. Sci. Engineering A327, 167–175.

Gupta, S. C., 2003. The classical Stefan problem: basic concepts, modelling and analysis. Elsevier, 385 pp.

Haas, C., February 1999. Ice tank investigations of the microstructure of artificial sea ice grown under different boundary conditions during INTERICE II. In: K. U. Evers, J. G., van Os, A. (Eds.), Proceedings of the HYDROLAB workshop. Hannover, Germany, pp. 107–113.

Hallett, J., 1963. The temperature dependence of the viscosity of supercooled water. Proceedings of the Physical Society 82, 1046–1050.

Hamberg, A., 1895. Studien über Meereis und Gletschereis (sea ice and glacier ice studies). Svenska Vetenskapsakademien, Bihang til Handlinger 21, 1–13.

Hansen, E. W., Gran, H. C., Sellevold, E. J., 1997. Heat of fusion and surface tension of solids confined in porous materials derived from a combined use of NMR and calorimetry. J. Phys. Chem. B 101, 7027–7032.

Hansen, G., Liu, S., Zhu, S.-Z., Hellawell, A., 2002. Dendritic array growth in the systems NH4Cl-H2O and (CH2CN)2-H2O: steady state measurements and analysis. J. Cryst. Growth 234, 731–739.

Happel, J., Brenner, J., 1986. Low Reynolds Number Hydrodynamics. M. Nijhoff, Dordrecht, Netherlands, 553 pp.

Hardy, S. C., 1977. A grain boundary groove measurement of the surface tension between ice and water. Philos. Mag. 35 (2), 471–484.

Hardy, S. C., Coriell, S. R., 1969. Morphological stability of cylindrical ice crystals. J. Crystal Growth 5, 329–337.

Hardy, S. C., Coriell, S. R., 1973. Surface tension and interfacial kinetics of ice crystals freezing and melting in sodium chloride solutions. J. Cryst. Growth 20, 292–300.

Hare, D. E., Sörensen, C. M., 1987. The density of supercooled water. II. Bulk samples cooled to the homogeneous nucleation limit. Journal of Chemical Physics 87 (8), 4840–4845.

Harned, H. S., Owen, B. B., 1958. The physical chemistry of electrolytic solutions. Amer. Chem. Soc. Monogr. Series. Reinhold Publishing, 803 pp.

Harrison, J. D., 1965a. Measurements of brine droplet migration in ice. J. Appl. Phys. 36 (12), 3811–3815.

Harrison, J. D., 1965b. Solute transpiration pores in ice. J. Appl. Phys. 36 (12), 326–327.

Harrison, J. D., Tiller, W. A., 1963. Ice interface morphology and texture developed during freezing. J. Applied Physics 34, 3349–3355.

Harrison, W. D., Raymond, C. F., 1976. Impurities and their distribution in temperate glacier ice. J. of Glac. 16 (74), 173–181.

Hartline, B. K., Lister, C. R., 1977. Thermal convection in a Hhele-Shaw cell. J. Fluid Mech. 79, 379–389.

Helland-Hansen, B., Nansen, F., 1909. The Norwegian Sea. Det Mallingske Boktrykkeri, Kristiania.

Hellawell, A., 1987. Structure and dynamics of partially solidified systems. Nijhoff, Ch. Local convective flows in partly solidfied systems, pp. 5–21.

Hellawell, A., Sarazin, J. R., Steube, R. S., 1993. Channel convection in partly solidified systems. Phil. Trans. R. Soc. Lond. A 345, 507–544.

Herring, J. R., 1963. Investigation of problems in thermal convection. J. Atmos. Sci. 20, 325–338.

Higdon, J. J. L., Ford, G. D., 1996. Permeability of three-dimensional models of fibrous porous media. J. Fluid Mech. 308, 341–361.

Hill, P. G., 1990. A unified fundamental equation for the thermodynamic properties of H_20. J.Phys. Chem. Ref. Data 19 (5), 1233–1274.

Hillig, W. B., 1958. The kinetics of freezing of ice in the direction perpendicular to the basal planes. In: Doremus, R. H., Roberts, B. W., Turnbull, D. (Eds.), Growth and Perfection of Crystals. Wiley and Sons, New York, pp. 350–360.

Hillig, W. B., 1998. Measurement of interfacial free energy for the ice/water system. J. Cryst. Growth 183, 463–468.

Hills, R. N., Loper, D. E., Roberts, P. H., 1982. A thermodynamic consistent model of a mushy zone. Q. J. Mech. Appl. Math. 36 (4), 505–539.

Hobbs, P., 1974. Ice Physics. Clarendon Press, Oxford, 837 pp.

Hoekstra, P., Osterkamp, T. E., Weeks, W. F., 1965. The migration of liquid inclusions in single ice crystals. J. Geophys. Res. 70 (20), 5035–5041.

Holloway, K. E., de Bruyn, J. R., 2005. Viscous fingering with a single fluid. Canad. J. Physics 83, 551–564.

Holten, V., Labetski, D. G., van Dongen, M. E. H., 2005. Homogenous nucleation of water between 200 and 240 K: New wave tube data and estimation of the tolman length. J. of chem. phys. 123, 104505–1.

Horton, C. W., Rogers, F. T., 1945. Convection currents in a porous medium. J. Appl. Physics 16, 367–370.

Horvay, G., Cahn, J. W., 1965. Dendritic and spheroidal growth. Acta Metall. 9, 695–705.

Howard, L. N., 1966. Convection at high Rayleigh Numbers. In: Görtler, H. (Ed.), Proc. of the Eleventh International Congress of Applied Mechanics. Munich, Germany, 1964, pp. 1109–1115.

Hruby, J., Holten, V., 2004. A two-structure model of thermodynamic properties and surface tension of supercooled water. In: 14th International Conference on the Properties of Water and Steam. Kyoto, pp. 241–246.

Huang, J., Bartell, L. S., 1995. Kinetics of homogeneous nucleation in the freezing of large water clusters. J. Phys. Chem. 99, 3924–3931.

Huang, W. D., Geng, X. G., Zhou, Y. H., 1993. Primary spacing selection of constrained dendritic growth. J. Cryst. Growth 134, 105–115.

Hunt, J. D., 1979. Keynote adress: cellular and primary dendrite spacings. In: Solidification and Casting of Metals. Metals Society London, Sheffield, pp. 3–9.

Hunt, J. D., 1991. A numerical analysis of dendritic and cellular growth of a pure material investigating the transition from array to isolated growth. Acta metall. mater. 39 (9), 2117–2133.

Hunt, J. D., 2001. Pattern formation in soldification. Sci. Techn. Adv. Mater. 2, 147–155.

Hunt, J. D., Lu, S. Z., 1996. Numerical modeling of cellular/dendritic array growth: spacing and structure predictions. Met. Mater. Trans. A 27 A, 611–623.

Hunt, J. D., McCartney, D. G., 1987. Numerical finite difference model for steady state array growth. Appl. Sci. Res. 44, 9–26.

Huppert, H. E., 1990. The fluid mechanics of solidification. J. Fluid Mech. 212, 209–240.

Huppert, H. E., Hallworth, M. A., 1993. Soldification of NH_4CL and NH_4BR from aqueous solutions contaminated by $CuSO_4$: the extinction of chimneys. J. Cryst. Growth 130, 495–506.

Huppert, H. E., Worster, M. G., 1985. Dynamic solidification of a binary melt. Nature 314, 703–707.

Hurle, D. T. J., 1969. Interface stability during the solidification of a stirred binary melt. J. Cryst. Growth 5, 162–166.

Hurle, D. T. J., Jakeman, E., Wheeler, A. A., 1982. Effect of solutal convection on the morphological stability of a binary alloy. J. Cryst. Growth 58, 163–179.

Hurle, D. T. J., Jakeman, E., Wheeler, A. A., 1983. Hydrodynamic stability of the melt during solidification of a binary alloy. Phys. Fluids 26, 624–626.

Hwang, I. G., Choi, C. K., 1996. An analysis of the onset of compositional convection in a binary melt soldified from below. J. Crystal Growth 162, 182–189.

Hwang, I. G., Choi, C. K., 2000. The onset of mushy-layer-mode instabilities during solidification in ammonium chloride solution. J. Crystal Growth 220, 326–335.

IAPWS, 1994. IAPWS release on surface tension of ordinary water substance. Tech. rep., International Association for the Properties of Water and Steam, London, Denmark.

IAPWS, 1996. Release on the IAPWS formulation 1995 for the thermodynamic properties of ordinary water substance for general and scientific use. Tech. rep., International Association for the Properties of Water and Steam, Fredericia, Denmark.

IAPWS, 1998. Revised release on the IAPS formulation 1985 for the thermal conductivity of ordinary water substance. Tech. rep., International Association for the Properties of Water and Steam, London, England.

IAPWS, 2003. Revised release on the IAPS formulation 1985 for the viscosity of ordinary water substance. Tech. rep., International Association for the Properties of Water and Steam, Vejle, Denmark.

ICT, 1929. International Critical Tables. Vol. 3. McGraw-Hill, New York.

Illiescu, D., Baker, I., Cullen, D., 2002. Preliminary microstructural and microchemical observations on pond and river accretion ice. Cold Reg. Sci. Techn. 35, 81–99.

Ivantsov, G. P., 1947. The temperature field around a spherical, cylindrical, or point crystal growing in a supercooled melt. Doklady Akademii Nauk SSSR 58, 567–569.

Jaccard, P. C., Levi, L., 1961. Segregation d' impuretes dans la glace. Z. Angew. Math. Phys. 12, 70–76.

Jamieson, D. T., Tudhope, J. S., 1970. Physical properties of seawater solutions: thermal conductivity. Desalination 8, 393–401.

Japikse, D., Jallouk, P. A., Winter, E. R. F., 1971. Single-phase transport processes in the closed thermosyphon. Int. J. Heat Mass Tranfer 14, 869–887.

Jeffery, C. A., Austin, P. H., 1997. Homogenous nucleation of supercooled water. J. Geophys. Res. 102 (D21), 25269–25279.

Jellinek, H. H., 1972. The ice interface. In: Horne, R. A. (Ed.), Water and aqueous solutions: structure, thermodynamics and transport processes. Wiley-Interscience, New York, pp. 65–107.

Jellinek, H. H. G., Gouda, V. K., 1969. Grain growth in polycrystalline ice. Phys. Status Solidi 31, 413–423.

Jenkins, D. R., 1990. Oscillatory instability in a model of directional solidification. J. Cryst. Growth 102, 481–490.

Jhaveri, B. S., Homsy, G. M., 1982. The onset of convection in fluid layers heated rapidly in a time-dependent manner. J. Fluid Mech. 114, 251–260.

Johnson, J. G., 1943. Studier av isen i Gullmarfjorden. Svenska hydrogr.-biolog. kommissionens skrifter, Ny serie: hydrografi 18, 1–21.

Jones, D. R. H., 1973a. The measurement of solid-liquid interfacial energies from the shapes of grain-boundary grooves. Phil. Mag. 27, 569–584.

Jones, D. R. H., 1973b. The temperature-gradient migration of liquid droplets through. J. Cryst. Growth 20, 145–151.

Jones, D.-R. H., 1974. The free energies of solid-liquid surfaces. J. Materials Sci. 9, 1–17.

Jones, D. R. H., Chadwick, G. A., 1971. Experimental measurement of solid-liquid interfacial energies: the ice-water-sodium chloride system. J. Cryst. Growth 11, 260–264.

Kachanov, M., Sevostianov, I., 2005. On quantitative characterization of microstructures and effective properties. Int. J. Solids Struct. 42 (42), 309–336.

Kaess, M., Magun, S., 1961. Zur Überhitzung am Phasenübergang fest-flüssig. Z. f. Kristallographie 116, 354–370.

Karma, A., Langer, J. S., 1984. Impurity effects in dendritic soldification. Phys. Rev. A 30, 3147–3155.

Karma, A., Lee, Y. H., Plapp, M., 2000. Three-dimensional dendrite-tip morphology at low undercooling. Phys. Rev. E 61 (4), 3996–4006.

Karma, A., Pelce, P., 1990. Stability of an array of deep cells in directional soldification. Phys. Rev. A 41 (12), 6741–6748.

Karma, A., Rappel, W.-J., 1998. Quantitative phase-field modeling of dendritic growth in two and three dimensions. Phys. Rev. E 57 (4), 4323–4349.

Katsaros, K., Liu, W. T., Businger, J. A., Tillman, J. E., 1977. Heat transport and thermal structure in the interfacial boundary layer measured in an open tank. J. Fluid Mech. 83 (2), 311–335.

Kaufmann, D. W., 1960. Sodium Chloride: The production and properties of salt and brine. Reinhold Publishing, Ch. Physical properties of sodium chloride in crystal, gas, and aqueous solution states, pp. 587–626.

Kawamura, T., 1988. Observations of the internal structure of sea ice by X Ray computed tomography. J. Geophys. Res. 93 (C3), 2343–2350.

Kawamura, T., Granskog, M. A., Ehn, J., Martma, T., Lindfors, A., Ishikawa, N., Shirasawa, K., Leppäranta, M., Vaikmäe, R., Dec. 2002. Study on brackish ice in the Gulf
of Finland. In: Ice in the Environment: Proceedings of the 16th IAHR International
Symposium on Ice. Int. Association of Hydraulic Engineering and Research, Dunedin,
New Zealand, pp. 165–171.

Kerr, R. C., Woods, A. W., Worster, M. G., Huppert, H. E., 1989. Disequilibrium and
macrosegregation during solidification of binary melt. Nature 340, 357–362.

Kerr, R. C., Woods, A. W., Worster, M. G., Huppert, H. E., 1990a. Solidification of an
alloy cooled from above part 1. Equilibrium growth. J. Fluid Mech. 216, 323–342.

Kerr, R. C., Woods, A. W., Worster, M. G., Huppert, H. E., 1990b. Solidification of an
alloy cooled from above part 2. Non-equilibrium interfacial kinetics. J. Fluid Mech.
217, 331–348.

Kerszberg, M., 1983a. Pattern formation in directional solidfication. Phys. Rev. B 27 (11),
6796–6810.

Kerszberg, M., 1983b. Pattern selection in directional solidfication. Phys. Rev. B 28 (1),
247–254.

Kessler, D. A., Koplik, J., Levine, H., 1988. Pattern selection in fingered growth phenomena. Adv. Physics 37 (3), 255–339.

Kessler, D. A., Levine, H., 1989. Steady-state cellular growth during directional solidification. Physical Rev. A 39 (6), 3041–3051.

Ketcham, W. M., Hobbs, P. V., 1967. The preferred orientation in the growth of ice from
the melt. J. Crystal Growth 1, 263–267.

Ketcham, W. M., Hobbs, P. V., 1969. An experimental determination of the surface
energies of ice. Phil. Mag. 19, 1161–1173.

Killawee, J. A., Fairchild, I. J., Tison, J.-L., Janssens, L., Lorrain, R., 1998. Segregation of
solutes and gases in experimental freezing of dilute solutions: Implications for natural
glacial systems. Geochimica and Cosmochim. Acta 62 (23/24), 3637–3655.

Kirgintsev, A. N., Shavinskii, B. M., 1969a. Directed crystallization of aqueous solutions of potassium chloride. Communication 1: crystallization without forced mixing.
Russian Chemical Bulletin 18 (7), 1329–1334.

Kirgintsev, A. N., Shavinskii, B. M., 1969b. Directed crystallization of aqueous solutions of potassium chloride. Communication 2: crystallization with forced mixing. Russian Chemical Bulletin 10 (8), 1528–1531.

Kiselev, S. B., Ely, J. F., 2001. Curvature effect on the physical boundary of metastable states in liquids. Physica A 299, 357–370.

Kiselev, S. B., Ely, J. F., 2002. Parametric crossover model and physical limit of stability in supercooled water. J. Chem. Phys. 116 (13), 5657–5664.

Kiselev, S. B., Ely, J. F., 2003. Physical limit of stability in supercooled D_2O and D_2O+H_2O mixtures. J. Chem. Phys. 118 (2), 680–689.

Knight, C. A., 1962a. Polygonization of aged sea ice. J. Geology 70, 240–246.

Knight, C. A., 1962b. Studies of arctic lake ice. J. Glaciol. 4 (33), 319–335.

Knight, C. A., Knight, N. C., 1972. Superheated ice: true compression fractures and fast internal melting. Science 10 (4061), 613–614.

Koch, L., 1945. The East Greenland ice. Vol. 130 of Meddelser om Grönland. C.- A. Reitzels.

Koch, W., 1924. Spezifisches Gewicht und spezifische Wärme der Volumeneinheit der Lösungen von Natrium-, Calcium,- und Magnesiumchlorid bei tiefen und mittleren Temperaturen. Zeitschr. ges. Kälte-Industr.. 31 (9), 105–108.

Koerber, C., 1988. Phenomena at the advancing ice-liquid interface: solutes, particles and biological cells. Quart. Rev. Biophys. 21 (2), 229–228.

Koo, K.-K., Ananth, R., Gill, W. N., 1991. Tip splitting in dendritic growth of ice crystals. Phys. Rev. A 44 (6), 3782–3790.

Koo, K.-K., Ananth, R., Gill, W. N., 1992. Thermal convection, morphological stability and the dendritic growth of crystals. AIChE Journal 38 (6), 945–954.

Koop, T., 2004. Homogeneous ice nucleation in water and aqueous solutions. Z. Phys. Chem. 218, 1231–1259.

Koop, T., Kapilashrami, A., Molina, L. T., Molina, M. J., 2000. Phase transitions of sea-salt/water mixtures at low temperatures: Implications for ozone chemistry in the polar marine boundary layer. J. Geophys. Res. 105 (D21), 26393–26402.

Kopczynski, P., Rappel, W.-J., Karma, A., 1996. Critical role in crystalline anisotropy in the stability of cellular array structures in directional solidification. Phys. Rev. Letters 77 (16), 3387–3390.

Koponen, A., Kataja, M., Timonen, J., 1997. Permeability and effective porosity of porous media. Phys. Ev. E 56 (3), 3319–3325.

Körber, C., Scheiwe, M. W., 1983. Observations on non-planar freezing of aqueous solutions. J. Cryst. Growth 61, 307–316.

Körber, C., Scheiwe, M. W., Wollhöver, K., 1983. Solute polarization during planar freezing of aqueous solutions. Int.J. Heat Mass Transf. 26 (8), 1241.1253.

Korosi, A., Fabuss, B. M., 1968. Viscosities of binary aqueous solutions of NaCl, KCl, Na2SO4 and MgSO4. Journal of Chemical Engineering Data 13, 548–552.

Kotler, G. R., Tarshis, L. R., 1968. On the dendritic growth of pure materials. J. Cryst. Growth 3-4, 603–610.

Kovacs, A., 1996. Sea ice: Part I. Bulk salinity versus flow thickness. CRREL Report 96-7, U.S. Army Cold Regions Research and Engineering Laboratory.

Krishnamurti, R., 1970. On the transition to turbulent convection. Part 2. The transtion to time-dependent flow. J. Fluid Mech. 42 (2), 309–320.

Krümmel, O., 1907. Handbuch der Ozeanographie. Vol. 1. J. Engelhorn, Stuttgart, 526 pp.

Kubelka, P., Prokscha, R., 1944. Eine neue Methode zur Bestimmung der Oberflächenspannung von Kristallen. Kolloid Zeitschr. 109 (2), 79–85.

Kubitchek, J. P., Weidman, P. D., 2003. Stability of a fluid-saturated porous medium heated from below by forced convection. Int. J. Heat Mass Transf. 46 (46), 3697–3705.

Kumai, M., Itagaki, K., 1953. Cinematographic study of ice crystal formation in water. J. Faculty of Science, Hokkaido, Japan Ser. 2 4, 235–246.

Kurowski, P., Cheveigne, S., Faivre, G., Guthmann, C., 1989. Cusp instability in cellular growth. J. Phys. France 50, 3007–3019.

Kurowski, P., Guthman, C., de Cheveigne, S., 1990. Shapes, wave length selection, and cellular-dendritic transition in directional soldification. Phys. Rev. A 42 (12), 7368–7376.

Kurz, W., Fisher, D. J., 1984. Fundamentals of Solidification. Trans Tech Publications, 247 pp.

Kusunoki, K., 1957. On the method of sampling of sea ice. In: 1st Proc. UNESCO Symposium on Physical Ocenaography. UNESCO, pp. 38–42.

Kvajic, G., Brajovic, V., 1970. Segregation of (134 Cs)+ impurity during growth of polycrystalline ice. Canad. J. Physics 48, 2877–2887.

Kvajic, G., Brajovic, V., 1971. Anisotropic segregation of K+ by dendritic ice crystals. J. Cryst. Growth 11, 73–76.

Kvajic, G., Brajovic, V., Pounder, E. R., 1971. Instability of a smoth-planar solid-liquid interface on an ice crystal growing from a melt. Can. J. Physics 49, 2636–2645.

Kvernvold, O., Tyvand, P. A., 1979. Nonlinear thermal convection in anisotropic porous media. J. Fluid Mech. 90, 609–624.

LaCombe, J. C., Koss, M. B., Corrigan, D. C., Lupulescu, A. O., Tennenhouse, L. A., Glicksman, M. E., 1999. Implications of the interface shape on steady state dendritic crystal growth. J. Cryst. Growth 206, 331–344.

Lake, R. A., Lewis, E. L., 1970. Salt rejection by sea ice during growth. J. Geophys. Res. 75 (3), 583–597.

Lamb, S. H., 1932. Hydrodynamics, 6th Edition. Dover Publications, New York, 738 pp.

Lan, C. W., Chang, Y. C., 2003. Efficient adaptive phase field simulation of directional solidification of a binary alloy. J. Cryst. Growth 250, 525–537.

Lange, M. A., 1988. Basic properties of Antarctic sea ice as revealed by textural anaylsis of cores. Ann. Glaciol. 10, 101.

Langer, J. S., 1980a. Eutectic solidfication and marginal stability. Phys. Rev. Letters 44 (15), 1023–1026.

Langer, J. S., 1980b. Instabilities and pattern formation in crystal growth. Rev. Mod. Phys. 52 (1), 1–28.

Langer, J. S., 1987. Chance and matter. North-Holland, New York, pp. 629–711.

Langer, J. S., 1989. Dendrites, viscous fingers, and the theory of pattern formation. Science 243, 4895.

Langer, J. S., Müller-Krumbhaar, H., 1978. Theory of dendritic growth - I. Elements of a stability analysis. Acta Metallurgica 26, 1681–1687.

Langer, J. S., Sekerka, R. F., Fujioka, T., 1978. Evidence for a universal law of dendritic growth rates. J. Crystal Growth 44, 414–418.

Langhorne, P., 1979. Crystal anisotropy in sea ice. Tech. rep., St. Johns, Newfoundland, Canada.

Langhorne, P. J., 1986. Alignment of crystals in sea ice under fluid motion. Cold Regions Sci. Techn. 12, 197–214.

Lapwood, E. R., 1948. Convection of a fluid in a porous medium. J. Cambridge Philos. Soc. 44, 508–521.

Lees, C. H., 1914. On the flow of viscous fluids through smooth circular pipes. Proc. Royal Soc. London A 91 (623), 46–53.

Leppäranta, M., 2005. The Drift of Sea Ice. Springer Praxis Books Geophysical Sciences. Springer, Berlin, 266 pp.

Levich, V. G., 1962. Physicochemical hydrodynamics. Prentice-Hall, Englewood Cliffs (first Russian edition 1952), 699 pp.

Levich, V. G., 1967. Theory of macroscopic kinetics of heterogenous and homogeneous-heterogeneous processes. Ann. Rev. Phys. Chemistr. 18, 153–176.

Levin, R. L., 1981. The freezing of finite domain aqueous solutions: solute redistribution. Int. J. Heat Mass Transfer 24 (9), 1443–1445.

Lewis, E. L., 1967. Heat flow through winter ice. In: Oura, H. (Ed.), Physics of snow and ice: International Conference on Low Temperature Science. Institute for Low Temperature Science, Sapporo, Japan, pp. 611–631.

Li, Q., Beckermann, C., 2002. Modeling of free dendritic growth of succinonitrile-acetone alloys with thermosolutal melt convection. J. Cryst. Growth 236, 482–498.

Light, B., Maykut, G. A., Grenfell, T. C., 2003. Effects of temperature on microstructure of first-year Arctic sea ice. J. Geophys. Res. 108 (C2), 3051–3066.

Lipton, J., Glicksman, M., Kurz, W., 1984. Dendritic growth into undercooled alloy melts. Mater. Sci. Engineeting 65 (1), 57–63.

Lister, C. R. B., 1974. On the penetration of water into hot rock. Geophys. J. R. Astron. Soc. 39, 465–509.

Liu, L. X., Kirkaldy, J. S., 1995. Thin film forced velocity cells and cellular dendrites I. Eexperiments. Acta Metall. Mater. 43 (8), 2891–2904.

Lock, G. S. H., 1990. The Growth and Decay of Ice. Cambridge University Press, Cambridge, 434 pp.

Lofgren, G., Weeks, W. F., 1969a. Effect of growth parameters on substructure spacing in NnaCl ice. Tech. Rep. CRREL Research Report 52, Cold Regions Research and Engineering Laboratory.

Lofgren, G., Weeks, W. F., 1969b. Effect of growth parameters on substructure spacing in NnaCl ice. J. Glaciol. 8 (52), 153–164.

Loper, D. E., Roberts, P. H., 2001. Mush-chimney convection. Stud. Appl. Math. 106, 187–227.

Lu, J. W., Chen, F., 1997. Assessment of mathematical models for the flow in directional solidification. J. Cryst. growth 171, 601–613.

Lu, S.-Z., Hunt, J. D., 1992. A numerical analysis of dendritic and cellular array growth: the sapcing adjustment mechanisms. J. Crystal Growth 123, 17–34.

Ma, D., 2002. Modeling of primary spacing selection in dendrite arrays during directional solidification. Metall. Mater. Trans. B 33, 223–233.

Mae, S., 1976. The freezing of small Tyndall figures in ice. J. Glaciol. 17, 111–116.

Maeno, N., 1967. Air bubble formation in ice crystals. In: Physics of Snow and ice. Vol. 1. Hokkaido University, Sapporo, Japan, pp. 207–218.

Mahler, E. G., Schlechter, R. S., 1970. The stability of a fluid layer with gas absorption. Chem. Eng. Science 25, 955–968.

Makkonen, L., 2000. Spacing in solidification of dendritic arrays. J. Crystal Growth 208, 772–778.

Maksym, T., Jeffries, M. O., 2000. A one-dimensional percolation model of flooding and snow ice formation on antarctic sea ice. J. Geophys. Res. 105 (C11), 26313–26331.

Malkus, W. V. R., 1954. Discrete transitions in turbulent convection. Proc. Roy. Soc. A 225, 185–195.

Malmgren, F., 1927. On the properties of sea ice. Vol. 1, No. 5 of The Norwegian North Polar Expedition with the Maud 1918-1925, Scientific Results. pp. 1–67, dissertation.

Malo, B. A., Baker, R. A., 1968. Cationic concentration by freezing. Vol. 73 of Advances in Chemistry. Amer. Chem. Soc., Ch. 8, pp. 149–163.

Marcet, A., 1819. On the specific gravity, and temperature of sea waters, in different parts of the ocean, and in particular seas; with some account of their saline contents. Phil. Trans. Royal Soc. London 109, 161–208.

Marion, G. M., Farren, R. E., 1999. Mineral solubilities in the Na-K-Mg-Ca-Cl-SO$_4$-% textscH$_2$O system: a re-evaulation of the sulfate chemistry. Geochim. Cosmochim. Acta 63 (9), 1305–1318.

Marion, G. M., Farren, R. E., Komrowski, A. J., 1999. Alternative pathways for seawater freezing. Cold Reg. Sci. Technol. 29, 259–266.

Martin, S., Drucker, R., Forth, M., 1995. A laboratory study of frost flower growth on the surface of young ice. J. Gephys. Res. 100 (C4), 7027–7036.

Martin, S., Yu, Y. Y., Drucker, R., 1996. The temperature dependence of frost flower growth on laboratory sea ice and the effect of frost flowers on infrared observations of the surface. J. Gephys. Res. 101 (C5), 12111–12125.

Matsuoka, Y., Kakumoto, M., Ishihara, K. N., Shingu, P. H., 2001. Desalination by macrosegregation. J. Japan Inst. Metals 65 (9), 791–794.

Maxworthy, T., 1989. Experimental study of interface stability in a Hele-shaw cell. Phys. Rev. A 39, 5863–5866.

Maykut, G. A., 1986. The surface heat and mass balance. Vol. 146 of NATO ASI Series. Plenum Press, Ch. 5, pp. 395–463, ed. by N. Untersteiner.

Maykut, G. A., Light, B., 1995. Refractive-index measurments in freezing sea-ice and sodium chloride brines. Applied Optics 34 (6), 950–961.

Maykut, G. A., Untersteiner, N., 1971. Some results from a time-dependent model of sea ice. J. Geophys. Res. 76 (6), 1550–1574.

McCay, T. D., McCay, M. H., Lowry, S. A., Smith, L. M., 1989. Convective instabilities during directional soldification. J. Thermophysics 3, 345–350.

McDonald, J. E., 1953. Homogeneous nucleation of supercooled water drops. J. Meteor. 10, 416–433.

McGuiness, M. J., Trodahl, H. J., Collins, K., Haskell, T. G., 1998. Non-linear thermal transport and brine convection in first-year sea ice. Ann. Glaciol. 27, 471–476.

McLaughlin, E., 1964. The thermal conductivity of liquids and dense gases. Chemical Review 64, 389–428.

McLean, J. W., Saffman, P. G., 1981. The effect of surface tension on the shape of fingers in a hele shaw cell. J. Fluid Mech. 102, 455–469.

Meese, D. A., 1989. The chemical and structural properties of sea ice in the southern Beaufort Sea. CRREL Report 89-25, Cold Regions Engineering Laboratory.

Meirmanov, A., 1992. The Stefan problem. Walter de Gruyter, Berlin, transl. from Russian by M. Neizgodka and A. Crowley, 244 pp.

Melnichenko, N. A., 1979. Study of the temperature-dependence of the brine content in sea ice by the pulsed NMR method. Oceanology 19 (5), 535–537.

Melnikov, I., 1995. An in situ experimental study of young ice formation on an Antarctic lead. J. Geophys. Res. C3 (100), 4673–4680.

Melnikov, I. A., 1997. The Arctic Sea Ice Ecosystem. Gordon and Breach, 204 pp.

Menzel, M. I., 2002. Multi-nuclear NMR in contaminated sea ice. Ph.D. thesis, RWTH Aachen.

Menzel, M. I., Han, S. I., Stapf, S., Blümich, B., 2000. NMR characterization of the pore structure and anisotropic self-diffusion in salt water ice. J. Magn. Reson. 143, 376–381.

Merchant, G. J., Davis, S. H., 1989. Directional solidification near minimum c_∞: two-dimensional isolas and multiple solutions. Phys. Rev. B 40 (16), 1140–1152.

Michaels, A. S., Brian, P. L. T., Sperry, P. R., 1966. Impurity effects on the basal plane solidification kinetics of supercooled water. J. Appl. Phys. 37 (13), 4649–4661.

Michel, B., 1978. Ice Mechanics. Les Presses De L'Université Laval, Québec, 499 pp.

Michel, B., Ramseier, R. O., 1971. Classification of river and lake ice. Can. Geotechnical J. 8, 36–45.

Millero, F. J., 1974. Seawater as a multicomponent electrolyte solution. Vol. Volume 5 of Marin Chemistry. Wiley-Interscience, Ch. I, pp. 3–80.

Millero, F. J., 1996. Chemical Oceanography. CRC Press, Boca Raton, pp. 469.

Millero, F. J., Leung, W. H., 1976. The thermodynamics of seawater at one atmosphere. Am. J. Sci. 276, 1035–1077.

Millero, F. J., Poisson, A., 1981. International one-atmosphere equation of state of sea-water. Deep-Sea Research 28 A, 625–629.

Millero, F. J., Sohn, M. L., 1992. Chemical Oceanography. CRC Press, Boca Raton, 531 pp.

Milosevic-Kvajic, M., Kvajic, G., Brajovic, V., 1973. Freezing paramters for ice monocrystals and the separation of basal platelets. Can. J. Phys. 51, 837–842.

Mironenko, M. V., Boitnott, G. E., Grant, S. A., Sletten, R. S., 2001. Experimental determination of the volumetric properties of NaCl solutions to 253 k. J. Phys. Chem. B 41 (18), 9909–9912.

Molho, P., Simon, A. J., Libchaber, A., 1990. Peclet number and crystal growth in a channel. Phys. Rev. A 42 (2), 904–910.

Mollendorf, J. C., Arif, H., Ajiniran, E. B., 1984. Developing flow and transport above a suddenly heated horizontal surface in water. Int. J. Heat Transfer 27 (2), 273–289.

Moss, E. L., 1878. Observations on Arctic sea-water and ice. Proc. Royal Soc. London 27, 544–559.

Muehlbauer, J. C., Sunderland, J. E., 1965. Heat conduction with freezing and melting. Appl. Mech. Rev. 18 (12), 951–959.

Muguruma, J., Higuchi, K., 1963. Glaciological studies on ice island T-3. J. Glaciol. 4 (36), 709–730.

Mullins, W. W., Sekerka, R. F., 1963. Morphological stability of a particle growing by diffusion or heat flow. J. Appl. Phys. 34 (2), 323–329.

Mullins, W. W., Sekerka, R. F., 1964. Stability of a planar interface during solidification of a dilute binary alloy. J. Appl. Phys. 35 (2), 444–451.

Muschol, M., Liu, D., Cummins, H. Z., 1992. Surface-tension-anisotropy measurements of succinonitrile and pivalic acid: comparison with microscopic solvability theory. Phys. Rev. A 46 (2), 1038–1050.

Nada, H., v. d. Eerden, J. P., Furukawa, Y., 2004. A clear observation of crystal growth of ice from water in a molecular dynamics simulation with a six-site potential model of h2o 266 (1-3), 297–302.

Nagashima, K., Furukawa, Y., 1997a. Nonequilibrium effect of anisotropic interface kientics on the directional growth of ice crystals. J. of Cryst. Growth 171, 577–585.

Nagashima, K., Furukawa, Y., 1997b. Solute distribution in front of an ice/water interface during directional growth of ice crystals and its relationship to interfacial patterns. J. Phys. Chem. 101 (32), 6174–6176.

Nagashima, K., Furukawa, Y., 2000a. Interferometric observation of the effects of gravity on the horizontal growth of ice crystals in a thin growth cell. Physica D 147, 177–186.

Nagashima, K., Furukawa, Y., 2000b. Time development of a solute diffusion field and morphological instability on a planar interface in the directional growth of crystals. J. Crystal Growth 209, 167–174.

Nairne, E., 1776. Experiments on water obtained from the melted ice of sea-water, to ascertain whether it be fresh or not. Phil. Trans. Royal Soc. London 66, 249–256.

Nakawo, M., Sinha, N. K., 1981. Growth rate and salinity profile of first-year sea ice in the high Arctic. J. Glaciol. 27 (96), 315–330.

Nakawo, M., Sinha, N. K., 1984. A note on brine layer spacing of first-year sea ice. Atmos.-Ocean 22 (2), 193–206.

Nakaya, U., 1956. Properties of single crystals of ice, revealed by internal melting. Research Paper 13, SIPRE.

Nelson, K. H., Thompson, T. G., 1954. Deposition of salts from seawater by frigid concentration. J. Marine Res. 13 (2), 166–182.

Nghiem, S. V., Kwok, R., Yueh, S. H., Gow, A. J., Perovich, D. K., Kong, J. A., Hsu, C. C., 1997. Evolution of polarimetric signatures of thin saline ice under constant growth. Radio Science 32 (1), 127–151.

Niedrauer, T. M., Martin, S., 1979. An experimental study of brine drainage and convection in young sea ice. J. Geophys. Res. 84 (C3), 1176–1186.

Nield, D. A., 1964. Surface tension and buoyancy effects in cellular convection. J. Fluid Mech. 19, 341–352.

Nield, D. A., 1967. The thermohaline Rayleigh-Jeffreys problem. J. Fluid Mech. 29 (3), 545–558.

Nield, D. A., 1968. Onset of thermohaline convection in a porous medium. Water Resources Res. 4 (3), 553.560.

Nield, D. A., 1975. The onset of transient convective instability. J. Fluid Mech. 71 (3), 441–454.

Nield, D. A., Bejan, A., 1999. Convection in Porous Media, 2nd Edition. Springer, 546 pp.

Notz, D., Wettlaufer, J. S., Worster, M. G., 2005. A non-destructive method for measuring the salinity and solid fraction of growing ice in situ. J. Glaciol. 51 (172), 159–166.

Oakes, C. H., Bodnar, R. J., Simonson, J. M., 1990. The system $NaCl-CaCl_2-H_2O$: I. the ice liquidus at 1 atm total pressure. Geochim. Cosmochim. Acta 54, 603–610.

O'Callaghan, M. G., Cravalho, E. G., Huggins, C. E., 1980. Instability of the planar freeze front during soldification of an aqueous binary solution. ASME J. Heat Transf. 102, 673–677.

Oldfield, W., 1973. Computer model studies of dendritic growth. Mater. Sci. Engin. 11, 211–218.

Onat, K., Grigull, U., 1970. The onset of convection in a horizontal fluid layer heated from below. Wärme- u. Stoffübertragung 3, 103–113.

Ono, N., 1967. Specific heat and heat of fusion of sea ice. In: Int. Conference on Low Temperature Science. Vol. 1. Ints. of Low Temperature Science, Sapporo, pp. 599–610.

Ono, N., 1968. Specific heat and heat of fusion of sea ice. In: Oura, H. (Ed.), Physics of snow and ice. Vol. 1. Inst. of Low Temperature Science, Hokkaido, Japan, pp. 599–610.

Ono, N., Kasai, T., 1985. Surface layer salinity of young sea ice. Ann. Glaciol. 6, 298–299.

Osipov, Y. A., Zheleznyi, B., Bondarenko, N. F., 1977. The shear viscosity of water supercooled to $-35°$. Russian Journal of Physical Chemistry 51 (5), 1264–1265.

Ozum, B., Kirwan, D. J., 1976. Impurities in ice crystals grown from stirred solutions. AICHE Symp. Ser. 72 (153), 1–.

Palm, E., Weber, J. E., Kvernvold, O., 1972. On steady convection in a porous medium. J. Fluid Mech. 54, 153–161.

Palosuo, E., 1961. Crystal structure of brackish and freshwater ice. In: Int. Assoc. Sci. Hydrol. Vol. 54. Gentbrugge, Belgium, pp. 9–14.

Palosuo, E., Sippola, M., 1963. Crystal orientation in salt-water ice. In: Kingery, W. D. (Ed.), Ice and Snow. M.I.T. Press, pp. 232–236.

Papadimitriou, S., Kennedy, H., Kattner, G., Dieckmann, G. S., Thomas, D. N., 2003. Experimental evidence for carbonate precipitation and CO2 degassing during sea ice formation. Geochim. Cosmochim. Acta 68 (8), 1749–1761.

Parrot, G. F., 1818. Über das Gefrieren des Salzwassers mit Rücksicht auf die Entstehung des Polar-Eises. Ann. d. Physik 57, 144–162.

Paterson, L., 1985. Fingering with miscible fluids in a Hele-Shaw-cell. Phys. Fluids 28, 26–30.

Pearson, J. R. A., 1958. On convection cells induced by surface tension. J. Fluid Mech. 4, 489–500.

Perey, F. G., Pounder, E. R., 1958. Crystal orientation in ice sheets. Can. J. Phys. 36, 494–502.

Perovich, D. K., Gow, A. J., 1991. A statistical description of the microstructure of young sea ice. J. Geophys. Res. 96 (C9), 16943–16953.

Perovich, D. K., Gow, A. J., 1996. A quantitative description of sea ice inclusions. J. Geophys. Res. 101 (C8), 18327–18343.

Perovich, D. K., Grenfell, T. C., 1981. Laboratory study of the optical properties of young sea ice. J. Glaciol. 27 (96), 331–346.

Perovich, D. K., Richter-Menge, J. A., 2000. Ice growth and solar heating in springtime leads. J. Geophys. Res. 105 (C3), 6541–6548.

Petrenko, V. F., Whitworth, R. W., 1999. Physics of Ice. Oxford University Press, 373 pp.

Petrich, C., Langhorne, P. J., Sun, Z. F., 2006. Modelling the interrelationships between permeability, effective porosity and total porosity in sea ice. Cold Reg. Sci. Techn. 44, 131–144.

Pettersson, O., 1883. On the properties of water and ice. Vol. II. Vega-expeditionens vetenskapliga jakttagelser, Stockholm, pp. 247–323.

Pettersson, O., 1907. On the influence of ice-melting upon oceanic circulation. Geograph. J. 30 (3), 273–295.

Phillips, O. M., 1991. Flow and Reactions in Permeable Rocks. Cambridge University Press, 285 pp.

Pitzer, K. S., Peiper, J. C., Bussy, R. H., 1984. Thermodynamic properties of aqueous sodium chloride solutions. J.Phys. Chem. Ref. Data 13 (1), 1–102.

Pocheau, A., Georgelin, M., 1999. Cell tip undercooling in directional solidification. J. Cryst. Growth 206, 215–229.

Pocheau, A., Georgelin, M., 2003. Cellular arrays in binary alloys: from geometry to stability. J. Cryst. Growth 250, 100–106.

Pounder, E. R., 1964. The Physics of Sea Ice. Pergamon Press, 151 pp.

Powell, R. W., 1958. Thermal conductivities and expansion coefficients of water and ice. Adv. Phys. 7, 277–297.

Prescott, P. J., Incropera, F. P., 1996. Convection heat and mass transfer in alloy solidification. Adv. Heat Transfer 28, 231–338.

Prevost, M., Gallez, D., 1986. Nonlinear rupture of thin free liquid films. J. Chem. Phys. 84 (7), 4043–4048.

Prodi, F., Levi, L., 1980. Aging of accreted ice. J. Atmosph. Sci. 37, 1375–1384.

Pruppacher, H. R., 1967. Interpretation of experimentally determined growth rates of ice crystals in supercooled water. J. Chem. Phys. 47, 1807–1813.

Pruppacher, H. R., Klett, J. D., 1997. Microphysics of clouds and precipitation, 2nd Edition. Vol. 18 of Atmospheric and oceanographic sciences library. Kluwer Acadamics Publ., 954 pp.

Quaresma, J. M. V., Santos, C. A., Garcia, A., 2000. Correlation between unsteady-state solidification conditions: dendrite spacings, and mechanical properties of Al-Cu alloys. Metall. Mater. Trans. A 31, 3167–3178.

Quincke, G., 1905a. The formation of ice and the grained structure of glaciers. Proc. Royal Soc. London A 76 (512), 431–439.

Quincke, G., 1905b. Über Eisbildung und Gletscherkorn. Annalen der Physik, Leipzig 18(11), 1–80.

Quincke, G., 1906. The transition from the liquid to the solid state and the foam-structure of matter. Proc. Royal Soc. London A 78 (521), 59–67.

Quintanilla, J., 2001. Measurement of the percolation threshold for fully penetrable disks of different radii. Phys. Rev. E 63, 061108–1.

Rae, J., 1874. On some physical properties of ice; on the transposition of boulders from below to above the ice; and on mammoth remains. Proc. Phys. Soc. London 1, 14–20.

Ragle, R. H., August 1963. Formation of lake ice in a temperate climate. Research report 107, Cold Regions Research and Engineering Laboratory, Hanover, New Hampshire.

Ramires, M. L. V., de Castro, C. A. N., Fareleia, J. M. N. A., Wakeham, W. A., 1994. Thermal conductivity of aqueous sodium chloride solutions. J. Chem. Eng. Data 39, 186–190.

Ramirez, J. C., Beckermann, C., 2003. Evaluation of a Rayleigh-number-based freckle criterion for Pb-Sn alloys and Ni-base superalloys. Metall. Mat. Trans. A 34, 1525–1536.

Ramirez, J. C., Beckermann, C., 2005. Examination of binary alloy free dendritic growth theories with a phase-field model. Acta Mater. 53, 1721–1736.

Ramprasad, N., Bennett, M. J., Brown, R. A., 1988. Wavelength dependence of cells of finite depth in directional solidification. phys. Rev. B 38, 583–592.

Randall, M., Bisson, C. S., 1920. The heat of solution and the partial molal heat content of the constitutents in aqueous solutions of sodium chloride. J. Amer. Chem. Soc. 42, 347–367.

Rasmussen, D. H., MacKenzie, A. P., 1972. Effect of solute on ice-solution interfacial free energy. In: Jellinek, H. H. G. (Ed.), Water structure at the water-polymer interface. Plenum Press, New York, pp. 126–145.

Rayleigh, L., 1879. On the capillary phenomena of jets. Proc. Royal Soc. London A 29, 71–97.

Rayleigh, L., 1916. On convective currents in a horizontal layer of fluid when the higher temperature is on the underside. Phil. Mag. 32, 529–546.

Reid, W. H., Harris, D. L., 1958. Some further results on the Benard problem. Phys. of Fluids 1 (2), 102–110.

Riahi, D. N., 2002. On nonlinear convection in mushy layers. Part 1. Oscillatary modes of convection. J. Fluid Mech. 467, 331–359.

Richardson, C., 1976. Phase relationships in sea ice as a function of temperature. J. Glaciolog. 17 (77), 507–519.

Richardson, C., Keller, E. E., 1966. The brine content of sea ice measured with a nuclear magnetic resonance spectrometer. J. Glaciol. 43 (43), 89–100.

Riedel, L., 1950. Wärmeleitfähigkeitsmessungen an kältetechnisch wichtigen Salzlösungen. Kältetechnik 4, 99–101.

Ringer, W. E., 1906. De verandringen in samenstelling van zeewater bij het bevriezen. Chemisch Weekblad 3 (15), 223–249.

Roberts, P. H., Loper, D. E., Roberts, M. F., 2003. Convective instability of a mushy layer - I: uniform permebaility. Geophys. Astrophys. Fluid Dyn. 97 (2), 97–134.

Robinson, R. A., Stokes, R. H., 1955. Electrolyte Solutions. The measurement and interpretation of conductance, chemical potential and diffusion in solutions of simple electrolytes. Butterworths, London.

Rocha, O. L., A-Siqueira, C., Garcia, A., 2003. Cellular/dendritic transition during unsteady-state unidirectional solidfication solidification of Ssn-Pb alloys. Mater. Sci. Engineering A 347, 59–69.

Rohatgi, P. K., Adams, C. M., 1967a. Freezing rate distributions during uniddirectional solidfication of solutions. Trans. Metall. Soc. AIME 239 (239), 850–857.

Rohatgi, P. K., Adams, C. M., 1967b. Ice-brine dendritic aggregate formed on freezing of aqueous solutions. J. Glaciol. 6 (47), 663–679.

Rohatgi, P. K., Brush, E. J., Jain, S. M., Adams, C. M., 1974. Denrditic structures produced on solidfication of multicomponent aqueous solutions. Mater. Sci. Engineer. 13, 3–18.

Rohrer, G. S., 2005. Influence of anisotropy on grain growth and coarsening. Ann. Rev. Mater. Res. 35, 99–126.

Rubinsky, B., Ikeda, M., 1985. A cryomicroscope using directional solidification for the controlled freezing of biological materials. Cryobiology 22, 55–68.

Rubinstein, L. I., 1971. The Stefan Problem. Vol. 27 of Translations of mathematical monographs. Amer. Mathem. Soc., translated by A. D. Solomon from the Russian, 419 pp.

Ruckenstein, E., Jain, R. K., 1974. Spontaneous rupture of thin liquid films. J. Chem. Soc., Faraday Trans. II 70, 132–147.

Ruedorff, F., 1861. Über das Gefrieren von Wasser aus Salzlösungen. Annalen d. Pys. 114, 63–81.

Ruedorff, F., 1872. Über das Gefrieren der Salzlösungen. Annalen d. Pys. 145, 599–622.

Rümelin, G., 1907. Über die Verdünnugswärme konzentrierter Lösungen. Z. phys. Chemie 58, 449–466.

Rutter, J. W., Chalmers, B., 1953. A prismatic substructure formed during solidification of metals. Can. J. of Phys. 31, 15–39.

Ryan, B. F., 1969. The growth of ice crystals parallel to the basal plane in supercooled water and supercooled metal fluoride solutions. J. Cryst. Growth 5, 284–288.

Ryvlin, A. Y., 1979. Method of forecasting flexural strength of ice cover. Problems of the Arctic and Antarctic 45, 99–108.

Saarloos, W., 2003. Front propagation into unstable states. Physics reports 386, 29–222.

Saffman, P. G., Taylor, G., 1958. The penetration of a fluid into a porous medium or Hele-Shaw cell containing a more viscous fluid. Proc. Roy. Soc. London Ser. A 245.

Saito, T., Ono, N., 1977. Percolation of sea ice. II: Brine drainage channels in young sea ice. Low Temp. Sci. A39, 127–132.

Sanderson, T. J. O., 1988. Ice Mechanics - Risks to Offshore Structures. Graham & Trotman, 253 pp.

Schlichting, H., 1965. Grenzschicht-Theorie. G. Braun, Karlsruhe, 736 pp.

Schoepf, W., 1992. Convection onset for a binary mixture in a porous medium and in a narrow cell: a comparison. J. Fluid Mech. 245, 263–278.

Schulson, E. M., 2001. Brittle failure of ice. Eng. fracture mech. 68, 1839–1887.

Schwarzacher, W., 1959. Pack-ice studies in the Arctic Ocean. J. Geophys. Res. 64 (12), 2357–2367.

Schwerdtfeger, P., 1963. The thermal properties of sea ice. J. Glaciol. 4 (36), 789–807.

Sciortino, F., Poole, P. H., Essmann, U., Stanley, H. E., 1997. Line of incompressibility maxima in the phase diagram of supercooled water. Phys. Rev. E 55 (1), 727–736.

Scoresby, W., 1815. On the Greenland and Polar Ice. Mem. Wernerian Society 2, 328–336.

Seidensticker, R. G., 1966. Comment on the paper by P. Hoekstra, Tt.E. Osterkamp and W. F. Weeks, 'The migration of liquid inclusions in single ice crystals'. J. Geophys. Res. 71 (8), 2180–2181.

Seidensticker, R. G., 1972. Partitioning of HCl in the water-ice system. J. Chem. Phys. 56 (6), 2853–2857.

Sekerka, R., 2004a. Equilibrium growth shapes of crystals: how do they differ and why should we care. Cryst. Res. Technol. 40 (4/5), 291–306.

Sekerka, R. F., 1965. A stability function for explicit evaluation of the Mullins-Sekerka interface stability criteron. J. Appl. Phys. 36 (1), 264–268.

Sekerka, R. F., 1967. A time-dependent theory of satbility of a planar surface during dilute binary alloy solidification. In: Peiser, H. S. (Ed.), Crystal Growth. Pergamon Press, Boston, 20-24 June 1966, pp. 691–702.

Sekerka, R. F., 1968. Morphological stability. J. Cryst. Growth 3-4, 71–81.

Sekerka, R. F., 2004b. Morphology: from sharp interface to phase field models. J. Crystal Growth 264, 530–540.

Selman, J. R., Tobias, C. W., 1978. Mass-transfer measurements by the limiting-current technique. Adv. Chem. Engineer. 10, 211–318.

Shapiro, L. H., Weeks, W. F., 1993. Influence of crystallographic and structural properties on the flexural strength of small sea ice beams. In: J. P. Dempsey, Z. P. Bazant, Y. D. S. R., Shyam, S. (Eds.), Ice Mechanics. Vol. 163. Am Soc. Mech. Engin., pp. 177–188.

Sharp, R. M., Hellawell, A., 1970. Solute distributions at non-planar, solid-liquid growth fronts. II. J. Cryst. Growth 6, 334–340.

Shesteperov, I. A., 1969. Observation of Tsurikov's formula. Okeanologiya 9 (4), 502–504.

Shibkov, A., Zheltov, M., Kolev, A., Kazakov, A., Leonov, A., 2005. Crossover from diffusion-limited to kinetics-limited growth of ice crystals. J. Cryst. Growth 285 (1-2), 215–227.

Shimada, W., Furukawa, Y., 1997. Pattern formation of ice crystals during free growth in supercooled water. J. Phys. Chem. B 101, 6171–6173.

Shumskii, P. A., 1955. Osnovy strukturnogo ledovedeniya (Principles of structural glaciology). Dover Publications, Inc., 1964 translated from Russian, 497 pp.

Siggia, E. D., 1994. High Rayleigh Number convection. Annu. Rev. Fluid. Mech. 26, 137–168.

Sinha, N. K., 1977. Technique for studying structure of sea ice. J. Glac. 18 (79), 315–323.

Sinha, N. K., 1978. Observation of basal dislocations in ice by etching and replicating. J. Glac. 21 (85), 385–395.

Sinha, N. K., Nakawo, M., 1981. Growth of first-year sea ice, Eclipse Sound, Baffin Island, Canada. Can, Geotechn. J. 18, 17–23.

Sinha, N. K., Zhan, C., 1996. Primary dendrite spacing of land-fast polar sea ice. J. Mater. Sci. Letters 15, 2118–2121.

Slack, G. A., 1980. Thermal conductivity of ice. Phys. Rev. B 22 (6), 3065–3071.

Smith, M. K., 1988. Thermal convection during soldification of a pure liquid with variable viscosity. J. Fluid Mech. 188, 547–570.

Smith, V. G., Tiller, W. A., Rutter, J. W., 1955. A mathematical analysis of solute redistribution during solidification. Can. J. Phys. 33, 723–744.

Snyder, D., Tait, S., 1998. A flow-front instability in viscous gravity currents. J. Fluid Mech. 369, 1–21.

Solomon, T. H., Hartley, R. R., Lee, A. T., 1999. Aggregation and chimney formation during solidification of ammonium chloride. Phys. Rev. E 60 (3), 3063–3071.

Somboonsuk, K., Mason, J. T., Trivedi, R., 1984. Interdendritic spacing: part I. Experimental studies. Metall. Transactions A 15, 967–975.

Souchez, R., Tison, J. L., Jouzel, J., 1988. Deuterium concentration and growth rate of Antarctic first-year ice. Geophys. Res. Letters 15 (12), 1385–1388.

Sparrow, E. M., Goldstein, R. J., Jonsson, V. K., 1964. Thermal stability in a horizontal fluid layer: effect of boundary conditions and nonlinear temperature profile. J. Fluid Mech. 18, 513–528.

Sparrow, E. M., Husar, R. B., Goldstein, R. J., 1970. Observations and other characteristcs of thermals. J. Fluid Mech. 41 (4), 793–800.

Speedy, R. J., 1987. Thermodynamic properties of suppercooled water at 1 atm. J. Phys. Chem. 91, 3354–3358.

Spencer, B. J., Huppert, H. E., 1998. On the soldification of dendritic arrays: Selection of the tip characteristics of slender needel crystals by array interactions. Acta Mater. 46 (8), 2645–2662.

Spencer, B. J., Huppert, H. H., 1999. The relationship between dendrite tip characteristics and dendrite spacings in alloy directional solidification. J. Crystal Growth 200, 287–296.

Spencer, R. J., Møller, N., Weare, J. H., 1990. The prediction of mineral solubilities in natural waters. Geochim. Cosmochim. Acta 54, 575–590.

Spinelli, J. E., Rosa, D. M., Ferreira, I. L., Garcia, A., 2004. Influence of melt convection on dendrtitic spacings of downward unsteady-state directionally solidified Al-Cu alloys. Mater. Sci. Engineering A 283, 271–282.

Stakelbeck, H., Plank, R., 1929. Über die Zähigkeit von Clornatrium-, Chlorkalzium- und Chlormagnesiumlösungen in Abhängigkeit von der Temperatur und Konzentration. Zeitschrift für die gesamte Kälte-Industrie 36, 105–112, 133–135.

Stander, E., 1984. Petrography of vertically sectioned first-year ice core. In: Int. Offshore Mechanics and Arctic Engineering Symp. Vol. 3. New Orleans, pp. 196–203.

Stander, E., 1985. Use of subgrains as paleostress indicators in first-year sea ice. In: Int. Conf. on Port and Ocean Engineering. Vol. 1. Narssarssuaq, Greenland, pp. 168–176.

Stander, E., Michel, B., 1989. The effect of fluid flow on the development of preferred orientations in sea ice: laboratory experiments. Cold Regions Sci. Tech. 17, 153–161.

Stauffer, D., 1991. Introduction to Percolation Theory, 2nd Edition. Taylor & Francis.

Stefan, J., 1889. Über die Theorie der Eisbildung, insbesondere über die Eisbildung im Polarmeere. Sitzungsberichte der kaiserlichen Akademie der Wissenschaften, Mathematisch-naturwissenschaftliche Classe, Abtheilung IIa 98, 965–983.

Steffen, K., Januar 1984. Oberflächentemperatur einer arktischen Polynya: North Water in winter. Ph.D. thesis, Eidgenössische Technische Hochschule Zürich, diss. ETH Nr. 7485.

Straub, J., Rosner, N., Grigull, U., 1980. Oberflächenspannung von leichtem und schwerem Wasser. Wärme- und Stoffübertragung 13, 241–252.

Surdo, A. L., Alzola, E. M., Millero, F. J., 1982. The $(p, \textsc{V}, \textsc{T})$ properties of conncentrated aqueous electrolytes. i densities and apparent molar volumes of nacl, na2so4, mgcl2 and mgso4 solutions from 0.1 mol/kg to saturation and from 273.15 to 323.15 k. J. Chem. Thermodynamics 14, 649.662.

Sverdrup, H. U., Johnson, M. W., Fleming, R. H., 1946. The oceans - their physics, chemistry and general biology. Prentice-Hall, New York, 1060 pp.

Tabata, T., Ono, N., 1962. On the crystallographic study of several kinds of ices. Low Temperature Science A 20, 199–214.

Taborek, P., 1985. Nucleation in emulsified supercooled water. Phys. Rev. B 32 (9), 5902–5906.

Tait, S., Jaupard, C., 1989. Compositional convection in viscous melts. Nature 338, 571–574.

Tait, S., Jaupard, C., 1992a. Compositional convection in a reactive crystalline mush and melt differentiation. J. Geophys. Res. 97 (B5), 6735–6756.

Tait, S., Jaupard, C., 1992b. The planform of compositional convection and chimney formation in a mushy layer. Nature 359, 406–408.

Takaki, T., Fukuoka, T., Tomita, Y., 2005. Phase-field simulation during directional solidification of a binary alloy using adaptive finite element method. J. Cryst. Growth 283, 263–278.

Tammann, G., Büchner, A., 1935. Die Unterkühlungsfähigkeit des Wassers und die lineare Kristallisaltionsgeschwindigkeit des Eises in wässrigen Lösungen. Z. f. anorgan. Chemie 222, 371–381.

Tavman, I. H., 1996. Effective thermal conductivity of granular porous materials. Int. Comm. Heat Mass Transf. 23 (2), 169–176.

Tennenhouse, L. A., Koss, M. B., LaCombe, J. C., Glicksman, M. E., 1997. Use of microgravity to interpret dendritic growth kinetics at small supercoolings. J. Cryst. Growth 174, 82–89.

Terwilliger, J. P., Dizio, S. F., 1970. Salt rejection phenomena in the freezing of saline solutions. Chem. Eng. Science 25, 1331–1349.

Thibert, E., Domine, F., 1997. Thermodynamics and kinetics of the solid solution of HCl in ice. J. Phys. Chem. 101, 3554–3565.

Thomas, D., Dieckmann, G. S., 2003. Sea Ice: An Introduction to its Physics, Chemistry, Biology and Geology. Blackwell, 402 pp.

Thompson, C. V., 1990. Grain growth in thin films. Ann. Rev. Mater. Sci. 20, 245–268.

Thurmond, V. L., Brass, G. W., August 1987. Geochemistry of freezing brines. Crrel report 87-13.

Tiller, W. A., 1962. Effect of grain boundaries on solute partitioning during prgressive solidification. J. Appl. Physics 33 (10), 3106–3107.

Tiller, W. A., 1991a. The science of crystallization: macroscopic phenomena and defect generation. Cambridge University Press, 484 pp.

Tiller, W. A., 1991b. The science of crystallization: microscopic interfacial phenomena. Cambridge University Press, 391 pp.

Tiller, W. A., Jackson, K. A., Rutter, J. W., Chalmers, B., 1953. The redistribution of solute atoms during solidification of metals. Acta Metall. 1, 428–437.

Tirmizi, S. H., Gill, W. N., 1987. Effect of natural convection on growth velocity and morphology of dendritic ice crystals. J. Crystal Growth 85, 488–502.

Tirmizi, S. H., Gill, W. N., 1989. Experimental investigation of the dynamics of spontaneous pattern formation during dendritic ice crystal growth. J. Crystal Growth 96, 277–292.

Tischenko, P. Y., 1994. A chemical model of seawater based on Pitzer's method. Oceanology 34 (1), 40–44.

Tison, J., Haas, C., Gowing, M. M., Sleewaegen, S., Bernard, A., 2002. Tank study of physico-chemical controls on gas content and composition during growth of young ice. J. Glac. 48 (161), 177–191.

Tison, J. L., Verbeke, V., 2001. Chlorinity/salinity distribution patterns in experimental granular sea ice. Ann. Glaciol. 33, 13–20.

Tolman, R. C., 1947. The effect of droplet size on surface tension. The journal of chemical physics 17, 333–337.

Trivedi, R., 1984. Interdendritic spacing: part II. A comparison of theory and experiment. Metall. Transactions A 15, 977–982.

Trivedi, R., Kurz, W., 1994. Dendritic growth. Int. Mater. Rev. 39 (2), 49–74.

Trivedi, R., Somboonsuk, K., 1984. Constrained dendritic growth and spacing. Mater. Sci. Engineering 65, 65–74.

Trivedi, R., Somboonsuk, K., 1985. Pattern formation during the directional solidfication of binary systems. Acta Metall. 33 (6), 1061–1068.

Trodahl, H. J., McGuiness, M. J., Langhorne, P. J., Collins, K., Pantoja, A. E., Smith, I. J., Haskell, T. G., 2000. Heat transport in McMurdo first-year sea ice. J. Geophys. Res. 105 (C5), 11347–11358.

Trodahl, H. J., Wilkinson, S. O. F., McGuiness, M. J., Haskell, T. G., 2001. Thermal conductivity of sea ice; dependence on temperature and depth. Gephys. Res. Lett. 28 (7), 1279–1282.

Tsurikov, V. L., 1939. Problema prochnosti morskogo l'da (the problem of ice strength). Severnyi Morskoi Put 16, 45–74.

Tsurikov, V. L., 1965. Formation of the ionic composition and salinity of sea ice. Okeanologiia 5, 463–472.

Tsurikov, V. L., 1974. Statistics of salt composition in sea ice. Okeanologiya 14 (3), 360–367.

Tsurikov, V. L., Tsurikova, A. P., 1972. The brine content of sea ice (statement of the problem). Oceanology 12 (5), 663–672.

Tucker, W. B., Gow, A. J., Richter, J. A., 1984. On small-scale horizontal variations of salinity in first-year ice. J. Geophys. Res. 89 (C10), 6505–6514.

Turnbull, D., 1950. Formation of crystal nuclei in liquid metals. J. Appl. Phys. 21, 1022–1028.

Turner, J. S., 1973. Buoyancy Effects in Fluids. Cambridge University Press, 367 pp.

Tyndall, J., 1858. On some physical properties of ice. Philos. Trans. Royal Soc. London 148, 211–229.

Tyndall, J., 1859. On the veined structure of glaciers, with observations upon white ice-seams, air-bubbles and dirt-bands, and remarks upon glaicer theories. Philos. Trans. Royal Soc. London 149, 279–307.

Tyndall, J., 1872. The forms of water in clouds and rivers, ice and glaciers. Appleton, New York.

Tyrrell, H. J. V., Harris, K. R., 1984. Diffusion in liquids. Butterworths, London, 448 pp.

Untersteiner, N., 1968. Natural desalination and equlibirium salinity profile of perennial sea ice. J. Geophys. Res. 73 (4), 1251–1257.

Untersteiner, N., 1986. The Geophysics of Sea Ice. Vol. 146 of NATO ASI Series, B: Physics. Plenum Press, New York, 1196 pp.

Vartak, B., Yeckel, A., Derby, J., 2005. On the validity of boundary layer analysis for flow and mass transfer caused by rotation during solution growth of large, single crystals. J. Cryst. Growth 283, 479–489.

Vitagliano, V., Lyons, P. A., 1956. Diffusion coefficients for aqueous solutions of sodium chloride and barium chloride. J. Amer. Chem. Soc. 78, 1549–1552.

Vitanov, N. K., 2000. Convective heat transport in a fluid layer of infinite Prandtl number: upper bounds for the case of rigid lower boundary conditions and stress-free upper boundary. Eur. Phys. J. B 15, 349–355.

Vlahakis, J. G., Barduhn, A. J., 1974. Growth rate of an ice crystal in flowing water and salt solutions. AIChE Journal 20 (3), 581–591.

Voorhees, P. W., 1992. Ostwald ripening of two phase mixtures. Ann. Rev. Mater. Sci. 22, 197–215.

Wadhams, P., 2001. Ice in the Ocean. CRC Press, 351 pp.

Wakatsuchi, M., 1974a. Experiments on the growth of sea ice and the rejection of brine. Low Temp. Sci. A32, 195–205.

Wakatsuchi, M., 1974b. On sea ice near Syowa Station, Antarctica. II. Salinity profile of sea ice. Low Temp. Sci. A40, 119–125.

Wakatsuchi, M., 1977. Experiments on haline convection induced by freezing. Low Temp. Sci. A35, 249–258.

Wakatsuchi, M., 1983. Brine exclusion processes from growing sea ice. Contrib. Inst. Low Temp. Sci. A33, 29–65.

Wakatsuchi, M., Kawamura, T., 1987. Formation process of brine drainage channels in sea ice. J. Geophys. Res. 92 (C5), 7195–7197.

Wakatsuchi, M., Ono, N., 1983. Measurements of salinity and volume of brine excluded from growing sea ice. J. Geophys. Res. 88 (C5), 2943–2951.

Wakatsuchi, M., Saito, T., 1985. On brine drainage channels of young sea ice. Ann. Glaciol. 6, 200–202.

Walker, D., 1859. Ice observations. Proc. Royal Soc. London 9, 609–611.

Warren, J. A., Langer, J. S., 1990. Stability of dendritic arrays. Phys. Rev. A 42 (6), 3518–3525.

Warren, J. A., Langer, J. S., 1993. Prediction of dendritic spacings in a directional-soldification experiment. Phys. Rev. E 47 (4), 2702–2712.

Weber, H., 1901. Die partiellen Differentialgleichungen der mathematischen Physik. Nach Riemann's Vorlesungen. Vol. 2. Vieweg und Sohn, Braunschweig, Ch. 6(45), Vordringen des Frostes, §45, 118-122.

Weber, J. E., 1977. Heat and salt transfer associated with formation of sea ice. Tellus 29, 151-160.

Weeks, W. F., 1958. The structure of sea ice: a progress report. In: Arctic sea ice. Vol. 598. Proc. Conf.on Arctic Sea Ice, Natl. Acad. Sci., pp. 96-98.

Weeks, W. F., 1961. Studies of salt ice I: The tensile strength of NaCl ice. Crrel research report 80, Cold Regions Research and Engineering Laboratory.

Weeks, W. F., 1998a. Growth conditions and the structure and properties of sea ice. Vol. 1. Department of Geophysics, University of Helsinki, Finland, pp. 25-104.

Weeks, W. F., 1998b. On the history of research on sea ice. Vol. 1. Department of Geophysics, University of Helsinki, Finland, pp. 1-24.

Weeks, W. F., Ackley, S. F., 1986. The geophysics of sea ice. In: The Geophysics of Sea Ice. Vol. 146 of NATO ASI Series. Plenum Press, pp. 9-164, ed. by N. Untersteiner.

Weeks, W. F., Anderson, D. L., 1958. An experimental study of strength of young ice. Trans. Amer. Geophys. Union 39 (4), 641-647.

Weeks, W. F., Assur, A., 1963. Structural control of the vertical variation of the strength of sea and salt ice. In: Kingery, W. D. (Ed.), Ice and snow. M.I.T. Press, pp. 258-276.

Weeks, W. F., Assur, A., 1964. Structural control of the vertical variation of the strength of sea and salt ice. Crrel research report 113, Cold Regions Research and Engineering Laboratory.

Weeks, W. F., Assur, A., 1967. The mechanical properties of sea ice. Monograph, Cold Regions Research and Engineering Laboratory.

Weeks, W. F., Cox, G. F. N., June 1974. Laboratory preparation of artificial sea and salt ice. Tech. Rep. SP 206, Cold Regions Research and Engineering Laboratory.

Weeks, W. F., Hamilton, W. L., 1962. Petrographic characteristics of young sea ice, Point Barrow, Alaska. The Amer. Mineralogist 47, 945-961.

Weeks, W. F., Lee, O. S., 1958. Observations on the physical properties of sea ice at Hopedale, Labrador. Arctic 11 (3), 134-155.

Weeks, W. F., Lee, O. S., 1962. Salinity distribution in young sea ice. Arctic 15 (2), 92–108.

Weeks, W. F., Lofgren, G., 1967. The effective solute distribution coefficient during the freezing of NaCl solutions. In: Physics of Snow and Ice. Vol. 1. Hokkaido University, Sapporo, Japan, pp. 579–597.

Weeks, W. F., Wettlaufer, J. S., 1996. Crystal orientations in floating ice sheets. In: The Johannes Weertman Symposium. The Minerals, Metals and Materials Society, pp. 337–350.

Weissenberger, J., Dieckmann, G., Gradinger, R., Spindler, M., 1992. Sea ice: A cast technique to examine and analyze brine pockets and channel structure. Limnol. Oceanogr. 37 (1), 179–183.

Wernick, J. H., 1956. Determination of diffusivities in liquid metals ny means of temperature-gradient zone melting. J. of chem. phys. 25 (1), 47–49.

Wettlaufer, J. S., 1992. Directional soldification of salt water: deep and shallow cells. Europhysics Letters 19 (4), 337–342.

Wettlaufer, J. S., 1998. Introduction to crystallization phenomena in natural and artificial sea ice. Vol. 1. Department of Geophysics, University of Helsinki, Finland, pp. 105–194.

Wettlaufer, J. S., Worster, M. G., Huppert, H. E., 1997. Natural convection during solidification of an alloy from above with application to the evolution of sea ice. J. Fluid Mech. 344, 291–316.

Wettlaufer, J. S., Worster, M. G., Huppert, H. E., 2000. Solidification of leads: theory, experiment and field observations. J. Geophys. Res. 105, 1123–1134.

Weyprecht, K., Wien, Germany 1879. Die Metamorphosen des Polareises. Österreich-Ungarische Arktische Expedition 1872-1874. Moritz Perles, 284 pp.

Wilcox, W., 1967. Mass transfer in fractional soldification. Ch. 3, pp. 47–112.

Wilcox, W. R., 1993. Transport phenomena in crystal growth from solution. Prog. Cryst. Growth and Chreact. 26, 153–194.

Wilkes, K. E., 1995. Onset of natural convection in a horizontal porous medium with mixed thermal boundary conditions. J. Heat Transfer ASME 117, 543–547.

Williams, K. L., Garrison, G. R., Mourad, P. D., 1992. Experimental investigation of growing and newly submerged sea ice including acoustic probing of the skeletal layer. J. Acoust. Soc.Am. 92 (4), 2075–2092.

Williamson, R. B., 1967. Morphology of ice solidified in undercooled water. In: Peiser, H. S. (Ed.), Crystal Growth. Pergamon Press, Oxford, pp. 739–743.

Wilson, L. ., 1978. On interpreting a quantity in the Burton, Prim and Slichter equation as a diffusion boundary layer thickness. J. Crystal Growth 44, 247–250.

Wilson, L. O., 1980. Analysis of microsegregation crystals. J. Cryst. Growth 48, 363–366.

Wintermantel, K., 1972. Über den Einfluss des Temperatur- und Konzentrationsfeldes auf die Kristallisation an gekühlten Flächen. Phd-thesis, Technische Hochschule Darmstadt, Fachbereich Maschinenbau.

Wintermantel, K., Kast, W., 1973. Wärme- und Stoffaustausch bei der Kristallisation an gekühlten Flächen: Teil I Versuchsanordnung und Ergebnisse bei ruhenden wässrigen Lösungen. Chemie-Ing.-Technik 45 (289), 728–731.

Wolfe, L. S., Hoult, D. P., 1974. Effects of oil under sea ice. J. Glaciol. 13 (69), 473–488.

Wollkind, D., Segel, L. A., 1970. A nonlinear stability analysis of the freezing of a dilute binary alloy. Phils. Trans. R. Soc. London A 268.

Wood, G. R., Walton, A. G., 1969. Kinetics of ice nucleation from water and electrolyte solutions. Research and development progress report, U.S. Department of the Interior.

Wood, G. R., Walton, A. G., 1970. Homogenous nucleation kinetics of ice from water. J. Appl. Phys. 41, 3027–3036.

Worster, M. G., 1986. Solidification of an alloy from the cooled boundary. J. Fluid Mech. 167, 481–501.

Worster, M. G., 1991. Natural convection in a mushy layer. J. Fluid Mech. 224, 335–359.

Worster, M. G., 1992. Instabilities of the liquid and mushy regions during soldification of alloys. J. Fluid Mech. 237, 649–669.

Worster, M. G., 1997. Convection in mushy layers. Annu. Rev. Fluid Mech. 29, 91–122.

Worster, M. G., 2000. Solidification of fluids. Cambridge University Press, Ch. 8, pp. 393–446.

Worster, M. G., Kerr, R., 1994. The transient behaviour of alloys solidified from below prior to the formation of chimneys. J. Fluid Mech. 269, 23–44.

Worster, M. G., Wettlaufer, J. S., 1997. Natural convection, solute trapping, and channel formation during solidification of saltwater. J. Phys. Chem. B 101, 6132–6136.

Xia, W., Thorpe, M. F., 1988. Percolation properties of random ellipses. Phys. Rev. A 38, 2650–2656.

Xu, J.-J., 1997. Interfacial theory of pattern formation: selection of dendritic growth and viscious fingering in Hele-Shaw flow. Springer series in synergetics. Springer, 305 pp.

Xu, J.-J., 2004. Dynamical theory of dendritic growth in convective flow. Kluwer Academic, 240 pp.

Xu, J.-J., Yu, D.-S., 2001. Further examinations of dendritic growth theories. J. Crystal Growth 222, 399–413.

Yang, W., Chen, W., Chang, K.-M., Mannan, S., de Barbadillo, J., 2001. Freckle criteria for the upward solidification of alloys. Metall. Mat. Trans. 32A, 397–406.

Yen, Y.-C., Cheng, K. C., Fukusako, S., 1991. Review of intrinsic thermophysical properties of snow, ice, sea ice, and frost. In: Zartling, Faussett (Eds.), International Symposium on Cold Regions Heat Transfer. Vol. 3. pp. 187–218.

Yi, Y., Sastry, A. M., 2004. Analytical approximation of the percolation threshold for overlapping spheres. Proc. R. Soc. Lond. A 460, 2353–2380.

Yosida, Z., 1967. Surface structure of ice crystal and its equilibrium form. In: Physics of Snow and ice. Vol. 1. Hokkaido University, Sapporo, Japan, pp. 1–18.

Zaytsev, I. D., Aseyev, G. G., 1992. Properties of aqueous solutions of electrolytes. CRC Press, Boca Raton, 1773 pp.

Zotikov, I. A., Zagorodnov, V. S., Raikovsky, J. V., 1980. Core drilling through the Ross Ice Shelf (Antarctica) confirmed basal freezing. Science 207, 1463–1464.

Zubov, N. N., 1943. L'dy Arktiki (Arctic Ice). Izdatel'stvo Glavsermorputi, Engl. Translation 1963 by U.S. Navy Oceanographic Office and American Meteorological Society, San Diego, Moscow, 490 pp.

Zukriegel, J., 1935. Cryologia Maris. Vol. 15 of Travaux Geographiques Tcheques. The Geohraphical Institute of the Charles IV. University.

Danke Takk Thankyou

The scientific work leading to the present thesis has been performed in an unconventional manner, more or loss freely evolving with time, not coupled to specific scientific programs, not hampered by the common bureaucratic overload. The following people and institutions have made this possible:

- I thank my parents Heidi and Reinhard Maus for the safety, patience and all love of this world. It is their harbour to which I always could return and from where I always could proceed again. Danke!

- During my time in Norway I have received considerable help and support from two professors, which made it possible that I could continue my research in Longyearbyen and Bergen. These are Peter Mosby Haugan and Tor Gammelsrød. Takk!

- Several institutions in Norway have hosted me kindly for quiet long periods as guest research fellow and provide me with many facilities. The Norwegian Polar Research Institute (NPI), the University Courses on Svalbard (UNIS) and the Geophysical Institute of the University Bergen (GFI) deserve my thanks for having strongly supported this independent piece of research. Takk!

The present work would neither have been possible without the freely available literature resources from libraries all over the world. My special thank goes to Berit Jacobsen at UNIS on Svalbard in the name of all librarians. I have met helpful people during my frequent visits of the libraries of the TUHH (Technische Universität Hamburg Harburg), the BSH (Bundesamt für Seefahrt und Hydrographie), and the ZMK (Zentrum für Meeres- und Klimaforschung) in Hamburg. In Bergen, Norway, I would like to mention the 'Realfagbiliotheket' (RBB) of the University Bergen and the library of Havforskningsinstitutt (HI). The Technical Information Library (TIB) Hannover was an important electronic source in many more difficult cases. I also profited from searching many other libraries' catalogues via SUBITO (in Germany) and BIBSYS (in Norway). The CRREL (Cold Regions Research and Engineering Laboratory) provided me with a lots of reports.

The sympathy of my friends has, of course, been a carrier of this work. Many have also offered me help in reading earlier manuscripts. I made little use of these offers, because I did not feel confident with the writing and results for a long time. In the end, the time was too short for doing so. Anne Linda Løhre, Karen Assmann and Jochen Reuder have read passages. This helped to improve them and, hopefully a bit, my English. Thank you!

For help with the field work in Adventfjorden on Svalbard and the analysis of many ice cores I would like to thank Cecilia and Magnus Bennett.

A large thank goes also to Bente Tverberg-Hellan, Rune Grimstad and Idar Hessevik, providing me with the necessary electronic working conditions and friendly help when the computer crashed.

During the final period of this work I was not a very social person. I thank my colleagues at the Geofysisk Institutt i Bergen for understanding this and the length of the final period.

Now, 'two weeks' are over and I am glad.